Energy Communities

Customer-Centered, Market-Driven, Welfare-Enhancing?

Energy Communities

Customer-Centered, Market-Driven, Welfare-Enhancing?

Edited by

Sabine Löbbe

Professor for Energy Economics, Reutlingen Center for Distributed Energy Systems and Energy Efficiency,
Reutlingen, Germany

Fereidoon Sioshansi

President, Menlo Energy Economics,
San Francisco, CA, United States

David Robinson

Senior Research Fellow,
Oxford Institute for Energy Studies,
Oxford, United Kingdom

ELSEVIER

ACADEMIC PRESS
An imprint of Elsevier

Academic Press is an imprint of Elsevier
125 London Wall, London EC2Y 5AS, United Kingdom
525 B Street, Suite 1650, San Diego, CA 92101, United States
50 Hampshire Street, 5th Floor, Cambridge, MA 02139, United States
The Boulevard, Langford Lane, Kidlington, Oxford OX5 1GB, United Kingdom

Notices
Knowledge and best practice in this field are constantly changing. As new research and experience broaden our understanding, changes in research methods, professional practices, or medical treatment may become necessary.

Practitioners and researchers must always rely on their own experience and knowledge in evaluating and using any information, methods, compounds, or experiments described herein. In using such information or methods they should be mindful of their own safety and the safety of others, including parties for whom they have a professional responsibility.

To the fullest extent of the law, neither the Publisher nor the authors, contributors, or editors, assume any liability for any injury and/or damage to persons or property as a matter of products liability, negligence or otherwise, or from any use or operation of any methods, products, instructions, or ideas contained in the material herein.

ISBN: 978-0-323-91135-1

For information on all Academic Press publications visit our website at https://www.elsevier.com/books-and-journals

Publisher: Charlotte Cockle
Acquisitions Editor: Graham Nisbet
Editorial Project Manager: Moises Carlo P. Catain
Production Project Manager: Kiruthika Govindaraju
Cover Designer: Greg Harris

Typeset by TNQ Technologies

Contents

Author biographies

Marina Artipoli is a Senior Analyst in the Infrastructure & Network, Regulatory & Antitrust Department at the Enel Group. She has experience in regulatory issues related to infrastructure and networks, energy supply, the resilience of networks, and commercial quality of electricity distribution service and connections.

Previously, Marina was involved in the regulation of electricity distribution, retailing, and unbundling of electricity distributors in Italy.

Marina graduated in Management, Innovation, and Internationalization of Businesses at La Sapienza University of Rome.

Jake Barnes is a Researcher at the Environmental Change Institute, University of Oxford. His research interests include the politics and sociology of innovation and change from community to national level.

Jake has worked with local governments and social enterprises in multiple projects to reflect, learn, and pursue progressive change and he has over 10 years of experience as a director of community enterprises.

He holds a BA (Hons) in International Relations and an MSc in Energy Policy and Sustainability from the University of Exeter, and a DPhil from the Science Policy Research Unit (SPRU), University of Sussex.

Thomas Bauwens is an Economist specializing in social and environmental sustainability. He is currently an Assistant Professor at the Copernicus Institute for Sustainable Development at Utrecht University where he leads research and teaches on circular economy, energy communities, and sustainability transitions.

Before joining the Copernicus Institute, he was a researcher at Oxford University and the Swiss Federal Institute of Technology in Lausanne.

He holds a PhD in Economics from HEC Management School, University of Liège, Belgium, and an MSc in Economics from the Economic School of Louvain, University of Louvain, Belgium.

Simona Benedettini has 10 years of experience as an energy economist. At present, she works for a major consulting firm and provides advice to private organizations and public institutions at the national and European level on a wide range of issues including electricity market design, network tariff regulation, renewable energy policies, and competition issues.

During her career, Simona has also worked on secondment at the Competition Department of the UK Financial Conduct Authority. She has published several articles on regulation and competition in the energy sector in academic journals and leading Italian newspapers.

She earned a PhD in Law and Economics from the University of Siena and has been a visiting scholar at the University of Illinois.

Darryl Biggar is a Special Economic Adviser at the Australian Competition and Consumer Commission (ACCC) and the Australian Energy Regulator (AER) where he provides advice on a range of competition and regulation matters, especially relating to the design of electricity markets and regulatory frameworks including the economic foundation of competition policy.

He is the coauthor of a textbook on the *Economics of Electricity Markets*. Prior to working at the ACCC, Dr. Biggar was an economist with the Competition Division of the OECD and has also worked for the New Zealand Treasury and as Lecturer in Law and Economics at University College London.

Darryl has a PhD in Economics from Stanford University and an MA in Mathematics from Cambridge University. He is originally from New Zealand.

Adriano Bisello is a Senior Researcher and Project Manager with the Institute for Renewable Energy at Eurac Research in Italy. Adriano's responsibilities range from local to European-funded projects in the field of smart cities and energy transition with a focus on multiple benefits and urban innovation.

His professional career started as a freelance urban and environmental planner, contributing to complex project design and strategic environmental assessments collaborating with several municipalities and engineering firms.

Adriano graduated in Urban and Regional Planning at the University IUAV of Venice and holds a PhD in Real Estate Economics and Appraisal from the University of Padua, Italy.

Tereza Borges is an Executive and Entrepreneur responsible for international sales with Lumenaza GmbH, primarily focused on the use of digital platforms as an enabler for the acceleration of the decentralization and decarbonization of the energy sector.

Before joining Lumenaza, she held leadership roles with the Great Britain smart metering program and cofounded n3rgy, offering API service to facilitate access to smart metering data in Great Britain. Prior to this, Tereza held regional and global leadership positions driving new business areas in large ICT corporations.

Tereza graduated in Electronic Engineering and holds an MSc in Automation from the Universidade Estadual de Campinas (Unicamp), Brazil.

Paulo Born is an independent consultant in utility regulation and electricity markets who regularly collaborates with colleagues at the Instituto Acende Brasil while working with a startup that supports low-income energy communities, and is a Senior Advisor to Roland Berger.

Previously, Paulo was a member of the Board of Directors of CCEE (Cámara de Comercializaçao de Energia Elétrica), the entity responsible for energy auctions, accounting and settlement of Brazil's electricity market; Director of Light, a power utility in Rio de Janeiro; and Vice President of Duke Energy International Brazil.

Paulo has a degree in Civil Engineering from UFPR, Brazil, and an MSc in Civil and Environmental Engineering from Cornell University, USA.

Sibylle Braungardt is a Senior Researcher and project manager in the Energy and Climate Division at Oeko-Institut in Freiburg, Germany. She is an internationally recognized expert in sustainable energy policy analysis with experience in research projects at the international, European, national, and community levels.

Prior to her current position at Oeko-Institut, she worked as a Senior Researcher at the Competence Center for Energy Policy and Energy Markets at Fraunhofer ISI.

She holds a PhD in Physics from the Institute of Photonic Sciences in Barcelona and an MSc in Environmental Sciences. Sibylle teaches energy policy at Freiburg University.

Pascale Château Terrisse is a Lecturer in Management at Toulouse INP ENSAT and a Researcher at INRAE UMR 1248 AGIR. She is developing research on social and solidarity economy organizations in various sectors, such as solidarity finance, social housing, renewable energy, and agriculture while studying management mechanisms to regulate economic and social tensions.

She is currently developing several research projects around multistakeholder cooperatives, focusing on their organizational stakes and practices.

Pascale holds a doctorate from the University of Paris-Est Marne La Vallée, France.

Valentina-Miriam Cittati is a Junior Researcher at the Institute for Renewable Energy of Eurac Research. She works on several European projects such as Interreg

ITA-AU Shelter, Impetus, Arv. Her research topics are energy communities, energy democracy, and social innovation.

She wrote a master thesis called *Energy Community and Policy: A New Network of Local Relationships. The Two Case Studies in* Piemonte. She presented her thesis at the Fifth Energy and Society Conference and wrote a paper that has been published by Energies journal.

Valentina-Miriam has a Bachelor's in Sociology at La Sapienza University of Rome. She holds an MA in European Studies on Society, Science, and Technology (University of Maastricht) and an MA in Management of Organizations and Territorial Networks (University of Trento).

Christian Chudoba is a Founder and the CEO of Lumenaza GmbH. He focuses on the development of flexible software and new business ideas that make sustainability commercially viable. Christian has substantial leadership experience in strategic business development and marketing.

Before founding Lumenaza, he held various management positions at Nokia Siemens Networks and Siemens.

He studied physics at the Freie Universität Berlin and received his PhD from Humboldt-Universität Berlin, before working at the Massachusetts Institute of Technology (MIT) in the Department of Electrical Engineering.

Toby D. Couture is Founder and Director of E3 Analytics, an independent renewable energy consultancy based in Berlin, Germany. He has extensive experience working in the sector and has advised government policymakers in over 40 countries worldwide on renewable energy policy, strategy, and investment.

Prior to founding E3 Analytics, Toby worked as Energy and Financial Markets Analyst at the National Renewable Energy Laboratory (NREL) in Colorado. Toby speaks French, English, German, and Spanish.

He is a previous recipient of the Fulbright Scholarship, has a Master's degree in Environmental Policy from the University of Moncton in Canada, and an MSc in Financial and Commercial Regulation from the London School of Economics in the UK.

Robert Crump is a junior trend scout at Trianel GmbH. His main interests are innovations in the energy and mobility sectors as well as sustainability and decarbonization. He is an Innovation Honors Class member of the TIME Research Area at RWTH Aachen University.

Robert graduated in Industrial Engineering at RWTH Aachen University and holds an MSc degree. In his studies at RWTH Aachen University, he specialized in Electrical Engineering as well as Innovation, Entrepreneurship, and Marketing.

Iñigo del Guayo is a Professor of Administrative Law at the University of Almería where he conducts research on a wide variety of issues related to energy law and regulation, about which he has published extensively and provided advice to public and private institutions.

Currently, he is the Chair of the Academic Advisory Group (AAG) of the Section on Energy, Environment, Natural Resources & Infrastructure Law (SEERIL) of the International Bar Association (IBA).

Iñigo has an LLM and a PhD from the University of Navarra and has been a visiting scholar of several European and US institutions on energy law.

Anita Del Pizzo is a Senior Researcher at Enel Foundation with expertise in distribution networks. She currently works on research projects regarding energy transition and electrification.

She joined Enel Group in 2007 in the Infrastructure and Networks Division working as a specialist on grid connection for retail customers in the liberalized market, and later on planning and management of distribution networks. She contributed to the development of the client application "Contatori on Demand," for which Enel received an award in the Business Performance category.

Anita holds a Master's degree (cum laude) in Electrical Engineering from the University of Naples Federico II.

Lea Diestelmeier is an Assistant Professor at the Groningen Centre of Energy Law and Sustainability in the Netherlands. Her research includes EU law, in particular EU energy law. Lea's research interest has moved toward energy communities, offering alternative small-scale means of energy supply. Beyond the implementation and operation of new technologies, she is also interested in how energy communities facilitate broader policy goals such as democratization and justice.

She teaches in the LLM program Energy and Climate Law and presents frequently at conferences and seminars.

Lea graduated from Groningen University, NL, and holds an LLM in Energy Law and a doctorate from that university.

Mariangela Di Napoli is the Head of Infrastructure & Networks, Regulatory & Antitrust Department at the Enel Group. She is responsible for the technical regulation for E-Distribuzione SpA, the main Italian DSO with more than 30 million customers. She manages the relationship with ARERA, the Italian regulator, on regulated activities such as metering, innovation and digitalization, unbundling, continuity of supply, the resilience of networks, and commercial quality of distribution service and connections.

Previously, she was the Executive Assistant to the Enel Group Infrastructure and Network Chief Operating Officer, after working in the utility sector for a large consultancy.

Mariangela holds an MBA and graduated in Management Engineering at the Federico II University of Naples.

Hajar El Karmouni is an Associate Professor in Management at University Paris-Est Créteil. She is focused on the organization of the social and solidarity economy, in particular on citizen collectives involved in the energy transition and alternative food systems. She is currently working on a research project on the accessibility of sustainable food systems.

She holds a PhD in Management from the University Paris-Est Marne La Vallée, France.

Mariëlle Feenstra is a Senior Researcher on gender and energy policy. Her contribution has been acknowledged both in academia and in practice by invitations to participate in policy formulation activities. She is a member of the ENGAGER network.

Mariëlle holds a PhD in Public Administration and Policy from the Centre of Sustainable Technology and Management of the University of Twente, the Netherlands.

Marco Galici is a PhD student at the University of Cagliari, working on power systems and distributed ledgers, where he studied during his research fellowship in 2018. He graduated in Electrical Engineering from the University of Cagliari in 2020.

Andrea Galliani is Deputy Director in the Wholesale Energy Markets and Environmental Sustainability Directorate of the Italian Regulatory Authority for Energy Networks and Environment (ARERA). He is involved in the innovation of regulation to allow the integration of renewable sources and distributed generation in the electricity system in Italy.

He also teaches exergetic analysis, state energy balances, and fundamentals of electricity regulation in the General Energetics course at the Polytechnic University of Milan.

Matthias Gebauer is the CEO of ECOSCOP Ltd., offering environmental and energy consulting to individuals, SME's, and public institutions for over 30 years. Since 2021, he has been working as Climate Protection Manager in the city of Trier.

He graduated in Political Sciences, Communication Science, and Social Psychology at the University of Munich. He then earned a European Diploma in Environmental Sciences at the Universities of Trier, Saarbrücken, Kaiserslautern, Metz, Nancy, Arlon, and Luxemburg in 1992.

Emilio Ghiani is an Associate Professor of Power Systems in the Department of Electrical and Electronic Engineering (DIEE). His research focuses on developing methodologies and tools for decision-making in planning and operating power distribution networks with significant penetration of distributed energy resources. This involves research to develop methods and tools for decision-making in planning and operating power distribution networks.

He is the author of more than 90 papers published in international journals or presented at national and international conferences.

He received his PhD in Electrical Engineering and Computer Science from the University of Cagliari.

Jean-Michel Glachant is the Director of the Florence School of Regulation and holds the Loyola de Palacio Chair at the European University Institute. He regularly advises the European Commission and energy regulators in Europe. His main research interests are the building of a common European energy regulation and policy, the achievement of the European internal market for energy and its decarbonization, and the industrial organization and market strategy of energy companies, all topics about which he has written extensively.

He serves as President-elect of the International Association for Energy Economics (IAEE) in 2022.

Jean-Michel received his PhD in Economics at La Sorbonne.

Daniel Gregetz is a Senior Project Engineer at SWT Stadtwerke Trier, the municipal utility in Trier, working in the department of Strategic Development – Power Generation. For the last 15 years, Daniel has implemented a number of projects on biogas, combined power and heat and solar energy, as well as the integration of renewable power and gas into the local energy system.

Before joining the utility in Trier, he was a project engineer at Fiorentini Deutschland GmbH and worked for HGC-Hamburg Gas Consult, an engineering firm.

Daniel graduated at the Trier University for Applied Sciences and holds a Diploma in Energy Supply Technology.

Rachel Guyet is the Director of the Master's program in Global Energy Transition and Governance with Centre International de Formation Européenne (CIFE) where she has been carrying out research projects on the local and social dimension of energy transition in Europe.

She is a member of the ENGAGER network. Before joining CIFE, she worked with Sciences Po, Paris, where she carried out research projects on energy poverty and social acceptance of energy transition and where she is still lecturing.

Rachel holds a PhD in Political Science from Institut d'Études Politiques of Grenoble, France.

André Hackbarth is a Senior Researcher and Lecturer at Reutlingen University, School of Engineering/Distributed Energy Systems and Energy Efficiency, Germany. His research interests include business models in the energy sector and consumer attitudes, preferences, and decision-making concerning energy-related behaviors and products.

Prior to his current position, he was a Research Associate at the Institute for Future Energy Consumer Needs and Behavior (FCN) at RWTH Aachen University, Germany.

He studied economics at Heidelberg University and holds a PhD from RWTH Aachen University.

Tina Hadler was a Product and Marketing Manager for Energy Communities at coneva, a startup founded in 2018 by SMA Solar Technologies before now heading the sales for Smart Metering Services at Theben, Germany. Focusing on smart metering and digital business models, Tina is responsible for shaping and defining products and services coneva offers to energy communities for which she has led various implementation projects in Germany.

Before joining coneva, she worked in management consulting focusing on smart metering and the digitization of the energy sector.

Tina holds an MSc degree in Technology and Management from the Technical University of Munich.

Florian Hanke is a PhD student at the European University Viadrina and a member of the transformational research cluster, where he received a scholarship from the Heinrich-Boell Foundation. Florian has coordinated a European research project on socially inclusive energy communities and energy poverty. Previously, he worked for a think tank and a social entrepreneurship accelerator.

Florian graduated in Business and Economics at the Berlin School of Economics and Law and holds an MA in European Studies from the European University Viadrina.

Paula Hansen is a Postdoctoral Researcher at the Environmental Change Institute, University of Oxford where she studies the operation and governance of energy communities as part of the Horizon2020-funded NEWCOMERS project. Paula's research centers around the interactions between humans and smart technologies and draws on commons governance, collective action, and systems theory. Previously, she worked on transdisciplinary energy innovation projects while working on her PhD research.

She holds a BSc in Business Management from King's College London, UK, an MSc in Sustainability Science and Policy from Maastricht University in the Netherlands, and a PhD from Curtin University in Australia.

Mohammad Reza Hesamzadeh is a faculty member at KTH, the Royal Institute of Technology, in Sweden and a Faculty Affiliate at the Program on Energy and Sustainable Development (PESD), Stanford University, and Research Affiliate at the German Institute for Economic Research (DIW Berlin).

Dr. Hesamzadeh has published books and papers on electricity market design; served as an Editor of IEEE Transactions on Power Systems and Guest Editor of other journals; and has been providing consulting services to private and government sectors.

He received his Docent from KTH Royal Insitute of Technology, Sweden, and PhD from the Swinburne University of Technology, Australia.

Achim Hill works as a Project Manager at the Environmental Campus Birkenfeld, a zero emission campus of the Trier University of Applied Sciences, focusing on energy and climate protection. He has experience in energy planning and optimization of buildings and technical facilities.

Before joining Trier University, he was the Managing Director of the Energy Agency for the region of Trier where he initiated a number of projects within the energy space from local to European level.

Achim gained a Diploma in Architecture at the University of Applied Sciences for Technology, Economics and Culture in Leipzig and a Master's degree in Economics at the University of Hagen, Germany.

Léon Francis Hirt is a PhD student at the Institute for Environmental Sciences, University of Geneva. His current research focuses on analyzing the dynamics of solar photovoltaic uptake in Switzerland, combining spatial energy analysis and insights from sociology.

Léon has a Bachelor's degree in Civil Engineering at the Swiss Federal Institute of Technology in Lausanne, a Master's degree in Environment and Climate Change at the University of Liverpool, and a Master's degree in Foundations and Practices of Sustainability at the University of Lausanne.

Richard Lee Hochstetler is a Director of Economic and Regulatory Issues at Instituto Acende Brasil, a think tank based in São Paulo focused on electric sector policy and regulation in Brazil and the organizer of the biennial Brazil Energy Frontiers conference.

He recently published a book on market design for the Brazilian electric power industry. Prior to joining Acende, he worked at the Federal Energy Regulatory Commission and was Partner at Tendências Consultoria Integrada, an economic consultancy in Brazil.

Richard obtained his BA in Economics from Goshen College and a Doctorate in Economics from the University of São Paulo, Brazil.

Simon Köppl is a Senior Researcher at FfE in Munich, leading a broad range of projects on the optimization of the electricity grid with a focus on distribution grids, the integration of flexibility in the energy system, and the impact of digitalization on the energy infrastructure and e-mobility.

Simon graduated in Electrical and Computer Engineering from the Technical University of Munich where he currently is a doctoral candidate.

Mark Kolesar is a Managing Principal at Kolesar Buchanan & Associates Ltd. and an affiliate consultant with Christensen Associates.

With over 30 years of experience in the regulated utilities sector, Mark has worked in regulation and public policy, external relations, marketing, strategy, business development, and mergers and acquisitions. He has over 20 years of corporate experience in the telecom sector, where he was Vice President, Economic Affairs at TELUS. He was a member of the Alberta Utilities Commission for 12 years, including 6 years as Vice-Chair and 2 years as Chair. Mark sits on the Advisory Board of the Center for Research in Regulated Industries at Rutgers University.

He has an Honors Degree in Philosophy and an MBA in Managerial Economics and Finance.

Maksym Koltunov is an independent Ukrainian researcher on energy economics with interest in the impact of energy communities on the welfare of towns and villages and smart city projects. Previously, he worked at the NGO "Skify" and different companies in Ukraine, the US, and China.

Maksym holds an MSc in Economics of Agribusiness from Kyiv National Economic University and a Joint European Master's degree in Comparative Local Development from the Corvinus University of Budapest, Regensburg University, and the University of Trento. He is an Erasmus Mundus alumni. Currently, Maksym is a PhD candidate in Circular Economy at the University of Trieste, Italy.

David Lellinger is a Climate Protection Manager of the city of Trier and currently involved in drafting Trier's first integrated climate protection scheme. His main interests are communal climate protection concepts, sustainable, low-tech architecture, building and district simulation, district heating, and photovoltaic systems.

He graduated at the Trier University of Applied Sciences with a Bachelor's in Engineering and a Master of Engineering in Energy Management.

Xin Li is a Research Fellow at the Norwich Business School, University of East Anglia, and Research Fellow at Oxford Climate Policy. His current research centers on innovative smart energy solutions for isolated communities in the EU, demand-side flexibility, and the energy transition in China.

Xin previously worked at the Oxford Institute for Energy Studies. He has published more than 20 papers in academic journals. His previous research focused on conceptualizing and modeling the interaction between human and environmental systems and modeling scenarios of future change.

Xin holds a PhD in Ecological Economics from the University of Leeds, UK.

Sabine Löbbe is a Professor at Reutlingen University, School of Engineering/Distributed Energy Systems and Energy Efficiency, Germany, and the Officer for Sustainability in the Extended President's Office of the university. Her consulting company advises utilities on strategy and business development and on organizational issues.

In her 25 years of professional experience before being appointed as professor, she was a Director for Strategy and Business Development at swb AG, Bremen and project manager at Arthur D. Little Inc. and at VSE AG, Saarbrücken.

She holds a doctorate in Business Administration from the Saarbrücken University and studied business administration in Trier, Saarbrücken, and EM Lyon, France.

Luca Lo Schiavo is the Deputy Director of the Infrastructure Regulation Directorate at the Italian Regulatory Authority for Energy Networks and Environment (ARERA), where he has worked since its establishment in 1997, and an expert on quality of service and innovation in the electricity sector.

He is a member of the Infrastructures Task Force of the European Agency for Cooperation of Energy Regulators (ACER) and of the Distribution Systems Working Group of the Council of European Energy Regulators (CEER). He has coauthored books on service quality regulation and has authored many regulatory papers. He won the International Confederation of Energy Regulators Award in 2012, and his paper, *Smart metering: an evolutionary perspective*, was recognized by the Energy Regional Regulatory Association in 2017.

Luca earned his MSc in Industrial Engineering from the Technical University of Milano and a specialization qualification in Public Policy Analysis at IDHEAP in Lausanne.

Reinhard Madlener is a Full Professor of Energy Economics and Management at the School of Business and Economics, RWTH Aachen University, Germany, and an Adjunct Professor at the Department of Industrial Economics and Technology Management, NTNU, Trondheim, Norway. His main research interests are in energy economics and policy, sustainable energy transition, smart grid economics and management, and energy technology diffusion.

He has served as the Senior Editor of Energy Policy and on numerous Editorial Boards of scientific journals. He is the Founder and current Director of the Institute for Future Energy Consumer Needs and Behavior at RWTH's interdisciplinary and integrated E.ON Energy Research Center.

He holds two Master's degrees and a PhD in the Social Sciences and Economics from Wirtschaftsuniversität Wien (WU) and a Postdoctoral Diploma in Economics from the Institute of Advanced Studies and Scientific Research, Vienna, Austria.

Marion Maignan is an Associate Professor in Management at the University Picardie Jules Verne in Amiens, France, working on socio-economic innovations related to sustainable food systems and renewable energies.

She holds a degree in Agricultural Economics from Montpellier Supagro, in France, in Philosophy from the Université Montpellier Paul Valéry, and a PhD in Management from the University Paris-Est Marne La Vallée, France.

Christoph Menke is a Full Professor for Energy Technology at the Trier University of Applied Sciences, Germany, and Distinguished Professor at the JGSEE/KMUTT, Bangkok.

His experience includes working in a number of international organizations including the World Bank, the Caribbean Development Bank, and the German International Cooperation (GIZ), and for the TATA Energy Research Institute in Bangalore, India. His main interests include energy technologies such as solar energy, cogeneration, energy policy, and planning.

He graduated in Mechanical Engineering at the Technical University of Braunschweig, Germany, with a PhD in Energy Planning from the same university.

Giuseppe Montesano was Deputy Director of Enel Foundation until 2021 where he undertook strategic studies and research focusing on energy transition and electrification. He is also the Vice-Chair of the Italian National Committee of the World Energy Council.

He has worked for Enel since 1988, developing renewable energy projects, designing environmental and low-carbon strategies, and advising on regulatory matters. He has held several managerial positions including Network and Tariff Regulation, Head of Environment, Head of European Regulation and Scenarios, and Head of European Energy Policy.

He graduated in Mechanical Engineering and specialized in Energy Transformation and Energy Systems.

Bruce Mountain is the Director of the Victoria Energy Policy Centre at Victoria University in Melbourne, Australia, where he leads a research team of economists, mathematicians, and data scientists. His main expertise is in regulatory and market economics applied to the electricity industry, with a particular interest in political economy.

Bruce has combined a career as a consultant in economics and policy with academic research and worked in Australia, South Africa, England, and France and has undertaken consultancy assignments in several other countries.

Bruce has a PhD in Economics from Victoria University and a Bachelor's and a Master's degree in Electrical Engineering from the University of Cape Town. He also qualified as a Chartered Management Accountant in England.

Mario Mureddu is a Postdoctoral Researcher at the University of Cagliari with interests spanning the application of complex systems, artificial intelligence (AI), and blockchain, to sustainable energy systems including integrating blockchain and AI in local and peer-to-peer energy markets for the renewable energy transition. Previously, he held a postdoctoral position at Jacobs University in Bremen.

He received a PhD in Physics from the University of Cagliari.

Carlo Papa serves as the Managing Director of Enel Foundation. He started his career in IBM Ireland and UK, was hired by Enel Distribuzione in 1997, and the year after joined the Corporate Strategic Planning Department. He founded the company's Corporate Venture Fund and led the Business Development practice in Russia and Ukraine. More recently, Carlo served as the Chief Innovation Officer of Enel Green Power and Vice President of ARISE, the UNISDR Private Sector Alliance for Disaster Resilient Societies. Actively involved as Fellow in the Aspen Institute and the German Marshall Fund of United States, he cooperates with two *pro bono* ventures: mentors4u.com and harambee-africa.org where he serves as Chairman of the Advisory Board.

He holds an MSc in Economics, a PhD in Management Engineering, and the TRIUM MBA jointly issued by NYU Stern, London School of Economics, and HEC.

Fabrizio Pilo is a Full Professor of Power Systems in the Department of Electrical and Electronic Engineering of the University of Cagliari. He has 30 years of experience in power distribution planning and development, is an IEEE Senior Member and CIGRE Distinguished Member, and a member of the CIRED Board of Directors. He has authored or coauthored more than 180 papers published in international journals or presented at national and international conferences.

He earned a PhD degree from the University of Pisa.

Vincenz Regener is a Research Associate at FfE in Munich where he works on future energy market designs and innovative energy concepts. Vincenz studied mechanical engineering at RWTH Aachen University and the Technical University of Munich, focusing on energy technology and mobility.

David Robinson is a Senior Research Fellow with the Oxford Institute for Energy Studies.

David was previously a Director and Board Member of two international economic consultancies, NERA and the Brattle Group. He has led academic and consulting projects on energy for over 30 years, most recently on market design and regulation to support consumer participation in electricity decarbonization in Europe, China, and Latin America.

David has a Doctorate in Economics from the University of Oxford, a Master's in Economic Policy and Planning from Erasmus University, and a Joint Honor's BA in Economics and Political Science from McGill University.

Nicolò Rossetto is a Research Fellow at the Florence School of Regulation where he studies and teaches mostly on electricity-related issues such as the establishment of a seamless transmission system in Europe, the digitalization of the electricity industry, and the future of energy communities.

Before joining FSR, Nicolò was external advisor to the World Bank and teaching assistant at the University of Pavia in Italy.

Nicolò holds a PhD in Economics, Law and Institutions from the Istituto Universitario di Studi Superiori of Pavia and a Master's degree in International Economics, Politics and Institutions from the University of Pavia.

Arianna Rossi is a Research Fellow of the Energy Department of the Politecnico di Milano and works with the Wholesale Energy Markets and Environmental Sustainability Directorate of the Italian Regulatory Authority for Energy Networks and Environment (ARERA). Arianna's main area of research is the innovation of regulation to allow integration of renewable energy sources and distributed generation in the Italian electricity system, with a focus on electric dispatching and ancillary services.

She graduated in Energy Engineering at the Politecnico di Milano.

Benjamin Schmid is a Postdoctoral Researcher at NUI Galway in Ireland with interests in energy citizenship, local energy governance, and configurative methodologies. Previously, he worked as a Research Associate at the Zurich University of Applied Sciences and as a doctoral student at the Swiss Federal Research Institute WSL.

Benjamin has a Bachelor's in Political Sciences from the University of Lucerne, a Master's in Sustainable Development from the University of Basel, and a PhD in Economic Geography from the University of Bern.

Rudolf Schöller was the Head of the department Strategic Development – Power Generation for the municipal utility in Trier, SWT Stadtwerke Trier, until his recent retirement.

For the last 35 years, Rudolf has accompanied many innovative projects in power grids, combined heat and power plants and renewable energies. His ongoing projects include an online grid controller and a pumped storage water plant to control the fluctuating production of solar and wind power with a capacity up to 900 MW.

Rudolf graduated from the Trier University for Applied Sciences and holds a Diploma in Electrical Engineering.

Mònica Serlavós is a Director at the Zoein Foundation, Co-President of the Swiss Association for Citizen Energy, and part-time lecturer at the University of Lausanne. Mònica's main interests include local energy and ecological transition and research-action methodology.

Earlier in her career, she led a research-action project in the Swiss French part of Switzerland. She founded, with several partners, the Swiss Association for Citizen Energy, the first national network for community energy initiatives in the country.

Mònica graduated in Political Sciences at Pompeu Fabra University in Barcelona, Spain, and holds a Master's in Development Studies from the Graduate Institute in Geneva, Switzerland, as well as a PhD in Environmental Sciences from the University of Lausanne.

Fereidoon Sioshansi is the founder and President of Menlo Energy Economics, a consulting firm advising clients on strategic issues in the electricity sector. He is the editor and publisher of *EEnergy Informer*, a monthly newsletter with international circulation.

His prior work experience includes working at So. Calif. Edison Co., EPRI, NERA, and Global Energy Decisions, acquired by ABB. This is his 14th edited or coedited volume since 2006 on different subjects including innovation and disruption at the grid's edge, consumer, prosumer, prosumagers, behind and beyond the meter, and variable generation, flexible demand.

He has a BS and MSc in Civil and Structural Engineering, an MSc and PhD in Economics from Purdue University.

Deyana Spasova is a Junior Researcher at the Climate and Energy Department of the Ecologic Institute in Berlin where she works on EU Member States' long-term planning toward climate neutrality and citizens' involvement in EU climate action. Previously, she worked at the Energy and Climate Division of Oeko-Institute in Freiburg.

She holds an MSc and MA in Global Studies from the Hochschule Karlsruhe in Germany, the National Chung Hsing University in Taiwan, and the Universidad de Monterrey in Mexico, as well as a Bachelor's degree in Journalism from the Robert Gordon University in Aberdeen, Scotland.

Elisabeth Springmann is a Research Associate at FfE in Munich where she focuses on the grid integration of e-mobility and the digitalization of distribution grid components as an instrument for grid management in an interconnected, smart energy system.

She graduated in Electrical and Computer Engineering at the Technical University of Munich.

Carlo Stagnaro is a Research Director of Istituto Bruno Leoni, a Milan-based think tank. Previously, he was Chief of the Minister's Technical Staff at the Italian Ministry of Economic Development.

He is a member of the Editorial Board of the journal Energia, a regular columnist to Il Foglio daily magazine, a member of the Institute of Economic Affairs' Academic Advisory Council, and Fellow of the Italian Observatory on Energy Poverty at the Padua University's Levi-Cases Center. His publications include *Power Cut? How the EU Is Pulling the Plug on Electricity Markets* (Institute of Economic Affairs, 2015) and *Contro il sovranismo economico* (Rizzoli, 2020) with Alberto Saravalle. He has published several articles on regulation and competition in the energy sector in academic journals and leading newspapers.

Carlo earned an MSc in Environmental Engineering from the University of Genoa and a PhD in Economics from IMT Alti Studi Lucca.

Teodora Stoyanova is a Coordinator of the Climate and Energy campaign of Greenpeace Bulgaria. For the last 7 years, she has been working on climate change, the need for energy transition, and transformation of the energy system in Bulgaria. Her areas of interest also include air quality and renewable energy, focusing on energy communities. She is a member of the team of the European Climate Foundation as a coordinator for Southeast Europe. Teodora lives and works in Varna, Bulgaria.

She holds a Bachelor's degree in Finance from Drexel University and a Master's degree in Environmental Management from the Yale School of the Environment in the United States.

Job Swens is a Consultant supporting municipalities, local communities, and companies with technical, legal, and governance solutions for renewable energy projects with a focus on the legal aspects of decentralized energy production, storage, and distribution. Job presently works for Spectral Energy, RvO (the Dutch governmental energy agency), and the municipalities of Amsterdam and The Hague, among others.

Previously, Job worked as an academic materials science researcher and as head of the R&D department at a switchgear company. Concerned about climate change, he started organizing conferences and moved to RvO. In 2008, he started his own consultancy J-OB – Joining Objectives for Sustainable World Energy Solutions.

Job graduated from the Delft University of Technology, NL, and holds an MSc in Applied Physics. He is currently finalizing his thesis for an MLL in Energy Law.

Matteo Troncia is a Postdoctoral Researcher at Universidad Pontificia Comillas. His research interests span complex systems, multicriteria methodologies, blockchain, and smart grid innovative applications.

He received a PhD in Industrial Engineering from the University of Cagliari.

Stefano F. Verde is an Assistant Professor in the Department of Economics and Statistics at the University of Siena, Italy. He is specialized in climate change economics and policy.

Before joining the University of Siena, he was the Deputy Director of the Florence School of Regulation – Climate, European University Institute, and an Adjunct Professor at the LUISS "Guido Carli" University, Rome. Stefano participated in several competitive research projects and authored several articles in international peer-reviewed journals.

He holds a BA in Economics from the University of Rome "Tor Vergata," an MSc, and a PhD in Economics from Trinity College Dublin, Ireland.

Andreas Weigand is a Senior Specialist for the decentralized energy system at Stadtwerke München (SWM), the municipal energy provider of Munich where he coordinates the company's activities in sector coupling and the integration of e-mobility into the distribution grid.

He has worked across the energy industry, starting his career as an engineer in a CCGT power station before setting up and marketing SWM's renewable energy portfolio across Europe in the trading department in Munich. He was part of the Steering Board and project leader of the power-to-heat field in "C/sells," a German research project on the cellular energy system.

R. Bruce Williamson is Commissioner Emeritus, Maine Public Utilities Commission, and Senior Fellow at the Public Utility Research Center at the University of Florida.

While being a Commissioner, he served on the Research Advisory Council of the Electric Power Research Institute (EPRI) and was Vice-Chair of the Federal Communication Commission's North American Numbering Council. Dr. Williamson also served as an Executive Board member of the Regional Greenhouse Gas Initiative (RGGI). Prior to becoming Commissioner, he worked for the University of Tennessee and for many years with SBC Communications, now part of AT&T. He began his career at Argonne National Laboratory and the Oak Ridge Institute for Energy Analysis.

He holds a PhD in Economics from the University of New Mexico, an MA degree in International Studies from the Korbel School, and a BA from Cornell University.

Foreword

For more than 30 years, Fereidoon Sioshansi has been testing our ability to follow the transformations that the electricity sector is facing and our understanding of these changes. Fereidoon also accelerated his efforts with "*Innovation & Disruption at the Grid's Edge*" in 2017, "*Consumer, Prosumer, Prosumager*" in 2019, "*Behind & Beyond the Meter*" in 2020, and "*Variable Generation, Flexible Demand*" in 2021. What's next? Fereidoon is now joined by Sabine Löbbe and David Robinson for a new volume on "*Energy Communities*," a topic bordering on a joke only a decade ago. Why not write a book on "*Amish utilities*"?

However, the truth is that the topic of the book is about many different things, except the Amish. The latest technology trends, roughly a decade old, are at work transforming the industry. First, the global push for wind and solar energy has resulted in low "levelized energy costs," putting distributed renewable resources on the agenda of many investors—be they individuals—owning more than 30% of all renewable energy capacity in Germany in 2019, and more than 50% of all wind capacity in Denmark in 2016—or professionals. Second, the deep digitalization of virtually everything, including communication with devices, ensures low transaction costs in many trading and sharing relations deemed impossible until recently, as described in 1970 by George Akerlof in his "*The Market for 'Lemons': Quality Uncertainty and the Market Mechanism.*"

These two strands of technology innovation have opened a new world in which many new things are possible. The question is: which ones?

The answer is not determined solely by technology. Neil Armstrong walked on the Moon on 20 July 1969: more than 50 years later, there is no viable business model for Moon trips. Technology, in other words, is only an "enabler." Investors and managers' plans, on the one hand, and consumers' preferences on the other hand, play a key role. However, they are not alone in influencing the energy decision-making process. Public policies and regulation are as effective as investors, managers, and consumers at making innovation viable in the energy sector.

Investors and managers usually interface with established utilities, or a few innovators creating new entrants and start-ups; but this time, lower economies of scale favor new entrants and not necessarily the established incumbents. Decarbonizing the energy sector also has a societal appeal which can call on new kinds of "societal entrepreneurs" to innovate. Bypassing the established businesses in a traditionally "vertically integrated" industry opens new appetites on a larger scale. Cooperatives, having always contested established businesses, are considering how to enter or to grow in this new field. Individuals and small entities such as small and medium enterprises or local authorities are receiving the same call. One day, maybe, all buildings with multiple owners or social housing units will join.

Of course, all of this has to be tested. Will energy consumers find direct action relevant? Will it be economically advantageous or socially and environmentally superior? And if so, for what kind of energy services? Joint energy procurement? Or energy generation and supply? Or energy delivery using communally owned grids? Or mainly for advanced digitalized interactions with wholesale and retail markets? These are important questions for which the jury is still out.

Currently, it is safe to say that new opportunities are emerging for collective action within communities. Elinor Ostrom, having got her Nobel Prize jointly with Oliver Williamson, used to gently mock him for focusing only on "*Market and Hierarchies*"; pointing out that humans had invented communities centuries ago, to deal with many common issues that markets and hierarchies are not well equipped to address. According to Elinor, one cannot hide that (i) addressing efficiently local externalities, or (ii) responding to preferences for locally differentiated goods, or (iii) simply grouping to reach an efficient scale of operation in production or trade are three alternative worlds of economic factors with economic rationale.

Nevertheless, having some economic rationale is not enough to make a living in the energy industry. The sector is deeply regulated, either to ensure markets can work despite the natural monopoly features of the grid, or to protect consumers from the outcome of markets, or to guarantee uninterrupted security of supply to everybody and the delivery of decarbonization targets. How do public policies and regulation consider the case of energy communities, their various forms, and the different services that they deliver? Are communities a new type of player in the energy sector, or another supplier, or a collective consumer, or a kind of integrated utility? Are consumers themselves the genuine nature of communities, or only the ones served by those? How should network monopolies, and their regulated metering and settlement processes treat and bill the communities and their members? The chapters of this book address many of these questions and more.

Collective action, where individuals or small players join to undertake a common initiative, is a fascinating issue. It is always of a dual nature. It can be addressed only as a particular case of economics, where particular economic factors play a bigger role. Or it can be seen as a particular area of the society, where particular societal factors play a bigger role. This dual nature of communities was at the core of Elinor Ostrom's research and remains relevant, proving a strong rationale for the task at hand.

Jean-Michel Glachant
Director
Florence School of Regulation
European University Institute
Florence, Italy

Introduction

Sabine Löbbe[1], *Fereidoon Sioshansi*[2], *David Robinson*[3]
[1] Reutlingen University, Germany, [2] Menlo Energy Economics, United States,
[3] The Oxford Institute for Energy Studies, United Kingdom

Traditionally, the power sector was organized from the top down: from large-scale and often distant generation to the end consumer, which was treated as the "load" to be served. The electricity system was "load following," with dispatchable generation providing the flexibility needed to meet unpredictable demand while maintaining security of supply and system stability.

In today's world, three challenges require a reevaluation of the consumer's role:

- Distributed energy resources (DERs)—including generation, storage, heat pumps, and devices behind end-consumer meters—are growing rapidly. For example, distributed solar photovoltaic capacity is projected to grow from 2018 to 2024 by 150% (IEA, 2019), which will contribute to net power flows changing direction.
- Most of the energy produced close to or on consumers' sites is renewable and intermittent; it is fed into the grid stochastically. The resulting fluctuation needs to be managed, by the grid, by other producers or storage providers, or by the consumers themselves. The consumer might take responsibility for stability and security of supply by evolving from being a "load" to being a prosumer, a prosumager, and ultimately a provider of system services (see Sioshansi, 2019).
- The development of DERs is essential for the energy transition and citizens will be central to the development of those resources. However, DERs will involve important consumer investment decisions and changes in consumer behavior.

For these and other reasons, consumers start to question their traditional role as passive "load," and regulation and policy must reflect the new role of active citizens. While the growth of DERs introduces complexity to the system, it also introduces new resources that can contribute to managing the same complexity.

Energy communities are one means to address these challenges, being themselves a possible institutional structure for organizing community energy, based on distributed resources and energy systems. As the World Economic Forum predicted (Burston, 2016), "in 2030, individual and community energy generation contributes to more than 50 percent of the energy mix in developed countries, up from less than 5 percent in 2016. [...] Politicians begin to see community energy ownership as the way to finally get citizens engaged in their energy future and to remove social and cultural barriers to the most affordable renewables."

To shape and support community energy, energy communities serve to integrate the consumer as an active part of the future energy system. Thus, as distributed generation, storage, and load management expand, the interest in energy communities increases. In customer-centered energy systems, just as with layers of an onion, the energy system needs to be redefined starting with the customer at the center. This transfers responsibilities to the end-consumer or to communities of consumers and those representing these stakeholders in the value chain. Worldwide, energy communities are emerging as part of the solution for a more sustainable, low-carbon, decentralized, resilient, and semi-independent energy systems.

Important challenges, however, need to be addressed to overcome hurdles to such a future. Energy communities have not been fully implemented at scale for a number of reasons including insufficient interest from consumers, technology that was not economically viable, and regulatory restrictions prohibiting viable business models from emerging. Moreover, as a number of chapters in this volume make clear, while the basic technology to do what is needed already exists, putting the pieces together and making them work at scale remains a challenge. However, as the political and regulatory support for energy communities grows, technology advances, and customers become more aware of and interested in local, ecologically sound alternatives and self-generation and consumption, the case for developing energy communities becomes more compelling. In this context, are energy communities going to play an important role in the future energy landscape? If so, what role, how, and when?

These questions inspired us to edit this book not knowing the answers when we started. The book presents a wide range of perspectives, some based on real case studies and others looking at emerging business models in different stages of maturity. Even more interesting, the volume reflects different—traditional and novel—perspectives and approaches from various disciplines and stakeholders, including policy makers, regulators, energy communities, technology providers, academics, as well as incumbent energy companies and new entrants providing services to energy communities.

As it turns out, the concept of energy communities is not new. On the contrary, they can be considered as the cradle of today's energy system, for electricity and gas. *Homo sapiens* probably formed communities when they settled—sharing all resources including energy, food, etc., within the community. In later stages, local leaders and politicians together with entrepreneurs ventured to bring energy services to communities, as Fig. 1 illustrates for the expansion of distribution lines to rural America. In many cases, the resulting organizations were the forebears of today's energy suppliers. However, to benefit from economies of scale and greater security of supply, most isolated local energy communities were eventually integrated into wider networks. Yet even today, many rural and municipal community-owned systems remain viable in the US, Europe, and elsewhere.

As Fig. 2 illustrates, today's energy communities tend to be "multipurpose," "multicontent," and "multistakeholder" systems. That is one reason why the concept of energy communities is so hard to nail down; it covers many different arrangements, with varying levels of interconnections to the rest of the energy system. This edited volume

Figure 1 Rural community-owned energy systems, called cooperatives, emerged in the US in the 1930s and remain viable to this day.

Source: Rural Electrification Administration, U.S. Department of Agriculture. "New Member-New Owner. REA Co-ops Are Member-Owned and Controlled." Special Collections, USDA National Agricultural Library. Accessed January 31, 2022, https://www.nal.usda.gov/exhibits/speccoll/items/show/215.

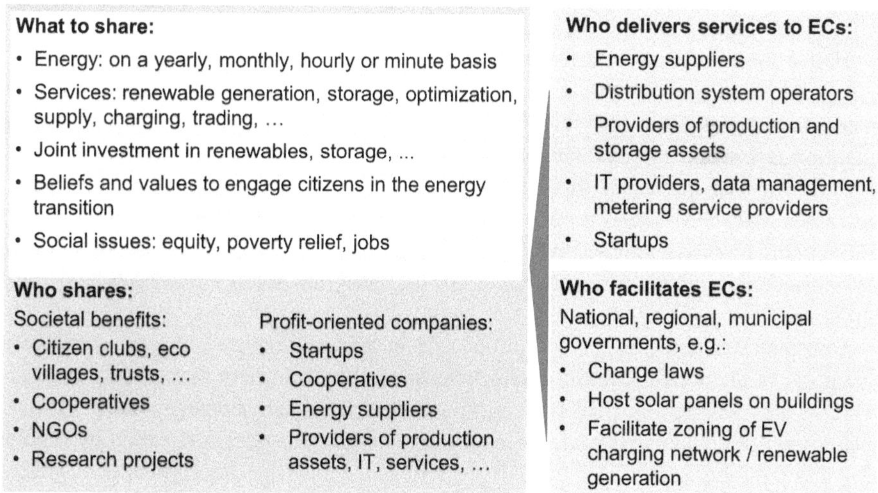

Figure 2 Overview of content, stakeholders, and service providers for energy communities.

explores how these communities can contribute to the transition toward a more decentralized distributed energy system, what political and regulatory frameworks currently exist and/or are needed to make energy communities viable, what technological solutions are promising, who controls which part of the energy system(s) and for what purpose, and how energy communities can become promising business models for different actors along the value chain.

Consequently, this volume addresses concepts and solutions covering all the purposes, groups of stakeholders, as well as the service providers identified in Fig. 2, and more. This is particularly important as there are significant differences of view on what exactly energy communities are and what they should aim to do. Moreover, they are developing for different reasons and with different motives around the world. The wide geographical coverage—Europe, North America, South America, and Australia—introduces different perspectives on the role of energy communities and how they are treated by policy makers and citizens. Without a doubt, recent European Union (EU) legislation actively supporting the creation of ECs has accelerated the move toward community energy in the EU. However, each Member State has significant leeway to decide how to transpose that legislation into national law. Although we see growing interest in other parts of the world, each country is different in terms of existing legislation, regulation, market design, and potential for energy communities to grow.

As reflected in the volume's title, to be sustainable and practical energy communities:

- Need to be customer-centered: they must attract energy customers (i.e., citizens) to participate actively in the EC or to buy their services—depending on the type of energy community;
- Should be market-driven: they should be integrated into the existing or the evolving future market design of the surrounding system; and
- Should be welfare-enhancing: defining "welfare enhancing" is a key issue, with some authors referring to the welfare of members of the energy community, while others define the concept in broader societal terms—welfare for members of the community and for nonmembers.

The book consists of 25 chapters by contributors from different disciplines and from different parts of the world.

In the book's **Foreword, Jean-Michel Glachant** notes that devoting a volume to energy communities might have looked like a joke only a decade ago. Today, however, such a volume is extremely welcome as energy communities are attracting increasing attention in Europe and elsewhere.

This growing interest is motivated by two trends: the rapid fall in the cost of renewables and the digitalization of virtually everything. Together, these two trends allow many new innovations in energy, including the emergence of collective action by individuals and entities such as small and medium-sized enterprises and local authorities.

Communities can play an important role in the production, allocation, and consumption of economic resources. Glachant has no doubts about that, but he wonders whether this will be the case in energy, a deeply regulated industry where not only technology or consumers' preferences matter, but also the decisions taken by regulators and policymakers.

The book is organized into four parts as outlined below.

Part 1: The concept of energy communities and their regulatory framework

This part provides an introduction to energy communities, including a taxonomy for understanding the different forms they take, as well as different perspectives on the legal, regulatory, and economic challenges they pose in different countries. These chapters raise important questions about the conditions under which energy communities are welfare-enhancing from a societal perspective, suggesting the need to establish and define a regulatory framework that has that perspective.

In **Chapter 1**, **A taxonomy of energy communities in liberalized energy systems, Nicolò Rossetto, Stefano Verde,** and **Thomas Bauwens** provide an overview of the extraordinarily multifaceted phenomenon of energy communities and offer a compass to navigate through the subsequent contributions in the volume.

The chapter introduces a pragmatic definition of energy communities as a heterogeneous set of collective actors performing one or more functions in the energy system, followed by a taxonomy of five categories of energy communities based on the main function performed and the level of maturity of their business models. For each category, the authors describe its business model, its strengths and weaknesses, and the policy and regulatory issues it raises.

The chapter's main contribution is to highlight the heterogeneity characterizing energy communities and the importance of properly distinguishing the various types of such organizations to make informed policy decisions.

In **Chapter 2, The EU policy framework for energy communities, Deyana Spasova** and **Sibylle Braungardt** provide an overview of the policy environment for energy communities in the EU.

The chapter compares the two legal definitions for energy communities introduced in the 2018 Renewable Energy Directive and the 2019 Electricity Directive, discusses relevant EU policy action, and offers insights into the development of energy communities in several EU Member States. The analysis illustrates that countries differ considerably with respect to their current deployment of energy communities and the legislative, social, and economic factors that influence their development.

The authors conclude that support is required for the implementation of the EU policy framework, particularly for Member States with few or no energy communities. The chapter also lays the foundation for the contributions that follow.

In **Chapter 3**, **Energy communities: A US regulatory perspective**, **Bruce Williamson** takes the perspective of a "typical" North American utility regulator in evaluating the introduction of energy communities, not *de novo*, but in service territories already adequately served by incumbent electricity utilities, both community owned and investor owned.

Examining the potential for welfare loss and welfare-improving possibilities of energy communities, the author notes that practical alternatives already exist in the many publicly owned systems in North America. These publicly owned entities—and some privately owned ones—are already experimenting with similar implementations, not by top-down directive in the EU mold, but driven by other incentives better suited to the North American legal, economic, social, and political structures.

The chapter's main contribution is a pragmatic regulatory evaluation of the EC concept, adjusting and molding it to fit the legislative and regulatory sensibilities across North America.

In **Chapter 4, Developing a legal framework for energy communities beyond energy law, Job Swens** and **Lea Diestelmeier** address the transposition of energy community–related EU legislation into national law, concluding that this requires more than the adaptation of energy legislation.

Based on experiences of the Dutch energy community "Schoonschip," the authors provide an inventory of legal barriers that energy communities need to overcome to fulfill their primary purpose to "provide environmental, economic, or social community benefits [...] rather than financial profits," as required by EU law. The authors analyze whether and how these barriers could be overcome in the Dutch legal framework.

The chapter's main conclusion is that a facilitating legal framework for energy communities requires a broader approach than implementing the specific provisions of the EU Directives into national energy legislation. This approach should emphasize the removal of the many barriers encountered by emerging energy communities.

In **Chapter 5, Alignment of energy community incentives with electricity system benefits in Spain, David Robinson** and **Iñigo del Guayo** assess whether the Spanish legislative, regulatory, and market framework for collective self-consumption will lower electricity system costs for all consumers, rather than shift costs to other consumers.

The chapter examines the national legislative context for the development of energy communities—including closed distribution systems, cooperatives, and community collective self-consumption—which is currently the main basis for community energy. It examines the alignment of the incentives for collective self-consumers with the interests of all consumers and recommends reforms to improve that alignment when transposing the EU legislation.

The main contribution of the chapter is to identify the legal and economic conditions that provide incentives for energy communities to lower overall system costs while also benefiting community members, offering a framework that may be applied to in other countries.

In **Chapter 6, The "virtual" model for collective self-consumption in Italy, Luca Lo Schiavo, Andrea Galliani,** and **Arianna Rossi** describe the pilot regulation that has been recently introduced in Italy to implement two configurations for collective action: collective renewable self-consumption and renewable energy communities. The authors point out that the current pilot regulation is transitional toward the full transposition of the EU Directive 2018/2001 in Italy.

The chapter provides a description of the model adopted by the Italian regulatory authority, called a "virtual model" because it permits collective self-consumption without the need for any new infrastructure, such as new distribution lines or meters.

The chapter's main contribution is to present the key characteristics of the Italian virtual model including its advantages, such as the customers' freedom to choose their retail energy supplier independently from participation in the collective self-consumption schemes.

In **Chapter 7, Energy communities: A North American perspective**, **Mark Kolesar** considers whether North America exhibits characteristics that can promote cooperative community-centric approaches to developing clean energy projects.

The chapter provides a survey of community energy projects in North America, drawing on business models that are emerging among energy communities elsewhere. It includes an overview of the policy framework in various North American jurisdictions as well as observations on the challenges faced and the models that appear to be most effective.

The author concludes that North Americans are motivated to engage in neighborhood-based renewable energy initiatives if given the opportunity and that a market-based approach that provides incentives for utilities and entrepreneurs to engage the public in these initiatives appears to be effective.

In **Chapter 8, Energy communities: Challenges for regulators and policy-makers, Darryl Biggar** and **Mohammad Reza Hesamzadeh** explore whether facilitating the creation of an energy community will improve overall economic welfare.

The chapter defines an energy community as "collective self-consumption"—that is, the ability to shift production and consumption within the group for billing purposes.

The authors show that there is a strong financial incentive to create such an energy community whenever the members of the group would otherwise face different retail prices at the margin. But as long as retail tariffs are inefficiently structured, creating an energy community exacerbates existing problems in the retail tariffs, leading to inefficient incentives for investment in generation and storage. The energy community potentially also undermines the ability of the local distribution network to efficiently price local congestion. The authors conclude that a tariff reform should precede energy communities.

Part 2: The appeal of energy communities to customers and citizens

This part expands on the role of citizens in energy communities. What entices citizens to engage in energy communities? What characteristics and attributes attract customers to want to become members? To what extent and how can energy communities contribute to engaging citizens in the energy transition? Finally, can energy communities enhance energy justice?

In **Chapter 9, What motivates private households to participate in energy communities? A literature review and German case study, André Hackbarth** and **Sabine Löbbe** explore preferences and motivations of private households regarding energy communities and discuss strategies for managers of such communities to increase participation of different customer groups.

The chapter reviews the literature and analyzes a survey of 1000 German households on this matter. It deduces pathways to support energy communities politically and strategies for organizations to attract potential participants by tailoring value propositions of energy communities to heterogeneous customer preferences.

The chapter's main contribution is to highlight the diversity of potential energy community participants. Aside from the economically driven majority, a substantial number of individuals can be attracted by social value propositions, e.g., environmental and community benefits or regionality. Also, prosumers, consumers, and citizens planning to invest in prosumption have significantly different preferences for energy community attributes.

In **Chapter 10, Community energy initiatives as a space for emerging new imaginaries? Experiences from Switzerland, Benjamin Schmid, Mònica Serlavós,** and **Léon Hirt** explore community energy initiatives through the lens of sociotechnical *imaginaries*, that is, collective and shared visions of desirable societies.

The chapter focuses on aspirations—views of desired future—and performances—enactment of the desired future—of community energy initiatives in Switzerland. It reveals the presence of alternative and plural sociotechnical imaginaries among community energy initiatives as well as tensions existing between aspirations and performances.

The chapter's main contribution is to illustrate that community energy can result in the emergence of new imaginaries that contribute to a sustainable energy transition, suggesting that such qualitative contributions of community energy should be considered in the design of policy instruments.

In **Chapter 11, The construction of a citizen-centered ecosystem for renewable energies in France**, **Pascale Château Terrisse, Hajar El Karmouni,** and **Marion Maignan** focus on an innovative multistakeholder cooperative supplying 100% renewable energy in France.

The chapter provides a description of the large ecosystem of this organization and how it provides multilevel and multisectoral models supporting the foundation for an ecological, economic, and social transition, while promoting the wide adoption and acceleration of the decentralized, community-based energy transition.

It identifies three key factors in ecosystems for renewable energies: a shared definition of the transition, important private and public financing dedicated to renewable energies, and the promotion of knowledge transfers on renewable energies. The authors recommend the growth of cooperatives as a means to restructure the energy ecosystem to be more local, democratic, citizen-centered, and decentralized.

In **Chapter 12, Energy communities' social role in a just energy transition**, **Florian Hanke**, **Rachel Guyet,** and **Mariëlle Feenstra** explore energy communities from the perspective of energy justice.

The chapter applies an energy justice framework to identify new opportunities for energy communities to contribute to a fair energy transition. It outlines energy communities' social role and illustrates how membership prerequisites can be a barrier to participation by vulnerable households.

The chapter's main contribution is to explore the barriers vulnerable groups face to become members of energy communities. It argues that socially inclusive energy transition depends on recognizing unequal participatory opportunities of different social groups. By examining existing best practice, the authors highlight the need for energy policy to recognize and reward the social welfare that energy communities create. To achieve this aim, they propose elements of an enabling framework to support energy communities' social role in contributing to a just transition.

Part 3: Enabling technologies, community design, and business models

This part demonstrates how technologies are applied in different ways to energy communities. The shape of community designs varies widely around the world and depends on the institutional, political, and regulatory framework. Within these very different environments, market participants—be they distribution network operators, energy suppliers, investors or IT providers—develop their business models to reflect their perspectives.

In **Chapter 13, Digitalization of peer-to-peer electricity trading within local energy communities, Emilio Ghiani, Mario Mureddu, Marco Galici, Matteo Troncia,** and **Fabrizio Pilo** point out that one of the motivating factors for developing energy communities is to create an opportunity for peer-to-peer trading among the members. Smart meters, Internet of Things, distributed ledgers, and digitalization enable energy communities based on renewable energy production to thrive.

The authors describe how local energy market participants can trade energy and provide services to the upstream distribution and transmission system operators and test the feasibility of their concepts using a laboratory-scale local energy community operated by P2P and blockchain technology.

The chapter's main conclusion is that, despite the availability of many technologies, an off-the-shelf solution for energy communities does not currently exist, which means that tailored solutions must be designed on a case-by-case basis. Standardization of components, interoperability, and cost reduction are critical requirements, as is the adaptation of national legislation and regulations, for the implementation of local energy communities.

In **Chapter 14, Enabling business models and grid stability: Case studies from Germany, Simon Köppl, Elisabeth Springmann, Vincenz Regener,** and **Andreas Weigand** present a concept that allows the marketing of small-scale flexibility capabilities that also enable their use for grid-stabilizing measures.

The chapter combines various concepts from living labs which demonstrate approaches that enable a cellular, participative, and diverse energy system to function. Inverting existing top-down approaches, locally and bottom-up organized concepts are designed to enable distributed responsibility for grid stability.

The chapter's main contribution is to create the organizational basis for citizen empowerment in the energy system. In contrast to the existing copperplate approach, this allows a synergistic interaction between community production, consumption, and the grid's capacities.

In **Chapter 15, The path to energy communities via local energy management and digital customer care, Tina Hadler** provides examples for practical implementation of energy communities as a platform-based business model.

The chapter describes the approach to energy communities from an energy service provider's perspective, providing a B2B2C portfolio based on digitization and decentralization of the energy transition in Germany and other European countries. A platform-based operating model must fulfill the regulatory requirements and the market needs of their clients. Several types of communities for different customer segments in various countries are examined to outline the implications for technical and process design.

The chapter's main contribution is to offer insights to how prosumers and consumers as members of a community can be connected, engaged, and incentivized to maximize the supply of locally produced energy from renewable sources.

In **Chapter 16, Governing energy communities: The role of actors and expertise in business model innovation, Jake Barnes** and **Paula Hansen** challenge the idea of single actors as the sole proprietors of energy community business models and articulate the role of actors and expertise in explaining what such business models achieve.

The chapter explores three ideal-typical governance arrangements of energy communities and examines their emergence and prevalence, their characteristics, and the business model archetypes achieved under each.

The chapter demonstrates that despite the seemingly wide range of activities open to energy communities, communities do not have a free hand when it comes to business model innovation, suggesting that specific business model archetypes are influenced by the governance arrangements they adopt, and the energy landscape they are embedded within.

In **Chapter 17, Grid-friendly clean energy communities and induced intracommunity cash flows through peer-to-peer trading, Reinhard Madlener** and **Robert Crump** examine the trade-offs between grid-friendliness and self-sufficiency of different types of clean energy communities (CECs) and intracommunity cash flows from local peer-to-peer (P2P) trading.

Using three versions of distribution grids—countryside, village, and suburban—the chapter examines residential load profiles, meteorological data, and the self-supply potential of distributed photovoltaics and wind power. Grid-friendliness is measured as changes in the community's net load gradient and the authors examine four options for CECs: improve energy efficiency; jointly purchase distributed PV systems; introduce P2P trading; and install a community-owned wind turbine.

The chapter's main conclusions are that smart combinations of measures help to enhance CEC grid friendliness, while cash flows generated from P2P trading are rather modest. Not surprisingly, the trade-offs between self-sufficiency and economic benefits to the members vary strongly depending on the CEC setup and the combination of measures.

In **Chapter 18**, **Italian energy communities from a DSO's perspective**, **Anita Del Pizzo, Giuseppe Montesano, Carlo Papa, Mariangela Di Napoli,** and **Marina Artipoli** highlight how the development of energy communities can benefit from technologically advanced distribution system operators (DSOs), which are able to provide a number of services and act as neutral market facilitators and innovators, ensuring reliability, resilience, cost effectiveness, and high service quality.

The chapter explores the nature of incentives, identifying their effects and ways to allocate subsidies consistently with the policy objective of reducing energy system costs. It also examines the criteria to define communities' boundaries and size.

The chapter's main conclusion is that the DSOs' role as neutral market facilitators can and should be to support energy communities in the simplest, most cost-effective and reliable way, while ensuring economic value to the energy system as a whole, helping the expansion of renewables and reducing the infrastructure footprint.

In **Chapter 19, Community energy design models in Brazil: From niches to mainstream, Richard Hochstetler** and **Paulo Born** focus on public policies that are most likely to foster effective energy communities, based on experiences with various models.

The chapter provides a comparative analysis of the different forms of energy communities that have emerged in Brazil, examines their objectives and how they have evolved over time, and evaluates how the regulatory environment can be improved to make them welfare-enhancing.

In the past, energy communities were only employed in remote or isolated regions, but recently, net metering programs have opened the door to widespread development of energy communities and consumer engagement. Energy communities have the potential to provide many benefits, but regulatory changes are necessary, including more granular retail pricing to induce efficient investment as well as policies to promote effective consumer engagement.

Part 4: Case studies and implementation

This part describes a few case studies illustrating how energy communities have been formed, how they function, and some of the remaining obstacles.

In **Chapter 20, Institutional and policy context of RECs in France and Italy: How to increase the welfare-enhancing capacity of the sector? Maksym Koltunov, Valentina Miriam Cittati,** and **Adriano Bisello** focus on factors that influence the deployment of renewable energy communities (RECs) and their welfare-enhancing capacity in France and Italy.

Using the quintuple helix innovation model, the chapter explains the current state of development, institutional forces and policy incentives affecting RECs, and includes case studies that reveal the benefits of RECs that adopt different business models.

The authors discuss how the circulation of knowledge among actors in five critical sectors—the market, the state, academia, civil society, and the interaction with the local natural environment—can stimulate innovation for RECs. They compare the conditions, stakeholders, and incentives in both countries.

In **Chapter 21, Energy communities in Europe: A review of the Danish and German experiences, Simona Benedettini** and **Carlo Stagnaro** review the early experience of citizen participation in and ownership of energy assets.

Both Denmark and Germany have adapted their national legislations to the requirements of EU directives concerning energy communities to some extent. Community energy has a long tradition in these countries, with private individuals owning 30 percent of renewable capacity for power production in 2019 in Germany, and 52% of wind power capacity in 2016 in Denmark.

The authors point out that the current experiences of community energy do not meet a stringent definition of energy community as defined by recent EU legislation. Moreover, they are often rooted in the past, and have historically enjoyed monetary or regulatory advantages. The future of energy communities in these countries, as well as in other EU member states, lies in emancipating energy communities from the protected environment where they have grown so far.

In **Chapter 22, Platform-based energy communities in Germany and their benefits and challenges, Christian Chudoba** and **Tereza Borges** draw on their experience from Germany to examine the advantages and challenges of using energy-as-a-service platforms to enable energy community models.

The chapter examines the historic context and underlines how platform-based approaches can remove barriers to innovation. A key point is the definition of the term energy service provider, which is fundamental for a new understanding of the energy market. The chapter provides an overview of the market drivers and the perspective of a broad range of actors, including traditional utilities and new entrants, supported by digital platforms that are reducing barriers to developing a diverse range of community models.

The chapter's main contribution is to articulate the potential of digital energy platforms as a foundation for community models and for the stacking of value.

In **Chapter 23, A community-based biomethane heat network: Case study from Trier, Christoph Menke**, **Achim Hill, Matthias Gebauer, Rudi Schöller, Daniel Gregetz,** and **David Lellinger** show how local stakeholders contribute to an energy community by joint planning of a local biomethane heat and power network in downtown Trier, the oldest town in Germany.

The project relies on biogas units upgraded to biomethane and used in local heat networks, which supply a hospital, a school, and retirement homes with low carbon heat and power through combined heat and power (CHP) units. Together with a new heat network using biomethane CHP units, it will supply the town hall, the theatre, and 800 buildings in the central quarter of Trier.

The authors document that the resulting heat network will reduce GHG emissions by roughly 90% compared to 1990, and serve potentially as a model for many downtown areas in Europe.

In **Chapter 24, Establishing energy communities in post-communist states: The case of Bulgaria, Toby D. Couture** and **Teodora Stoyanova** explore the challenges that former communist countries in Eastern and Central Europe face in their attempts to encourage the development of citizen-owned renewable energy projects, with a focus on Bulgaria.

These challenges include a complicated historical relationship to collective ownership, as well as the relatively low levels of household income. Low levels of household savings make it difficult for citizens to pool their financial resources and invest in common, community-based projects even when the willingness is there.

The chapter argues that different interventions are likely to be required to encourage the growth of energy communities in countries like Bulgaria compared to other jurisdictions, and that the greatest needs are to provide direct financial support, including grants and cofinancing, as well as to secure municipal involvement and investment.

In **Chapter 25, Sustainable island energy systems: A case study of Tilos island, Greece, Xin Li** explores the TILOS project that demonstrates the potential to integrate local small-scale battery storage systems, renewable energy technologies, smart meters, and demand side measures. The project's success illustrates the potential for green energy transitions on a small and isolated island.

The chapter examines the experience and lessons learned from the implementation of the project. During the first year of operation, renewable energy rose from 0 to over 50 percent of energy consumption. This success reflects the design and operation of the energy system and active participation of stakeholders, the local residents.

The main conclusion of the chapter is that engagement of the locals led to a high level of acceptance of the new power system. This suggests that effective communication and participation are critical to the success of all community energy projects.

In the book's **Epilogue, Bruce Mountain** examines the differing definitions of energy communities and responses to the question of whether they are—or can be—citizen-centered, market-driven, and welfare-enhancing, a reference to the title of the volume.

Rhetorically asking what motivated the editors to compile such a volume in the first place, the short answer is that with recent advances in technology, the distributed production, storage, and trading of electricity is now not only possible but economically attractive—and this offers potential for increasing levels of individual energy self-sufficiency as well as various forms of communal distributed production and electricity trade.

The Epilogue's main insight is that while in many ways today's energy communities might be taking us back to a place from whence we first came, we are now motivated to make better use of local renewable resources while avoiding some of the inefficiencies of the existing electricity infrastructure.

References

Burston, J., 2016. 5 Predictions for Energy in 2030. World Economic Forum. https://www.weforum.org/agenda/2016/11/5-predictions-for-energy-in-2030.

IEA, 2019. Renewables 2019: Analysis and Forecasts to 2024. International Energy Agency, Paris. https://doi.org/10.1787/b3911209-en.

Sioshansi, F. (Ed.), 2019. Consumer, Prosumer, Prosumager – How Service Innovations will Disrupt the Utility Business Model. Academic Press, London, UK.

Part One

The concept of energy communities and their regulatory framework

A taxonomy of energy communities in liberalized energy systems

1

Nicolò Rossetto[1], *Stefano F. Verde*[2] *and Thomas Bauwens*[3]
[1]Florence School of Regulation, Florence, Italy; [2]Department of Economics and Statistics, University of Siena, Siena, Italy; [3]Utrecht University, Utrecht, Netherlands

1. Introduction

Energy communities are a heterogeneous phenomenon in liberalized energy systems. They are not completely new, as they used to play a marginal role, both in Europe and North America.[1] However, it is only in the past few years that they have received increasing attention and started to leave the niches to which they were relegated for decades.[2] This development is not only quantitative but also qualitative: beyond the increase in their number and size, energy communities are also diversifying and taking on new activities.

Community-driven initiatives are today particularly visible in the electricity sector. In countries such as Denmark and Germany, the development of generation technologies that are reasonably efficient and easy to operate on a small scale and the implementation of public policies supporting the massive deployment of renewable energy sources (RESs) have contributed to the emergence of energy communities primarily involved in the generation of electricity (Bauwens et al., 2016). The digitalization of electricity infrastructure and the electrification of end uses such as transport and heating are now offering further opportunities. For instance, community-driven initiatives are increasingly observed which involve the aggregation of distributed energy resources (DERs), the establishment of microgrids, and the provision of shared mobility (THEMA and Multiconsultant, 2018; Tounquet et al., 2019; Verde and Rossetto, 2020).

As several chapters in this book show, energy communities differ in many respects. These include not only the functions performed but also the maturity level of their business models, the geographical scope of their operations, the number and characteristics of their members, their governance, and the legal form adopted. This

[1] The most notable exception is represented by the cooperatives that played an important role in the electrification of rural areas in Europe and the United States, mostly during the first half of the 20th century.

[2] Statistics on energy communities are hard to find and often not easily comparable across countries. In Europe, REScoop.eu gathers about 1900 renewable energy cooperatives with roughly 1.25 million individual members (https://www.rescoop.eu/). REScoop.eu estimates that these are only about half the renewable energy cooperatives active in Europe.

Energy Communities. https://doi.org/10.1016/B978-0-323-91135-1.00004-3

heterogeneity is a defining feature of energy communities as well as one of the reasons why it is difficult to provide a straightforward and fully satisfactory definition of what they are. It is also a source of frequent confusion in public debate and a major obstacle to sound and constructive policymaking since different types of energy communities present different advantages and disadvantages to their members, provide different costs and benefits to the rest of society, and pose different policy and regulatory challenges.

Energy communities can be classified in different ways and the existing literature has already discussed some of them. In order to provide a pragmatic compass for practitioners, policymakers, and regulators to navigate through the many types of energy communities that are observed nowadays in liberalized energy systems, this chapter adopts two classifying criteria: the main function performed by the energy community and the level of maturity of the business model adopted. The first criterion specifies how a community interacts with the rest of the energy sector, while the second distinguishes between more established and more innovative communities. The combination of these two criteria clusters energy communities in five categories that are characterized by different sets of strengths and weaknesses, and which raise different policy and regulatory issues. A description of these categories and the different types of communities they include provides an overview of the key issues currently debated in public discourse and can guide the reader through the next chapters of the book, which tend to focus on specific types of energy communities.

In addition to this introduction this chapter consists of five sections:

- Section 2 discusses the concept of energy community in greater depth and highlights how, beyond a very few core characteristics, energy communities are highly heterogeneous;
- Section 3 provides a taxonomy of energy communities based on the main function(s) they perform and the level of development of their business models;
- Section 4 illustrates the types of energy communities that have a longer tradition and can be nowadays considered well established; and
- Section 5 completes the overview of the taxonomy by discussing the most recent types of energy communities that are emerging due to the increasing digitalization and electrification of energy systems, followed by the chapter's conclusions.

2. A heterogeneous set of collective actors

There is no consensus in the academic literature and public discourse on what precisely energy communities are (Bauwens et al., 2022). The multidimensionality of the terms "community" and "energy" implies that different people may have different things in mind when they refer to energy communities. Nonetheless, by examining the initiatives that normally receive this label in the world of practice, it is possible to agree on the fact that, broadly speaking, energy communities constitute a heterogeneous set of collective actors performing one or more functions in the energy sector. This is the pragmatic definition used in this chapter. A short discussion will be sufficient to clarify its main elements.

First, energy communities are collective actors that gather households, small- and medium-sized enterprises (SMEs), and other organizations, often of small size, such as local public bodies and nonprofit associations. The members of an energy community have something in common. This can be a place or an interest (Moroni et al., 2019). For instance, members can all be located in the same neighborhood or they can all be concerned about the implications of climate change. Community members are also typically small or at most medium-sized energy consumers with no or a limited professional stake in the energy sector. Because of these commonalities, the members of a community tend to see each other as peers and community governance is normally democratic and horizontal rather than centralized and hierarchical.

Second, energy communities perform one or more functions in the energy sector. Households, SMEs and other organizations establish a new energy community or join an existing one because they are willing to take direct action in the field of energy. Instead of delegating the provision of energy to traditional utilities, they aspire to jointly satisfy their energy-related needs and preferences such as consuming electricity generated from renewable or local energy sources. By joining their resources and coordinating their actions, the members of an energy community can overcome some of the obstacles that often frustrate individual consumers eager to play an active role in energy. Collective action can be, for instance, more cost-effective than individual initiatives because it allows the exploitation of higher economies of scale and scope. Collective action can also generate new sources of value by creating new relations among community members and providing relational goods (Verde and Rossetto, 2020).[3] Importantly, community members are generally not merely passive beneficiaries of collective action; they also tend to be directly involved in it. Spontaneous and widespread participation, for instance, via voluntary work or attendance at general assemblies, is a key feature of energy communities (Walker and Devine-Wright, 2008).

Third, energy communities are heterogeneous. Beyond the core characteristics mentioned above, energy communities may differ from one another in several respects. They can differ in terms of size and internal homogeneity: some communities may have only a few tens of members, all of them being very similar to each other, while other communities may be formed of thousands of members with very different backgrounds and socio-demographics. The dominant motivations among the community members and the prevalent goals of their collective action can also differ from one community to another (Bauwens, 2016). In one case, it could be the provision of electricity in a rural area neglected by investor-owned utilities, while in another case it could be a desire to contribute to an accelerated transition to a low-carbon economy. Differences are also possible in terms of the activities performed and the geographical level at which collective action takes place. Although electricity generation from RES is a frequently mentioned activity, energy communities often perform other functions

[3] "Relational goods" refer to forms of social interaction of a non-instrumental nature and in which the non-anonymity of the parties involved constitutes a value. Examples include social approval, solidarity, friendship, and the satisfaction of having complied with a social norm (Bauwens, 2019).

as well. In some cases, their outreach is somewhat small (e.g., a neighborhood or a village), while in others communities operate at the regional or even national level. Energy communities may also present different degrees of participation and be more or less egalitarian in their governance structure. While in some communities every member has exactly the same powers, in other communities, it is possible to have members that have a stronger position and more influence on collective decisions than others. This can also be the result of the various legal forms that a community can adopt. Although the cooperative model is a frequently selected option, other possibilities exist.

3. A taxonomy to inform policy and regulatory debates

There are many possible ways to classify energy communities and no single taxonomy is perfect. Depending on the context or the purpose, some classifications can suit better than others. The academic literature already offers some examples in this regard (Gui and MacGill, 2018; Moroni et al., 2019; Braunholtz-Speight et al., 2020; Reis et al., 2021). These taxonomies have their advantages and relevance. However, some of them remain arguably at an excessively high level of abstraction, making it difficult to identify the concrete implications of different energy communities for the organization and regulation of the energy sector. Other taxonomies, in contrast, consider too many dimensions, mirroring too closely the multifaceted reality of community energy and failing to provide a sufficient level of synthesis. A different perspective is needed to provide practitioners, policymakers, and regulators working in liberalized energy systems with a comprehensive but simple compass to navigate through the many forms that energy communities can assume and understand what policy and regulatory issues they typically raise. The main function performed and the level of maturity of the business model adopted by an energy community are useful in this regard.

The main function performed refers to the primary activity or set of activities within the energy sector that a community is focused on.[4] Observation of energy communities operating in liberalized energy systems suggests that communities can perform several different functions (Ofgem, 2017; THEMA and Multiconsultant, 2018; Tounquet et al., 2019; Caramizaru and Uihlein, 2020). They can provide energy consumers and prosumers with support, produce and supply energy, deliver energy via transmission and distribution networks, or deploy and manage DERs.[5] Depending on the main function performed—i.e., on what it does from the energy sector point of view—an energy community will interact with the rest of the sector in a specific way and play

[4] The main function performed by an energy community should not be confused with the motivation(s) of the community itself and its members. The former refers to the object of collective action, while the latter to the objective that collective action wants to achieve. See the chapter by Hackbarth and Löbbe for more on motivations to participate in energy communities.

[5] Over time, observations of new cases of energy communities may lead to reassessing the four main functions identified in this chapter.

a certain role. The impact of its emergence and development will be different, as will be the way it is affected by existing and prospective public policies and regulation.

The level of maturity of the business model refers to the possibility for energy communities to adopt different business models, some of them more mature than others, when performing a specific function within the energy sector. Experience with some business models is consolidated in the sense that community-driven initiatives adopting them are numerous and sometimes date back several years or even several decades (although this does not imply that they play a quantitatively large role in the energy sector nowadays). These types of energy communities have proved to work, at least under certain circumstances, and can be legitimately called "well established." In contrast, experience with other business models is much less consolidated. Concrete initiatives are less common and more recent—they are no more than three or four years old, tend to involve few people or limited geographical areas, and are often related to research or pilot projects led by universities and innovative firms. It is usually not fully clear whether they can work and under what circumstances. The energy communities implementing these models act as innovators and, thus, we refer to them as "new kids on the block." Depending on the level of maturity, information available on the different types of energy communities is more or less abundant, their strengths and weaknesses are more or less understood, as are the policy and regulatory issues they raise.

By combining the main function performed and the level of maturity of the business model, five categories or families of energy communities can be identified (see Table 1.1). Each category can contain one or more types of communities which share the same function and level of maturity but are somewhat different in their other dimensions (e.g., the value proposition for members or the revenue stream).[6]

The next two sections of this chapter provide an overview of the categories identified and the types of energy communities they contain, addressing first those that are well established and then the new kids on the block. This overview includes descriptions of the community business models, their strengths and weaknesses, and the policy and regulatory issues raised.

4. Well-established energy communities

Collective action in the energy sector is not a recent development and several categories of well-established energy communities exist. They perform one of the three following main functions: provide energy consumers and prosumers with support; produce and supply energy;[7] and deliver energy via networks. Multiple types of communities implementing different mature business models exist in the case of the first two

[6] Three combinations of the two criteria are empty, since no observed type of energy community can be assigned to them, at least for the time being.

[7] In this chapter, energy generation and production are considered synonyms. The same applies to energy retail and supply.

Table 1.1 A taxonomy of energy communities.

	Level of maturity of the business model	
Main function performed	**Well-established communities**	**New kids on the block**
Support provision to energy consumers and prosumers	• Joint purchasing groups ➢ DERs ➢ Collective switching • Assistance providers ➢ Energy services ➢ Facilitated access to finance	• Not observed
Energy production and supply	• Community energy producers • Community energy retailers ➢ Asset light ➢ Integrated with production	• Energy sharing communities ➢ Collective self-consumption ➢ Peer-to-peer energy exchange
Energy delivery via networks	• Utility cooperatives	• Not observed
Deployment and management of DERs	• Not observed	• Shared electric mobility community providers • Community aggregators • Community microgrids

functions. These are joint purchasing groups and assistance providers, and community energy producers and retailers. In the case of energy delivery via networks, the most common community type is the utility cooperative.

4.1 Joint purchasing groups and assistance providers

4.1.1 Business model description

Energy communities can focus on providing support to energy consumers and prosumers. This support can take two forms. First, community members can aggregate their demand for DERs or energy supply and collectively look for better bargains with technology vendors or energy retailers. This type of community, which can be labeled a joint purchasing group or a collective switching mechanism, acts as an asset-light intermediary for its members, who retain full control of their purchasing decisions and contract relations. The community exploits its larger scale and possibly the specific skills of some members to get lower prices for the assets or the energy ultimately purchased by its members.

Second, community members can receive technical and financial assistance regarding energy-related issues. This type of community, which can be labeled an assistance provider, offers various energy services such as performing building energy audits and implementing energy efficiency enhancing solutions (replacing inefficient heating systems, insulating buildings, installing distributed generation behind the meter, etc.). A network of technicians, often members of the community, and company partners is normally available to do this. This type of community can also provide members with financing for their investments in DERs or energy efficiency solutions. In cases where the community pursues a strong social goal, such as tackling energy poverty, this assistance can be given to people in need at reduced prices or freely. This type of community may also conduct campaigns and carry out educational initiatives for the purpose of raising community members' awareness of energy issues.

4.1.2 Strengths and weaknesses

Energy communities belonging to this category often operate at the local level and rely on existing social relations that promote trust and cooperation. However, they may also be active on a broader scale by gathering individuals with a common interest, such as the willingness to act autonomously in the energy sphere, to save on their bills or to contribute to a more sustainable energy system. These communities can take advantage of monetary and non-monetary economies of scale compared to initiatives by individual energy consumers and prosumers. Still, they often remain quite small and may face tough competition from dedicated solutions provided by traditional commercial actors like energy utilities and energy service companies.

4.1.3 Policy and regulatory issues

Joint purchasing groups and community assistance providers can foster the active involvement of costumers in the transition to a more sustainable energy system. Their impact may be particularly relevant for less affluent community members, who may struggle to afford to invest money in more efficient energy appliances and dwellings, not to mention distributed generation. They can also facilitate coordinated initiatives among the inhabitants of multi-apartment buildings. However, the benefits are not purely economic or environmental as these initiatives can also promote a sense of community and social ties between the members. From a regulatory perspective, these energy communities do not raise significant problems as they address deregulated activities and do not disrupt the existing use of public energy networks. The consumers and prosumers they support will continue to individually interact with the rest of the energy system. On the contrary, these communities may reduce the need for public bodies to focus on implementing customer protection and empowerment.

4.2 Community energy producers and retailers

4.2.1 Business model description

Energy communities can focus on producing and supplying energy to their members and to non-member energy customers. This can take two forms. First, a community can invest in new energy generation capacity, often but not necessarily RES-based. The energy produced from community assets, in most cases electricity,[8] is physically injected into the public grid and typically sold to an off-taker via a bilateral contract, or more rarely directly on the wholesale spot market. Revenue from the sale of energy is used to remunerate community members, finance investments in new capacity, or support other social or environmental community initiatives. This type of community, which can be labeled a community energy producer, is often local—it gathers people living in a village or a valley to exploit local primary energy sources. However, it can also operate over a larger geographical area involving members spread over an entire country and building generation assets that can be quite distant.

Second, a community can provide energy to its members, who then become customers of the community. The energy supplied often presents some specific features, for instance, it can be generated entirely from RES, and can be either purchased by the community from other producers or produced by directly-owned community assets. In the former case, the community can be labeled an asset-light community energy retailer, which conveys the preferences and needs of its members to external energy producers. In the latter case, the community also invests in its own generation capacity and closes the circle from production to consumption. It can then be labeled a community energy retailer integrated with production.[9] Community energy retailers, integrated or not, tend to be less geographically concentrated than community energy producers and to attract members at the regional or national level.[10]

4.2.2 Strengths and weaknesses

Since the last decade of the 20th century, these two types of energy communities have gradually gained importance in Europe and the United States due to the significant development of renewables and the support policies implemented by several governments (Wierling et al., 2018). They have also benefited from the liberalization of retail markets and the possibility for households and other small consumers to change supplier.

Community energy producers can leverage the geographically dispersed availability of RESs and their suitability for decentralized exploitation by many small- and medium-sized plants. Communities are particularly suited to operate at this scale, say in the 100 kW–5 MW range, which is normally more efficient than that in which individual households or small businesses can act, say in the 1–200 kW range.

[8] Some energy communities produce heat or biogas, which are injected in local district heating networks or natural gas pipelines. See the chapters by Benedettini and Stagnaro and by Menke et al.

[9] In France, Enercoop is an example of a community energy retailer integrated with production. See the chapter by Château Terrisse et al.

[10] In Europe, there are now a few community energy retailers that have a strong regional dimension. Ecopower, for instance, has most of its members located in Flanders (Belgium), while Som Energia is centered in Catalonia (Spain). Nevertheless, this regional dimension does not prevent people from other parts of Belgium or Spain joining them. In the United States, a community retailer-like institution, Community Choice Aggregation (CCA), normally gathers the residential customers of a municipality or a county (see the chapter by Kolesar for more information about the North American experience with energy communities).

Obviously, utility-scale plants are even more efficient, but for certain types of renewables, the efficiency gap is not so large as it is for fossil fuels. Vis-à-vis individual prosumers, energy communities also have the possibility to select better sites in terms of primary energy endowment and to invest in a portfolio of technologies and sites, which may make their energy generation and the related revenue stream more stable over time. Finally, especially when the community operates at the local level, investment in energy generation assets by community producers has the advantage of expanding the ownership of the initiative. By involving all the people that can be affected by the realization of a project in the decision-making process, communities can more easily raise awareness, incorporate individual concerns, and gradually reach consensus on how to implement the project and allocate its costs and benefits in a satisfactory manner. Local opposition to new infrastructure, a serious problem that often slows down permitting procedures and completely blocks RES projects, can then be better dealt with.

Community energy retailers rely on the possibility for energy customers to choose their supplier and on a large enough share of them valuing energy with specific features and possibly paying premium prices for it. Where they have close relations with a community energy producer or they have their own generation assets, community retailers can benefit from synergies between production and supply. In particular, they can procure the electricity needed to supply their customers at a more stable and possibly lower price. By doing this they can also strengthen the sense of autonomy and purpose that community members often look for.

Despite the advantages mentioned above, energy communities involved in producing and supplying energy also present some important weaknesses. Although they are more efficient than individual prosumers, community energy producers often struggle to achieve a level of costs that is competitive with the usual price of energy on wholesale markets. The limited amount of technical skill and financial capital that communities can mobilize, compared to that available to traditional energy companies, means that they often have to resort to support from professional third parties (e.g., wholesale traders and balance responsible parties) and may not invest in large enough plants. Long and uncertain permitting and authorization procedures necessary to build and connect new plants constitute a relevant barrier, especially for communities that rotate around a single project: failure to get the permits and authorizations needed in a reasonable timeframe may demotivate members and possibly lead to community disbandment. On top of that, the very same democratic and horizontal governance that characterizes these communities can reduce the speed at which they make decisions and their willingness to seize new but risky opportunities. In the last few years, a phase-out of community-friendly RES support mechanisms, such as feed-in tariffs (FiTs) for solar PV and onshore wind energy, has been followed by a significant slowdown in the establishment of new renewable energy cooperatives in several European countries. These developments confirm how difficult it is for community-driven initiatives to invest in energy generation and grow quickly without dedicated policy measures (Herbes et al., 2017, 2021; Braunholtz-Speight et al., 2020).

Energy retailing is a relatively low-margin activity subject to intense competition which is quite demanding in terms of administrative processes and obligations, as in many liberalized energy systems retailers represent the main interface between the final customer and the rest of the sector, and are responsible for a series of activities such as tax and levy collection. As a result of these obligations, energy communities willing to be involved in supplying energy must deal with many additional tasks which normally require the support of professional third parties and generate new costs, which are particularly significant when spread over a thin customer base.

Finally, when they achieve a larger scale, community energy producers and retailers can be subject to internal tensions. They may start attracting people more interested in financial returns rather than collective action per se or the non-economic goals originally pursued by the community. The level of participation of their members may diminish. At the same time, the need for an efficient and effective decision-making process may require a change in the governance of the community and the adoption of a more hierarchical structure. As a result, some of the distinctive characteristics of community energy may start to fade away (Bauwens and Devine-Wright, 2018; Bauwens, 2019).

4.2.3 Policy and regulatory issues

Community energy producers and retailers play an important role in the democratization of the energy transition (Morris and Jungjohann, 2016). They offer people and small organizations an opportunity to be directly involved in the move to a different, more decentralized and less carbon-intense energy system, and to benefit from it. From a public policy perspective, they help mobilize private capital and address problems of local acceptability of new RES projects. They are also actors that can relaunch economic activity in rural and sparsely populated areas, thereby fighting their impoverishment and depopulation (Taylor, 2019).

However, from a legal and regulatory point of view, they raise some issues, for which clear solutions have not yet always emerged (De Almeida et al., 2021). An important challenge in this regard is how to ensure fair and non-discriminatory treatment of energy communities as new participants in liberalized energy markets. Given the specific characteristics of energy communities, for instance the fact that they typically qualify as non-commercial actors, it could be argued that non-discriminatory treatment implies dedicated rules and a favorable regulatory framework (Roberts, 2020). On the other hand, regulators and policymakers should avoid providing unjustified advantages that penalize competition or final customers. Similarly, it could be argued that the specificities of community initiatives should be taken into account when developing new RES support mechanisms, in order to ensure the concrete possibility for communities to access public support and effectively allow democratization of the energy transition.[11] At the same time, of course, these rules should not translate into unnecessary costs that ratepayers have to bear to finance the decarbonization of the energy system.

[11] In the EU, Directive (EU) 2018/2001 of the European Parliament and of the Council of 18 December 2018 on the promotion of the use of energy from renewable sources (recast) foresees an obligation for Member States to ensure competition for public support on an equal footing (art. 22.7).

4.3 Utility cooperatives

4.3.1 Description of the business model

Energy communities can focus on delivering energy via networks. This normally takes the form of local utility cooperatives in which households, businesses, and other organizations gather and invest their resources to build the infrastructure necessary to connect individual customers to energy generation plants or to transmission networks carrying energy from distant producers.[12] This type of energy community has a long tradition which dates back to the first half of the 20th century and justifies the frequent epithet of historical energy cooperatives.[13] At that time, large rural and mountain areas of Europe and North America still had no modern energy infrastructure, and private and publicly owned energy companies were generally unwilling or unable to provide people and businesses with access to electricity, gas or heat, normally because of unsatisfactory profit expectations or limited available capital. As a result, in several cases, it was local cooperatives that took the initiative and invested in generation, distribution, and sometimes even in transmission assets.[14] Electrification in numerous Alpine valleys and in wide parts of inland US occurred in this way.

More recently, dissatisfaction with the results of liberalization and privatization of local public services such as water and electricity distribution have fueled a process of "remunicipalization," in which local municipalities or groups of citizens claim back the infrastructure for the provision of those services. This phenomenon, although quantitatively somewhat limited, provides the basis for the emergence of new communities focused on the delivery of energy via networks at the local level.[15]

4.3.2 Strengths and weaknesses

Utility cooperatives blossomed decades ago because of a lack of competitive alternatives for the procurement of energy, in particular electricity. On the one hand, investor-owned and publicly owned utilities were not interested or able to supply people and businesses located in sparsely populated or isolated areas. On the other hand, the cost of producing electricity individually was well beyond the reach of most people and businesses. In this context, collective action represented a good solution. Benefiting from close social relations and the trust existing among local community members, cooperatives were able to mobilize local scarce resources and invest in

[12] Both in Europe and the United States municipal utilities have sometimes performed a role similar to that of historical cooperatives. However, although local governments traditionally play a major role in the decision-making processes in these utilities and indirectly channel the interests and preferences of people and businesses belonging to the local community, there is no consensus on whether municipal utilities are a type of energy community or not.

[13] In the United States, the expression rural electric cooperative is common.

[14] Many of the 830+ rural electric cooperatives existing in the United States jointly own 60+ so-called generation and transmission cooperatives. These cooperatives of cooperatives invest in large power plants and transmission lines or buy bulk power on behalf of their members (Imgarten, 2016).

[15] Elektrizitätswerke Schönau (EWS) in Germany is an example of a cooperative of citizens who purchased the local electricity distribution grid (https://www.ews-schoenau.de/).

long-lasting physical assets. Patient capital provided by the members was essential to build hydropower stations and electric distribution lines with very long pay-back periods. By considering the preferences and needs of the users during system planning, and by ensuring a fair distribution of costs and benefits, utility cooperatives could and can still today successfully target issues of local acceptability that often afflict the construction of new energy infrastructure.

However, their local nature and small size are also a cause of weakness for this category of energy communities insofar as they prevent them operating at a larger and more efficient scale. Such downside is compounded by current changes in electricity distribution systems that call for significant new investment, not only in "iron and copper" but also in new digital technologies, and require new skills and competences, which utility cooperatives may struggle to acquire or contract (Rossetto and Reif, 2021). Merging or federating with similar entities can be a way out for them, but this is also likely to water down their local community dimension.

4.3.3 Policy and regulatory issues

Regarding their implications for the energy system, utility cooperatives played an important role in ensuring access to modern forms of energy and exploiting local energy sources, often renewable-based, in areas where neither private initiatives nor state-owned companies were present. However, with the completion of basic energy infrastructure almost everywhere in developed countries and the provision of energy access to basically everybody, the rationale for this type of community is now reduced. Moreover, the importance that liberalized energy systems attribute to unbundling competitive activities (e.g., energy generation and supply) from those with monopoly characteristics (e.g., distribution) challenges the business model of these communities, which are often vertically integrated from generation to distribution and supply. Ensuring that old and new utility cooperatives provide high-quality distribution services and avoid discrimination among network users is an issue that regulators are sometimes worried about (Thema and Multiconsultant, 2018; CEER, 2019).

5. New kids on the block

Recent trends in energy systems, such as decentralization, digitalization, and electrification, open new spaces on which collective action can focus and which enable the appearance of innovative energy communities. Four types of energy communities can be identified. While energy sharing communities deal with the production and supply of energy in a new way, shared electric mobility providers, community aggregators, and community microgrids address the deployment and management of DERs in different ways.

5.1 Energy sharing communities

5.1.1 Business model description

The development of distributed generation technologies such as rooftop solar PV and the most recent waves of digitalization in the electricity system allow communities to be involved in energy sharing, a set of business models which has attracted much attention in public and regulatory discourse over the past three or four years, and which has been associated, although not always in a consistent way, with the concepts of collective self-consumption (CSC) and peer-to-peer (P2P) energy exchange (Brown et al., 2019; REScoop.eu and ClientEarth, 2020).[16] These two expressions can be used to differentiate two types of communities belonging to this category.

Energy communities performing CSC invest in community generation assets—typically solar PV plants—and possibly also in storage devices, and use the energy output to cover, either physically or virtually, the energy consumption of their members, at least partially. CSC can be done in a strict sense when both the community assets and the members are all behind the same connection point to the public grid, or it can be virtual if the community's assets and members are connected to different points in the public grid, which they then make use of in order to share the energy generated.[17] In CSC communities, individual members continue to have their own energy suppliers that take care of their residual demand when there is not enough energy produced by the community's assets. Members also continue to be fully autonomous in terms of when and how much energy to consume on their premises.

Energy communities enabling P2P energy implement the concept of energy sharing in a slightly different way. In this case, the community gathers a group of individual consumers, prosumers and prosumagers,[18] and allows them to exchange the energy injected into the grid by their individually owned assets, typically rooftop solar PV or electric vehicles, possibly on conditions that are individually chosen by the members and reflect their personal preferences (e.g., price, origin, and destination of the energy). The community may also invest in community assets and attribute shares of them to individual members, but its key task remains to provide, directly or via a third party, the digital infrastructure on which energy exchange can occur.[19] This infrastructure represents a fundamental "matching loop" as it reduces the otherwise exorbitant transaction costs for the peers (Glachant and Rossetto, 2021). P2P energy exchange can take place at the local level, for instance, within the same building or neighborhood, or it can

[16] According to the European legislation, groups of active customers, jointly acting renewable self-consumers, citizen energy communities, and renewable energy communities are entitled to share energy by means of collective self-consumption and peer-to-peer energy trading. See the chapter by Spasova and Braungardt for additional information.

[17] Partagélec Pénestin in France is an example of a community implementing virtual CSC (Verde and Rossetto, 2020).

[18] A prosumager is a consumer who produces and stores (at least part of) the energy she consumes by means, for example, of a solar PV panel and a battery installed on her premises.

[19] The Beehive Project by Enova Energy Community in Australia is an example of a community implementing P2P energy exchange (https://www.enovaenergy.com.au/shared-community-battery).

be supra-local. In this case, peers are connected to different parts of the same distribution grid or even to different distribution grids.

5.1.2 Strengths and weaknesses

Communities focused on energy sharing are today particularly attractive since the cost of generating energy at the distribution level can be competitive with the retail rates final customers usually face. If they are allowed to share the production from their individual or community generation assets, members of a community can put together their different load patterns and are likely to obtain higher self-consumption rates that increase savings on their energy bills. However, a community involved in energy sharing can address not only the economic interests of its members but also their frequent desire to be more independent from traditional players, feel part of a group of peers, and contribute to the decarbonization of the energy system. Individuals that do not have the space or financial resources to afford renewable energy investments are likely to be particularly empowered, as are people that like the idea of choosing with whom they exchange energy and the price they get. In an energy sharing community, these people can have the possibility of rewarding producers they know above market prices or even donate any surplus energy to others they want to help. Close social relations and digital tools enabling energy flow monitoring, information sharing, and rating by members all support the coordination of collective action in this type of communities.

In terms of weaknesses, communities focused on energy sharing are dependent on a series of regulatory decisions and on a general evolution of the regulatory framework that fully legitimize this relatively new type of business model. Indeed, the possibility that small and non-professional entities could share, directly or via a community, their energy over the public grid was generally not contemplated when energy markets were restructured in the 1990s and 2000s. Depending on the specific decisions that regulators make, for instance, regarding the space and time interval within which CSC can be performed or regarding the network charges and levies applied to shared energy, the economics of energy sharing can significantly change, as distributed generation is often cheaper than full retail rates but is rarely competitive with the price of the kWh traded at the wholesale level. If the economics are not particularly good, it could then be more difficult for a community to agree on a sustainable way of sharing the benefits and to attract enough members with sufficiently diversified generation and load profiles (Abada et al., 2020; Wilkinson et al., 2020).

5.1.3 Policy and regulatory issues

Although this new type of energy community is not yet quantitatively so important, it is the one which regulators and policymakers nowadays mostly focus on, at least in Europe (see the chapters by Robinson and del Guayo, Lo Schiavo et al., as well as Biggar and Hesamzadeh).

From a policy point of view, the promotion of energy sharing via the emergence of CSC and P2P energy communities is considered a potentially important way to support a further and more efficient decarbonization of the energy system. After the phase-out of FiTs for small- and medium-sized RES-based plants, allowing energy sharing by households and small organizations provides prosumers with a set of new business models which may be cost-effective for the system (if consumption is geographically close to production, the energy system may benefit from reduced energy losses and lower network capacity needs). As mentioned above, allowing energy sharing also looks like an inclusive policy, since it offers the possibility for less affluent people to access renewable energy and benefit from its deployment. Interestingly, some energy distribution system operators (DSOs) seem eager to facilitate the emergence of this type of energy community and support their activities (see the chapter by Del Pizzo et al.). On the other hand, energy sharing and its implementation in communities raise various regulatory issues that relate to both network and energy market regulation (CEER, 2019; REScoop.eu and Client Earth, 2020). This is not the place to address them; a list of some key questions that deserve attention will suffice:

- What should be the spatial and temporal boundaries within which energy can be collectively self-consumed?
- What network charges and other levies should apply to the energy shared within a community?
- To what economic regulation should DSOs be subject in order to incentivize them to support energy sharing communities?
- Should community members be free to leave a community without an exit charge?
- What should be the rights and duties of the energy suppliers of the final customers involved in energy sharing?
- Should all members of a P2P energy community have the same energy supplier or should they be allowed to have different suppliers? How should the various interactions be settled in the second case?

The reader is invited to look at other chapters in this book for more detailed assessments of these issues.

5.2 Shared mobility providers, community aggregators and microgrids

5.2.1 Business model description

The most recent waves of digitalization reaching distribution networks and customers' premises, the increasing decentralization of energy production, and the electrification of final energy uses like transport provide communities with the opportunity to become involved in the deployment and management of DERs, both on the production and consumption sides. Concrete cases of communities performing this new function are still limited in number and the policy and regulatory framework is sometimes still under development. Nevertheless, various initiatives have been launched in the past few years and have attracted the attention of households, SMEs, urban developers, digital start-ups and other organizations willing to experiment with new technologies and explore new energy solutions and business models (THEMA and Multiconsultant, 2018; Tounquet et al., 2019; CEER, 2019; REScoop.eu, 2021). At least three types of energy communities performing this function can be observed nowadays: shared electric mobility community providers, community aggregators and community microgrids.

A community providing shared electric mobility purchases a fleet of electric cars or other vehicles (e-bikes, e-scoters) and makes them available to its members. The community then manages the use of the vehicles via a digital application and can also take care of their recharging by installing and operating a few recharging points in the geographical area of interest.[20]

By actively managing and combining the load and generation of its members, a community aggregator unleashes the flexibility potential of energy consumers, prosumers and prosumagers, and more effectively integrate renewables and other technologies, such as electric vehicles and domestic batteries, in the energy system. By aggregating its members, an activity that can be carried out at both the local and supra-local level, an energy community can increase the share of energy produced that is collectively self-consumed or internally exchanged. It can also directly participate in energy markets and sell ancillary services to system operators and other market parties.

A community microgrid goes one step further by establishing its own small-scale "delivery loop" physically linking all the members, their resources, and those installed by the community itself. It then operates this space in an integrated way with the aim of satisfying the energy needs of its members while minimizing the cost of the energy imported from the public network to which the microgrid is connected. A typical example of this is a new residential area endowed with generation resources (e.g., rooftop solar PV and heat pumps), storage (e.g., batteries and heat tankers), and highly efficient and controllable loads (e.g., heating and cooling systems, smart lighting, and EV charging), in which households are not directly connected to the public distribution grid but to a grid developed on private land that covers their demand for energy services with local resources, as far as it is efficient to do so, and fills any residual gap with energy from the public network.[21]

5.2.2 Strengths and weaknesses

Community providers of shared electric mobility can leverage on a good understanding of the needs and preferences of their members (e.g., where and when they need mobility services, preferences for electricity-based mobility, and green energy). They can also benefit from patient capital provided by members or an external source of funding such as public grants and loans. This can support the gradual uptake of the shared mobility service and the related community. On the other hand, these energy communities may suffer because of the capital intensity of this type of activity (purchasing a few electric cars can exhaust the community's available financial resources), the need to employ a digital platform to manage the service, and the competition from

[20] Examples are Som Mobilitat in Spain (https://www.sommobilitat.coop) and Partago in Belgium (https://www.partago.be/).

[21] An example is Trent Basin in Great Britain (https://www.trentbasin.co.uk/).

alternative modes of transport (e.g., buses and trains) or offers from commercial entities. These potential weaknesses may explain why energy communities of this type are established at the local level but then try to develop relations with similar initiatives, with the aim of reducing the cost of the ICT and ensuring the possibility for their members of benefitting from services provided by other communities elsewhere.[22]

Energy communities that become involved in the aggregation business can expand the benefits that their members derive from their individual or collective investments in generation, storage and controllable loads (by improving the self-consumption rate or by selling the flexibility of community assets, a community aggregator can generate additional economic value for its members). These communities also benefit from the social relations that link the various members in order to recruit their DERs more easily. After all, individuals tend to be less reluctant to give control of their energy usage and the related data to someone else if they trust him or her. By being able to recruit more resources, a community aggregator is likely to better carry out its activity. On the other hand, aggregating DERs and operating in energy markets is not easy, particularly when residential loads are involved. In some countries, this business model is not yet possible at all because of existing regulation. And even when it is possible, rules and market conditions can be challenging, and significant technical and market expertise may be required (ACER, 2021). This may explain the difficulty that community initiatives face when they try to expand into the aggregation business and the importance of partnering with skilled professional actors, which on the other hand may reduce the autonomy and bottom-up nature of the community.

Energy communities that operate in a fully integrated way with the local grid and the resources connected to it benefit from the increasing possibility of monitoring and controlling energy flows at a very granular level. This allows optimization of local resources and a reduced use of energy from the large public network.[23] However, they also face some important challenges. In particular, they have to support significant investments at the beginning of their operation, while their continued viability may be dependent on a long-term commitment to the community by its members. A few members wishing to leave the community may generate additional or unexpected costs and possibly make efficient exploitation of local resources more difficult to achieve. The success of a microgrid can then be fragile.

5.2.3 *Policy and regulatory issues*

The emerging category of energy communities involved in the deployment and management of DERs can play a relevant role in the context of deep decarbonization of energy systems that most countries around the world are committed to by the middle

[22] An example of cooperation between shared mobility community providers is The Mobility Factory SCE (https://www.themobilityfactory.eu/). Established in 2018 by eight European cooperative enterprises, it developed an e-car sharing platform for the exclusive use of its members, which can be adjusted to reflect their specific needs and preferences.

[23] Cost savings may derive not only from a reduced amount of energy import but also from the possibility of importing energy when it is cheaper (e.g., at night) or of reducing the connection capacity with the public grid and paying lower network charges.

of the 21st century. Deep decarbonization requires a growing use of intermittent RESs such as wind and sun and of electricity as the reference energy carrier for transport and heating. Integrating DERs in an efficient and effective manner will be fundamental. Energy communities providing shared electric mobility, aggregation, and microgrid services can help in this regard and support the achievement of public climate goals.

Community providers of shared mobility can foster the early adoption of new forms of more sustainable mobility, particularly among users that cannot afford to buy them privately or that are not reached by commercial actors. From a regulatory point of view, these communities do not pose specific issues since they deal with a non-regulated service. However, coordination with the planning of local public transport, the expansion of the local electricity distribution grid and the EV recharging infrastructure may increase the net value of these initiatives for society.

By offering energy system operators flexibility or achieving an optimal use of the various energy resources at the local level, community aggregators and microgrids may lead to reduced system operating costs (e.g., lower energy losses and redispatching) and to a lower demand for network capacity over time (see the chapters by Koppl et al. and Hadler). They may also offer solutions that improve the reliability of the energy service for people who live in remote and sparsely populated areas (see the chapter by Lee). System resilience can also be strengthened by the possibility for microgrids to operate in islanding mode during conditions of stress for the system and eventually support its return to normal operation after a major outage (Fox-Penner, 2020).

However, it is not clear whether the proliferation of community microgrids is necessarily a positive development.[24] Energy regulators are concerned that small private networks mushrooming at the edge of the public distribution grid can increase the overall cost of distributing energy and cause an unfair and inefficient shift of some stranded costs to consumers that are not part of these communities (CEER, 2019). Existing DSOs may be illegitimately hurt by the undermining of their monopoly rights, and the existence of a single entity that controls the local network and the assets connected to it runs against the fundamental principles that guided the liberalization of the industry in the 1990s and 2000s (i.e., the unbundling of network activities from generation and supply). Several questions then emerge:

- How can market neutrality be ensured within a microgrid?
- How can the fundamental rights of final energy customers be guaranteed, in particular their right to freely choose their suppliers and to receive a service of a sufficiently high quality?
- How can data privacy and data access to all the eligible parties be ensured?

These are all questions that community microgrids raise and for which a consensus answer is still awaited.

[24] Although aggregation raises some unsettled issues, its performance by a community does not pose specific concerns to energy regulators (CEER, 2019).

6. Conclusions

Energy communities represent an extraordinarily multifaceted socio-technical phenomenon that has acquired increasing importance over the first two decades of the 21st century. Confined for a long period of time to a few niches, the most relevant case being that of utility cooperatives in rural or mountain areas, energy communities are nowadays growing in number, expanding in new territories and differentiating in terms of the range of services they offer, their goals, memberships and business models. The decentralization of energy production, the new waves of digitalization targeting distribution grids and customer's premises, and the electrification of end uses support this trend with the result that several types of energy communities can be observed already today in the liberalized energy systems of Europe and North America.

Energy communities can perform different main functions and adopt business models that are more or less mature. Some communities focus on the provision of support to energy consumers and prosumers, and others on the production and supply of energy, most often renewable-based. Some communities concentrate on delivering energy via networks, and others on deploying and managing DERs. Some communities have well-established business models and have proved to flourish when certain conditions are met, while others have a much shorter track-record and it is still unclear if and how they can thrive.

Distinguishing the various categories and types of energy communities is important to better understand the way they interact with the rest of the energy system, how their development impacts the existing organization and regulation of the sector, and how they are or may be affected by the choices made by policymakers and regulators. Treating the phenomenon of energy communities as an indistinct monolith confuses public discussion on their benefits, drawbacks, and the necessity—or not—of adopting specific policies and rules to steer and support their uptake.

Energy communities allow individual households, SMEs, local public bodies and other organizations to take the initiative in the energy space and contribute to the transition to a more sustainable system. They offer an important opportunity to engage energy customers and overcome significant obstacles to decarbonization, such as opposition to the construction of new infrastructure and the intermittent availability of some RESs. Public policies and sector regulation are therefore expected to intervene and promote the establishment and functioning of new energy communities. However, given the heterogeneity of these collective actors, the measures for implementation are not always the same. Specific arrangements may help the uptake of certain types of communities, while they may be irrelevant or even counterproductive for others. A taxonomy that distinguishes the various forms of community energy on the basis of the function performed and the level of maturity represents a useful tool to facilitate discussion and adoption of the best measures.

Finally, it is essential to remember that not all energy communities have the same positive impact on society. In some cases, the benefits of collective action can be strictly confined to those directly involved, with the rest of the energy system and society mostly bearing the costs. In these cases, support by regulation or other public policies is not warranted. Clearly identifying these cases is important and again the taxonomy proposed in this chapter may be of help.

Acknowledgment

The authors would like to thank Jean-Michel Glachant, Valerie Reif and Tim Schittekatte for their comments and feedback on previous versions of the text. They would also like to warmly thank the editors of the book for their feedback and patience. Any error or mistake remains the sole responsibility of the authors.

Bibliography

Abada, I., Ehrenmann, A., Lambin, X., 2020. On the viability of energy communities. Energy J. 41 (1), 113–150.

ACER, 2021. Market Monitoring Report 2020—Electricity Wholesale Market Volume, Agency for the Cooperation of Energy Regulators, 30 October.

Bauwens, T., 2016. Explaining the diversity of motivations behind community renewable energy. Energy Pol. 93 (June), 278–290.

Bauwens, T., 2019. Analyzing the determinants of the size of investments by community renewable energy members: findings and policy implications from Flanders. Energy Pol. 129, 841–852.

Bauwens, T., Gotchev, B., Holstenkamp, L., 2016. What drives the development of community energy in Europe? The case of wind power cooperatives. Energy Res. Soc. Sci. 13 (March), 136–147.

Bauwens, T., Devine-Wright, P., 2018. Positive energies? An empirical study of community energy participation and attitudes to renewable energy. Energy Pol. 118, 612–625.

Bauwens, T., Schraven, D., Drewing, E., Radtke, J., Holstenkamp, L., Gotchev, B., Yildiz, O., 2022. Conceptualizing community in energy systems: a systematic review of 183 definitions. Renew. Sustain. Energy Rev. 156 (March), 111999.

Braunholtz-Speight, T., Sharmina, M., Manderson, E., McLachlan, C., Hannon, M., Hardy, J., Mander, S., 2020. Business models and financial characteristics of community energy in the UK. Nat. Energy 5, 169–177.

Brown, D., Hall, S., Davis, M.E., 2019. Prosumers in the post subsidy era: an exploration of new prosumer business models in the UK. Energy Pol. 135 (December), 110984.

Caramizaru, A., Uihlein, A., 2020. Energy Communities: An Overview of Energy and Social Innovation, JRC Science for Policy Report. Joint Research Centre.

CEER, 2019. Regulatory Aspects of Self-Consumption and Energy Communities. Council of European Energy Regulators. Ref: C18-CRM9_DST-05-03, 25 June.

De Almeida, L., Cappelli, V., Klaussmann, N., van der Soest, H., 2021. Peer-to-Peer Trading and Energy Community in the Electricity Market—Analysing the Literature on Law and Regulation and Looking Ahead to Future Challenges. EUI. Working Papers, RSC 2021/35, March.

Fox-Penner, P., 2020. Power after Carbon: Building a Clean, Resilient Grid. Harvard University Press.

Glachant, J.-M., Rossetto, N., 2021. New transactions in electricity: peer-to-peer and peer-to-x. Econ. Energy Environ. Policy 10 (2), 41–55.
Gui, E.M., MacGill, I., 2018. Typology of future clean energy communities: an exploratory structure, opportunities, and challenges. Energy Res. Social Sci. 35, 94–107.
Herbes, C., Brummer, V., Rognli, J., Blazejewski, S., Gericke, N., 2017. Responding to policy change: new business models for renewable energy cooperatives—barriers perceived by cooperatives' members. Energy Pol. 109, 82–95.
Herbes, C., Rilling, B., Holstenkamp, L., 2021. Ready for new business models? Human and social capital in the management of renewable energy cooperatives. Energy Pol. 156, 112417.
Imgarten, J., 2016. Generation and Transmission Cooperatives: Going beyond Buying and Selling Electricity, 19 August, available at: https://www.cooperative.com/publications/ccc/Pages/Generation-and-Transmission-Cooperatives-Going-Beyond-Buying-and-Selling-Electricity-151.aspx.
Moroni, S., Alberti, V., Antoniucci, V., Bisello, A., 2019. Energy communities in the transition to a low-carbon future: a taxonomical approach and some policy dilemmas. J. Environ. Manag. 236, 45–53.
Morris, C., Jungjohann, A., 2016. Energy Democracy: Germany's Energiewende to Renewables. Palgrave Macmillan.
Ofgem, 2017. Local Energy in a Transforming Energy System, Ofgem's Future Insights Series. Office of Gas and Electricity Markets (January).
Reis, I., Gonçalves, I., Lopes, M., Antunes, C.H., 2021. Business models for energy communities: a review of key issues and trends. Renew. Sustain. Energy Rev. 144, 111013.
REScoop.eu, 2021. Flexibility Services for Energy Cooperatives. An Overview of Possible Flexibility-Based Services Using Residential Equipment Control. REScoop.eu, Brussels.
REScoop.eu, ClientEarth, 2020. Energy Communities under the Clean Energy Package. Transposition Guidance. REScoop.eu, Brussels.
Roberts, J., 2020. Power to the people? Implications of the Clean Energy Package for the role of community ownership in Europe's energy transition. Rev. Eur. Comparative Int. Environ. Law 29 (2), 232–244.
Rossetto, N., Reif, V., 2021. Digitalization of the electricity infrastructure: a key enabler for the decentralization and decarbonization of the power sector. In: Montero, J., Finger, M. (Eds.), A Modern Guide to the Digitalization of Infrastructure, pp. 217–265.
Taylor, K., 2019. Governing the Wind Energy Commons: Renewable Energy and Community Development. West Virginia University Press.
THEMA, Multiconsultant, 2018. Descriptive Study of Local Energy Communities. Commissioned by NVE. THEMA Report 2018-20, December.
Tounquet, F., De Vos, L., Abada, I., Kielichowska, I., Klessmann, C., 2019. Energy Communities in the European Union—Revised Final Report. ASSET Project, May.
Verde, S., Rossetto, N., 2020. The Future of Renewable Energy Communities in the EU. An Investigation at the Time of the Clean Energy Package. EUI, August.
Walker, G., Devine-Wright, P., 2008. Community renewable energy: what should it mean? Energy Pol. 36 (2), 497–500.
Wierling, A., Schwanitz, V.J., Zeiß, J.P., Bout, C., Candelise, C., Gilcrease, W., Sterling Gregg, J., 2018. Statistical evidence on the role of energy cooperatives for the energy transition in European countries. Sustainability 10, 3339.
Wilkinson, S., Hojckova, K., Eon, C., Morrison, G.M., Sandén, B., 2020. Is peer-to-peer electricity trading empowering users? Evidence on motivations and roles in a prosumer business model trial in Australia. Energy Res. Social Sci. 66, 101500.

The EU policy framework for energy communities

2

Deyana Spasova [1] *and Sibylle Braungardt* [2]
[1]Ecologic Institute, Berlin, Germany; [2]Öko-Institut e.V, Freiburg, Germany

1. Introduction

The European Union (EU) has been methodically building and continuously enhancing a framework of climate and energy policies, with the predominant aim of facilitating the transition toward cleaner energy and achieving its Paris Agreement commitments for reducing greenhouse gas emissions globally. This framework culminated in the introduction of the European Green Deal with its action plan for Europe to become the first continent with no net emissions of greenhouse gases through a multilayered transition. Under the Green Deal, the EU is also strengthening its climate and energy policy framework for 2030—first introduced in the Clean energy for all Europeans package in 2016. The package consists of eight legislative acts, focused on enhancing energy efficiency, setting ambitious goals for renewable energy (RE) utilization, improving the governance of the Energy Union and giving consumers more rights and upgrading electricity markets (European Commission, 2019).

Within the framework of the Clean energy for all Europeans package, the European Commission also acknowledges the important role of energy consumers in achieving climate neutrality. A focal point of the package is its objective for consumers to become more directly involved in the energy market, or in other words—for consumers to evolve into "prosumers." The term "prosumer" was first introduced in a 1981 book[1] by Alvin Toffler, and although its meaning has since been embodied in a variety of phrases (as, for example, "producer and consumer," "proactive consumer," "provider and consumer," and "professional consumer" (Kotilainen, 2019)), its essence remains that of the active consumer.

Two directives of the Clean energy for all Europeans package confirm the dedication of the EU to strengthening the position of citizens in the energy market. The recast Renewable Energy Directive[2] (RED II) of December 2018 and the revised Electricity Directive[3] of June 2019 introduce three types of energy prosumption for the first time. They focus not only on ensuring that citizens are able to be proactive in the energy sector, but also on establishing a support framework for those who wish to collaborate in their aim to be proactive. The two directives present the following concepts:

[1] Toffler, Alvin (1980). The third wave (Vol. 484). New York: Bantam books.
[2] Directive (EU) 2018/2001.
[3] Directive (EU) 2019/944.

Energy Communities. https://doi.org/10.1016/B978-0-323-91135-1.00022-5

- "Renewables self-consumers";
- "Renewable energy communities"; and
- "Citizen energy communities."

While energy prosumerism and communal energy are not new concepts in Europe (from the so-called "drying cooperatives"[4] in Germany to wind power communities in Denmark, prosumers have a long history on the continent), what is new is their only very recent introduction into EU legislation.

1.1 Renewable Energy Directive

The RED II is the 2018 revision of the 2009 Renewable Energy Directive,[5] the former published as part of the Clean energy for all Europeans package. Among other additions, the recast directive establishes a common framework for the promotion of RE in the EU until 2030. Member States (MSs) shall collectively ensure that the share of energy from renewable sources in the Union's gross final consumption of energy in 2030 is at least 32%. Furthermore, Article 21 and Article 22 provide legal definitions, respectively, of renewables self-consumers and renewable energy communities (RECs).

Article 21 obliges MSs to ensure that consumers are entitled to become renewables' self-consumers, who may generate RE, including for their own consumption, as well as store and sell their excess renewable generation. MSs are also to see to it that self-consumers, located in the same building, are able to engage jointly in these activities. Moreover, an enabling framework for the promotion of renewables self-consumption is to be established in every MS, in which the potential and incentives, as well as the financial and regulatory barriers for the concept's development, are to be assessed.

While Article 21 focuses on the activity of individual and joint renewables self-consumption, Article 22 defines the rights and obligations of "renewable energy communities" (RECs), which are described as legal entities in Article 2 of the same directive. According to Article 22, MSs shall ensure that RECs are entitled to produce, consume, store, and sell RE, including through renewables power purchase agreements; to share RE produced by units owned by the REC; and access all suitable energy markets directly or through aggregation. MSs must also guarantee that final customers have the right to join a REC without being subject to unfounded or unfair terms. As in the case of self-consumers, Member countries are obliged to assess existing barriers and perspectives for the development of RECs in their territories, create an enabling framework for their progress, and transpose the RED II into national law by June 30, 2021.

As part of its Fit for 55 package, the European Commission proposed a revision of the RED II on July 14, 2021. However, the revision "does not plan to change provisions on energy communities as the new rules are still being implemented in the

[4] Holstenkamp, Lars (2012). Ansätze einer Systematisierung von Energiegenossenschaften. https://doi.org/10.13140/RG.2.2.36503.68005.

[5] Directive (EU) 2009/28.

MSs and the Commission has started nonlegislative actions to foster the roll-out of energy communities" (European Commission, 2021).

1.2 Electricity Directive

The recast Electricity Directive entered into force in 2019, substituting its predecessor from 2012. In the revised version, the European Commission acknowledges the effectiveness and cost-efficiency of community energy in meeting citizens' expectations for their empowerment as electricity consumers. It also introduces the "citizen energy community" (CEC) as a legal entity (Article 2.11), to be established on an open, voluntary basis and with the core objective "to provide environmental, economic, or social community benefits to its members or shareholders or to the local areas where it operates rather than to generate financial profits" (Article 2). Article 16 of the directive obligates MSs to ensure the free (including cross-border) participation of citizens in such a community, and their open access to all electricity markets. MSs are required to compile these facilitating regulatory conditions into an enabling framework for the development of CECs and transpose them into national law by December 31, 2020.

The balance of this chapter is organized as follows:

- Section 2 provides a comparison between RECs and CECs based on key elements;
- Section 3 examines other relevant policy developments; and
- Section 4 discusses energy communities in EU MSs, followed by the chapter's conclusions in Section 5.

2. RECs versus CECs

At first glance, the RED II and the Electricity Directive appear almost identical, as the structures of the two legislations mirror one another. Nevertheless, the two directives differ in key elements with regard to energy communities. In its proposal for a revision of the RED II from July 2021, the European Commission stated that stakeholders have called for a clarification of the definitions of RECs and CECs. The following text examines, compares, and summarizes the main features of RECs and CECs as originally defined.

2.1 Activities

RECs, as well as CECs, can engage in electricity generation, consumption of self-generated electricity, sharing of self-generated electricity within the community, storage, sales, and aggregation within the community, as well as participation in markets outside the community.

The Electricity Directive introduces the possibility for CECs to become distribution system operators (DSOs) either under the general regime or as closed DSOs. CECs are also entitled to engage in energy efficiency services or charging services for electric vehicles or to provide other energy services to members or shareholders.

The RED II provides the opportunity for RECs to advance energy efficiency at household level and to help fight energy poverty through reduced consumption and lower supply tariffs.

2.2 Purpose

An REC, as well as a CEC, is "to provide environmental, economic, or social community benefits for its shareholders or members or for the local areas where it operates, rather than financial profits" (Directive (EU) 2018/2001, Article 2.16, and Directive (EU) 2019/944, Article 2.11).

2.3 Participation

Participation in both RECs and CECs shall be open and voluntary, including cross-border participation. The Electricity Directive specifically points out that members or shareholders of a CEC are entitled to leave the community. The RED II states that members maintain their rights or obligations as final customers.

2.4 Energy sources

The RED II does not impose a limitation on the type of energy that RECs can utilize, while the Electricity Directive describes CECs as specifically focused around electricity. However, it is important to note that the Electricity Directive also does not specifically point out that the electricity produced, consumed, shared, stored, etc., by a CEC must necessarily come from renewable sources. While it is difficult to imagine that citizens could operate a community nuclear or coal power plant, the failure of the Electricity Directive to clearly define the energy source of CECs leaves room for interpretation as to whether energy from nuclear power or fossil fuels (e.g., gas cogeneration) could be allowed.

2.5 Control

The control for a CEC is entrusted to members or shareholders who are natural persons, local authorities, including municipalities and small enterprises. Decision-making powers within a CEC are only given to members or shareholders that are not engaged in large-scale commercial activity and for which the energy sector is not a primary area of economic activity.

An REC, on the other hand, is controlled by shareholders or members, located in the proximity of the RE project. These can include natural persons, micro, small or medium-sized enterprises, or local authorities, including municipalities. According to the RED II, "renewable energy communities should be capable of remaining autonomous from individual members and other traditional market actors that participate in the community as members or shareholders, or who cooperate through other means such as investment" (Directive (EU) 2018/2001, (71)).

2.6 Geographical scope

The control of an REC being entrusted to people in close proximity to the project provides RECs with a smaller geographical scope for action. In this respect, "RECs can generally be seen as a subset, or type, of CEC" (Roberts et al., 2019), but it can also be concluded that there is the possibility for RECs to be controlled by more types of enterprises than CECs.

The Electricity Directive does not define geographical constraints for CECs and electricity sharing enables consumers to be supplied with energy without being in physical proximity to the generating installation.

2.7 Rights and obligations

According to the Electricity Directive, CECs have rights and obligations such as freedom of contract, the right to switch supplier, the responsibilities of the DSO, the rules on network charges, and balancing obligations. The RED II asserts that RECs maintain the rights and obligations of the energy community members as customers.

2.8 Possible forms of entity

RECs should be able to take any form of entity as long as they exercise the rights and are subject to the obligations specified in the EU legislation. In the case of CECs, a similar statement about rights and obligations is included in the Electricity Directive, although the text goes into more specifics by listing a few examples of entities that a CEC can take the form of, including associations, cooperatives, partnerships, nonprofit organizations, and small- or medium-sized enterprises.

2.9 Support schemes

The RED II provides RECs with the opportunity to participate in available support schemes on an equal footing with large participants. Under the directive, support schemes include, but are not limited to, investment aid, tax exemptions or reductions, tax refunds and simplified permitting procedures for small RE installations.

The Electricity Directive does not specifically mention such possibilities for CECs. However, the directive ensures that, as CECs can engage in electricity sharing, incentives shall be provided without prejudice to applicable network charges, tariffs, and levies in accordance with a transparent cost–benefit analysis of distributed energy resources developed by the competent national authorities.

Both RECs and CECs can benefit from incentives for collective self-consumption.

2.10 Ownership of the electricity distribution network

The Electricity Directive entitles CECs to own, establish, purchase, or lease distribution networks and to autonomously manage them. In contrast, RECs can use either a public network or a private distribution network in the ownership of its members.

Table 2.1 provides an overview of the key elements of RECs and CECs.

Table 2.1 Key elements of RECs and CECs.

Element	RECs	CECs
Activities	Electricity generation, consumption of self-generated electricity, sharing of self-generated electricity within the community, storage, sales, and aggregation.	
	Opportunity to advance energy efficiency at household level and to help fight energy poverty through reduced consumption and lower supply tariffs.	Possible to become distribution system operators either under the general regime or as closed distribution system operators; entitled to engage in energy efficiency services or charging services for electric vehicles or to provide other energy services to members or shareholders.
Purpose	To provide environmental, economic, or social community benefits for its shareholders or members or for the local areas where it operates, rather than financial profits.	
Participation	Open and voluntary and including cross-border participation.	
	–	The Electricity Directive specifically points out that members or shareholders of a CEC are entitled to leave the community.
Energy sources	Renewable energy sources	Electricity. The Electricity Directive does not specifically exclude any electricity generation technology, meaning that potentially renewables, nuclear power, or fossil fuels may be used.

Table 2.1 Key elements of RECs and CECs.—cont'd

Element	RECs	CECs
Control	Shareholders or members, located in the proximity of the RE project, including natural persons, micro-, small- or medium-sized enterprises or local authorities, including municipalities. Should be capable of remaining autonomous from individual members and other traditional market actors.	Members or shareholders who are natural persons, local authorities, including municipalities and small enterprises. Decision-making powers are only given to members or shareholders that are not engaged in large-scale commercial activity and for which the energy sector is not a primary area of economic activity.
Geographical scope	Close proximity to the project.	No geographical constraints; consumers can be supplied with energy without being in physical proximity to the generating installation.
Rights and obligations	Maintain the rights and obligations of the energy community members as customers.	Maintain the rights and obligations such as freedom of contract, the right to switch supplier, the responsibilities of the distribution system operator, the rules on network charges, and balancing obligations.
Possible forms of entity	Any form of entity as long as it exercises rights and is subject to obligations defined by the EU legislation.	Any form of entity, including associations, cooperatives, partnerships, nonprofit organizations, and small- or medium-sized enterprises, as long as it exercises rights and is subject to obligations defined by EU legislation.
Support schemes	Can benefit from incentives for collective self-consumption.	
	Can participate in available support schemes (investment aid, tax exemptions or reductions, tax refunds, and simplified permitting procedures for small RE installations) on an equal footing with large participants.	No specifically mentioned support schemes for a CEC, however; since CECs can engage in electricity sharing, they can also benefit from related incentives.
Ownership of the electricity distribution network	Can use either a public network or a private distribution network in the ownership of its members.	Entitled to own, establish, purchase, or lease distribution networks and to autonomously manage them as DSO.

3. Other relevant policy developments

The RED II and the Electricity Directive are the two legislative acts, which give a legal status to community energy initiatives. However, these initiatives and their importance for the energy transition have been mentioned in several other recent EU policy actions.

The Commission's *EU Strategy for Energy System Integration*[6] from July 2020, for example, proposes an action plan for a faster transition toward a more integrated energy system. In the communication on the strategy, the Commission refers to energy communities as an important part of this integration, including as a means to further develop the use of local biomass energy by farmers.

Another European strategy—the *Renovation Wave*[7]— launched in October 2020, aims to trigger an EU-wide buildings' renovation initiative, in order to tackle persistent energy efficiency barriers and to initiate new investment connected to digitalization and growth. One of the main objectives of the Renovation Wave is to expedite the deployment of renewables from local sources and the integration of energy systems at neighborhood level, including local energy generation and storage (European Commission, 2020b). According to the European Commission,

> *Energy communities generate, consume, store and sell energy, and can offer tools for the most vulnerable citizens to lift them out of energy poverty. To exploit their untapped potential as active players in the energy system, the Commission will look closely into the implementation of the Electricity Market Directive and the Renewable Energy Directive and use concerted actions to support their progressive creation and diffusion across MS. The Commission will further explore how to promote energy communities and disseminate good practices*
>
> *(European Commission, 2020b, p. 18).*

Several projects under the EU's Horizon 2020 program have aimed, among other things, to identify good practices regarding energy communities and to disseminate them. For example, the *COME RES - Community Energy for the uptake of renewables in the electricity sector. Connecting long-term visions with short-term actions*[8] project focuses on advancing RECs as defined in the RED II. The *Collective Action Models for the Energy Transition and Social Innovation (COMETS)*[9] project examines the role of citizen-driven initiatives, such as energy communities, cooperatives, and purchasing groups and their contribution to the energy transition.

[6] COM/2020/299 final.
[7] COM/2020/662 final.
[8] See COME RES (2021).
[9] Available online at: http://www.comets-project.eu.

In December 2020, the Commission also presented the *European Climate Pact*[10]—part of the Green Deal initiatives—which acknowledges the role for communities and in particular those dealing with RE. The aim of the pact is to give citizens, communities, civil society, and other stakeholders, a voice and an active role in climate action. In the communication establishing the pact, the European Commission not only recognizes the necessity for more societal and communal involvement in tackling climate issues, but also specifically gives attention to the contribution of community energy projects to increasing the use of renewables and fostering participation and active engagement of citizens.

4. Energy communities in European Member States

There were more than 2400 energy communities in Europe at the beginning of 2015 and their current number is estimated to be around 3000 (Vansintjan, 2015, 2018). It is important to note here that as EU legislation on energy communities was introduced later than these estimations, the approximate number of communities is not limited to RECs and CECs (as defined by the RED II and the Electricity Directive), but includes other forms of community energy as well. A 2020 study, published by the European Commission, refers to over 3500 RE cooperatives in Europe—a cooperative model, which the authors call "a type of energy community" (Caramizaru and Uihlein, 2020, p. 20).

Community energy initiatives predominate in Northern and Western Europe, and in particular in Germany, Denmark, the Netherlands, and the United Kingdom, in terms of both historical presence and number (Yildiz et al., 2015; Bauwens et al., 2016). Germany, for example, has a long tradition of citizen participation in the energy sector, dating back to early 20th century electricity cooperatives (Holstenkamp, 2012; Yildiz et al., 2015). In present times, community energy and citizen participation are integral components of the strategy for the German energy transition (Energiewende), which 90% of Germans strongly support (Fraune, 2015; Agora Energiewende and Renewable Energy Institute, 2017). The endorsement of decentralized RE as part of the "Energiewende" also meant a boom in energy communities in the country (Gancheva et al., 2018). Presently, there are more than 1700 energy communities in Germany—the most in any European country (Kahla et al., 2017).

In some Southern countries, such as Italy,[11] energy communities are presently gaining momentum as an emerging concept and are becoming more widespread, although the Italian province of South Tyrol has historically been associated with energy cooperatives and a large number of community initiatives (Magnani and Osti, 2016).

[10] COM/2020/788 final.

[11] The chapters by Lo Schiavo et al., Del Pizzo et al., and Koltunov et al., provide more detailed analyses of energy communities in Italy.

In Eastern Europe,[12] community energy is not well developed, and in some regions, communal initiatives are practically nonexistent. An analysis of energy communities in 16 postsocialist European nations from 2020 found "no relevant project" in 10 of them (Capellán-Pérez et al., 2020, p. 7). In Bulgaria, for instance, community energy is not present and there are various barriers for its successful development, including legislative, economic, and social issues (Spasova and Braungardt, 2021).

4.1 Belgium

The government of the Wallonia region in Belgium introduced the concept of "communautés energie renouvelable" (RECs) in a Decree of April 30, 2019. In the Decree, RECs are described as legal entities, entitled to produce, consume, share, store and sell electricity from renewable sources in a local perimeter and which are entitled to reduced policy-related charges (Hannoset et al., 2019). The decree, although more limited (it only refers to electricity), is consistent with the EU policy framework for energy communities and "was already in its final adoption stages when the RED II was officially adopted" (Hannoset et al., 2019, p. 51). The timing makes Belgium a particularly interesting case, as it resulted in the RED II being transposed early into national law.

Belgium's National Energy and Climate Plan[13] (NECP) also provides a detailed action plan for the development of renewables self-consumption and RECs, although it does not specifically include energy cooperatives, which are already well established in the country (e.g., Ecopower). The plan refers to a new regulatory framework for renewables self-consumption projects within a single building, measures for the establishment of a support framework for renewable heating and cooling, a system of incentives specifically targeted at residential consumers for the deployment of heat pumps and boilers, the creation of a "Heating Decree," and the establishment of a renewable obligation in new buildings.

The Horizon 2020 COME RES project examined the transposition and implementation progress of the RED II in several MSs, including Belgium. It concluded that the definition of RECs in the Belgian decree is similar to the definition of RECs in the RED II, although further regulations of RECs' elements and a definition, which does not allow interpretations, are necessary. Moreover, the project determined that participation of Belgian customers in RECs is open to all, and the degree allows them to freely produce, consume, store, sell, and share RE. However, as of December 2020, the Belgian government had not carried out an assessment of the barriers for RECs and the decree does not contain an enabling framework or support schemes for the specificities of energy communities.

[12] For an analysis of energy communities in Bulgaria, see the chapter by Couture and Stoyanova.

[13] The NECPs of all EU Member States are available at: https://ec.europa.eu/energy/topics/energy-strategy/national-energy-climate-plans_en.

4.2 Germany[14]

The large number of energy communities in Germany is a reflection of a long history of consumer participation and cooperative models, as well as Germans' trust in community energy initiatives, which they associate with "civic engagement, participation, and empowerment in energy matters" (Yildiz et al., 2015, p. 68) and a strong support framework.

There are two variations of energy communities in Germany—"Energiegenossenschaften" (energy cooperatives) and "Bürgerenergiegesellschaften" (citizen energy cooperatives). A legal definition of citizen energy cooperatives was formulated for the first time in the 2017 Renewable Energy Act (EEG). Such cooperatives must consist of no less than 10 people as voting members or shareholders, each holding no more than 10% of the voting rights, while member citizens are entitled to at least 51% of the voting rights (Bundesministerium der Justiz und für Verbraucherschutz, 2021).

A type of REC, popular in Germany, is bioenergy villages, which must cover at least 50% of their heat and electricity demand with locally produced biomass and if necessary other renewables.[15] The first such project (Brohmann et al., 2006) was launched in 2006. Since then, bioenergy villages have gained significant popularity around the country and are now a well-known concept, which continues its expansion. At least 50% of homeowners in the village need to be willing to participate in order for a bioenergy village to be created. Currently, there are 170 bioenergy villages in Germany, while 42 are on their way to becoming one.[16] Bioenergy villages are not legally, legislatively, or regulatorily distinguished from energy communities.

The European Commission regards the legal framework for RECs in Germany as "already well established" (European Commission, 2020a, p. 23). The NECP of Germany outlines the special privileges of energy communities in regards to the funding of onshore wind energy and provides a regulatory framework for the development of RECs (3.1.2.v.6), which covers the open participation of consumers in RECs and the free access of RECs to existing support schemes.

As of December 2020, the transposition of the RED II in Germany is weak (COME RES, 2021). A legal definition of RECs in line with the one of the RED II does not exist. According to the EEG, German final customers are entitled to participate in "citizen energy companies," the scope of which is limited to wind energy only, while their range of eligible participants is broader than in the case of RECs. There is no precise definition of the activities that citizen energy companies are entitled to undertake, and energy sharing is essentially undefined and thus impossible. No government entity has carried out an assessment of the barriers for or the potential of RECs and there is no

[14] The chapters by Köppl et al. Benedettini and Stagnaro, Chudoba and Borges, and Menke et al., provide more detailed analyses of energy communities in Germany.

[15] According to Fachagentur Nachwachsende Rohstoffe e. V.: Bioenergiedörfer. Bundesministerium für Ernährung, Landwirtschaft und Verbraucherschutz. Available online at https://bioenergiedorf.fnr.de/bioenergiedoerfer/liste.

[16] According to a database by the German Agency for Renewable Resources (FNR). Available online at: https://bioenergiedorf.fnr.de/bioenergiedoerfer/liste.

enabling framework in line with the RED II for such communities. Concrete privileges are also only given to onshore wind energy projects.

Moreover, joint self-supply with electricity from within a building as described in the RED II has been found to be currently impossible to undertake in Germany, as according to the EEG, both RE system operators and end consumers are required to be the same legal entity, preventing several people from joining together as "jointly acting self-suppliers" (Institut für ZukunftsEnergie-und Stoffstromsysteme, 2021). According to the same report, the development of energy communities in Germany is possible, since energy cooperatives already meet the requirements, defined in the RED II. However, as such cooperatives cannot operate energy sharing among their members without becoming an energy supplier with all associated obligations, energy communities can still not fully (and freely) operate within the scope of the RED II (Institut für ZukunftsEnergie-und Stoffstromsysteme, 2021).

4.3 Denmark[17]

Denmark is one of the European pioneers in the development of energy communities, as the first wind cooperatives in the country emerged in the late 1970s, formed by collective antinuclear networks (Mey and Diesendorf, 2018; Wierling et al., 2018). Since then, Danish policies have continuously supported the development of community energy initiatives, beginning with tax-free grants for individuals and cooperatives in 1979, the recognition of a long-term perspective for communal wind energy projects in 1981, and a mandatory 20% ownership share for people living close to new wind energy installations in 2009 (Mey and Diesendorf, 2018; Gorroño-Albizu et al., 2019).

Energy communities in Denmark do not necessarily correspond to the definitions in the RED II and the Electricity Directive. Community energy in Denmark is common in the form of citizen ownership of wind turbines and as ownership in the country's most common heat supply system—district heating (DH)—usually in the form of municipal companies in big cities or consumer-owned cooperatives in rural areas. The dominance of citizen ownership in the Danish DH is the direct consequence of DH systems being by law prohibited from making profit (Gorroño-Albizu et al., 2019). As DH companies are still profitable, however, profit is subsequently returned to customers in the form of reduced heat rates (Chittum and Østergaard, 2014). This makes investments in DH systems unappealing to commercial investors and appealing to consumers (Gorroño-Albizu et al., 2019). In 2016, there were 407 DH companies in Denmark and 341 of them were consumer cooperatives (Gorroño-Albizu et al., 2019; Danish Utility Regulator, 2016).

[17] The chapter by Benedettini and Stagnaro provides more a detailed analysis of energy communities in Denmark.

4.4 The Netherlands[18]

Similar to the cases of Germany and Denmark, the development of community energy initiatives in the Netherlands is connected to the 1970s oil crisis and the realization of Dutch society that the country is highly dependent on foreign imports of fossil fuels (Oteman et al., 2014; Proka et al., 2018; Vansintjan 2015; Kooij et al., 2018). The first grassroots initiatives in the Netherlands—the wind cooperatives—were established in the early 1980s, and community energy initiatives have since grown to more than 300 in 2017 (Koirala et al., 2018; Proka et al., 2018). Despite a rapid development of RECs in recent years, an economically oriented rationale, limited support from institutions, and a strong fossil fuel lobby have continuously restricted the growth of energy communities (Oteman et al., 2014).

According to the Dutch NECP, renewable self-consumers are supported by a VAT reimbursement on the purchase of solar panels and a "net-metering scheme." Based on the scheme, renewable electricity, fed back into the grid, is deducted from the electricity purchased from the grid. This means that as long as the electricity purchased from the grid is offset by electricity fed back into the grid, small-scale consumers are exempt from paying supply costs, energy tax, surcharge for sustainable energy, or VAT. Until recently, Dutch citizens could benefit from a postcode scheme, which was replaced by a kWh-related subsidy in 2021. Additionally, in December 2020, the European Commission approved the Renewable Energy Transition Incentive Scheme (SDE++), broadening the scope of its predecessor SDE+. The new scheme will operate until 2025 and will use an auction process to award subsidies to RE technologies depending on the amount of avoided emissions. The scope of the incentives comprises a variety of categories, including renewable energy production projects, cogeneration, heat pumps, and compost heating. Tax benefits are also available for Dutch energy customers. The country's NECP also contains plans to focus on simplifying administrative procedures for energy communities.

A legal definition of RECs, in line with the RED II, exists in the draft of the new Dutch Energy Law (COME RES, 2021). The draft foresees entitlements for customers to participate freely in an REC, as well as to produce, consume, store, sell, and share RE. The Dutch government also aims to establish nondiscriminatory frameworks for fair competition and demand response and storage, including through aggregation. Although a governmental assessment of barriers has not been carried out yet, the establishment of an enabling framework had been planned for late 2020.

4.5 Spain[19]

Renewable electricity self-consumption in Spain is regulated by the Royal Decree 244/2019 of April 5, 2019. In the decree, self-consumption is described as the individual or joint consumption of electricity coming from generation installations close to and

[18] The chapter by Swens and Diestelmeier provides a more detailed analysis of a specific energy community in The Netherlands.

[19] The chapter by Robinson and del Guyao provides more detailed analyses of energy communities in Spain.

associated with consumption installations. Economic compensations attached to different schemes are also updated, along with easing the regulatory process for small-scale producers. RECs in particular are given a legal definition in the Royal Decree law 23/2020 of June 23, 2020 (COME RES, 2021).

A key measure to promote RE in Spain's NECP is Measure 1.13 "Local energy communities." According to the plan, the term "local energy communities" encompasses both the description of RECs in the RED II and of CECs in the Electricity Directive, as well as energy cooperatives (e.g., Som Energia), which are popular in the country (Pellicer-Sifres et al., 2020). Among the proposed mechanisms to promote these communities are the creation of a legislative framework, the elimination of barriers, the promotion of demonstration projects, training and capacity-building programs for members, and the possible creation of a support office for financing and technical assistance. Another measure, aiming to strengthen the proactive role of citizens in decarbonization, is built around, among others, mechanisms to promote the diversity of actors in the electricity market, collective financing strategies, and improving the cooperation between citizens and municipalities.

Spain is one of the countries which, according to the COME RES project, has made good progress in transposing the RED II in regards to energy communities. Spanish RECs are given a legal definition, entirely compliant with the one in the RED II. However, customers' entitlements are not fully addressed and are not yet formally in national legislation. There is also no enabling framework, encoded into Spanish law, and no support schemes in line with the RED II.

4.6 Greece[20]

Greece is a European trailblazer regarding the introduction of enabling legislation for energy communities and a report counts 409 community energy initiatives (not limited to RECs and CECs) in the country (Electra Energy, 2020). The necessity to alleviate energy poverty in the country, as well as a fragmented and confusing legal framework on community energy, resulted in the initiation of the 2018 new law (L. 4513/2018) on energy communities (Douvitsa, 2018). Reflecting on the contents of the Proposal for the EU directive on the promotion of the use of energy from renewable sources[21] and the Proposal for a directive on common rules for the internal market in electricity[22] from 2016, the law addresses tackling energy poverty and enabling the green transition of Greece through supporting the creation of energy communities. Building on the two directives, the law describes the activities that an energy community can engage in (producing, storing, consuming, and selling electric or thermal or cooling energy), as well as optional activities such as attracting capital for investment within the district of the community, handling funding programs, and supporting vulnerable citizens in

[20] The chapter by Li provides a more detailed analysis of one island community energy project in Greece.
[21] COM (2016) 767.
[22] COM (2016) 864.

sustainability matters (Douvitsa, 2018). Other key points, such as membership and cooperative share requirements, are also discussed in the law.

At the time of its introduction, the law was welcomed and praised as innovatory in Europe. Larissa, Thessaloniki, and Athens were reported to be preparing to use virtual net-metering and to be developing plans that would provide free solar energy to energy-poor households (Gancheva, et al., 2018; Electra Energy, 2020). However, in February 2021, the European federation of energy cooperatives REScoop.EU published a report[23] together with Greenpeace Greece, WWF Greece, and Electra Energy, which calls for a change in the Greek institutional framework for the support and development of energy communities. According to the report, some of the issues in Greece include complex administrative procedures, lack of governmental support and funding for private investors, and adjustments in Greek legislation inconsistent with the country's NECP and EU legislation. Market participation in Greece has also been condemned as unequal in the report, as private investors are said to have "hijacked" the advantages of legislation, aimed at local energy initiatives, resulting in the Ministry blocking all incentives for energy communities, which as of 2022 "are to compete with private investors in bids to ensure operational reinforcement for renewable energy projects" (REScoop.EU, 2021).

5. Conclusions

The introduction of legislative approaches to support community energy at EU level provides an opportunity for strengthening its role in the energy transition across the EU MSs. Establishing definitions and provisions for energy communities in EU legislation and the ongoing activities to encourage such initiatives can act as an important driver.

However, the various MSs differ largely with respect to the current deployment and role of energy communities as well as their national legislative frameworks addressing and affecting these communities. While energy communities have a long tradition in MSs such as Denmark and Germany, community energy plays a minor role in Eastern European countries.

In order to ensure a successful implementation of the EU legislation and to support the deployment of community energy across the continent, support needs to be provided for the implementation of the EU requirements, particularly for the MSs that currently lag behind with the deployment of community energy. Barriers, including a lack of experts and know-how, high percentages of energy poverty, low gross disposable income of households, a lack of a regulatory framework, no financial or technical support, administrative burdens, and social implications, have all been found to be detrimental to the development of energy communities, yet are widespread in countries lagging behind (Spasova and Braungardt, 2021). The chapter by Couture and

[23] See REScoop.EU (2021).

Stoyanova provides a case study of Bulgaria—an example of the many countries that face challenges regarding the implementation of EU legislation on energy communities and their general development. It also discusses key barriers and offers recommendations for better supporting energy communities' deployment.

However, transposing EU legislation into national law is not enough—a coordination of transparent national policies is also necessary. In that sense, assessing the experiences of countries, successful in energy communities' deployment, and their transferability would be highly beneficial for struggling MSs.

Nevertheless, the question remains to what extent MSs with already successful legislation on energy communities should change it, in order for it to be in line with EU directives. As discussed in the case of Germany, for example, national, time-proven strategies such as energy cooperatives (and the legislation that governs them), which are already successfully operating and developing further, will need to be adjusted so as to align with EU legislation, although the model has proven to be effective in the deployment of community energy in the country. The following chapters provide a detailed analysis of regulatory and policy frameworks, including case studies from various countries and the challenges that these countries face when implementing both national and EU legislation on energy communities.

Bibliography

Agora Energiewende and Renewable Energy Institute, 2017. 10 Q & A on the German Energiewende. Available online at: https://www.renewable-ei.org/en/images/pdf/20170303/REI_Report_20170303_10FAQ_GermanyEnergiewende_EN.PDF. (Accessed 16 April 2021).

Bauwens, T., Gotchev, B., Holstenkamp, L., 2016. What drives the development of community energy in Europe? The case of wind power cooperatives. Energy Res. Social Sci. 13, 136–147. https://doi.org/10.1016/j.erss.2015.12.016.

Brohmann, B., Fritsche, U., H€unecke, K., 2006. Case Study. The Bioenergy Village J€uhnde. In Report by €Oko-Institute eV for Create Acceptance Work Package.

Bundesministerium der Justiz und für Verbraucherschutz, 2021. Gesetz für den Ausbau erneuerbarer Energien. Available online at: http://www.gesetze-im-internet.de/eeg_2014/EEG_2021.pdf. (Accessed 16 April 2021).

Capellán-Pérez, I., Johanisova, N., Young, J., Kunze, C., 2020. Is community energy really non-existent in post-socialist Europe? Examining recent trends in 16 countries. Energy Res. Social Sci. 61, 101348. https://doi.org/10.1016/j.erss.2019.101348.

Caramizaru, E., Uihlein, A., 2020. Energy Communities: An Overview of Energy and Social Innovation. European Commission. JRC Science for Policy Report.

Chittum, A., Østergaard, P.A., 2014. How Danish communal heat planning empowers municipalities and benefits individual consumers. Energy Pol. 74, 465–474. https://doi.org/10.1016/j.enpol.2014.08.001.

COME RES, 2021. Assessment Report on Technical, Legal, Institutional and Policy Conditions. Deliverable 2.1. Available online at: https://come-res.eu/fileadmin/user_upload/Resources/Deliverables/COME_RES_D2.1__Assessment_report_FINAL.pdf. (Accessed 21 May 2021).

Danish Utility Regulator, 2016. Energitilsynets Fjernvarnmestatistik. December 2016, Valby, 2016.
Douvitsa, I., 2018. The new law on energy communities in Greece. Cooperativismo e Economía Social 40 (2017–2018), 31–58.
Electra Energy, 2020. Mapping of Energy Communities in Greece. WP1, 1.2 (Athens).
European Commission, 2019. Clean Energy for All Europeans Package. Available online at: https://op.europa.eu/s/ptis. (Accessed 3 August 2021).
European Commission, 2020a. Commission Staff Working Document. Assessment of the Final National Energy and Climate Plan of Germany. SWD(2020) 904 final. Brussels, 14.10.2020.
European Commission, 2020b. Communication from the Commission to the European Parliament, the Council, the European Economic and Social Committee and the Committee of the Regions a Renovation Wave for Europe—Greening Our Buildings, Creating Jobs, Improving Lives. COM(2020) 662 final. Brussels, 14.10.2020.
European Commission, 2021. Proposal for a Directive of the European Parliament and of the Council Amending Directive (EU) 2018/2001 of the European Parliament and of the Council, Regulation (EU) 2018/1999 of the European Parliament and of the Council and Directive 98/70/EC of the European Parliament and of the Council as Regards the Promotion of Energy from Renewable Sources, and Repealing Council Directive (EU) 2015/652. COM/2021/557 final. Brussels, 14.7.2021.
Fraune, C., 2015. Gender matters. Women, renewable energy, and citizen participation in Germany. Energy Res. Social Sci. 7, 55–65. https://doi.org/10.1016/j.erss.2015.02.005.
Gancheva, M., O'Brien, S., Crook, N., Monteiro, C., 2018. Models of Local Energy Ownership and the Role of Local Energy Communities in Energy Transition in Europe. European Committee of the Regions, Bruxelles, Belgique.
Gorroño-Albizu, L., Sperling, K., Djørup, S., 2019. The past, present and uncertain future of community energy in Denmark: critically reviewing and conceptualising citizen ownership. Energy Res. Social Sci. 57, 101231. https://doi.org/10.1016/j.erss.2019.101231.
Hannoset, A., Peeters, L., Tuerk, A., 2019. Energy Communities in the EU—Task Force Energy Communities. D3.12.d. Bridge Horizon 2020.
Holstenkamp, L., 2012. Ansätze einer Systematisierung von Energiegenossenschaften.
Institut für ZukunftsEnergie- und Stoffstromsysteme, 2021. Stand der Umsetzung der RED II-Richtlinie in Deutschland mit Blick auf die Bürgerenergie. Kurzstudie.
Kahla, F., Holstenkamp, L., Müller, J.R., Degenhart, H., 2017. Entwicklung und Stand von Bürgerenergiegesellschaften und Energiegenossenschaften in Deutschland.
Koirala, B.P., Araghi, Y., Kroesen, M., Ghorbani, A., Hakvoort, R.A., Herder, P.M., 2018. Trust, awareness, and independence: insights from a socio-psychological factor analysis of citizen knowledge and participation in community energy systems. Energy Res. Social Sci. 38, 33–40. https://doi.org/10.1016/j.erss.2018.01.009.
Kooij, H.-J., Oteman, M., Veenman, S., Sperling, K., Magnusson, D., Palm, J., Hvelplund, F., 2018. Between grassroots and treetops: community power and institutional dependence in the renewable energy sector in Denmark, Sweden and The Netherlands. Energy Res. Social Sci. 37, 52–64. https://doi.org/10.1016/j.erss.2017.09.019.
Kotilainen, K., 2019. Energy prosumers' role in the sustainable energy system. In: Leal Filho, W., Azul, A., Brandli, L., Özuyar, P., Wall, T. (Eds.), Affordable and Clean Energy. Encyclopedia of the UN Sustainable Development Goals. Springer, Cham. https://doi.org/10.1007/978-3-319-71057-0_11-1.
Magnani, N., Osti, G., 2016. Does civil society matter? Challenges and strategies of grassroots initiatives in Italy's energy transition. Energy Res. Social Sci. 13, 148–157. https://doi.org/10.1016/j.erss.2015.12.012.

Mey, F., Diesendorf, M., 2018. Who owns an energy transition? Strategic action fields and community wind energy in Denmark. Energy Future 35, 108–117. https://doi.org/10.1016/j.erss.2017.10.044.

Oteman, M., Wiering, M., Helderman, J.-K., 2014. The institutional space of community initiatives for renewable energy: a comparative case study of The Netherlands, Germany and Denmark. Energy Sustain. Soc. 4, 11. https://doi.org/10.1186/2192-0567-4-11.

Pellicer-Sifres, V., Belda-Miquel, S., Cuesta-Fernandez, I., Boni, A., 2020. Learning, transformative action, and grassroots innovation: insights from the Spanish energy cooperative Som Energia. Energy Res. Social Sci. 42, 100–111. https://doi.org/10.1016/j.erss.2018.03.001.

Proka, A., Loorbach, D., Hisschemöller, M., 2018. Leading from the niche: insights from a strategic dialogue of renewable energy cooperatives in The Netherlands. Sustainability 10 (11), 4106. https://doi.org/10.3390/su10114106.

REScoopEU, 2021. Executive Summary: Development of Community Energy in Greece under Pressure. Available online at: https://www.rescoop.eu/news-and-events/news/executive-summary-development-of-community-energy-in-greece-under-pressure. (Accessed 19 April 2021).

Roberts, J., Frieden, D., d'Herbemont, S., 2019. Energy Community Definitions. Deliverable Developed under the Scope of the COMPILE Project: Integrating Community Power in Energy Islands (H2020-824424).

Spasova, D., Braungardt, S., 2021. Building a common support framework in differing realities—conditions for renewable energy communities in Germany and Bulgaria. Energies 14 (15), 4693. https://doi.org/10.3390/en14154693.

Vansintjan, D., 2015. The Energy Transition to Energy Democracy. Power to the People. REScoop.eu. Available online at: http://www.collective-action.info/sites/default/files/webmaster/_PUB_The-energy-transition-to-energy-democracy.pdf. (Accessed 16 April 2021).

Vansintjan, D., 2018. Where Will Citizens Fit within Europe's Energy Transition? EURACTIV MEDIA NETWORK BV. Available online at: https://www.euractiv.com/section/energy/opinion/where-will-citizens-fit-within-europes-energy-transition/. (Accessed 16 April 2021).

Wierling, A., Schwanitz, V.J., Zeiß, J.P., Bout, C., Candelise, C., Gilcrease, W., Gregg, J.S., 2018. Statistical evidence on the role of energy cooperatives for the energy transition in European countries. Sustainability 10 (9), 3339. https://doi.org/10.3390/su10093339.

Yildiz, Ö., Rommel, J., Debor, S., Holstenkamp, L., Mey, F., Müller, J., et al., 2015. Renewable energy cooperatives as gatekeepers or facilitators? Recent developments in Germany and a multidisciplinary research agenda. Energy Res. Social Sci. 6. https://doi.org/10.1016/j.erss.2014.12.001.

Further reading

Local Energy Scotland, 2015. Community Renewables Meets 2020 Target. Available online at: https://www.localenergy.scot/news-and-events/2015/october/community-renewables-meets-2020-target/. (Accessed 19 April 2021).

Scottish Government, 2017. Scottish Energy Strategy: The Future of Energy in Scotland. The Scottish Government.

Scottish Government, 2021. Energy Strategy: Position Statement. Available online at: https://www.gov.scot/publications/scotlands-energy-strategy-position-statement/pages/3/. (Accessed 19 April 2021).

Energy communities: a U.S. regulatory perspective

3

R. Bruce Williamson [1,2]
[1]Maine Public Utilities Commission, Hallowell, ME, United States; [2]Public Utility Research Center, University of Florida, Gainesville, FL, United States

1. Introduction

This chapter offers an overview of the possibilities of implementing energy community (EC) deployment in North American utility regulatory jurisdictions at the state and province level.[1] In both the EU and North America, legislators and regulators see the need for appropriate response to the energy—environment—climate challenges of our time. ECs are one such response, and in a variety of forms, are appearing in North American practice.[2] ECs are expected to move the central tendency of our electricity supply and demand balance closer to the consumer and the community, and away from the large centralized generation systems of the past. This expectation is thought to be significant, although in the historical or institutional sense the attention to grassroots, community, or small municipal energy systems has a long history.[3] In the 1890—1920 nascency of electric utilities, small town and community-owned utilities were the major form of ownership, some eventually selling to investor-owned operations where advantages of specialized expertise, cost, and scale existed, or combining in operation with other public systems. As will be discussed in more detail, the signal difference for this conversation today in implementing ECs is that there is

[1] The standard disclaimers apply that nothing in this discussion should be construed as legal counsel or advisement, and the views expressed here are those of the author alone.
[2] See the chapter by Kolesar in this volume.
[3] EIA (2019). "In smaller cities and towns, local governments began setting up their own electric distribution utilities. By the late 1800s and early 1900s, municipalities ran most utilities, and more than 3000 existed by 1923. Later, technological improvements in generation and transmission made smaller plants uneconomical, and many cities sold their equipment and transferred their customers to IOUs." There are at present some 2770 municipal, county, and cooperative utilities in the US, not counting the Federal- and State-owned public power systems; in Canada there are at least 364 smaller municipal utilities, several larger municipals, as well as six provincially owned public power systems. Canadian data downloaded Nov. 1, 2021 from https://www.thecanadianencyclopedia.ca/en/article/electric-utilities. Priest (1993) writes that municipalities also experimented extensively in the 19th century and into the early part of the 20th century with franchising utility services, at times to more than one franchisee simultaneously. These experiments had uneven results, and were abandoned.

Energy Communities. https://doi.org/10.1016/B978-0-323-91135-1.00013-4

nearly universal coverage of the populations in North America by either publicly owned or private investor owned electric utilities.[4] In other words, ECs would not be *de novo* operations in unserved areas of North America, but created out of incumbent service territories.

The concept of an EC refers to community-owned and operated renewable energy installations, although in areas of North America, the terms "distributed generation" (DG) and "microgrid" may be used to suggest a commonality of resource type, technical design, or modularity, rather than special emphasis on community attributes built around a transaction-oriented energy production and delivery core. There are now many ways to configure and power ECs with renewable generation, two-way power flow capability and demand side flexibility.[5] Regulators at the state and provincial levels in North America are already in proceedings to consider DG and microgrid developments that include communities.[6] Although it is the case that regulatory commissions traditionally have little say about the operations and pricing of publicly owned utilities, it so happens that publicly owned systems are also expanding member-owned renewable generation and consumption portfolios in their own service areas.[7] Where privately owned utility incumbents are involved, however, the regulator must apply the "public interest" regulatory standard that accompanies the statutorily assigned duty of "ensuring safe, adequate, and reliable service at just and reasonable rates."

The chapter is organized as follows:

- Section 2 describes the North American regulatory and policy framework;
- Section 3 evaluates the impacts of EC market entry in already-served markets; and
- Section 4 examines the central tenets of this book, whether ECs can be "market driven" and "welfare enhancing," followed by the chapter's conclusions.

[4] The term "public power" includes all town and municipally owned, county-owned, State-owned, Provincially owned, and Federally owned electric service utilities, in contradistinction to investor-owned electric utilities. The North American term "Public Utility" is confusing, as it can mean either "publicly owned" or "investor owned" depending upon the context.

[5] The microscale balancing possible between supply coupled with demand side flexibility enables a smaller overall energy footprint for the community. This could allow for lower average costs of operation by minimizing requirements for storage and generation, and the need for imported power. The more this can be done without incentives or external subsidy, the more market-driven and welfare enhancing the EC investment will be.

[6] For one example from 2018 by a privately owned utility working with a community in Chicago, see "ComEd Approved to Build One of First Microgrid Clusters in the Nation," downloaded Nov. 10, 2021 from https://www.comed.com/News/Pages/NewsReleases/2018_02_28.aspx.

[7] Publicly owned utilities and cooperatives have only light-handed regulatory oversight when asked for, or when legislatively assigned duties or levies (such as contributions for efficiency and low income assistance programs) are applied. The light touch oversight of publicly owned utilities goes back to the U.S. Supreme Court's 1921 Springfield case, which affirmed the "right of municipal governments to sell electricity to private consumers, without regulation by state public utilities commissions." Springfield Gas & Elec. Co. v. City of Springfield, 257 U.S. 66 (1921). See APPA (2016), 34. Clarification in personal correspondence and additional information from interview with Paul Zummo, Director, Policy Research and Analysis, American Public Power Association, Nov 8, 2021, Louisville, KY. Several unregulated publicly owned community transformation projects (City of Holyoke, Seattle City Light, Sacramento Municipal Utility District among others) are described on the American Public Power Association website at https://www.publicpower.org/periodical/article/optimizing-utility-transformation, viewed Nov. 11, 2021.

2. The regulator, the legislator, and the North American regulatory policy framework

When it comes to converting ideas into legislation and regulatory action that will authorize EC entry into existing energy service markets, the devil is in the details. There may be expectation by some observers that North American regulatory commissions[8] at the state, provincial, and federal levels could unilaterally modify the laws to favor EC deployment at the local level. If North America had no utility providers or infrastructure in place, this might be possible as a "greenfield" or *de novo* situation, but that is clearly not the case. North American and European practices in regard to regulatory legislation and regulatory commissions are somewhat different, and further, utility enterprises, whether publicly or investor owned, have ongoing utility operations with either an explicit or implicit obligation to serve. There are also long-recognized Federal and State/Province separations of power and areas of responsibility, and protections against takings of private property without legal due process.

Moreover, elected legislators create the actual public utility laws set in statute, and are as much "regulators" as the administrative regulatory agencies referred to as regulatory commissions. The latter do not make policy, which is the purview of the elected legislators, but for over a century regulatory commissions have been the place where evidence can be examined in detail. Regulatory commissions are also where full legal due process is available for all parties[9] to regulatory proceedings, which legislatures have neither the time nor the expertise to provide.[10]

Consider now the regulator's initial acquaintance with policy for EC advancement in North America. Suppose that EC developers seek unrestricted entry into existing service territories of both publicly owned and investor owned utilities. These utilities are closely integrated with grid scale transmission in organized, regional markets. The physics of electron flow dictate that in sufficient numbers and placement, ECs could have far-reaching effects—some beneficial, some not so—on energy costs, locational marginal prices, incumbent utility providers, transmission and distribution grid reliability and costs, electrification efforts, vehicle integration, and consumers and businesses whether members of an EC or not. If a petition by an EC developer for a Certificate of Public Convenience and Necessity (CPCN) came before a hypothetical utility regulatory commission, the regulatory commission's legal staff and

[8] Variously named Corporation Commissions, Public Service Commissions, Public Utilities Commissions, Public Utilities Boards, Energy and Utilities Boards, Utility and Review Boards, Department of Public Service, and Department of Public Utilities. When speaking of the utility regulatory role in North America it is worth noting that North American state and provincial jurisdictions also empower consumer or public advocates to represent broad consumer interests before Utilities Commissions, and although public advocates are not "regulators," per se, they participate in and help shape many electricity regulatory outcomes in North America. State Attorneys-General may also be called on to enforce consumer protections or business practices that fall under general trade protections and outside a utility commission's jurisdiction.

[9] The term "party" in a legal proceeding refers to persons or entities with recognized standing to participate in a proceeding as a result of being sufficiently affected by the matter at issue.

[10] Public Utilities Department (1915). See also Kahn (1988) and Schwartz et al. (1985) for the background and emergence of core concepts in utilities regulation.

Commissioners would look to the legal foundation in the state or province statute and rule, and perhaps law court rulings of a pertinent nature. Preferably, the EC developer has community agreement, located a suitable site, received the physical siting permit, received all necessary environmental review, and environmental permitting and approvals.[11] The utility regulator is expected to comply with what the legislature has set in statute and required of the applicant; these requirements may be specific in a qualitative sense, but not always simple to define or measure.[12] Often technical conferences with interested parties or public witness hearings may be part of due process to allow parties and the public an opportunity to comment. This would be a simplified hypothetical approach in many North American regulatory jurisdictions for an EC petition.[13] What then is the problem? The friction between parties to our hypothetical case can occur when the chosen EC site is selected within utility service territory that is:

- Already franchised by the legislature to an investor owned utility which provides adequate, reliable service to the same customers at reasonable rates; or
- Within the service boundaries of an existing community/municipal utility adequately serving their citizen-owners.

To the first point, State/Provincial legislatures have the ability to modify or even terminate the franchises of investor-owned utilities under exceptional circumstances. This inevitably impacts legal contracts and infrastructure investments purpose-built under the utility franchise obligation to serve. An action to modify or terminate the investor owned utility franchise almost certainly faces contentious valuation proceedings and law court challenges for the taking of private property without appropriate compensation.[14] State laws vary,[15] but the length of such proceedings and the

[11] Statute may specify whether environmental reviews or other approvals may occur before, during, or after the application for CPCN is made. There have been situations where developers of community energy programs ran into protracted siting disagreements with a town's elected Board of Supervisors that were not resolved, thus mooting the petition for CPCN to operate.

[12] As an example of the specificity of a State Legislature providing guidance to regulators with a transmission line CPCN in Maine, Title 35-A M.R.S §3132 part 6 states "In determining public need, the commission shall, at a minimum, take into account economics, reliability, public health and safety, scenic, historic and recreational values, state renewable energy generation goals, the proximity of the proposed transmission line to inhabited dwellings and nontransmission alternatives to construction of the transmission line, including energy conservation, distributed generation, or load management." Where proceedings take longer tends to be where the legislative statute does not anticipate the scope, specifically prohibits some aspects of the petitioner's application, contains ambiguity that situates the request between statutory criteria, leaves doubt whether the public interest is served, or draws significant opposition to development impacts.

[13] The Interstate Renewable Energy Council and Vote Solar Initiative in the US have published "Model Program Rules" rules for legislatures which explicitly waive regulatory commission oversight of energy communities and from treatment as utilities. State and provincial legislatures may accept this advocacy position and in doing so create an unregulated class of citizen-owned energy providing entities distinct from already lightly regulated cooperatives and municipal utilities. See Appendix B of US Department of Energy (2011).

[14] 5th and 14th Amendments to the US Constitution, the latter, restricting the powers of state government to take private property without due process and fair compensation. See Sullivan (n.d.).

[15] APPA (2016) 28.

uncertainty of the outcome has led some legislatures to find different, and sometimes indirect means and smaller scale applications or pilots to set larger policy objectives in motion.[16] Regardless of the method, legislatures typically rely on utility regulatory commissions to sort out the details and provide legal due process for the parties involved if the parties have not already filed separately in a law court of jurisdiction.

Independent of legislative action to terminate or modify the monopoly franchises of privately owned electric utilities, cities and towns may "municipalize" existing investor owned utility infrastructure within their own municipal boundary, thereby acquiring privately owned infrastructure that is already in use serving municipal customers.[17] There have been 20 successful municipalizations since 2000 in the United States, and others are under consideration.[18] In both these instances, even if there are legitimate public interest reasons to promote market entry and EC deployment by takeover of the investor owned incumbent utility's infrastructure and customer base, there is still the requirement of fair valuation and just compensation paid to the incumbent utility owner for the infrastructure acquired by the municipality or acquired with franchise modification or termination of the franchise by the legislature of the State or Province.[19]

The second bulleted point above highlights risks to existing cooperative and other citizen-owned utilities by policy-supported EC developments seeking to locate *within* the customer base of the cooperative. It may seem hypothetical, but as one narrowly avoided example will show, subsidized entry of competitors within the existing service areas of incumbent citizen-owned municipal utilities and electric cooperatives raises equity, fairness, and sustainability issues that are directly relevant to the thesis of this volume, as well as being wasteful of societal resources. Such an instance was eventually headed off after the committee of jurisdiction in the 2019 Session of the Maine State Legislature had moved ahead with a bill that would allow microgrid

[16] See APPA (2016) 28. States have community-based renewable energy programs and subscription-based citizen ownership of distributed generation (primarily solar PV). Most states require their investor-owned utilities to buy a certain portion of their electricity from green sources, so that renewable systems have firm buyers, and placing the state's utilities in a position of being marketers of renewable power.

[17] Under the U.S. Constitution's 10th Amendment, the powers not given to the Federal government are reserved to the states, and include the so-called "police powers" which enable a state or municipality to regulate behavior and enforce order within their territory for the betterment of the health, safety, morals, and general welfare of their inhabitants, which could include under "general welfare" the creation of municipal utilities. Downloaded Nov. 1, 2021 from https://en.wikipedia.org/wiki/Police_power_%28United_States_constitutional_law%29.

[18] APPA (2016). See also Public Utility Research Center (2018), 83. In recent years a few publicly owned utilities have either been sold to other publicly owned utilities, or less often, "demunicipalized," i.e., privatized.

[19] APPA (2016), 28 et. seq.

developers to own up to 10 MW of generation capacity within any community in the state.[20] Moreover, the EC developers would also qualify for state subsidies and were guaranteed purchase of their exported power by other utilities, effectively shifting costs and risks onto nonmembers. Allowing subsidized microgrids of 10 MW to be established in the footprint of existing consumer owned utilities with system loads of 15, 17, or 19 MW (as was the case in Maine) would quickly bankrupt the citizen-owned utilities. Enfranchising a new class of supply within and to the detriment of a preexisting, socially empowered, citizen-owned electric utility should be a quintessential equity issue for policy makers.

In brief, the introduction of ECs into areas already served by utility providers and the extent to which the public interest is served by doing so will raise significant issues for the regulator. Several of these key issues will be addressed next.

3. Evaluating EC entry in already served markets

As noted above, the case of *de novo* entry of EC operations into unserved areas presents no problem for regulators. It is a problem for the regulator if by policy ECs are granted unrestricted entry into already existing electricity markets, particularly if accompanied by presumption at the outset that ECs are "market driven" and "welfare enhancing." For example, if unrestricted market entry requires cost shifts that incentivize ECs but not other grid customers, there should be a public interest concern about welfare impacts on customers and impacts on the safe, reliable and affordable operation of any incumbent utility, public or private. An EC developer may assert the existence of "collective benefit"[21] for EC members and more generally, for nonmembers and society, but that claim must be unpacked carefully to understand all components and their implications. In this context,[22] the regulator must come to terms with:

- Cost causality, how costs will be recovered and from whom, how risks will be assigned, and extent of tax incentive or cost subsidy required from the state;
- How decarbonization goals are cost effectively facilitated by EC operations;
- How wholesale and retail electricity markets may be affected as ECs grow in number and extend their scope of operations over time; and
- How EC members, nonmembers, and grid users are affected by possible technical and economic externalities created by ECs.

The points are addressed in turn below.

[20] The bill, LD 13, "An Act to Allow Microgrids That Are In the Public Interest," eventually died without passage at the close of the 129th Legislature's Second Session on Nov. 16, 2020. The public testimony to the Energy, Utility, and Technology Joint Legislative Committee by Ms Debra Hart, who represented Maine's cooperative and municipal utilities at the hearing of Jan. 31, 2019, is paraphrased here. Downloaded Nov. 5, 2021 from https://legislature.maine.gov/bills/display_ps.asp?snum=129&paper=HP0014&PID=1456.

[21] The chapter by Kolesar quotes Walker and Devine−Wright (2008).

[22] This discussion is confined to EC penetration of existing electricity markets.

Direct subsidization of renewable energy projects is not uncommonly blessed by legislatures in statute, and usually implemented by levying assessments on customer utility bills. In the case of net metering arrangements, they may also exempt that customer group from some or all transmission and distribution (T&D) costs that pay for grid usage.[23] The principle of cost causality applied here is that all grid customers rely on grid use, whether receiving power only or exporting as well as receiving power, and all grid customers should bear a fair share of the common costs. This means that an exemption or discount for one group of customers (lowering cost to the favored group) will be paid for by another group (who will consequently face higher costs) to make up the missing revenue required to maintain the common cost of the grid. To the extent that EC member owners produce power and self-consume entirely on an "islanded" community-owned network, there is no impact on the grid.[24] But to the extent that ECs must use the grid for occasional or regularized power trades,[25] any waiver from paying their fair share of grid utility infrastructure shifts the unpaid grid costs to those grid customers who do not belong to the EC.[26]

Financial risks might also be shifted to nonmembers in some of the financing arrangements for the EC. If tax subsidies or levies are used to offset investment expense with credits or exemptions, the financial risk for lenders who back EC development is transferred from EC members to nonmembers and, indeed, to all other citizens not benefitting from the credit or exemption. Moreover, there is an opportunity cost: the more that EC revenue streams benefit from government entitlements and support, the more investment flows into the EC sector, drawing societal resources away from potentially more beneficial activities. Some of this is intended with legislation for climate action, but it may come on top of already existing subsidies through renewable energy standards. Ultimately, the layering of subsidy and cost shifts among citizens and customer classes will increase market distortions, will be politically difficult to unwind later, and will reduce social welfare.

To the first point above, these are quite obviously not Pareto-optimal outcomes[27]—there are clear winners and losers—which violates the sense of collective benefit. In

[23] Commonly solar PV-owning customers or communities, or possibly other renewables, such as wind (see chapter by Benedettini and Stagnaro).

[24] "Islanded" meaning electrically separated or "at will" separable from the grid. There will be an impact on incumbent electricity suppliers, who would see a disappearance from usual load obligations with a self-supplying or demand responding microgrid or EC. Microgrids and ECs have the ability to cooptimize supply and load behind the meter, but unless continually islanded, may import or export power as needed through the incumbent utility or distribution operator.

[25] See chapter by Madlener and Crump on peer-to-peer trading and formal bilateral power purchase agreements (PPAs); and the use by wind communities of an aggregator to market power and optimize trading, in the chapter by Benedettini and Stagnaro.

[26] Arguments are made to justify feed in tariffs or net metering exemptions from the grid provider's T&D costs, but these are not economic arguments, and have their source in providing extra policy incentives for renewable energy development. Other arguments, which are economic in nature, are those that attempt to disentangle or rebalance the multipart tariffs and the alignment with system fixed, variable, and marginal costs.

[27] At its simplest, the economic concept of Pareto-optimal solutions is that at some desired point of trade-offs no one can be made better off without someone else being made worse off.

principle, there is only one wallet in the room, so to speak, and it belongs to the taxpayer—customer—ratepayer. If utility bills are the favored means to collect levies for various programs,[28] there is an adverse income effect[29] that particularly harms citizens and communities with lower incomes or on fixed pensions. These residents of limited means are forced to pay a relatively higher price per kilowatt-hour than an EC member who may pay nothing for the grid connection and may be financed by an investment loan that gives cost-lowering tax credits or exemptions for members. In all, government subsidies inevitably distort markets and redistribute income among grid participants by shifting costs and risks, and all else equal, tend to lessen any collective benefit or societal gain that may be claimed from EC expansion.

The second point on the list is the matter of whether decarbonization goals are being accomplished cost effectively with application of limited societal resources. A commitment of societal resources such as labor, capital, energy, materials, etc., may ultimately accomplish decarbonization, but might be done wastefully if a least cost approach is not used. If resources, e.g., capital, are applied to excess, it will detract from the accomplishment of other essential societal goals. Finding the least cost solution to achieve a given societal goal such as decarbonization is a standard regulatory method for maximizing the social welfare that can be achieved for a given expenditure of resources. The chapter by Robinson and del Guayo takes this approach to address whether the production of cleaner kilowatt hours and decarbonization will be accomplished with a least cost strategy. Suffice it to say here that a policy construct which incentivizes renewables to the point of economic oversupply cannot be, by definition, the least cost approach to decarbonization, and will not maximize social welfare, yet it has been observed in practice.[30]

[28] See Howe and Reeb (1999). The use of the utility bill to collect taxes or other levies has to do with the low price elasticity for utility service, which reduces tax avoidance when bundled with utility payments. The chapter by Robinson and del Guayo points out that the typical Spanish monthly electric bill is comprised of 46% taxes and levies. Levies include cross-subsidies for "renewables, cogeneration, and residues." These amounts "reflect the government's ability to collect revenue to pay for public policies."

[29] An income effect in economics is the change in overall purchasing power resulting from the change in the price of a good or service; here, the increased price of electricity to nonmembers due to the built-in levies (which flow to subsidized EC members, in our example) lowers the total purchasing power of all customers, but is particularly burdensome to those with lower incomes and on fixed incomes, and decreases social welfare.

[30] Incentivized distributed generation (DG) developments in New England's ISO-NE area are leading to a predictable oversupply as new DG entrants find the economic rents attractive. By a recent estimate, Maine will soon have more than twice the generation capacity it needs. Combined with similar incentive programs in the five other states in New England, the excess generation being sold on the grid will depress wholesale electricity prices, which in turn will raise the requirement for capacity payments to dispatchable generators (or storage) to be available when intermittent resources cannot produce. While spot electricity is cheapened, the customers in New England will pay substantially more in capacity payments on their electric bills. The problem is aggravated if demand response and efficiency programs continue to shrink peak load and overall load. The province of Ontario also recently acted to reduce an oversupply situation when it canceled 758 additional renewable projects stimulated by an overly generous feed in tariff (FIT). The renewable power benefitting from the FIT was replacing the province's abundant low-cost hydropower. Downloaded Nov 5, 2021 from https://www.instituteforenergyresearch.org/international-issues/canadian-provinces-fighting-back-against-trudeaus-carbon-tax/.

The third point concerns the grid externality effects (positive or negative) of EC deployment. There are externalities from an increasing number of ECs participating on the larger grid without measures taken to appropriately assign costs and evaluate the individual and collective grid impacts, if any.[31] One consideration is that an ecosystem of ECs, enabling trade of exported and imported power as their internal loads and self-generation vary, can impact grid stability if their choice of siting and scale will necessitate upgrades to ensure safe and reliable operation for all grid users.[32] How those upgrade costs are shared among ECs, or perhaps paid by the Regional Transmission Operator (RTO) or the Distribution System Operator (DSO)—who then assess all grid customers—should be examined for fairness in application. There will be welfare gains to the extent that EC developers take responsibility for necessary upgrades to the grid caused by their decisions, and once in operation, provide critical grid support. The fair approach is to ensure costs are assigned to those who impose the costs, so that grid reliability is maintained, and all users of the grid can benefit. Without this approach, ECs might shift their costs to non-EC customers and other grid users and thereby lessen overall social welfare.

Another consideration is that an ecosystem of ECs can impact the cost-effective operation of electricity markets. Sioshansi et al. (2008) observe that the process for unit commitment from either centrally committed or self-committed market control structures is economically imperfect. Centrally committed operations suffer from incentive compatibility problems, while self-committed operations create economic inefficiencies from uncoordinated actions. In either extreme there is potential for exercise of market power by participants, which may distort pricing for all grid customers and add to the difficulty of reliability planning for the RTO/DSO.[33] Reliability requirements are met on the supply side by retaining dispatchable fast-ramping capacity reserves to be available when called. The more reserves that are needed because self-committed EC operations are not coordinated with the system operator, the higher that capacity payments to the reserve units will need to be, and the higher the eventual cost of electricity to customers. It is a genuine benefit to the RTO/DSO, however, if certain requirements for capacity reserves and ancillary services can be met in part with EC operations. These capabilities with EC coordination could actually lower overall costs of grid operations and enhance overall social welfare.

One last consideration on externalities with EC deployment is that ECs necessarily become part of the larger grid's response to extreme events and natural disasters. It is a matter of rendering mutual assistance when needed. An RTO/DSO may need to coordinate with all grid-attached resources, capable of islanding or not. Modern interdependencies among many community systems elevate the problem of large-scale outages,

[31] Potential grid stability impacts (not cost shifts) are considered by the Eastern Interconnection Planning Cooperative (EIPC) and similar organizations bringing system operators together with incumbent utilities, both publicly and privately owned, to ensure policymakers make well-informed decisions that affect grid stability. See EIPC (2021).

[32] Sopher (2013). See also the discussion of the DSO perspective in the chapter by Del Pizzo et al.

[33] Clearly, there is a role for an independent market monitor to ensure customers of the grid are not harmed by exercises of supplier market power.

which can quickly cascade to impact drinking water supplies, schools, public safety, health care, and essential food supplies for a larger region. ECs themselves might experience extreme events and require emergency assistance from the grid. Inevitably, ECs will find themselves part of national critical infrastructure. This is somewhat at odds with the ideal of energy self-determination and freedom from large utility operators, an idea mentioned by respondents in the chapter on EC members by Hackbarth and Löbbe, in this volume. Nonetheless, EC member-owners ultimately should see the advantage to being good grid citizens in crises, even if this means the RTO/DSO has visibility into EC operations, for the public safety and overall social welfare of the state or region.[34]

4. Can ECs be "market driven" and "welfare enhancing"?

The answer is a matter of degree.[35] Simply put, ECs can be both market driven and welfare enhancing, to be discussed in this section.

4.1 Market driven?

Regulators should be keen to prevent cream skimming[36] by EC developers eager to enter the incumbent's franchise market. ECs, as new market entrants, are unlikely to be tasked with the legislative and regulatory obligations faced by the electric utility incumbent. The requirements placed on incumbents include the obligation to serve all customers, just and reasonable pricing, allowances to favor low income customers, obligations to buy green power (sometimes priced above-market), and perhaps most critically, being unable to exit the market or portions of the market when continued service there is unremunerative. Arriving without those obligations, new EC entrants are in a strategic position to cream skim the best areas in the incumbent's service territory, leaving the electric utility incumbent in a steadily worsening situation with low density, lower income, and/or higher cost service areas. Economic analysis shows that inducing the exit of the incumbent by occupying the best possible areas to serve is very

[34] Federal Energy Regulatory Commission (2021). Very recently, the severe weather-induced outages impacting water, gas, and electricity supplies in Texas in February 2021 are certain to spur actions to strengthen FERC and ERCOT visibility into market participant capabilities and arrangements, including demand response arrangements, to prevent a repeat occurrence.

[35] The thesis of this volume also asks if ECs are "customer centric," a presumption we can grant by definition. Like ECs, publicly owned utilities, municipals, and cooperatively owned utilities are by definition and institutionally "customer centric."

[36] The effort by a competing firm to take the highest margin, lowest cost-to-serve customers from another firm and avoiding the lower margin, higher cost-of-service customers. In regulated utilities, incumbent utilities have an obligation to serve all customers in a given service territory without discrimination, making it easy for unrestricted competitors to select the most profitable customers or areas only. Kahn writes at length about the impacts on regulated and deregulated industries. Kahn (1988) II. 221–246.

similar to the strategy that an unfettered incumbent would take in deterring new entry.[37] The shoe is simply on the other foot.

In fact, the market entry of EC competition can and should be done in a way that minimizes economic welfare loss to society as a whole by minimizing the creation of stranded assets resulting from legislated policy changes that unevenly favors entrants over existing service providers.[38] The right approach to introducing healthy competition minimizes social costs by avoiding a revenue-cost "death spiral" of the incumbent, who is unable to raise rates, and unable to abandon the low density, highest cost service areas. It would be difficult to claim that EC deployments are market driven to the extent that ECs are loosed upon regulated markets without thought for minimizing welfare loss as the market continues to evolve. If electricity markets are to be deregulated, then it should occur with unsubsidized, economically competitive entrants with the most cost-advantageous technology and operations. Rethinking both incumbent and new entrant obligations, combined with tariff reform, can do much to redress the imbalance and reduce welfare loss from cream skimming actions by market entrants or anti-competitive behavior by incumbents, and could ensure that EC introductions are market driven.

4.2 Welfare enhancing?

Can ECs be welfare enhancing? As explained earlier, there are ways to ensure net societal gains from EC deployment with welfare improvements associated with minimizing stranded costs, preventing cream skimming, minimizing cost shifting and use of cross subsidies and aiding grid stability. The choice of optimal policy and agreement on actions gets more difficult, however, when the concept of welfare enhancement itself is expanded to include broadly stated socio–cultural–ecological priorities and goals.

Moreover, in the present context of EC infrastructure investments, welfare gain or loss in evaluating welfare enhancement will be directly influenced by the sustainability of the EC investment itself and the resources it consumes over the life of the EC. At a

[37] Tirole (1990), 324.

[38] "Economic bypass" in existing markets opened to greater competition is the consequence of cost-lowering technologies allowing the adopters to compete successfully on price/quality against incumbents and other competitors without need for incentives and subsidy. There is a net welfare gain to society as older, higher cost technologies are phased out and replaced by newer more cost-effective technologies. A contrasting example would be the "uneconomic bypass" that occurred for telecom network providers by unburdened market entrants, which occurred in the telecom industry's deregulation in North America. Entrants were allowed to carry traffic through to customers without paying certain levies the incumbent network owner was required to pay. For the present discussion of electric utilities, in some North American markets, incumbent investor-owned utilities are also allowed to adopt new cost-lowering technologies identical to what an EC could own—and they are in some cases doing so. In other North American markets ("restructured markets"), incumbent utilities are forbidden to own generation. Where utilities are restricted from owning generation but retain the obligation to serve and the inability to exit the market, the unrestricted entry of ECs who have no obligation to serve is the prescription for a "death spiral" outcome, in which incumbent shareholders lose their utility investment, and EC nonmembers, i.e., all other grid customers, are left without service, for a significant overall societal welfare loss.

minimum, enterprises that are not sustainable in a business and economic sense will waste society's resources and investor time and effort. In failure they provide only a low or even negative return on society's investment in ECs, relative to investments that otherwise could have been made to decarbonize our energy systems and mitigate climate change.

The EC planners' problem is necessarily more difficult when the socio–cultural–ecological aspects of sustainable welfare enhancement are considered. The epistemological challenge is that these aspects are difficult to identify, difficult to agree on, and difficult to prioritize across citizenry, circumstance, and time. Lacking consistent, accurate, and reliable ways of measurement, planners may never be sure if the resources applied were the right resources and in the right amounts at the right time. The likely root of the problem will come from diversity in EC member-owner preference structures when operations begin.[39] Kenneth Arrow's theorem on preference ranking for collective decision-making comes to mind—here, settling on what the priorities should be,[40] but here also on the societal resources needed. In practice, it is conceivable that targets will vary among ECs, across regions, member-owners, time, and even from one national political leader to the next. It may be very difficult to verify progress of the overall investment. The more one looks at the challenge of sustainable welfare enhancement, it is at minimum a resource problem, an information problem, a coordination problem, and a measurement problem, all of which are laid at the door of EC member-owners to resolve with a very limited tool kit. The measurement of business, economic, environmental, and decarbonization progress seems simple by comparison.

5. Conclusions

In 1921, US Supreme Court Justice Louis Brandeis, remarking on the course of regulatory and antitrust development, said, "It is one of the happy incidents of the federal system that a single courageous state may, if its citizens choose, serve as a laboratory, and try novel social and economic experiments without risk to the rest of the country."[41] Faced by the clear need to decarbonize our energy systems, there is an increasing amount of utility and community experimentation in North American states and provinces. There is significant progress in application of distributed resource technology, efficiency measures, and integration of demand side management to reduce generation requirements with both publicly owned and investor-owned electric utilities.

It also helps to take the longer view on the important matter of sustainability of these investments, and an example from outside North America may be a guide.

[39] The chapter by Hackbarth and Löbbe addresses ex ante attitudes of potential EC participants. *Ex post* preferences may differ once experience is gained.

[40] See Arrow (1950). Recent research on the imperfect nature of most if not all voting methodologies suggest a similar methodological dilemma.

[41] *New State Ice Co. v. Liebmann*, 285 U.S. 262 (1932). See Schwartz et al. (1985), 57.

The chapter in this volume by Benedettini and Stagnaro describes Denmark's long experience with renewable-based communities, and it provides food for thought on the matter of ECs as sustainably customer-centered, market-driven, and welfare-enhancing contributors. The Danish experience with citizen wind communities was as progressive and future oriented as anyone could imagine at the time. Nonetheless, in fewer than three decades, it was obsoleted by political change, technological advance, and market change. A later wave of energy market opening in Denmark brought more cost-effective wind generation and greater scale economies, and hastened the decline of most of the original wind communities. The lesson is stark: the one constant that cannot be addressed completely by our designs today is the bruising certainty of change over time. Within a matter of one or two - decades, today's favored designs[42] may be impractical or uninteresting, consumers and owners may be frustrated by artificial intelligence and second guessing of human plans, and operation and maintenance costs may prove unsustainable. EC systems may expand or merge with other ECs to achieve better scale economies,[43] or seek to reduce overhead costs by outsourcing to aggregators.[44] And perhaps most significantly, the national enabling statutes can be superseded by different national priorities, better knowledge, new approaches, or shifted policy emphasis.

None of this is to say the regulatory and policy effort to bring ECs into mature markets is not worth the candle. There are a number of ways that ECs can add welfare enhancing benefits to their communities and regions, even without considering the social and cultural good that may be possible within the individual EC operation. ECs should be tried, tested, measured, and learned from over the next 5 years.[45] With that knowledge and experience, more can be implemented quickly where needed—in the EU or in North America or elsewhere. It will take a steady effort by legislators and regulators to minimize the use of subsidies, taxes and costs shifts, and to pursue market and tariff reforms that maximize societal benefits from a more competitive and innovative energy market. The pursuit of sustainable change for a globally decarbonized future requires real commitment—and candor—to ensure that limited resources are wisely applied. The alternative, that of adding cost shifters and distortions into energy markets, and favoring this party or that according to political whim rather than experience and evidence, is a road to economic welfare loss and societal decline.

[42] Digital technologies have on three occasions since the 1990s overpromised on and underdelivered on grid benefits. See Slayton (2013).

[43] A process not unlike how the mergers of smaller utility systems for scale economies built most of the utility companies we have today in North America.

[44] According to the chapter by Benedettini and Stagnaro, the Danish wind communities eventually outsourced their power sales to Vindenergie, an aggregator.

[45] Learning from the member-owner experience over extended time periods of behind-the-meter EC supply balanced with flexible demand response for a given scale of operation would be highly useful in sizing additional deployments and setting realistic expectations.

Bibliography

American Public Power Association, 2016. Public Power for Your Community: Local Control. Local Priorities. A Stronger Local Economy. Arlington, Virginia. Downloaded October 30, 2021 from. https://www.publicpower.org/municipalization.

Arrow, K.J., 1950. A difficulty in the concept of social welfare. J. Polit. Econ. (58), 328−346.

Briggerman, A., Costinescu, R., Bond, A., 2012. Survey of State Municipalization Laws. Prepared for the American Public Power Association by Duncan & Allen, Washington, DC. Downloaded October 30, 2021 from. https://www-static.bouldercolorado.gov/docs/Survey_of_Municipalization_Laws_-_Duncan_and_Allen__FINAL___00027359_-1-201504271104.pdf.

Eastern Interconnection Planning Cooperative, 2021. Planning the Grid for a Renewable Future. Downloaded November 14, 2021 from. https://insidelines.pjm.com/new-eipc-paper-outlines-needs-of-the-grid-with-increased-renewables/.

Energy Information Administration, 2019. Investor-owned Utilities Served 72% of U.S. Electricity Customers in 2017. Downloaded October 28, 2021 from. https://www.eia.gov/todayinenergy/detail.php?id=40913.

Federal Energy Regulatory Commission, 2021. FERC, NERC to Open Joint Inquiry into 2021 Cold Weather Grid Operations. Downloaded October 17, 2021 from. https://www.ferc.gov/news-events/news/ferc-nerc-open-joint-inquiry-2021-cold-weather-grid-operations.

Howe, E.T., Reeb, D.J., 1999. State and local electric utility taxes: evolutionary taxation of a deregulating monopoly. Am. J. Econ. Sociol. 58 (1), 115−128.

Kahn, A., 1988. The Economics of Regulation Principles and Institutions. MIT Press, Cambridge, Massachusetts.

Kahn, A., 1998. Letting Go: Deregulating the Process of Deregulation. MSU Public Utilities Papers, East Lansing, MI.

Kahn, A., 2004. Lessons from Deregulation: Telecommunications and Airlines After the Crunch. AEI-Brookings Joint Center for Regulatory Studies, Washington, DC.

Priest, G., 1993. The origins of utility regulation and the 'theories of regulation' debate. J. Law Econ. 36 (1), 289−323. Part 2.

Public Utilities Department, 1915. First Annual Report of the Public Utilities Commission, State of Maine, for the Year Ending October 31, 1915. Sentinel Publishing Co, Waterville, Maine.

Public Utility Research Center, 2018. Valuing Municipal Utilities – the Case of the Potential Sale of JEA in Jacksonville. University of Florida, Gainesville, Florida.

Schwartz, L., Flynn, J.J., First, H., 1985. Free Enterprise and Economic Organization: Government Regulation, sixth ed. Foundation Press, Mineola, New York.

Sioshansi, F.P. (Ed.), 2008. Competitive Electricity Markets: Design, Implementation, Performance. Elsevier, New York.

Sioshansi, R., Oren, S., O'Neill, R., 2008. The cost of anarchy in self-commitment-based electricity markets. In: Sioshansi, F.P. (Ed.), Competitive Electricity Markets: Design, Implementation, Performance. Elsevier, New York.

Slayton, R., 2013. Efficient, secure, green: digital utopianism and the challenge of making the electrical grid smart. Inf. Cult. 48 (4), 448−478.

Sopher, P., 2013. Lessons learned from Germany's Energiewende. Renew. Energy Law Pol. Rev. 6 (2), 99−112. Downloaded on October 11, 2021 from. https://www.jstor.org/stable/26256443.

Sullivan, E.J., n.d. A Brief History of the Takings Clause. Washington University, St. Louis, Missouri. Downloaded October 31, 2021 from https://landuselaw.wustl.edu/Articles/Brief_Hx_Taking.htm.

Tirole, J., 1990. The Theory of Industrial Organization. MIT Press, Cambridge, Massachusetts.

U.S. Department of Energy, 2011. A Guide to Community Solar: Utility, Private and Non-profit Project Development. D0E/G0-102011-3189. Washington, DC. January 2011. Downloaded September 30, 2021 from. https://www1.eere.energy.gov/solar/pdfs/47692.pdf.

Developing a legal framework for energy communities beyond energy law

4

Job Swens[1] *and Lea Diestelmeier*[2]
[1]J-OB—Joining Objectives, Rotterdam, Netherlands; [2]Groningen Centre of Energy Law and Sustainability, University of Groningen, Groningen, Netherlands

1. Introduction

With the reform of the legal framework governing the European Union (EU) energy sector in 2018 and 2019, known as the *"Clean Energy for All Europeans"*—Package (CEP) (EU Commission, 2019), the EU legislature acknowledges the need to empower consumers and for the first time addresses "citizens" in particular to accelerate the transition toward a low-carbon energy sector. With this focus on consumers and citizens, the EU energy transition is set to be wider than a shift of energy sources, entailing also the inclusion of individuals and new organizational forms of distributing the costs and benefits of this transition. One of the most prominent novel legal concepts that has been adopted under the CEP is the "energy community." In fact, two different definitions for energy communities have been introduced, namely the "Renewable Energy Community" (REC) under the Directive on the promotion of renewable energy sources, and the "Citizen Energy Community" (CEC) under the Directive on common rules for the internal market for electricity.[1] A large variety of energy communities already existed in EU member states prior to the adoption of those Directives (EU Commission Joint Research Center et al., 2020), but with the adoption of the Directives national law-makers have to revise their legal frameworks to ensure that the requirements of the Directives are met and that the further development of energy communities is explicitly facilitated and promoted.

While the Directives leave a large degree of discretion to the member states to determine the exact legal form of energy communities (e.g., cooperative, partnership, association, nonprofit organization, etc.) and the activities they may carry out (e.g., production, consumption, selling, storing, and additionally for CECs distribution system operation, energy efficiency services, charging services, etc.), the Directives take a prescriptive approach in the purpose that energy communities have to fulfill: energy communities have to *"provide environmental, economic, or social community benefits*

[1] Directive (EU) 2018/2001 on the promotion of the use of energy from renewable sources [2018] OJ L328/82 (hereinafter "Renewable Energy Directive"), article 2(16); Directive (EU) 2019/944/of 5 June 2019 concerning common rules for the internal market for electricity [2019] OJ L158/125 (hereinafter "Electricity Directive"), article 2(11).

Energy Communities. https://doi.org/10.1016/B978-0-323-91135-1.00019-5

for its shareholders or members or for the local areas where it operates, rather than financial profits."[2] Essentially, this provision establishes the distinct feature of energy communities compared to "conventional" entities (e.g., production and supply companies and distribution system operators [DSOs]) which strive for profit margins or efficiency in a liberalized energy market.

The question of what distinguishes energy communities from other actors in the energy sector has been investigated by long-standing research on energy communities which proposes to categorize energy communities in two dimensions: a process dimension, which determines *who* is involved and *who* exercises influence, and an outcome dimension, which determines *how* outcomes of an initiative are spatially and socially distributed, and *who* benefits in economic and social terms (Walker and Devine-Wright, 2008). Incorporating those distinctive elements in a legal framework for energy communities is key for the concept to thrive and indeed facilitate civic actors to participate in the energy transition, as envisioned by the EU (Savaresi, 2019). National law-makers will thus need to find a balance between integrating energy communities in a liberalized market setting and enabling them to achieve wider goals, i.e., *"environmental, economic, or social community benefits"* (Diestelmeier, 2021).

This chapter aims at contributing to the understanding of the balancing act between market liberalization and community benefits when developing a legal framework for energy communities. It does so by drawing insights from the Netherlands at two levels: "on the ground" experiences from a small scale but progressive energy community located in Amsterdam (*"Schoonschip"*) and the legal framework and thus the "scope of action" for energy communities. The guiding question is how *"Schoonschip"* implements community benefits and whether and how this is facilitated by the legal framework applicable to energy communities in the Netherlands. While this specific case may not be read as a general model for adjusting legal frameworks in other jurisdictions, it addresses a number of general aspects of energy community legislation and can inform law and policymakers and various stakeholders in energy communities about the need to introduce a reform of energy sector legislation which expands the liberalized market narrative. The chapter is organized as follows:

- Section 2 outlines the purpose and roles ascribed to energy communities as established by EU legislation;
- Section 3 presents the case study of the energy community *"Schoonschip";*
- Section 4 presents the legislation allowing for "regulatory sandboxing" for energy communities in the Netherlands and experiences of *"Schoonschip"* with this legislation;
- Section 5 discusses how the experiences of *"Schoonschip"* can inform the transposition of the EU provisions on energy communities into national legal frameworks; and
- Section 6 concludes.

[2] Electricity Directive, Article 2.11b; Renewable Energy Directive, Article 2.16c.

2. EU law on the purpose of energy communities

The EU legal framework obliges the member states to develop a *"facilitating legal framework for energy communities."* Yet, the Directives leave a considerable level of discretion to the member states concerning the exact implementation and definition of the legal form, and the roles for energy communities. While variety in transpositions is not per se negative, as it reflects the different circumstances at member state level, it is important that the adapted legal framework enables an energy community to fulfill the primary purpose, the provision of *"environmental, economic, or social community benefits for its shareholders or members or for the local areas where it operates."* This suggests that energy communities have to serve a much broader role than merely producing, distributing, and supplying energy; that is to incorporate a societal value. In other words, a community not *for* energy, but *through* joint activities related to energy.

While the Directives do not provide any further guidance on how this purpose of energy communities should be fulfilled, recital 43 of the Electricity Directive and recitals 67 and 71 of the Renewable Energy Directive present more concrete ideas for the roles energy communities are expected to take. The following list provides an overview of the concrete roles ascribed to energy communities as presented in these recitals:

- Community energy offers an inclusive option for all consumers to have a direct stake in producing, consuming, or sharing energy;
- Community energy initiatives focus primarily on providing affordable energy for their members or shareholders rather than on prioritizing profit-making;
- By directly engaging with consumers, community energy initiatives demonstrate their potential to facilitate the uptake of new technologies and consumption patterns, including smart distribution grids and demand response, in an integrated manner;
- Community energy can also advance energy efficiency at household level and help fight energy poverty through reduced consumption and lower supply tariffs; and
- Community energy enables certain groups of household customers to participate in the electricity markets, who otherwise might not have been able to do so (specifically referring to vulnerable customers and tenants).

This suggests that the EU legislature on the one hand ascribes societal and even social goals to energy communities, and on the other hand expects these communities to participate in a competitive market (see chapter by Hanke et al.). This potential tension might leave member states puzzled concerning the implementation in their national legal frameworks. That this is indeed a challenge is illustrated in Sections 3 and 4 which present the case of the energy community *"Schoonschip"* in the Netherlands and how the legal framework both facilitated and obstructed this initiative.

3. The energy community *"Schoonschip"* in Amsterdam

While *"Schoonschip"* can be classified as an energy community, the initiative is not limited to energy-related activities. The origin of *"Schoonschip"* dates back to 2008, when a group of citizens of Amsterdam wanted to demonstrate an alternative way of housing and living following two main objectives, namely incorporating circular economy approaches in the use (and reuse) of resources (e.g., energy, water, nutrients) and fostering a community along common social values of that group. The implementation of those objectives was (and is) challenging. However, as the group is very small, starting with 8 and growing to 46 households, innovative technical solutions could be developed and implemented without facing extremely high investment costs. Inspired by a single houseboat which was equipped to be largely self-sufficient, the group had the ambition to develop a floating neighborhood consisting of several houseboats which would be connected by a jetty (see Fig. 4.1). This also inspired the name *"Schoonschip"* which on one hand means *"clean ship"* but also refers to the Dutch saying "*schoon schip maken (make clean ship)*," which means as much as "*come clean.*"

The idea was to connect the houseboats to a single community grid, which would have one connection point to the public grid. Behind this connection point, the community intended to operate its own local smart microgrid with the help of an energy management system. This system had already been tested at a property close to the site where *"Schoonschip"* was to be developed. At this test site, called *"De Ceuvel,"* a microgrid is now up and running connecting 17 offices and a restaurant. The energy management system balances supply and demand within the test site, electricity is exchanged between the different offices, and the restaurant and payments are settled via a blockchain token

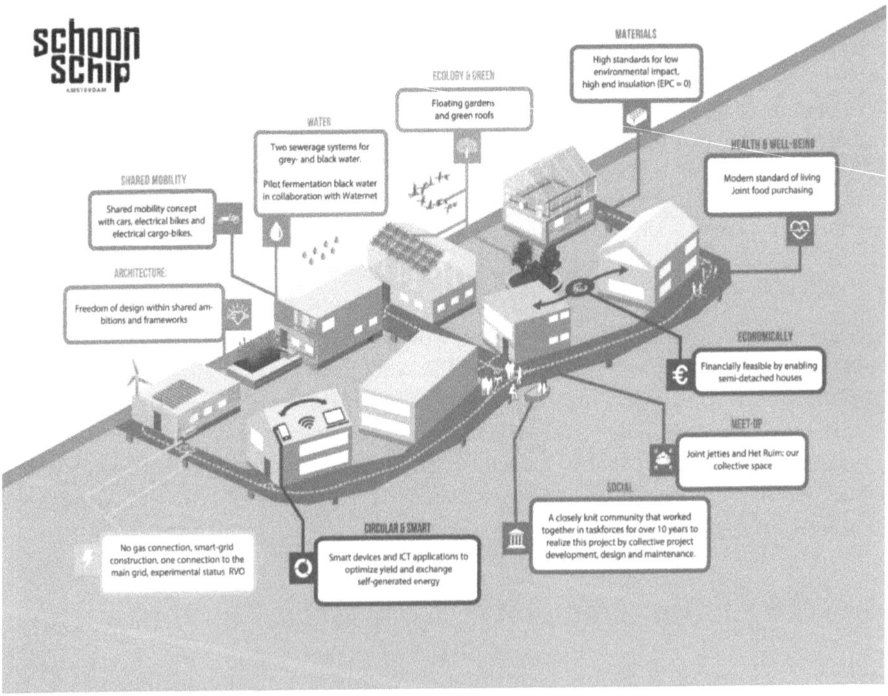

Figure 4.1 Schematic overview of the original design for "Schoonschip."
Source: https://schoonschipamsterdam.org; © Metabolic, Amsterdam AND © Schoonschip, Amsterdam.

called *"Jouliette."* In this way, a "peer-to-peer" supply system was established. As the technology was successfully tested, the *"Schoonschip"* community was confident they would be able to realize their objectives of a more sustainable and social (energy) community by sharing resources and achieving a larger degree of self-sufficiency.

"De Ceuvel" however is a single site, single owner project, where regulations regarding grid operation and supply are not applicable as all "grid" management and electricity "supply" activities take place "behind the meter," within the installation of *"De Ceuvel."* *"Schoonschip"* on the other hand would be a multisite project, connecting different dwellings of different owners via a grid, and would thus be subject to these regulations. The implementation of the objectives therefore not only required technological feasibility, but also a facilitating organizational framework. In the context of a liberalized energy sector, regulation strictly prescribes the roles and tasks of market parties, such as production and supply companies, on the one hand, and network-related activities on the other hand. A facilitating organizational framework thus requires legislation which would allow the community to carry out the tasks it wanted to assume. For the case of *"Schoonschip"* this set of tasks would include the operation of grid infrastructure (coordinating consumption, storage, production), providing metering services within the community, "sharing" electricity within the community, and feeding any surplus electricity into the public grid system. Additionally, the community wanted to offer aggregated flexibility services to the DSOs of the public grid system, so they would also need to assume the role of aggregating flexibility within the community and being able to access all electricity trading markets. Of course, while the capacity of the community *"Schoonschip"* is small, the main aim is to live and showcase an alternative way of organizing the production and use of resources (most prominently, energy), to the conventional liberalized market setting, combining sustainability objectives and social values within one community. In the taxonomy developed in the chapter by Rossetto et al., *"Schoonschip"* would be considered a microgrid, with commonly managed local production, joint consumption, and aggregation.

After 10 years of preparation, including technological developments and legal struggles, the construction of the floating neighborhood envisioned by the community *"Schoonschip"* started in 2018 and was finally completed in 2021 (see Fig. 4.2). The community now consists of 46 dwellings located on 30 ships (14 single dwelling ships and 16 double dwelling ships), housing 144 inhabitants. The average energy consumption of the dwellings in the community turns out to be around 2.500 kWh per year, which is well below the expected value, considering the extra electricity consumption for electric heating. From the total community consumption, 92% is produced by 500 solar panels located on the roofs of the dwellings. To overcome temporary shortage or overproduction of electricity, 30 batteries (one per ship), with a total capacity of approximately 300 kWh/125 kW, were installed.

The energy management system, which optimizes production, storage, and demand every 5 seconds, enables *"Schoonschip"* to shave consumption and production peaks to well below 100 kW. Due to this peak shaving, *"Schoonschip"* could be connected to the common grid with a low power connection with a capacity of just 100 kW, approximately one sixth of a standard DSO connection for 46 dwellings. *"Schoonschip"* therewith shows that local energy management can seriously contribute to increase

Figure 4.2 Recent photo of the "Schoonschip" community, as finally established in 2020. Source: https://schoonschipamsterdam.org; © Schoonschip.

direct consumption of locally produced renewable energy and mitigate problems of distribution and transmission capacity shortages and local congestion.

Nowadays, "*Schoonschip*" is recognized as an exemplary energy community, contributing to both an increase of consumption of renewable energy through local storage and reducing the pressure on the distribution and transmission systems through local peak shaving and congestion mitigation.

4. In search of novel legislation for energy communities

As the ambitions of "*Schoonschip*" were very much future oriented and did not fit in the regulatory setting of the energy sector, implementation seemed to be impossible from a legal perspective. Yet, the development of "*Schoonschip*" took place at a time when discussions on how to organize a more sustainable society and economy reached a higher and broader governance level in the Netherlands. In this context, the 2013 "*Energy Agreement for Sustainable Growth*" (hereafter the "Energy Agreement") concluded by the Dutch government, with various (industrial) stakeholders and nongovernmental organizations, acknowledged the need for sustainable growth and consequently, among others, the development of new organizational forms for energy production and supply.[3]

[3] The formulation of the "Energy Agreement" was initiated by the Social and Economic Council of the Netherlands ("SER"), which is an advisory body to the government and in which representative organizations of employers and employees as well as independent experts work together to reach agreement on key social and economic issues. The SER advises the Dutch government and Parliament on social and economic policy. It also facilitates agreements such as the "Energieakkoord voor duurzame groei," Kamerstukken II 2012/13 30,196, nr. 202. See for more information https://www.ser.nl/nl/thema/energie-en-duurzaamheid/energieakkoord.

The Energy Agreement emphasized the potential of decentralized generation of renewable energy sources to contribute to the goal of increasing the overall share of renewable energy sources in gross final consumption of energy in the country. Alarmed by the problems experienced by "*Schoonschip*" and a few other early energy communities, the existing legal framework was recognized as one of the main obstacles for the further development of decentralized energy projects.[4] This led to the adoption of a motion by the Dutch Parliament to use the provision in the Dutch Electricity Act (hereinafter Electricity Act) to "deviate from the Electricity Act, by means of experiment."[5] In order to implement this motion and investigate the legal obstacles further, the government adopted a Decree which allowed for exemptions of specified provisions in the national Electricity Act for projects which aim at contributing to decentralized renewable electricity production and improve the efficiency in grid operations. Essentially, this Decree was establishing a "regulatory sandbox" for energy communities. The following subsection outlines the main provisions of this Decree.

4.1 The experimentation decree for decentralized sustainable electricity production

The Decree allowing for the regulatory sandbox, called the "*Experiments Decentralized Sustainable Electricity Production*" Decree[6] (hereafter "Experimentation Decree" or "the Decree"), entered into force in 2015. Its main aim was to allow for exemptions from specific provisions of the national Electricity Act to investigate whether new organizational forms would accelerate decentralized renewable electricity generation and more efficient use of the grid infrastructure.[7] Essentially, the Decree allows small customers organized as a cooperative or an owners' association not only to engage in collective generation of renewable electricity, but also to become suppliers and, under certain conditions, even system operators. These options were in line with the ambitions of "*Schoonschip,*" especially the options which allowed for innovative and autonomous system operation and "peer-to-peer supply."

The Experimentation Decree allowed for two sorts of projects to be set up by owners' associations or cooperatives which are determined by their size in terms of connected customers, namely "project grids" and "large experiments." "Project grids" have one connection to the public distribution system, are located within a geographically delineated area, and have a maximum of 500 connected customers.[8] "Project grids" may autonomously construct, own, and operate the local grid. "Large experiments," on the other hand, take place within the service area of a locally assigned

[4] Energieakkoord voor duurzame groei (Energy Agreement), 83.

[5] Elektriciteitswet 1998 (Dutch Electricity Act), art. 7a.

[6] Besluit van 28 februari 2015, houdende het bij wege van experiment afwijken van de Elektriciteitswet 1998 voor decentrale opwekking van duurzame elektriciteit (hereinafter Experimentation Decree), Stb. 2015, 99.

[7] Nota van Toelichting Besluiten Experimemten (hereinafter Explanatory Memorandum Experimentation Decree), K.St. 34627 nr.48, Attachment 881510, para. 1.2.

[8] Experimentation Decree, art. 1.

DSO and have a maximum of 10,000 connected customers. As they are not confined to a specific site, "large experiments" can also operate virtually, within the service area of a DSO. Contrary to "project grids," the asset management of the grid remained the responsibility of the DSO.[9] As part of the exemption, both types of projects would automatically be granted a license to supply to small customers, which is normally only granted to larger energy supply companies, and both types of projects could determine their own network tariffs, without being subject to prior approval by the Dutch National Regulatory Authority which only had to approve the method of calculation and not the tariffs themselves.[10] As the Decree was set up as a "regulatory sandbox," the number and the duration of projects was limited to 10 "project grids" and 10 "large experiments" annually with a duration of 10 years.

Initially, the Experimentation Decree was set to be open for 4 years (2015–2019) with a possibility for an extension of another 4 years. However, based on a negative advice by the Council of State, the Decree was not extended.[11] One of the main points of criticism of the Council of State was that the Decree could be conflicting with EU energy law and in particular with the Electricity Directive of 2019, which had not yet been transposed into the Dutch law. Questionably, the Council of State argued that the transposition of EU law would make another "regulatory sandbox" for this purpose redundant. The next section outlines the experiences of the exemplary energy community *"Schoonschip"* with the Experimentation Decree. Essentially, this section looks for the insights that have been developed in the "regulatory sandbox" and whether these could facilitate the development of a legal framework enabling energy communities to contribute to establish a more sustainable and social community.

4.2 Experiences with the "experimentation decree": insights from "Schoonschip"

While the Experimentation Decree surely provided favorable regulatory conditions for the realization of *"Schoonschip,"* in the course of the development of the project, several other legal obstacles caused delays and forced the project to deviate from its original plans. Identifying and understanding these obstacles is essential for developing facilitating national legal frameworks for energy communities in the course of the transposition of the EU legal provisions. This section presents the most prominent obstacles, which might also be relevant for other community projects.

Despite the fact that *"Schoonschip"* community obtained the right to supply to its members in the context of being recognized as a "licensed supplier" under the Experimentation Decree, "peer-to-peer" energy supply was still not possible because the supply license did not apply to the individual members, but rather only to the entity *"Schoonschip."* While in this case the issue was resolved by applying "peer-to-energy

9 Explanatory Memorandum Experimentation Decree, para 2.3.3.

10 Experimentation Decree, article 12.

11 See letter from the Minister of Economic Affairs and Climate to the Upper House of Parliament of 10 December 2020 as published in Kamerstuk, 34627, nr. F.

community-to-peer supply," it becomes clear that the term "peer-to-peer" supply remains an unclear concept from an (energy) law perspective and needs to be defined with the transposition of the EU Directives.

"Schoonschip" also experienced problems resulting from the fact that consumers retained their right to switch their supplier, i.e., individual members could switch to an energy supplier other than the community. This was problematic as *"Schoonschip"* was developed as a "project grid" and was planning to construct, own, and operate its own grid. Potentially switching members implied complications for the calculation of the grid operation costs and network tariff. The core principle of a liberalized market, i.e., "free choice of supplier," seems to cause uncertainty for the development of energy communities. Another obstacle resulted from the real estate policy of Amsterdam, which required considering the heat pumps as part of the houseboats where they were installed. This conflicted with the objective of *"Schoonschip"* to create a separate asset for all heat pumps in order control and manage production and demand efficiently by using the heat pumps as smart storage facilities, because the economic ownership was assigned to the owners of the houseboats by accession.

Two other significant problems in the realization of the project were related to energy tax allocation. The Environmental Tax Act determines that an energy tax is due for electricity supplied to a connection to a grid, whether self- or jointly generated or purchased from a commercial electricity supplier.[12] This provision is one of the problems faced by most early energy communities. The initial idea was that the electricity produced and directly consumed within the "project grid" of "*Schoonschip*" would not be charged any energy tax individually. However, the tax authority ruled that a "project grid" under the Experimentation Decree should be recognized as a "grid" for the Environmental Tax Act, and that thus energy tax was due for any electricity exchanged within the *"Schoonschip"* community. This had a significant effect on the financial feasibility of the project. Also, double-charging tax for electricity storage constituted a problem for the project. The electricity stored in batteries installed at *"Schoonschip"* was subject to energy tax twice: when charging the battery and when subsequently supplying to one of the houseboats. Following a debate in parliament, the secretary of state of the ministry of finance agreed to solve this matter by January 1, 2021, in the course of the transposition of the EU Electricity Directive.[13] So far, however, the announced amendment has not yet entered into force and *"Schoonschip"* is being charged twice for stored electricity.

The following section discusses whether and to which extent these experiences have been taken into consideration when drafting new legislation for the energy sector, and in particular the provisions for energy communities.

[12] Wet Belastingen op Milieugrondslag (Environmental Tax Act), article 50 (1).

[13] Letter of the Minister to the Parliament regarding the prevention of double energy taxation. See: https://www.rijksoverheid.nl/documenten/kamerstukken/2019/07/05/kamerbrief-over-het-voorkomen-van-dubele-energiebelasting.

5. The way forward: transposing EU law on energy communities in the Netherlands

As mentioned above, the Netherlands has not yet transposed the Renewable Energy Directive and the Electricity Directive,[14] but is currently reforming the entire energy sector legislation by not only revising current laws, but also merging the Gas Act and the Electricity Act into one Energy Act. Of course, this could be a great moment to develop a legal framework for energy communities by transposing the EU-relevant provisions, and by learning from the experiences of energy communities and in particular those with an exemption under the Experimentation Decree (Diestelmeier and Swens, 2021).

In November 2021, a second draft of the Energy Bill was presented. Instead of differentiating between the EU legal concepts REC and CEC, the Dutch legislator chose to only define "Energy Community" (in Dutch: *"energiegemeenschap"*) as follows:

> *Energy Community: legal person which, for the benefit of its members or shareholders, conducts activities in the energy market and has as its purpose to provide environmental, economic, or social benefits to its members or shareholders or the local areas where it operates, and not to make profit.*[15]

Following the specific provisions from EU legislation for the RECs, the draft adds that energy communities developing renewable energy projects may decide to (1) also include medium-sized enterprises and/or (2) limit the effective control to those members and shareholders who are located in "close proximity" of the renewable energy projects of the community. In contrast to the EU provisions on energy communities, the draft Bill does not prescribe the specific activities that energy communities should be entitled to carry out. The Explanatory Memorandum to the draft Bill justifies this approach by the assumption that it is, in principle, irrelevant who conducts a specific activity. From that point of view, it is more relevant to define activities and the conditions under which they can be carried out than to define actors with the activities they should be allowed to carry out. Potentially, this approach does not limit energy communities to any specific set of tasks and thus appears even more favorable for energy communities than the EU legislation. However, in fact, it does not, in particular, facilitate energy communities. Especially, considering that energy communities are supposed to fulfill a broader role than the conventional activities (production, supply, distribution) by incorporating social objectives (as outlined in Section 2), it might be necessary to develop a framework specifically supporting energy communities.

Drawing from the experiences from the *"Schoonschip"* community, three core activities are identified, namely grid operation, energy sharing, and energy supply.

[14] The deadline for the transposition of the Renewable Energy Directive was 31 December 2020 and for the Electricity Directive 30 June 2021.

[15] Concept Wetsvoorstel Energiewet dd 17-11-2021 (hereinafter Draft Energy Bill), article 1.1, p 1.

Contrary to the approach in the Experimental Decree, the Dutch legislator did not include the option for energy communities to operate or own a grid system, as it considered it "not necessary" and argued that grid operation should, for, among other, safety reasons, be carried out by professional and experienced entities like DSOs. This choice deprives energy communities of the possibility to use local balancing and peak shaving to design a lighter and more efficient grid and reduce local grid costs. It is at least curious that the permission to operate or own a grid, which was one of the options specifically investigated under the Experimentation Decree, has not been incorporated in the draft Bill, as it appeared to have had a very positive contribution to the functioning of the energy communities under the Decree.

The draft Bill does not define or even mention "sharing" of energy. Instead of sharing, "mutual supply" is implemented by providing an exemption from the general rule that supply of energy to small (household) end-users is prohibited without a supply license. This requirement, which appeared to be a barrier for energy communities, functions as a consumer protection measure, ensuring that supply companies comply with certain standards and providing the assurance that in case of bankruptcy of a licensed supplier another licensed supplier will take over. The draft Bill provides the option for "mutual supply" by allowing energy communities to supply electricity to their members or shareholders if the electricity is generated by that legal entity.[16] However, this energy community is only allowed to supply to their members or shareholders if the supply can be allocated to that energy community. So far this can only be achieved via what is called "Secondary Allocation Points" (SAPs). This would require a splitting of the consumer's installation into one part that is supplied by the primary (licensed) supplier and a second part, supplied by the community. In addition, purchasing an SAP and the splitting require a significant investment which would constitute an extra barrier to "mutually supply" energy among the participants of an energy community.

The draft Energy Bill follows the Directives and includes both concepts, CECs and RECs. However, it does so in a rather limited way: the draft Bill does not propose a facilitating legal framework for energy communities, but rather sticks to the minimum requirements. Furthermore, while the approach to focus on activities rather than actors does not create specific obstacles for energy communities to engage in an activity in the energy sector, it does restrict the scope of energy communities to "yet another market actor" in the liberalized market. Furthermore, the ambition of the EU Directives that energy communities could also play a role in the social dimension, by including vulnerable customers or mitigate energy poverty, is not further developed or facilitated.

Table 4.1 provides an overview of the extent to which the different "required rights" for energy communities, such as "*Schoonschip,*" to operate effectively are covered under the different legal frameworks. The draft Energy Bill (Concept NL law in the table) appears to include all the provisions needed for such energy communities. The barriers

[16] Draft Energy Bill, article 2.19, sub 2.

Table 4.1 Required rights for energy communities to operate.

Provision \ Legislation	Current NL law	Reg. Sandbox	EU-CEP	Concept NL law
Operate the local grid	X	√	Opt.	X[a]
Provide metering services within community	√[b]	√	√	√[b]
Share electricity within the community	√[c]	√	√	√[d]
Coordinate consumption, storage, and generation dispatch	√	√	√	√[b]
Feed electricity into the public grid	√	√	√	√
Aggregate flexibility within the community	?	√	√	√[b]
Have access to all electricity markets	√	√	√	√[e]

Notes: Where X = not covered by legislation; √ = covered by legislation; opt. = optional.[17].
[a]Possibly no longer required due to other provisions.
[b]Requires second set of metering equipment for sufficient resolution.
[c]Sharing not defined; considered mutual supply, license required.
[d]Sharing not defined; considered mutual supply, one allocation per supplier required.
[e]Only applicable for larger or aggregated power and larger or aggregated volumes.

however, as addressed above, lie in the specific requirements, as specified in the notes to this table, or in other legislation, for instance, on taxation.

6. Conclusion

This chapter sets out to provide concrete ideas on how to develop a "facilitating legal framework for energy communities," allowing them to fulfill their purpose as formulated in the EU Directives. The case study of *"Schoonschip"* shows that a facilitating legal framework for energy communities needs to extend beyond a few selected provisions in energy law, and also needs to include other fields of law such as, for example, tenancy law, land registry regulations, (energy) tax law, and corporate law. This means that for the transposition of the EU Directives and the provisions on energy communities, other laws might need to be adjusted. This is also in line with the Directives which require an "enabling legal framework" for energy communities, which should extend beyond the energy sector. This finding is of special importance considering that the purpose of energy communities, as established in EU legislation, is not only related to the "classical" activities in the energy supply chain, such as production, distribution, and supply, but also includes environmental and social objectives. Especially, the latter—social—objective is emphasized in the EU

[17] Optional here relates to the provision in Electricity Directive, art. 16,4 stating that "Member States ***may*** decide to grant citizen energy communities the right to manage distribution networks in their area of operation."

provisions and entails social inclusion, affordability, empowerment, and protection against energy poverty (see Section 2). Considering that the main aims of the Electricity Directive and the Renewable Energy Directive are, respectively, to promote a liberalized market setting and to accelerate of implementation of energy from RESs, the social objectives of energy communities might be overlooked when implementing the provisions. The case study of *"Schoonschip"* showed that for an effective transposition of the Electricity Directive and the Renewable Energy Directive, achieving the purpose of energy communities, it is not sufficient to "simply" allow energy communities to participate in the market, for example, by awarding a supply license and granting the right to operate a system, but also requires careful consideration for which purpose those activities are carried out by an energy community. In other words, energy communities are not supply companies nor DSOs, but fulfill (parts of) those functions for another purpose along with different roles than the conventional undertakings in the sector. This chapter also showed that much is already happening "on the ground" while legislation is being drafted.

The case of *"Schoonschip"* provides an example of an energy community which seems to closely implement the concept of energy communities as it is set out in the EU Directives. Though "*Schoonschip*" might be considered a relatively small community, the majority of the obstacles faced by this community will be faced by larger communities equally. The fact that *"Schoonschip"* is locally restricted and physically confined to a closed smart grid does not set it apart from larger communities. Also, larger communities of consumers connected to the public grid will face the same problems with, for example, peer-to-peer energy supply, energy sharing, storage, and real estate definition. Studying and learning from existing initiatives, and in particular, those developed under regulatory sandbox schemes, is key for national law-makers when transposing the EU legal provisions on energy communities.

References

Diestelmeier, L., 2021. Citizen energy communities as a vehicle for a just energy transition in the EU—challenges for the transposition. Oil Gas Energy Law 19 (1). https://www.ogel.org/article.asp?key=3945.

Diestelmeier, L., Swens, J., 2021. Energy communities in The Netherlands: learning from local energy initiatives. In: Roggenkamp, M.M., Banet, C. (Eds.), European Energy Law Report XIV. Intersentia.

EU Commission, 2019. Clean Energy for All Europeans, (European Union, 2019). https://doi.org/10.2833/9937.

EU Commission Joint Research Centre, Caramizaru, A., Uihlein, A., 2020. Energy Communities: An Overview of Energy and Social Innovation. Publications Office of the European Union. https://doi.org/10.2760/180576.

Savaresi, A., 2019. The rise of community energy from grassroots to mainstream: the role of law and policy. J. Environ. Law 31 (3), 487−510. https://doi.org/10.1093/jel/eqz006.

Walker, G., Devine-Wright, P., 2008. Community renewable energy: what should it mean? Energy Pol. 36 (2), 497−500. https://doi.org/10.1016/j.enpol.2007.10.019.

Alignment of energy community incentives with electricity system benefits in Spain

5

David Robinson [1] *and Iñigo del Guayo* [2]
[1]Oxford Institute for Energy Studies, Oxford, United Kingdom; [2]University of Almería, Almería, Spain

1. Introduction

The question addressed in this chapter is whether the current Spanish framework of legislation, regulation, and markets aligns the incentives to create energy communities (ECs) with the interests of *all* consumers on the electricity system, not just members of the communities.

Spain has not yet transposed EU legislation pertaining to ECs, and there are consequently no operating communities in Spain that correspond precisely to the EU definition of Citizen Energy Communities (CECs) and Renewable Energy Communities (RECs). However, there are two other kinds of ECs that can operate today under Spanish legislation: Collective Self-Consumption (CSC)[1] and Closed Distribution Systems (CDS). CSC legislation is designed to allow consumers to collectively self-consume the electricity they generate; in Spain, these consumers are usually organized as cooperatives. CDS legislation, on the other hand, is designed to enable electro-intensive industries to lower their electricity costs. Both are different from the EU versions of ECs. However, because CSCs are similar to the EU version of RECs, this chapter concentrates on CSC communities.

There are too many gaps in the information available to be definitive about our assessment of alignment. Nevertheless, there is enough information to draw the conclusion that reforms are required to provide efficient price signals to ECs, incentivizing them to contribute to lowering system costs, not just community members' costs. Furthermore, all agents in the power system, including individual consumers, aggregators, ECs, generators, and storage providers, should receive the same efficient price signals, along with nondiscriminatory regulatory treatment, thereby encouraging more effective competition.

The chapter has five sections, in addition to this introduction:

- Section 2 introduces the analytical framework and the cost components of Spain's electricity system final prices;
- Section 3 summarizes current and relevant Spanish legislation;

[1] EU legislation uses the term "jointly acting renewables self-consumers." Spanish legislation refers to "*auto-consumo colectivo*." They are fundamentally the same concept.

Energy Communities. https://doi.org/10.1016/B978-0-323-91135-1.00008-0

- Section 4 analyzes features of Spain's legislation, regulation, and markets that support alignment between EC incentives and the interests of all electricity consumers;
- Section 5 analyzes the features that do not support that alignment; and
- Section 6 offers a preliminary assessment of alignment and makes policy recommendations to improve it.

2. Analytical framework

The analytical framework used in this chapter focuses on two aspects of welfare enhancement: economic efficiency and distributional impact.[2] To the extent that communities and their members have incentives to invest and operate in ways that benefit their members as well as consumers outside the community, because they lower system-wide costs, the relationship between communities and the system is considered to be well aligned. If the framework has the effect of raising overall system costs, or benefiting the communities at the expense of electricity consumers outside the communities, the framework is considered to be not well aligned. The focus is on the relationship between the community and the wider system, not on the relationship between the community and its members.

The aim of public policy related to ECs should be to ensure, inter alia, that they help to lower overall costs of developing a decarbonized electricity system and do not shift costs to nonmembers of the communities. Fig. 5.1 below summarizes the main categories of cost for a residential consumer in the Spanish electricity sector in early

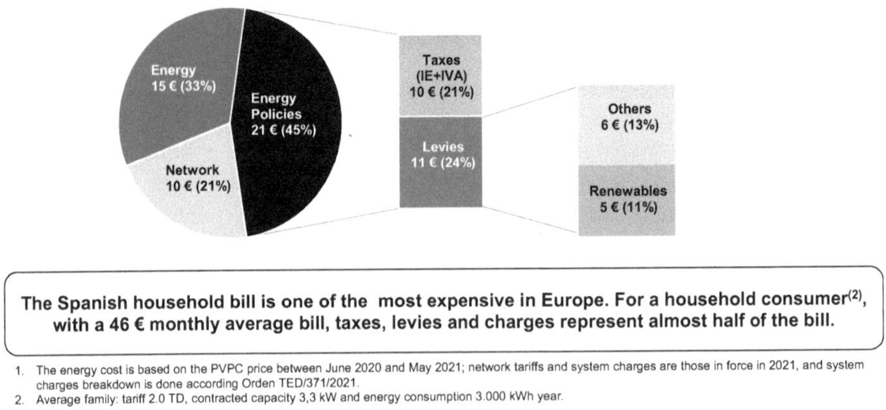

Figure 5.1 Costs of the Spanish electricity system represented in residential tariff.
Source: Endesa.

[2] For more on efficiency objectives, see Robinson (2021). This chapter does not examine social welfare in the broader sense; see Hausman (2019) for a critique of the standard microeconomic model of welfare.

2021 (and for many years before then), before Spanish wholesale electricity prices spiked upward.[3]

- Generation and retail supply are open to competition and referred to in the figure as energy costs; they account for about a third of a residential consumer's electricity bill. To a large extent, these are variable costs of the system in the short term; so a reduction in energy consumption corresponds to a reduction in system costs, while more consumption will raise those costs.
- Transmission and distribution networks are regulated and account for 21% of the residential consumer's bill. A very large share of these costs is fixed in the short term. A change in consumption will not usually have an important effect on network costs in the short term, although there are times when changing consumption will significantly increase or reduce congestion on the system, potentially affecting future investment requirements.
- Public policy costs (levies and taxes) are 45% of the bill. These are not related to the cost of regulated and competitive businesses, but rather reflect the government's ability to collect revenue to pay for public policies. In particular, levies amount to over €10 billion per annum, and about 24% of a residential consumer's bill. The largest levy element funds historic investments in renewables, cogeneration, and residues. Between regulated business, taxes and policy levies, governments determine over 2/3 of the final price paid by residential consumers.

The aim of this chapter is to examine how, under current regulation, ECs may increase or decrease the cost of electricity supply (regulated and competitive businesses) and shift those costs and public policy costs to nonmembers.

3. Current legislation

Before introducing the legislation that applies to collective self-consumption and coops, this section provides some legislative backdrop related to the initial steps toward transposing EU legislation and some indication of anticipated future legislative developments.

3.1 Preparation for the transposition of EU legislation

Although Spain has not transposed the EU legislation on ECs, there is Spanish legislation that paves the way for transposition.

Closed distribution systems (CDS). Royal Decree-law no 20/2018, of 7 December[4] introduced legal measures to guarantee a "fair" energy transition for electro-intensive industries, to enable them to remain competitive in international markets and to thereby avoid carbon leakage. It introduced the possibility of closed distribution

[3] As result of the spike in wholesale electricity prices, which rose from €30–60/MWh in the first quarter to over €200/MWh in October 2021, the Spanish Government introduced a series of temporary measures to reduce the impact on final consumers, especially over 10 million consumers on a default price for small consumers (PVPCs), which reflects hourly wholesale prices. These measures are not considered here.

[4] Spanish Official Bulletin (here in after referred as to BOE) of 296, of 8 December, 2018.

systems and addressed the issue of state aid for those industries. At the end of 2020, the Government approved a new Statute for electro-intensive consumers, through Royal Decree no 1106, of December 15, 2020.[5] There are clear legal differences between closed distribution systems and ECs (for example, closed distribution systems are not obliged to be incorporated), but the combination of closed distribution systems with electro-intensive consumers may serve as a first step toward the creation and development of Citizen Energy Communities (CECs). There is a Draft Royal Decree developing the procedure and the requirements that must be met for the granting of administrative authorization for closed electricity distribution networks. In that draft, consumers may become a cooperative, suggesting that in future laws cooperatives may be the preferred legal entities to incorporate ECs.[6]

Special remuneration for generation by nonprofit energy communities. Act no 6/2018, of 3 July, introduced a new paragraph within Final Disposition no 3 of the 2013 Electricity Act. This Disposition empowers the Government to pass secondary legislation related to renewable energies owned by ECs and to offer special remuneration for these energies. The provision below has not yet been developed further. However, in September 2021 the Government announced its intention to provide €100 million to support the creation of ECs within the framework of Spain's Plan for Recovery, Transformation, and Resilience (PRTR).[7]

> *"The Government is empowered so that all those generation facilities owned by energy communities, understanding these as nonprofit organizations, individuals, or small and medium-sized companies whose majority shareholders or members are individuals, local or provincial entities, or also other small and medium-sized companies, may have a special remuneration treatment as an essential vehicle for their necessary permanence in the generation market."*

Definition of renewable energy communities. Until very recently, Spanish legislation did not recognize "renewable energy communities" as defined by Article 2 (16) of the Renewable Energy Source (RES) 2018 Directive.[8] However, Royal Decree-law no 23/2020 of 23 June amended the 2013 Electricity Act and added three new agents to the electricity system: (i) storage, (ii) independent aggregators, and (iii) Renewable Energy Communities (RECs). The latter definition is almost the same as the one contained in the RES Directive. Of course, the mere introduction of a definition in the 2013 Electricity Act is not enough to promote RECs or enable their creation.

[5] BOE no 328, of December 17, 2020.

[6] https://energia.gob.es/es-ES/Participacion/Paginas/Index.aspx.

[7] https://www.idae.es/noticias/el-gobierno-anuncia-100-millones-en-ayudas-para-impulsar-las-comunidades-energeticas. See Ministerial Order no TED-1446-2021, of 22 December 2021, that defines the framework for the conferral of financial incentives to pilot projects involving energy communities (EU "Implementa" Program) under the Spanish Recovery, Transformation, and Resilience Plan (Spanish Official Bulletin of 24 December 2021).

[8] Directive (EU) 2018/2001 of the European Parliament and of the Council of 11 December 2018 on the promotion of the use of energy from renewable sources (recast): OJ L 328, of 21 December 2018.

Much more must be regulated and decided by the Spanish Government. Moreover, a definition and further rules on CECs are still lacking.

Government guidance on creating energy communities. Some guidance on the creation of ECs can be found in the Spanish National Energy and Climate Plan (PNIEC).[9] The PNIEC provides some relevant provisions and foresees that by promoting such communities there will be an increase in the number of agents in the electricity sector, as well as in the number of participative projects in renewable energies generation and throughout the electricity system.[10] The PNIEC underlines some main differences between the two communities defined in EU legislation (RECs and CECs). It also aims more generally to promote ECs. In order to achieve that goal, it foresees several mechanisms for action to be carried out by the Ministry for Ecological Transition and the Demographic Challenge. Although these mechanisms are broadly positive for the development of ECs, none of them has been implemented so far; as always, the devil will be in the details. In implementing EU legislation in Spain, it is expected that the Government will take into account the Guide to Promote Energy Communities, published in March 2019 by IDAE (2019), which proposes some detailed regulatory measures.[11]

Renewable energy communities mentioned in existing Spanish legislation on auctions. The 2013 Electricity Act introduced a new system for the promotion of renewable energy sources. New electricity generation capacity making use of renewable sources could benefit from public support following periodic central government auctions. A revised central auctions system was put into force with Royal Decree-Law no 23, of June 23, 2020. Among others goals, it aims to promote the role of renewable ECs. With that objective in mind, it empowers the Government to take into account the special conditions of these communities so that they can compete in renewable auctions on a level playing field, or benefit from the remuneration the auctions offer.[12] Royal Decree no 960 of November 3, 2020, which develops Royal Decree-law no 23 of June 23, 2020, provides the economic framework for the governance of the production of electricity by means of renewable energies.[13] Articles 3 and 8 (Section 13) of that Royal Decree regulate the award of economic support for renewable energies through auctions. However, the first two auctions for renewable generation capacity, which took place on January 26, 2021, and October 19, 2021, respectively, did not foresee any special treatment of ECs.[14] This omission reflects a more general lack of proper regulation in force, which is an obstacle to the development of ECs. To provide another example, the long awaited Royal Decree no 1183/2020 of December 29, 2020 that regulates access and connection to the distribution and transport networks, a

[9] Plan Nacional Integrado de Energía y Clima 2021–30 (PNIEC), 20th January 2020.

[10] Pages 9, 14, and 92, PNIEC. Measure no 1.13 of the PNIEC expressly refers to energy communities.

[11] IDAE is the *Instituto para la Diversificación y el Ahorro Energético*, part of the Spanish Ministry for Ecological Transition and the Demographic Challenge.

[12] New Article 7 bis, paragraph third, of the 2013 Electricity Act.

[13] BOE no 291, of 4th November 2020.

[14] See Ministerial Order no TED/1161/2020 of 4 December 2020.

key component of the legal framework for renewable energies, does not contain a single reference to ECs.[15]

Public consultation on energy communities. In late 2020, the Spanish Government opened a window for public consultation to obtain input for a future Royal Decree on ECs. The Spanish Association for Energy Law (AEDEN using the Spanish initials) argues that the process of developing legislation has begun but that there is much more to do to enable ECs to be part of a new flexible decarbonized energy system. There was also a government call for the expression of interest to identify mechanisms to foster the emergence of local ECs, as an instrument for economic recovery following the pandemic.

Long-Term Decarbonization Strategy. The Spanish Government approved the Long-Term Decarbonization Strategy,[16] where a reference is made to renewable communities when dealing with the role of citizens in the process.

Law on Climate Change and Energy Transition. To end this overview of current legislation and of plans related to ECs, it is worth noting that the recent Act on Climate Change and the Energy Transition contains no provision for Energy Communities. This is curious, since this law intends to transpose other pieces of recent EU legislation. The explanation appears to be rooted in a lack of political consensus about ECs within the coalition Government.[17]

Energy Communities and the PRTR. As mentioned earlier in this section, the Government has announced its intention to dedicate funds from the Spanish PRTR to support ECs. In particular, it has plans to spend €100 million on three lines of support. The first would familiarize people and organizations with the concept of an EC and help to identify future partners and members. The second would support planning and the constitution of ECs. The third, based on competitive bidding, would finance projects involving renewable energy, energy efficiency, and electric mobility. In addition, the Government has identified other programs financed by the PRTR that will benefit ECs, including the MOVES III program to support electric mobility and incentives for self-consumption of renewable energy. As always, the devil is in the detail, but it is clear that the Government is making an effort to demonstrate its commitment to ECs and to encourage citizen participation in the transition. The first call for aid was made through two IDAE resolutions of 12 January 2022, amounting to up to 40 million euros, for small and medium sized initiatives.[18]

3.2 Energy communities today: collective self-consumption and coops

The remainder of this section summarizes the current legislation that is being used as the basis for creating ECs. It uses the term EC here to refer to communities in Spain,

[15] BOE no 340, of 30th December 2020.

[16] Estrategia a largo plazo para un economía española moderna, competitiva y climáticamente neutra en 2050, November 2020 (point 7.1).

[17] Act no 7, of the 20th May 2021, on Climate Change and Energy Transition (BOE no 121, of 21st May 2021).

[18] Spanish Official Bulletin no 12, of 14 January, 2022.

typically organized as coops that may exercise the right to collective self-consumption. In other words, existing cooperatives and new ones are using existing legislation to build ECs before EU legislation for RECs and CECs has been transposed.

Collective self-consumption. Royal Decree-law no 15/2018, of 5 October[19] introduced legal changes to allow all Spanish consumers to produce electricity and to consume the electricity they produce, including collective self-consumption among several consumers with an "interior network," as, for instance, neighbors in an apartment building. That law states that the Government will further develop the concept of "interior networks," where the key feature is that generation is close to consumption. Today, in addition to sharing generation within an internal network, it is possible to share generation and consumption when they are connected through the low-voltage network within 500 m, or connected to the low-voltage network that shares the same substation, or the generation and consumption are within the same property registry number, as defined by the first 14 digits.

This allows self-consumption by a group of consumers in such a way that the networks required for self-consumption are no longer considered necessarily as an asset belonging to the local distribution company. This legislation opens the door to self-generation by individual consumers and by collectives that are within the vicinity of the generating assets.

Collective self-consumption, developed by Royal Decree no 244/2019, of 5 April,[20] allows various consumers from the same community (community of owners, a neighborhood, an industrial area, etc.) to collectively benefit from the same generation facilities, located in the community area. Although there is some conceptual similarity between collective self-consumption and the EU definition of ECs (particularly in the form of RECs), there are clear differences, since the group of consumers engaged in collective self-consumption cannot engage in certain other electricity activities allowed for ECs, such as aggregation and selling in wholesale markets.[21] Although collective self-consumption can be carried out by either residential or industrial consumers, there are a number of limits within Royal Decree no 244/2019, which make self-consumption less attractive for industrial consumers (for example, distance or other limits between producers and consumers). The implicit idea appears to be that collective self-consumption of industrial consumers will take place under the application of laws related to closed distribution systems.

The Spanish regulation on collective self-consumption could be a first step toward the creation of EU-style ECs, but not necessarily. There is a fundamental distinction between an REC, as defined by EU legislation, and a group of consumers who self-produce, store, and share energy. Whereas financial gains cannot be the main purpose of ECs under EU law, one of the aims of collective self-consumption is to obtain an economic benefit. Collective self-consumers cannot engage in any other activity other than the production, storage, consumption, and the sale of surplus electricity to the

[19] BOE no 242, of 6 October 2018.

[20] BOE no 83, of 6th April 2019.

[21] The chapter by Spasova and Braungardt compares the functions that RECs and CECs can engage in.

grid. On the other hand, once created and recognized as such according to EU legislation, ECs will be able to involve themselves in a variety of other activities, for instance, aggregation, power purchase agreements (PPAs) and participation in wholesale markets.[22] Furthermore, whereas ECs must become legal entities with legal personality under EU legislation, that is not the case for a group of self-consumers under Spanish law.

Cooperatives. As regards the legal form for CSC, there is a wide range of choices. A 2020 report published by the European Commission estimated that there were 33 EC initiatives in Spain (EC 2020). The cooperative is currently the preferred form. Energy coops have a long history in Spain, with many over 100 years old. Union Renovables,[23] for instance, includes about 20 cooperatives with a municipal or regional coverage and more than 85,000 members. Som Energía[24] has 75,000 members, refers to itself as the Green Energy Coop, and has activities throughout Spain. Sapiens Energía is another group that is building ECs in a number of municipalities, principally in the region of Valencia.[25] Other cooperatives, including those in Enercoop Grupo[26], are also in the Valencia region.

The general regulatory framework for cooperatives is Act no 27, of July 16, 1999. Cooperatives are a suitable legal structure to support renewable ECs in Spain, especially because of their "fit" with local government, in accordance with the Local Organization Act 1985. Municipal governments are allowed to participate in coops and therefore in CSC ECs.[27] There are several types of cooperatives and one of them is the so-called "mixed cooperative," which refers to a cooperative where stakeholders have a share of the capital of the cooperative but do not participate in the activities of the cooperative itself. In other words, it will be the community, in the form of a cooperative that will operate as foreseen in the EU Directives. The agent entitled to operate in electricity markets will be the cooperative itself, not each member of the cooperative. However, under EU legislation, all consumers currently have the right to choose their retail supplier. This also applies to members of a coop for the energy needs not supplied by the coop.

[22] Article 22 of the 2018 RES Directive refers to the functions of RECs: "produce, consume, store, and sell renewable energy, including through renewables power purchase agreements, access all suitable energy markets both directly or through aggregation in a nondiscriminatory manner, other commercial energy services, access to finance and information."

[23] http://www.unionrenovables.coop.

[24] https://www.somenergia.coop/es.

[25] https://sapiensenergia.es.

[26] https://www.grupoenercoop.es.

[27] Act No 7 of 2 April 1985, as modified by Act No 27, of the 27 December 2013. The Local Act 1985 has been developed by several RDs and by the Regulation of the Local Organization Act 1985 (1986), approved by RLD No 781 of 18 April 1986.

4. Factors potentially favoring alignment

There are at least five ways the current Spanish framework could support the creation of ECs that lower system-wide costs and member costs, while not shifting costs to non-members (see also Box 5.1). The aim is to create the conditions that would have this effect over time, thereby reducing the costs of the energy transition.

4.1 Improving competitive pressures on the system

ECs may be able to strengthen competitive pressures throughout Spain. First, self-generation competes with large-scale generation. Sometimes, this is treated simply as lower demand in the wholesale market, leaving the concentration index in that market unchanged. But looked at in terms of total generation (distributed plus central system), we are witnessing increased competition. Furthermore, ECs may purchase electricity on Power Purchase Agreements (PPA) from companies outside the community. Second, ECs are competing with retail suppliers, who buy from the wholesale market on behalf of final consumers. Our understanding is that all members of the ECs have the freedom to leave the community and members have the right to purchase from the retail supplier of choice for energy not supplied within the community. Provided the playing field is level, these competitive options in generation and retail supply introduce incentives for the community itself and for external suppliers and generators to innovate, improve the quality and range of services, lower prices, and generally respond to consumer preferences.

However, the playing field in Spain does not currently support effective competition. For collective self-consumption, there is no legislation that explicitly permits participation in all organized markets for energy or ancillary services, but nor does it explicitly prohibit that participation. There is a legal principle by which individuals and/or companies can operate in whatever economic activities they want, save when this is expressly prohibited by law. However, due to the complex existing regulation of the electricity sector and the rules for trading, one may argue that the reverse principle applies: agents can operate in a specific electricity business only when they have been allowed to do so and the regulations make their participation viable. In particular,

Box 5.1 Five ways ECs can potentially benefit the electricity system and all consumers

1. ECs can improve competitive pressures.
2. ECs can lower the costs of the *existing* system.
3. ECs can reduce the costs of the *future* system.
4. ECs can reduce total system carbon emissions at least cost.
5. ECs can benefit their members and society as a whole, in other ways.

there are no clear regulations for the participation of independent aggregators in Spanish electricity markets and this absence also affects ECs, which are in this respect akin to independent aggregators.[28] In future, when EU legislation is transposed, ECs must be empowered to participate in all markets, both existing markets and new ones.

Leaders of ECs and their advisors have reported to the authors a barrier to entry related to the difficulty (e.g., delays) in obtaining authorizations for collective self-consumption from local distribution system operators (DSOs). This may reflect a general problem of administrative delays in obtaining authorisations in Spain.

In short, the potential for effective competition could and should be improved, in particular by enabling access for ECs to all external markets, ensuring fair and efficient access to networks and speeding up administrative approvals.

4.2 *Incentives to reduce costs of the existing system*

ECs and all other agents system should have incentives to reduce incremental costs both for themselves and for the existing system. For that condition to exist, prices paid by these communities should reflect the incremental costs of the network, system services, and energy. In that case, when the community reduces the energy bought from the existing system by self-consuming its own generation, it reduces system costs and should be compensated for doing so. The compensation should reflect reductions in network losses and in costs of operation and maintenance, system services (e.g., balancing), and energy. However, it is important to distinguish the parts of the system where costs are reduced from those where costs may be unchanged or rise, and to be clear over what period of time the change in costs may occur.

Fig. 5.2 helps to distinguish where building new generation may raise transmission system costs and where it may lower them. It maps areas of Spain by reference to the difference between generation and demand.[29] In areas with more generation than demand (i.e., red areas (gray in print version), including Cadiz in the south), building more local generation does not lower transmission network costs; it may raise the costs, requiring more network capacity to move the generation to demand. In areas where demand is greater than generation (i.e., blue areas (dark gray in print version), including Barcelona and Madrid), building local generation will lower transmission losses and other costs. For instance, an EC in Madrid, Barcelona, or Valencia will lower losses but not in Cadiz or Tarragona.

This figure is not precise enough to assess the impact of new generation (i.e., building an EC) on the distribution network. But it does illustrate the need to be specific about network costs before determining to what extent an EC will lower costs of

[28] Currently, retail supply companies may participate as demand aggregators in certain balancing markets, but independent aggregators (not integrated with supply) and energy communities are not permitted to do so. Furthermore, neither kind of aggregator is authorized to sell into other existing or proposed wholesale energy markets. For instance, an early Draft Order creating a capacity market in the Spanish electricity system excluded independent aggregators from said market, an exclusion which would affect energy communities, in their capacity as aggregators: https://energia.gob.es/es-ES/Participacion/Paginas/Index.aspx.

[29] For more information on this approach, see Tevar et al. (2019, 2021).

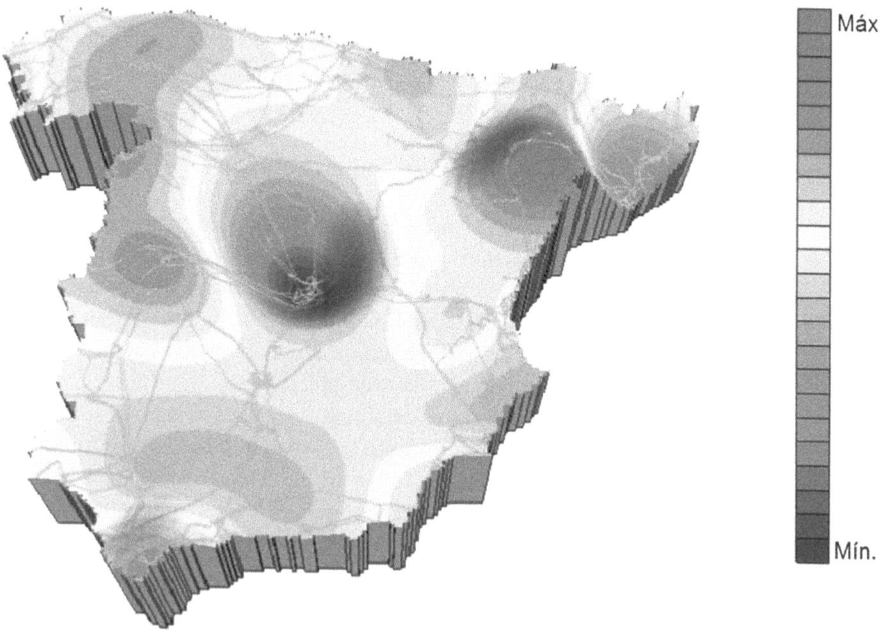

Figure 5.2 Spain 2020 horizon planning, average generation - average demand (MW). Source: Own elaboration based on REE graphics as produced in (Gómez-Expósito et al., 2018).

the system, both in the short term and in the longer term, when new investment may be needed or avoided. In some cases, the cost reduction may be substantial and in other cases very little or even negative.

This raises at least two problems related to Spanish regulation:

- Under current Spanish legislation, individual or collective autoconsumers save the variable component of the regulated access tariff for each unit of energy they generate themselves and do not buy from the system. They also save the variable cost of energy, including the variable price of wholesale energy, balancing costs, capacity payments, and estimated losses.[30] As explained later in Section 5, this saving for the autoconsumer is significantly more than the resulting reduction in the system's costs. Although this is an incentive to create an EC, it is an inefficient signal that may raise overall costs and involves passing costs to non-members of the EC.
- Locational differences are not recognized by Spanish network tariffs, which are uniform nationally. Hence, compensation does not reflect the underlying economic conditions, in particular the balance of generation versus demand. The absence of locational differences introduces inefficiency, encouraging investment in generation where there is

[30] The regulator (CNMC) apparently did not take into account the effect of avoided network charges when they calculated the network tariffs; the implication would be that there will be some missing money. Apparently, they set up a working group to study this.

already an excess to local needs, thereby raising the costs of the network to transmit the generation elsewhere. It may also not provide adequate incentives to build generation near to demand.

4.3 Incentives to reduce costs of the future system

An EC may be able to reduce future investment and operating costs elsewhere in the system. On the one hand, the penetration of renewable generation and distributed energy resources (DERs) in ECs may reduce the need for additional large-scale generation, storage, and high voltage networks. DER includes not only generation, but also storage and smart electrical devices that can be programmed to respond to price signals and compete with conventional resources in wholesale markets.

On the other hand, whether this penetration actually reduces system cost depends on the comparative costs of the alternatives. Large-scale solar and wind is significantly less expensive than small-scale renewables, even when one takes into account the avoided network costs of distant large-scale generation. The implication is that small-scale solar generation in an EC is likely to be more expensive than solar generation built at larger scale. To be economically viable without permanent subsidies, ECs need to minimize the disadvantage through larger-scale generation projects and to find ways to lower their costs and create new sources of revenue, for instance, using the DER to offer flexibility services in local and wholesale markets. Many of the additional revenue sources depend on changes in the overall policy, regulatory, and market framework, reinforcing the view that the system benefits of ECs will only emerge over time and will require reform.

We can see the beginning of the needed reform in the access tariffs introduced in June 2021. They offer an economic incentive for autoconsumption by small consumers—with contracted capacity below 15 kW—at times of network peak. By discouraging demand at peak times (denoted in Fig. 5.3 by the overlap between the hours of most sun and the period of the "high" spot energy price of €62.4/MWh[31]), these tariffs reduce peak congestion on the grid and the need to invest in grid reinforcement in areas where the grid is congested. However, because this is a uniform national tariff that ignores local conditions, it means that uncongested grids remain on the system and have to be funded by all users.

The tariff is also an inducement for autoconsumption because the peak access charge occurs at times of day when rooftop solar will be operating, as reflected in Fig. 5.3. Because the tariffs favor individual and collective autoconsumers and lower congestion on grids when congestion is a problem, one could argue that they align the incentives of ECs with the wider interests of the system. However, as explained above, the alignment is far from perfect due to the fact that it ignores local differences. Furthermore, Section 5 explains how the design of these tariffs introduces incentives for ECs that effectively shift costs to nonmembers of the EC.

[31] Recently, spot energy prices have reached levels four times higher.

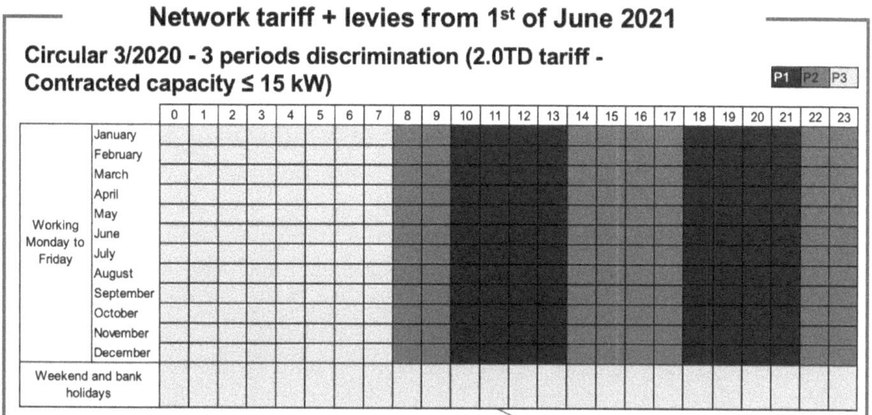

62,4 €/MWh

PV Production

	1	2	3	4	5	6	7	8	9	10	11	12	13	14	15	16	17	18	19	20	21	22	23	24
January	0	0	0	0	0	0	0	0,1	0,2	0,3	0,4	0,5	0,4	0,3	0,2	0,1	0	0	0	0	0	0	0	0
February	0	0	0	0	0	0	0	0,2	0,3	0,5	0,6	0,6	0,6	0,5	0,3	0,2	0	0	0	0	0	0	0	0
March	0	0	0	0	0	0	0,1	0,3	0,4	0,6	0,6	0,7	0,6	0,6	0,4	0,3	0,1	0	0	0	0	0	0	0
April	0	0	0	0	0	0,1	0,2	0,4	0,5	0,6	0,7	0,8	0,7	0,6	0,5	0,4	0,2	0,1	0	0	0	0	0	0
May	0	0	0	0	0	0,1	0,3	0,4	0,6	0,7	0,8	0,9	0,8	0,7	0,6	0,4	0,3	0,1	0	0	0	0	0	0
June	0	0	0	0	0	0,2	0,3	0,5	0,6	0,8	0,9	0,9	0,9	0,8	0,6	0,5	0,3	0,2	0	0	0	0	0	0
July	0	0	0	0	0	0,2	0,3	0,5	0,7	0,8	0,9	1	0,9	0,8	0,7	0,5	0,3	0,2	0	0	0	0	0	0
August	0	0	0	0	0	0,1	0,3	0,4	0,6	0,7	0,8	0,9	0,8	0,7	0,6	0,4	0,3	0,1	0	0	0	0	0	0
September	0	0	0	0	0	0	0,2	0,3	0,5	0,6	0,7	0,8	0,7	0,6	0,5	0,3	0,2	0	0	0	0	0	0	0
October	0	0	0	0	0	0	0,1	0,2	0,4	0,5	0,6	0,6	0,6	0,5	0,4	0,2	0,1	0	0	0	0	0	0	0
November	0	0	0	0	0	0	0	0,1	0,2	0,4	0,4	0,5	0,4	0,4	0,2	0,1	0	0	0	0	0	0	0	0
December	0	0	0	0	0	0	0	0,1	0,2	0,3	0,4	0,4	0,4	0,3	0,2	0,1	0	0	0	0	0	0	0	0

Figure 5.3 Network tariff and levies charged through access tariffs.
Source: Endesa.

There is another potential contribution for the system from the expansion of DERs, including those developed by ECs. Overall, the penetration of intermittent renewable energy increases the fragility of the electricity system. As conventional power plants leave the system, many of the ancillary and balancing services they provided need to be replaced to maintain system stability. Some of these—frequency control, reserves, black-start—can be provided by new resources that are provided at large scale and connected to the transmission grid; the grids themselves may need to be upgraded and extended in order to connect these large-scale resources to markets. However, ECs and other commercial configurations (e.g., aggregators) will soon be in a position to combine multiple DERs through digital platforms that potentially act as Virtual Power Plants (VPPs). These VPPs—even when they do not provide all of the services a generator can provide—compete with large-scale generation and storage resources connected to the transmission network (see Mancarella, 2020). The system will need both distributed and centralized resources, with distributed resources increasingly important to ensure resilience. Indeed, the provision of ancillary services may be the most important basis for ECs to reduce system costs and to earn additional revenues.

It is important to explore in detail how ECs in specific locations could lower system costs by contributing balancing services, secondary and tertiary reserves, and black start capability (see Canca et al., 2018).

A good example of this future potential from VPP platforms is the use of electric vehicle batteries to provide ancillary and reserve services, through smart charging (Grid to Vehicle, G2V) and smart discharging (Vehicle to Grid, V2G). Since consumers buy vehicles as a means of transport, and on average only drive them 10%–20% of the time, the batteries offer energy storage at incremental costs that are very low by comparison to storage facilities that are built solely to provide energy and ancillary services to the system. DERs are especially relevant for short-term flexibility.

It is too early to say whether these potential net benefits for the system will be realized by ECs. However, it is clear that they will only be realized if Spanish legislation clarifies the role of ECs and creates markets in which they (as well as consumers and aggregators) can participate effectively.

4.4 Incentives to reduce total system carbon emissions

When ECs consume or store the power they generate, their members rely less on large-scale electricity generation. This has no noticeable effect on incremental carbon emissions when the system energy at the margin is also renewable; it is simply displacing one renewable source by another. However, if the community is able to invest in storage and demand management and use the resulting flexibility so that the community stores renewables when they are not required and consumes (or sells) that stored renewable energy when the marginal energy on the system is generated from fossil fuels, then the community is contributing to a reduction of system wide emissions. Today, ECs in Spain, like Sapiens, are investing in storage with precisely that objective. However, whether this is the most economic means of integrating renewables into the system will depend on the circumstances and the alternatives.

4.5 Other beneficial effects

ECs potentially deliver many other beneficial effects to their members and to society at large. The most important ones include citizen participation in and control over key decisions affecting their lives. To the extent that ECs enable and encourage citizens to be engaged in taking these decisions and to feel that they are directly benefiting from them, communities almost certainly enhance political and social support for policies aimed at promoting the energy transition. Furthermore, if these communities are able—as intended—to create local employment and address poverty among members, they have the potential to make other important social contributions. The potential alignment in these respects is very positive, and the relevance of these other contributions will be case specific. In their chapter, Swens and Diestelmeier consider ECs as a means to achieve a wider set of objectives. The resulting benefits, if realized, are additional to the economic objectives studied in this chapter.

5. Factors that potentially do not favor alignment

This section analyzes five factors that potentially either raise system-wide costs or shift costs to nonmembers and therefore do not favor alignment between the EC members and the interests of nonmembers (see also Box 2).

5.1 Incentives that raise system costs and shift them to nonmembers of EC

As mentioned earlier, incentives for self-consumption of renewable electricity can potentially lead to higher system costs because of the reduced economies of scale associated with local renewable generation. In Spain, for instance, the January 2021 auctions for large-scale renewables yielded an average price of solar PV of about €25/MWh. Estimates for solar PV in ECs are typically higher than that. When networks are already built, the cost difference between large- and small-scale generation is very likely to be much greater than the short-term savings in network costs (mainly losses) in areas where generation is already greater than demand. Consequently, the net benefits for the system as a whole of more self-generation, instead of buying energy from the system, will not be evident in the short term. However, in areas where generation is less than demand, the net benefits for the system may be large in the short term, and the current tariff system may not fully compensate the self-generator.

In Spain, there are a number of implicit economic incentives supporting self-consumption. Some of these are transparent fiscal incentives, for instance, a reduction in real estate tax, which can be as high as 50% for 5 years, to promote solar panels, or up to 95% of the tax on construction, installation, and works, in addition to personal income tax deductions.[32] These explicit subsidies have no direct short-term economic impact within the power sector because the subsidies are paid from general taxation, not by other electricity consumers or companies. However, it is not clear whether these

Box 5.2 Five ways ECs may not benefit wider system interests

1. Economic incentives to self-generate may raise the cost of the energy transition and shift costs to nonmembers of the EC.
2. Tariffs based on "contracted capacity" are an incentive to create an EC to reduce contracted capacity, potentially raising costs for nonmembers of the community.
3. Different prices for exporting self-generation and for buying from the system could be an incentive to create communities to arbitrage the difference, with no welfare benefit.
4. Subsidies that are hidden and not time-limited will raise costs of the transition.
5. Market design and governance issues may discourage the realization of the potential benefits of ECs that were outlined in Section 4.

[32] https://elpais.com/sociedad/2021-04-30/autoconsumo-solar-la-hucha-para-ahorrar-con-las-nuevas-tarifas-electricas.html.

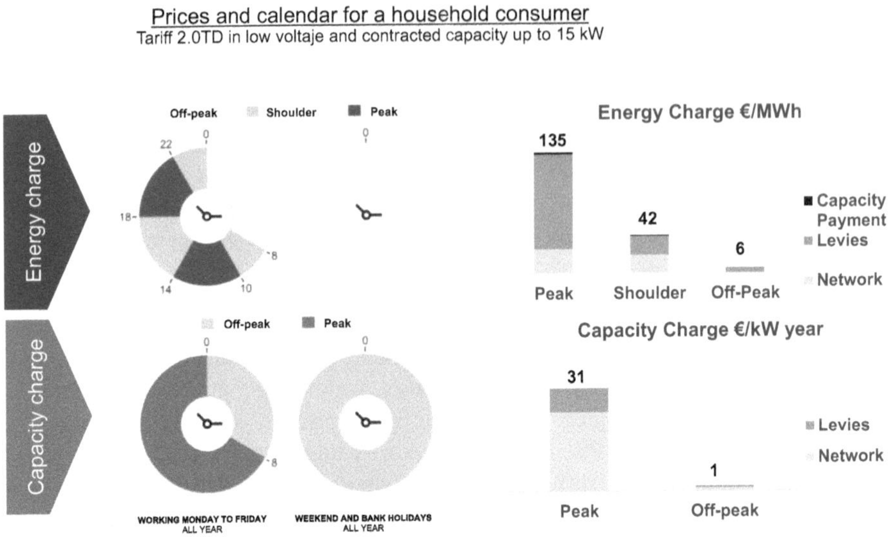

Figure 5.4 Time of use tariffs for residential consumers.
Source: Endesa

subsidies are an efficient use of public funds. Arguably, they are more justified in the areas where there is a need for more local generation to balance demand.

There is another incentive for autoconsumption that implies shifting costs from community members to nonmembers. This is related to autoconsumers not paying the variable component of the electricity tariff. As explained in Section 4, this policy is sensible as long as the variable tariff component reflects the true variable costs of the system, in particular, reduced network losses. However, in Spain, the variable component of the access tariff includes costs that are not variable, in particular, policy levies. Fig. 5.4 summarizes the access tariff structure. In particular, levies to finance policy costs are by far the largest share of the Energy Charge in the access tariff. Individual and collective self-consumers avoid the levies for every unit of electricity they generate, rather than buy from the system. But the costs reflected in the levies do not disappear; they must be recovered from other electricity consumers. So, by reducing consumption of energy bought from the system, the EC effectively policy shifts those costs to nonmembers.

5.2 *Incentives to reduce contracted capacity may shift costs to nonmembers*

The Spanish access tariffs include an important fixed component related to the amount of capacity the consumer contracts, which is supposed to reflect the consumer's maximum demand. It covers some of the fixed cost of networks as well as a number of policy costs. Under the recently introduced tariff regime, residential consumers are authorized to select two levels of contracted capacity, corresponding to peak and off-peak periods.[33] While there may well be an economic case for reducing over time the total level of contracted capacity, care is needed in the interim to avoid shifting regulated costs of the existing network to nonmembers of the EC.

A tariff design based on contracted capacity provides an incentive for consumers to create ECs because the members of the EC will be able to reduce the total contracted capacity and, hence, what they pay for access to the network; this can lead to shifting certain costs to nonmembers. This problem applies especially to all large energy consumers, including Citizen Energy Communities (CECs). This is not an economic efficiency problem, but rather a concern related to the need to rebalance the tariffs, essentially increasing the payment per kW. It is important to explain the concern that policy will need to address.

Contracted capacity in Spain reflects the capacity needed to meet a consumer's peak demand. Because consumers have different demand patterns, the required system capacity to meet the peak for all consumers will be less than the sum of the required capacities for each consumer because of the nonsimultaneous nature of consumer demand. If 10 consumers each require 10 kW, this amounts to 100 kW; due to the nonsimultaneity factor, the regulation requires that the system be designed to meet a combined demand of 75 kW. The cost of the 75 kW is then allocated to consumers on the basis of their individual peak demand (i.e., 10 kW $\times$ 10 $=$ 100). However, if the consumers combine in an EC, they will contract for 75 kW. Furthermore, because of their size, they may be offered a more attractive tariff. In this case, system costs cannot be recovered without redesigning tariffs, probably raising the tariff per kW at the expense of consumers who are not in ECs and who are likely to include many people from households who do not have the resources to invest in solar panels.

The distributional effect is a temporary result of the tariff design. One aim of network tariff design should be to leave consumers indifferent between joining a community or not (as far as their network tariff payment is concerned), except to the extent that any network tariff savings for the consumer reflect avoided costs for the system, for instance, lower losses or avoided network investment.

[33] See three "decisions" (circulares) of the Spanish regulator (CNMC): Circular 2/2019, of November 12, which establishes the methodology for calculating the rate of financial remuneration for transmission and distribution of electricity, regasification, natural gas transportation, and distribution; Circular 5/2019, of December 5, which establishes the methodology for calculating the remuneration of electricity transmission; Circular 3/2020, of January 15, which establishes the methodology for calculating electricity transmission and distribution tolls (as modified by Circulares no 7/2020, of 22 July, and no 3/2021, of 17 March). Among other legal norms, see also Resolution of the CNMC, of the 18 March 2021, establishing access tariffs to transport and distribution networks, as from the 1 June 2021 (BOE no 70, of 23 March 2021); Resolution of the General Directorate for Energy Policy and Mines, of 28 April 2021, establishing the content of electricity bills for suppliers of last resort (BOE no 103, of 30 April 2021); and Ministerial Order no TED/371/2021, of 19 April, 2021, establishing the prices of the costs of the systems and applicable capacity payments, as from the 1st June 2021 (BOE no 96, of 22 April 2021).

5.3 Arbitrage between export and import prices

Where consumers pay more for the energy they buy from the system than they receive for energy they sell to it, there may be potential to arbitrage through the creation of an EC, lowering community costs without improving welfare for the system as whole. The generator will earn more selling to the community than when selling to the system, and buyers will pay less for energy generated in the community than when they buy energy from the system. The community members benefit, but there is no benefit to the rest of the system.[34] Indeed, there is likely to be an overall loss of welfare if the economic cost of self-generation in the community is higher than the system generation being displaced.

Whether this is happening in Spain is a moot point, and will depend on the way communities are regulated in the future. At first sight, the current legislation seems to deal with this correctly, through "net billing," which uses the same hourly wholesale energy price for net energy sales by the individual to the system, as the price that applies to energy purchased from the system on the default tariff (i.e., the PVPC) for small residential consumers. Those conditions apply to about 10 million residential consumers today.

However, as explained earlier, autoconsumers avoid not only the variable costs included in the variable component of the tariffs. They also avoid various other policy and fixed system costs that will not decline when autoconsumers stop paying for them. So while the export and import price of energy are the same for consumers on the default tariff, there are additional benefits from autogeneration. The implication is when the EC does not pay these costs, they will be shifted to others.

5.4 Implicit subsidies

There are many reasons why governments may wish to support the creation of ECs, especially over the longer term. Indeed, EU legislation requires them to do so. These could well justify direct subsidies, for instance, through reduced local property taxation or direct subsidies paid by government. However, under the current arrangements in Spain, some important subsidies are hidden, mainly in the design of tariffs. Avoiding the payment of policy levies and fixed system costs inevitably means shifting costs to other consumers.[35]

5.5 Market design and governance

There are a number of electricity market design and governance features in Spain that may act as barriers to welfare-enhancing trade between the EC and the system. One of the most important challenges is gaining access to existing wholesale energy and ancillary markets. Legislation must ensure that consumers and their representatives

[34] This issue is covered in detail in the chapter by Biggar and Hesamzadeh.

[35] Lo Schiavo et al. explain in their chapter how the Italian regulator has designed a regulatory pilot that pays an explicit subsidy (related to kWhs that are self-generated) to the energy community, for a limited period.

(aggregators and energy communities) have access to these markets. Furthermore, these markets were originally designed for large-scale generation and not with DERs in mind. They will therefore need to be reformed to reflect and encourage the participation of these new resources.

A second challenge is to create local markets for congestion and energy and ensure that ECs have access to these markets, as buyers and sellers. ECs are likely to be able to contribute especially to balance supply and demand at the local level and to assist in managing distribution network congestion. Spain has been planning for a few years to carry out local market pilots in multiple cities but, for a variety of reasons, there has been very little progress to date. It would be sensible to include ECs in these pilots.

A related barrier is the governance of networks and local markets. In Spain and most other European countries, the owner of the distribution network is also the operator of the distribution system (DSO), which manages congestion and operations on the distribution network. To manage distribution system congestion in future, it is important to select the most economic combination of resources, choosing among the resources owned by the grid itself (and its affiliated supply, aggregator, and generator companies) and other operators, including independent aggregators, final consumers, storage companies, and ECs. These markets need to be transparent and managed in a way that is free from any conflicting interest. In Spain, regulatory or structural reform is needed to ensure that the DSO acts with neutrality in selecting the providers of flexibility services through a competitive process.

6. Assessment and recommendations

This chapter has identified a framework for assessing to what extent ECs have incentives that align the interests of their members with the interests of nonmembers. To illustrate the approach, it refers not to the EU definition of ECs, but rather to communities that fit within existing Spanish legislation for collective self-consumption.

6.1 Assessment

Today, there are features of the Spanish regulatory framework that favor the creation of collective self-consumption communities. However, most of these involve economic incentives (e.g., avoidance of policy costs) that do not reduce the short-run costs of the system or that shift costs to nonmembers. Uniform national tariffs may not adequately reward ECs in areas where there is a shortage of generation to meet local demand, and may overreward them in areas where there are no economic benefits for the system. Overall, the incentives may raise the long-term costs of the system by encouraging inefficiently small renewable generation units. Furthermore, there are barriers to the creation of these communities, notably the difficulties they experience in gaining authorization to connect to the distribution network and to participate in wholesale markets. The current framework is neither efficient nor fair.

Even with an appropriate framework—recommendations for which are discussed below—it is difficult for collective self-consumption in Spain to lower system costs in the short term. The full economic benefits of ECs for the system will only emerge over time if the communities reduce their generation cost disadvantage, notably by gaining economies of scale, and create new sources of revenue, in particular drawing on their flexibility to sell into balancing and ancillary service markets. This will only be possible with reforms to legislation, regulation, and markets.

6.2 Recommendations

It is premature to draw firm conclusions about reform, but we would recommend attention to the following issues when transposing EU legislation for ECs in Spain:

1. ECs will initially require financial support, which should not come at the expense of electricity consumers outside the EC, but rather as direct and transparent subsidies from government and that are limited in time. However, it is important to analyze the efficacy of public policies in meeting their objectives.
2. Future Spanish regulation should give all ECs stronger incentives to reduce their costs and to create new sources of revenue, while ensuring they receive nondiscriminatory treatment with respect to network and market access.
   a. To help them reduce costs, legislation should enable sharing and trading (e.g., PPA) over longer distances than currently allowed in order to take advantage of the economic benefits of larger-scale energy and storage facilities. This can be done by extending the geographical limits of RECs and by supporting the creation of CECs, which allow trading over much wider geographical areas.
   b. To increase potential income for ECs, it is important to create local energy and congestion markets, and to guarantee EC access to all new and existing wholesale markets, whether local or system-wide. Within the EC, it will be essential to develop flexibility, in particular relying on storage and demand flexibility, to help integrate renewable generation, whether produced inside or outside the EC.
   c. Future network tariff design should leave consumers indifferent between buying electricity as individuals or through an EC (at least with respect to the payment of the network tariff), except to the extent that the community lowers system costs, in which case community members should be compensated accordingly. Future network tariffs for all consumers should provide strong incentives to use the networks when they are uncongested.
   d. Moving away from uniform national tariffs would enable providing efficient locational signals, encouraging local generation and ECs in areas where this would lower overall generation and network costs.
3. The current legislation provides financial incentives to self-consume, individually or as a community, by enabling the self-consumer to avoid paying certain taxes, network charges, or policy levies. This implies shifting system costs to others outside the community. Future legislation should ensure that the total collected taxes, network charges, and policy levies are unaffected by the decisions to join an EC. More generally, policy should not shift system costs from EC members to nonmembers. An important option to consider is passing policy levies - especially legacy costs - to the central government budget (Robinson, 2019).
4. To ensure nondiscriminatory access to networks and local markets for energy and congestion management, legislation should ensure that the DSO treats all resources without discrimination with respect to their ownership. It should also ensure that DSOs have incentives to facilitate the connection of ECs.
5. Future legislation will need to ensure that members of an EC are free to leave (subject to agreed exit fees) and to choose the retail supplier for energy not provided by the EC.

In transposing EU legislation for Renewable Energy Communities and Citizen Energy Communities, it is important to reform Spanish legislation in the direction suggested in this conclusion. Otherwise, there is serious risk that these communities will depend indefinitely on financial support from consumers outside the community and that they could raise the cost of the energy transition. There is also a risk of balkanizing the electricity system, further increasing costs. By contrast, we anticipate that with the right legislation and incentives, ECs in Spain will help to accelerate the energy transition.

These recommendations apply equally for other new participants in the electricity markets, in particular those that are motivated by financial gain: aggregators, closed distribution systems, and consumers with DERs. The goal—and the case for policy support—is to enable all these new actors to build their capability to compete, eventually without subsidies. Efficient price signals and a level playing field should apply to all existing and new participants, allowing them to compete effectively, benefiting all consumers and helping to reduce the costs of the energy transition.

Bibliography

Canca, D., Arcos-Vargas, Á., Núñez, F., 2018. Blackout Risk Mitigation by Using Medium Size Gas Turbines. Energy 148, 32–48.

Caramizaru, E., Uihlein, A., 2020. Energy Communities: An Overview of Energy and Social Innovation, EUR 30083 EN. Publications Office of the European Union, Luxembourg.

Gómez-Expósito, A., Arcos-Vargas, A., Maza-Ortega, J.M., Rosendo-Macías, J.A., Alvarez-Cordero, G., Carillo-Aparicio, S., González-García, T., 2018. City-friendly Smart Network Technologies and Infrastructures: the Spanish Experience. In: Proceedings of the IEEE, 106(4). IEEE, pp. 626–660.

Hausman, D., 2019. Enhancing Welfare without a Theory of Welfare. Behavioural Public Policy. Cambridge University Press, 2019.

IDAE, March 2019. Guía para el Desarrollo de Instrumentos de Fomento de Comunidades Energéticas Locales. Gobierno de España.

Mancarella, P., September 2020. Electricity Grid Fragility and Resilience in a Future Net-Zero-Carbon Economy. Oxford Energy Forum. Issue 124. https://www.oxfordenergy.org/wpcms/wp-content/uploads/2020/09/OEF124.pdf.

Robinson, D., November 2019. Prices Behind the Meter: Efficient Signals to Support Decarbonization. Oxford Institute for Energy Studies.

Robinson, D., 2021. Energy Communities and Flexible Demand. In: Sioshansi, F. (Ed.), Generation, Flexible Demand. Academic Press (Elsevier).

Tevar, G., Gómez-Expósito, A., Arcos-Vargas, A., Rodríguez-Montañés, M., 2019. Influence of Rooftop PV Generation on Net Demand, Losses and Network Congestions: a Case Study. Int. J. Electr. Power Energy Syst. 106, 68–86.

Tévar-Bartolomé, G., Gómez-Expósito, A., Arcos-Vargas, A., Rodríguez-Montañés, M., 2021. Network Impact of Increasing Distributive PV Hosting: a Utility-scale Case Study. Sol. Energy 217, 173–186.

The "virtual" model for collective self-consumption in Italy

6

Luca Lo Schiavo [1,*], *Andrea Galliani* [1,*] *and Arianna Rossi* [2]
[1]ARERA, Italian Regulatory Authority, Milan, Italy; [2]Politecnico di Milano, Milan, Italy

1. Introduction

The Clean Energy for all Europeans Package (CEP) establishes the European legislative and regulatory framework for achieving European energy objectives for 2030, as a step to full decarbonization by 2050. Among the many innovations provided in the CEP, important novelties include the introduction of the concept of "jointly acting self-consumers" and of "energy communities," in order to involve citizens in the transformation of the energy sector. For example, the participation of citizens and local authorities in energy communities might raise local acceptance of renewable energy, thereby encouraging local investment and the participation by citizens in the broader energy transition.

With the growing importance of self-consumption, and, above all, of renewable self-consumption, it is necessary to establish a regulatory framework which allows the development of self-consumption not only for individual prosumers but also for extended configurations, characterized by multiple users, for example, citizens living in apartment buildings.

Self-consumption is the consumption of electricity produced at the same time on the same site or nearby. It does not depend on power plants' sources and technologies, grid access, or specific corporate structures. Therefore, it is perfectly compatible also with the case of the jointly acting self-consumers, i.e., a group of at least two self-consumers located in the same building or multiapartment block.

On the other hand, energy communities are not necessarily limited to self-consumption because they have the wider goal of promoting social cooperation among participants which can lead to facilitating investments consistent with system transformation, e.g., renewables, energy efficiency. Moreover, energy communities might facilitate the collective purchase or sale of electricity, simplifying participation in wholesale energy and ancillary services markets. From this perspective, energy communities might be a useful instrument to combat energy poverty, which is one of the most significant social themes of citizen communities. This chapter is limited to the analysis of the self-consumption aspect of energy communities.

* The opinions expressed in this chapter are the personal opinions of the authors and do not necessarily represent the official position of the Authority nor commit the Authority to any course of action in the future.

Energy Communities. https://doi.org/10.1016/B978-0-323-91135-1.00017-1

In Italy, before the first introduction of "jointly acting self-consumers" and of "energy communities", several configurations had already been defined to enhance self-consumption. In a nutshell, there are a lot of definitions of self-consumption which could be traced back to the so-called *one-to-one configuration*: one customer and one producer that manage a power plant (the customer and the producer, in some cases, can also be two different parties). In addition, the *closed distribution system configurations* are defined in accordance with EU Directive 2009/72/CE (EU, 2009; repealed by Directive 2019/944 (EU, 2019a)). In Italy, in existing "one-to-one" configurations, self-consumption has already reached about 28 TWh, of which almost 21% is attributable to renewables.

The possibility to be jointly acting renewable self-consumers and to establish Renewable Energy Community (REC) collective schemes was introduced in Italian legislation for the first time with article 42bis of decree-law 162/2019 (Italian President, 2019; confirmed by law 8/2020 (Italian President, 2020)), that explicitly anticipates, on the matter of collective self-consumption, the full transposition of Directive EU 2018/2001 (RED II) (EU, 2019b). According to this decree-law, both jointly acting renewable self-consumers and RECs have the goal of sharing electricity as it is self-consumed. The Italian Regulatory Authority for Energy, Networks and Environment (ARERA, or the Authority) is mandated by this piece of legislation to define all regulatory aspects.

This chapter focuses on the Italian regulatory model, defined by the Authority with resolution 318/2020 (ARERA, 2020a), adopted in Italy to implement the collective self-consumption schemes foreseen by the decree-law 2019/162. The regulatory model allows users who are part of the new configurations to maintain their rights (including that of choosing their own retail supplier) to enter and exit these new configurations, and use self-consumed electricity according to the benefits derived, without any need for technical interventions, such as building new lines or installing new meters. The model allows also the separate management of the support scheme (i.e., a policy decision to achieve the EU target and the energy transition) separately from the economic value of self-consumption (which, instead, is related to the actual economic benefit of self-consumption) from the perspective of the electricity system.

This chapter consists of three sections in addition to the Introduction.

- Section 2 introduces the European and Italian regulatory frameworks for collective self-consumption and energy communities and
- Section 3 describes the regulatory model for collective self-consumption recently implemented in Italy; followed by the chapter's conclusions.

2. Legislative and regulatory framework

2.1 European framework

The mechanism outlined at the European level requires each Member State to contribute to the achievement of the common objectives by setting national targets.

In addition to preparing an Integrated National Energy and Climate Plan (NECP), Member States are also asked to implement, among others, Directive 2018/2001 (EU, 2019b) and 2019/944.

The first of these directives introduces the two concepts analyzed in this chapter, "jointly acting renewables self-consumers" and "renewable energy communities":

- Jointly acting renewables self-consumers (JARSC) are defined in Article 2(15) as a group of at least two jointly acting renewables self-consumers who are located in the same building or multiapartment block. A renewables self-consumer is defined in Article 2(14) as *"a final customer operating within its premises located within confined boundaries or, where permitted by a Member State, within other premises, who generates renewable electricity for its own consumption, and who may store or sell self-generated renewable electricity, provided that, for a non-household renewables self-consumer, those activities do not constitute its primary commercial or professional activity."*
- A REC is defined in Article 2(16) as an autonomous legal entity based on open and voluntary participation and is effectively controlled by shareholders or members that are located in the proximity of the renewable energy projects that are owned and developed by that legal entity. The shareholders or members of RECs are natural persons, small- or medium-sized enterprises, or local authorities, including municipalities. The primary purpose of RECs is to provide environmental, economical, or social community benefits for its shareholders or members or for the local areas where it operates, rather than financial profits.

Article 2(46) of directive 2019/944, on the other hand, introduces the concept of Citizens Energy Community (CEC) as a legal entity, based on voluntary and open participation, whose primary purpose is to provide environmental, economical, or social community benefits to its members or shareholders or to the local areas where it operates rather than to generate financial profits. The concept of CEC is very similar to that of the REC but differs in some respects. One difference is that the REC may include any kind of renewable energy, whereas the CEC refers solely to electricity.[1]

In the cases of REC and CEC, member States shall ensure that participation is open and voluntary and that their shareholders and members do not lose their rights and obligations as household customers or active customers. In the framework of full liberalization of energy retail markets, article 4 of Directive 2019/944 states that all customers shall be able to purchase electricity from the supplier of their choice.

2.2 New legislative framework in Italy

With article 42bis of the decree-law 162/19 (converted into law 8/2020), the Italian government established a transitional regime in force until the transposition of CEP Directives, with the aim of collecting data and learning from field experience. Thus, this is a transitional regime, which anticipates the transposition of the European Directives.

This transitional regime introduces the concept of jointly acting renewable self-consumers and RECs, defining the perimeter and the permissible power plants in these

[1] The chapter by Spasova and Braungardt compares and contrasts RECs and CECs.

configurations. The transitional regime does not anticipate all the configurations of collective self-consumption foreseen by CEP, but rather only those related to renewables. In particular, art. 42bis of the decree-law 162/19 foresees that JARSC and RECs, in the transitional regime, are enabled to produce electricity by means of Renewable Energy Source (RES) power plants with capacity not above 200 kW in operation after the date of entry into force of the law converting decree-law 162/19 (i.e., from March 1, 2020) and within 60 days following the date of entry into force of the law transposing Directive 2018/2001. It must be noted that the Directive 2018/2001 does not foresee constraints related to size or date of entry of operation of the RES power plants, so these limits will probably be removed in the Italian legislation when the transposition of CEP Directives is completed.

Moreover, JARSC must be located within the same building or multiapartment block, as defined by Directive 2018/2001. By contrast, under the Italian transitional legislation for RECs, customers and producers must be connected to Low-Voltage (LV) grids connected to the same transformer substation MV/LV: the limited perimeter derives from the fact that, in the transitional regime, RECs have the goal of sharing electricity as it is self-consumed.

According to decree-law 162/19, consumers belonging to the collective self-consumption configurations fully retain their rights as final customers. In fact, each single customer shall be able to choose its own supplier. Moreover, he or she shall be able to opt-in and opt-out of the collective scheme (jointly acting renewable self-customers or REC): there shall be no restrictions on members joining the collective scheme and then leaving it. Participants regulate the relationships by means of private law contracts which identify a delegated subject (the so-called "contact person"), who is responsible for the allocation of the shared electricity. Participants may also delegate the management of the payment to retail companies and of the collection of RES-related incentives from the public agency for renewables, the GSE—Gestore dei Servizi Energetici S.p.A.

Moreover, according to decree-law 162/19, participants of REC or JARSC schemes must use the public grid (i.e., the public distribution network, at low and medium voltage, to which consumers and generating units belonging the configuration are connected) to share with each other the electricity produced by the RES plants belonging to the configuration. In this way, they avoid having to build or own private electricity grids. In fact, a key principle underlying this transitional regime is to make the best possible and efficient use of the existing public grid, avoiding the construction of new private grids where they are not needed. For the same reason, RECs and JARSC cannot buy or lease the local public grid.

In each hour, the shared electricity is defined as the minimum between the electricity produced and injected into the public grid by the RES plants belonging to the collective scheme and the electricity withdrawn from the public grid by the participating customers. In Italy, each customer has its own separate meter, so it is possible to distinguish the energy withdrawn or fed into the grid by each single participant in the configuration (REC or JARSC) and by consumers who are not participants.

Finally, art. 42bis of the decree-law 162/19 foresees that the Minister of Economic Development shall define the unitary value of an incentive tariff (in €/MWh) for the remuneration of RES power plants installed within the allowed configurations. Separately, the Italian regulatory authority shall identify the value of the regulated tariff components which are not technically applicable to shared electricity and not charged to participants for the electricity they share. The logic for not charging for these tariff components is that the electricity is instantaneously self-consumed in the same portion of the local public grid and comparable to physical self-consumption on site.

This explicit separation between the support scheme (a political choice) and the valuation of self-consumption (a regulatory choice, as it is inherently related to cost-reflectiveness) had been advocated by the Authority itself in several hearings at Parliament (ARERA, 2019; ARERA, 2020b) in order to avoid implicit (hidden) incentives.

On the one hand, the valuation of self-consumption for the electricity consumed on the same site where it is produced (either instantaneously or via storage systems, regardless of who is operating on the same confined site, and regardless of the power plant energy source) should be defined to reflect the actual economic benefit (i.e., avoided cost) of self-consumption for the electricity grid. This benefit is due to reductions in electricity transit costs and grid losses. On the other hand, support schemes for the electricity produced only from RES power plants are necessary to achieve the ambitious European target on decarbonization and RES penetration; the EU target is for RES to meet 32% of gross final energy consumption by 2030 (even more ambitious targets are under discussion within the "Fit for 55" package) and for the EU to become the world first climate neutral continent by 2050. Thus, the methodology for the valuation of system benefits from collective self-consumption is entrusted to the Authority, while the Italian Government defines the incentive for RES power plants embedded in new self-consumption schemes.

The incentive was defined by the Minister of Economic Development, with ministerial decree September 16, 2020 (Italian Minister of Economic Development, 2020). It is equal to 100 €/MWh for the electricity self-consumed among JARSCs, whereas it is equal to 110 €/MWh for the electricity shared within the RECs. This support scheme is granted for 20 years, for both collective schemes. This incentive value is like the one actually applied in Italy for small-size photovoltaic power plants and it has been defined assuming that, as expected, photovoltaic systems will be mostly used in JARSCs and RECs.

3. The Italian pilot regulation for collective self-consumption

After wide public consultation (consultation paper 112/2020; ARERA, 2020c), with the resolution 318/2020, the Authority launched the pilot regulation related to shared electricity for self-consumption by REC and JARSC.

The pilot regulation defined by the Authority is based on a regulatory model that is called a "virtual model" because it allows the enhancement of collective self-consumption without the need for any new infrastructural interventions like new direct lines or new meters. Participants stay connected to their point of delivery and the electrical energy is shared using the public distribution grid. It is a "virtual model" with the aim of enhancing real self-consumption.

In addition to avoiding the need for new infrastructure, the virtual model guarantees transparency and flexibility for all electricity consumers wishing to take part in collective initiatives, including the safeguarding of their rights as final customers, and it is easily adaptable for any future needs. Customer rights are guaranteed since each participant can modify its own choices, both in relation to the participation in the collective self-consumption configuration (opt-in and opt-out) and in relation to its choice of the supplier for all the electricity withdrawn from the public grid (be it shared or not). Last but not least, as already described in previous paragraphs, the virtual model separates the economic benefits associated with self-consumption and the explicit incentives, avoiding hidden, implicit, incentives.

From a practical point of view, the virtual model foresees that each member of the collective scheme has to pay the traditional electricity bill for energy withdrawn from the public grid and then receives a «cash-back » corresponding to the shared electricity. Such a cash-back represents the recognized value of self-consumption. The value of the cash-back is computed on a monthly basis by GSE and it is different for JARSCs and for RECs.

The cash-back has to be defined on the basis of grid benefits (avoided costs) resulting from self-consumption. These have been identified by the Authority as the reduction of grid losses and of grid costs covered through the variable part of the transmission and distribution tariff[2]. In Italy, the distribution network tariff is completely capacity driven (in euro/kW) for households, and the volumetric part is extremely small (covering costs of residual High Voltage (HV) lines belonging to distribution networks) and for simplicity is applied only to LV and MV business customers. On the contrary, the Italian transmission network tariff is fully volumetric for LV and MV customers, while it is mostly capacity driven for HV customers. The transmission and distribution tariffs only include grid costs, without other levies. Benefits related to energy dispatch costs are currently not demonstrable: an in-depth study on further benefits of collective self-consumption on dispatching cost is ongoing. No benefits are related to general system charges as they are independent of self-consumption: they are applied to all electricity withdrawn from the public grid and not returned.

If, instead of implementing a virtual regulatory model, other solutions had been adopted (for example, the possibility of building private electricity grids to which the general system charges would not have been applied), it would not have been possible to attribute to the self-consumption the avoided system costs. Furthermore, the implicit incentives that would have derived from rewarding self-consumption on a new private network (for example, the avoidance of the general system charges) could have promoted the creation of private configurations in the absence of real systemic efficiency.

[2] Details about transmission, distribution, and metering costs are available at www.arera.it/allegati/docs/19/568-19com.pdf.

The cash-back is paid by GSE to the "contact person" who is responsible for the allocation of the received cash flows among the participants in the configuration. The nature of the contact person has been identified by the Authority for both configurations. For jointly acting renewables self-consumers, the contact person can be either the legal representative of the building (or of multiapartment block) or the producer that manages one or more generating units belonging to the configuration. For a REC, the contact person is the community itself as it is already a legal entity.

In addition to the value of the self-consumed electricity, the contact person receives from the GSE the incentive defined by the Minister of Economic Development (100 or 110 €/MWh depending on the type of collective scheme) for the shared electricity.

In this way, the political incentive and the valuation of self-consumption remain separate but they are received together by the same the contact person, simplifying the cash flow management of the configuration itself. This also enables payment of the incentive only for the agreed duration, while the cash-back payment for self-consumed electricity may be granted for longer. The allocation of the received cash flows among the participants in the configuration is made according to a private agreement.

Detailed elements are provided below to better explain how the regulatory model works, with some numerical examples. Fig. 6.1 shows a simplified chart that describes the way the shared energy is determined within the same multiflat building (i.e., a JARSC configuration; the REC one works the same way as far as shared energy identification is concerned).

As an example:

- A is a generator unit (for example, a photovoltaic plant but other renewable plants may be admitted, and also storage) that injects its own overall production into the public grid;
- B is a prosumer because it produces when its power plant is available, consumes part of its production on site, and, at any given moment, injects the excess production into the public grid and takes the missing energy from the public grid;

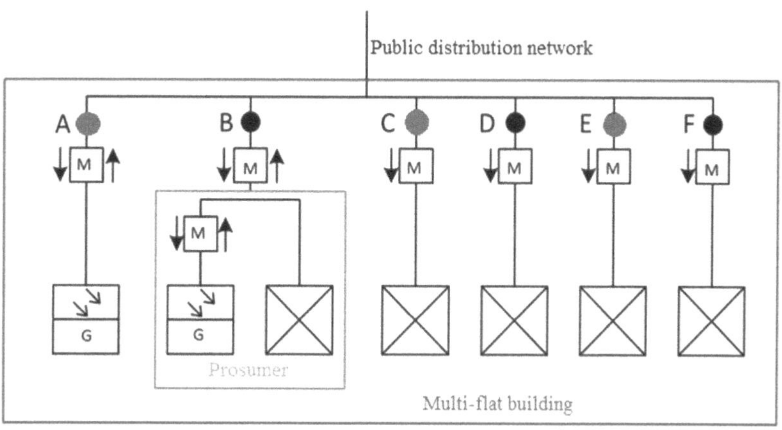

Figure 6.1 Shared energy among participants of a collective self-consumption scheme.

- C, D, E, and F are consumers who withdraw all their electricity from the public grid; and
- "M" identifies a meter managed by the Distribution System Operator (DSO).[3]

Now, imagine that only A, C, and E decide to participate to the JARSC configuration, whereas B, D, and F choose not to participate.

From an operational point of view, the contact person must submit an application to GSE, which checks that all requirements are met and, if so, concludes a contract. In order to compute the electricity to be valued (and incentivized), GSE receives from DSOs metering data on electricity withdrawn from and fed into the grid through each connection point that is part of the configuration (points A, C, and E in Fig. 6.1). However, in Italy, validated hourly measures are not yet available for all the connection points, because the first generation of smart meters typically used a time-band framework[4] for delivery points up to 55 kW of capacity; the substitution of these meters with second generation (2G) ones, which provide hourly metering data for all delivery points, is ongoing. The smart meter replacement campaign from first generation to second generation has reached 20 million customers, i.e., more than 50% of the Italian electricity customer base (update: June 2021; Wired, 2021).

On the basis of hourly measures received from DSOs, GSE calculates the shared electricity which, for each hour h, is the following:

$$Shared\ electricity\ (h) = \min_h \begin{cases} \sum_{i=1}^{n} Eout_{i,h} \\ \sum_{i=1}^{n} Ein_{i,h} \end{cases}$$

where

- i is the i-th participant of the configuration (in red (light gray in print version) in Fig. 6.1);
- $Eout_{i,h}$ is the energy withdrawn from the grid in a given hour h by the i-th participant; and
- $Ein_{i,h}$ is the energy fed into the grid in a given hour h by the i-th participant.

Only for shared electricity, GSE pays to the contact person the valuation of self-consumed electricity and the incentive.

A numerical example in Box 6.1 may be useful.

As stated above, the value of the incentive defined by the Minister of Economic Development, with ministerial decree September 16, 2020, is 100 €/MWh for JARSC or 110 €/MWh for REC. In Annex A, the valuation of self-consumed electricity as defined by the Italian Authority is explained, distinguishing the case of JARSC and the case of RECS.

[3] All meters (M) are able to meter separately electricity withdrawn from the network and electricity injected into the network.

[4] 3 time bands are used: peak, from 8.00 a.m. to 7 p.m. on weekdays; shoulder, from 7 to 8 a.m. and from 7 to 11 p.m. on weekdays, and from 7 a.m. to 11 p.m. on Saturdays; off-peak all the remaining hours and days.

Box 6.1 Numerical example

Consider three cases:

- In case 1, in an hour: A feeds 1 kWh into the public grid, C takes 7 kWh from the grid, and E takes 3 kWh from the grid, as measured by the DSO;
- In case 2, in an hour: A feeds 9 kWh into the public grid, C takes 7 kWh from the grid, and E takes 3 kWh from the grid, as measured by the DSO; and
- In case 3, in an hour: A feeds 11 kWh into the public grid, C takes 7 kWh from the grid, and E takes 3 kWh from the grid, as measured by the DSO.

In the virtual model, expenditure for withdrawal and "cash-back" for self-consumption are kept fully separate. The total injections and the total withdrawals may be sold/bought from the supplier chosen, separately, by A, C, and E; so they can each choose their own supplier and may accept the best offer operating on the market: as a result, the electricity supplier for community members is not necessarily the community.

In all the three cases, C and E pay their suppliers for 10 kWh. Therefore, the expenditure for energy withdrawn from the public system is exactly the same for all three cases. For the same hour, A is paid by a supplier for 1 kWh in case 1, 9 kWh in case 2, and 11 kWh in case 3. Finally, the REC or JARSC scheme has a shared electricity equal to 1 kWh in case 1, 9 kWh in case 2, and 10 kWh in case 3, for which GSE pays to the contact person the valuation (avoided system cost) of self-consumed electricity and the incentive.

So, if we look at the cash-back due to self-consumption benefit and RES incentives, in case A the cash-back is proportional to 1, in case 2 it's proportional to 9, and in the third case it's proportional to 10. This demonstrates that the virtual model enhances real collective self-consumption.

Take note that no compensation for a given hour is allowed for energy consumed in different hours, so there is no "net metering".

4. Conclusions

No results are yet available as the first JARSCs and RECs are being implemented. On the basis of the very first estimates currently available, there could be requests for the construction of 1100 photovoltaic plants within JARSCs and RECS, for a total capacity of 24 MW. A total production of about 13 GWh is estimated, of which 8 GWh could be self-consumed, implying an injection of 5 GWh to the system.

Although the estimates of capacity and energy may appear low, this "virtual" regulatory pilot, being implemented by the Italian Authority until the full transposition of CEP directives, will likely prove interesting for the valuation of self-consumed electricity in collective schemes.

On the customer side, the virtual model ensures flexibility for all participants and safeguards their rights as final electricity customers in the framework of a fully liberalized retail energy market. Customers belonging to the same collective self-consumption configuration may opt-in and opt-out and are not obliged to be supplied by the same retail company: indeed, each participant can freely change its own electricity supplier at any time, fully independently from the collective self-consumption group to which he or she belongs. In other words, the electricity supplier for community members is not necessarily the community itself. This ensures that each customer and each producer maintain their rights and obligations, as required by the Directive 2018/2001.

On the grid side, the virtual model is efficient because it allows the enhancement of collective self-consumption without the need for any infrastructure interventions or new meters: no private (and potentially inefficient) grids or microgrids are necessary.

On the system side, the virtual model makes it possible to separate the valuation of self-consumed energy from the incentive, which is conceptually different. This could make it possible to apply the valuation of self-consumed energy to configurations with existing plants, limiting the incentive to self-consumption deriving from new plants that require economic support.

Being still a transitional measure, this model cannot be considered yet as the complete "regulation of energy communities," since these communities can provide many other services in addition to self-consumption. For instance, energy communities may develop initiatives for fighting fuel poverty or for increasing energy efficiency in buildings, as described, e.g., in the chapters by Robinson and del Guyao, Ghiani et al., and Del Pizzo et al.

For all the above reasons, as an interim measure prior to transposing the EU Directives, the virtual model offers features that encourage real collective self-consumption without jeopardizing full liberalization of electricity retail market or the integrity of distribution public networks. Moreover, it has the advantage of allowing a correct separation between the role of the independent regulatory Authority and the role of the Government, respectively, in valuing and in promoting self-consumption. Some versions of these features are likely to be incorporated in the final legislation, no doubt reflecting the experience from the pilot regulation.

Annex A

Valuation of self-consumed electricity: the case of jointly acting renewable self-consumers

In this case, the valuation of self-consumed electricity is related to the avoidance of the grid costs covered through the variable part of the transmission and distribution tariff, and to the reduction of grid losses for energy withdrawn by customers at a lower (or at least equal) voltage level than the one of the generating units belonging to the JARSC configuration (limited to the reduction of grid losses not already recognized in current regulation).

Therefore, the valuation (i.e., reflecting the avoided system cost) of the self-consumed electricity (C_{SE}), for each month, in the JARSC is equal to:

$$C_{SE} = T + L$$

The first term, T, represents a proxy of the reduction of grid costs due to self-consumption:

$$T = E_{SF} * C_{SE,m}$$

- E_{SF} is the sum, for a month, of hourly self-consumed electricity which is the minimum, for each hour, between the overall injected energy into the system by each power plant in the JARSC and the overall withdrawn energy from the system; and
- $C_{SE,m}$ is the sum of the variable parts (in €/kWh) of the transmission tariff (called $TRAS_E$ in Italy) for low-voltage customers and the highest value of the variable component of the distribution tariff for "other electrical uses" in low-voltage grid. These values are defined yearly by the Authority. In 2021, they are respectively equal to 0.794 c€/kWh and 0.062 c€/kWh. Thus, $C_{SE,m}$, the sum of these two variables costs, is equal to 0.856 c€/kWh in 2021.

The second term, L, instead, is an estimate of the reduction of costs related to grid losses due to self-consumption:

$$L = \sum_{i,h} \left(E_{SF,i} * c_{nl,i} * P_z\right)_h$$

where

- i is the voltage level;
- h is the hour, varying from 1 to the number of hours in the month;
- $E_{SF,i}$ is the minimum, for each hour, between the overall injected energy into the network by each power plant (starting from the ones connected to the low-voltage distribution grid) and the overall energy withdrawn from connection points to the network with voltage level equal to or lower than the voltage level to which are connected the generating units belonging to the configuration;
- c_{nl} is the coefficient for the avoided grid losses, equal to 1.2% for shared electricity produced by generating units connected to medium-voltage distribution grid, or 2.6% for electricity produced by generating units connected to the low-voltage distribution grid. These are only the losses avoided on the grid at the voltage level of the generation plant in the JARSC configuration. In fact, in the current Italian regulation, the reduction of grid losses attributable to self-consumption is already recognized through the increase of the quantity of electricity injected into the low- and medium-voltage grid, in the same manner as for any other distributed generator. More specifically, the electricity fed into the low and medium voltage grids is already increased by applying a percentage coefficient that takes into account the avoided grid losses, set equal to the technical losses from the very high voltage and up to the transformer between the voltage level in question and the level of higher voltage (the losses of the grid having the same voltage level to which the system is connected are not included). So, as the electricity injected into the grid is already increased by applying a percentage coefficient (and, as analyzed in the example, that increased electricity is sold to a

seller), the term "L" only needs to recognize avoided network losses not already recognized; and

- P_z is the hourly energy price formed on the day-ahead energy market in the market zone in which the configuration is located.

Valuation of self-consumed electricity: the case of renewable energy communities

The contribution to the valuation of the self-consumed electricity (C_{SE}), for each month, is equal to:

$$C_{SE} = E_{SF} * C_{SE,m}$$

In this case, the benefits of self-consumed electricity within an REC are only related to the grid costs covered through the variable part of the transmission and distribution tariff. Losses on the grid at the voltage level of the production plant are not avoided, because the grid is used, unlike in the case of buildings or multiapartment blocks.

Bibliography

ARERA, 2019. Position Paper 12th March 2019, 94/2019/I/com. Milan.

ARERA, 2020a. Resolution 4th August 2020, 318/2020/R/eel. Milan.

ARERA, 2020b. Position Paper 21st May 2020, 175/2020/I/com. Milan.

ARERA, 2020c. Consultation Document 1st April 2020, 112/2020/R/eel. Milan.

EU, 2009. The European Parliament and the Council of European Union. "Directive 2009/72/EC", Brussels.

EU, 2019a. The European Parliament and the Council of European Union. "Directive 2019/944", Brussels.

EU, 2019b. The European Parliament and the Council of European Union. "Directive 2018/2001", Brussels.

Italian Minister of Economic Development, Ministerial Decree 16th September 2020, 2020. Rome.

Italian President of the Republic, Decree-Law 30th December 2019, N. 162, 2019. Rome.

Italian President of the Republic, Law 28th February 2020, N. 8, 2020. Rome.

Wired, 2021. Open Meter, il progetto di sostituzione avanza e supera quota 20 milioni. www.wired.it/scienza/energia/2021/05/14/open-meter-enel-quota-20-milioni.

Energy communities: a North American perspective

7

Mark Kolesar[1,2]
[1]Kolesar Buchanan and Associates, Calgary, AB, Canada; [2]Christensen Associates, Madison, WI, United States

1. Introduction

In North America as elsewhere, policy makers and regulators are wrestling not only with how to achieve rapid and substantial GHG reduction targets, but how to do so in ways that attract and maintain broad public engagement and acceptance. Getting general support for broad policy objectives for limiting GHGs is relatively easy. Mitigation and adaptation measures aimed at climate protection are becoming generally accepted as necessary. But obstacles often arise when individuals are asked to participate in achieving specific objectives.

By way of example, in 2015 Alberta, Canada established a goal of phasing out coal and obtaining 30% of Alberta's energy from renewable sources by 2030. The Renewable Energy Program proposed to add 5000 MW of renewable capacity, using an auction to solicit long-term power purchase agreements (PPAs). Among the successful bidders was a project by EDP Renewables SH Project GP Ltd. for the 248 MW Sharp Hills Wind Project.[1]

The hotly contested project required approval of the Alberta Utilities Commission.[2] Among the issues raised were allegations of unfair negotiating with landowners; concerns about visual impacts; adverse environmental effects; noise; reduced property values; health and safety implications; aeronautical effects; and agricultural and land use effects. None of this was unexpected. However, the undercurrent to all of these issues was divisions in the community between those who opposed and those who supported the project ostensibly because supporters would profit from having turbines on their land.

By comparison, in 2018, the recast Renewable Energy Directive 2018/2001/EU (RED II) of the European Commission came into force as part of the Clean Energy for All Europeans policy.[3] Cornerstones of the EU policy include promoting local

[1] Refer to Alberta Utilities Commission Proceeding 23066: https://www.auc.ab.ca/regulatory_documents/ProceedingDocuments/2018/23066-D01-2018.pdf#search=Proceeding%2023066.

[2] The author was Chair of the Alberta Utilities Commission at the time and chaired the proceeding panel that adjudicated the application for approval and licensing.

[3] Refer to https://ec.europa.eu/energy/topics/renewable-energy/renewable-energy-directive/overview_en.

Energy Communities. https://doi.org/10.1016/B978-0-323-91135-1.00020-1

energy generation, consumption, and trading and embracing the concept of energy communities.[4] Specifically, the EU is aiming to have some 16% of its electricity generation from cooperative community-oriented projects by 2050. This would require about half of Europe's households to engage in some form of renewable generation, the majority of which is expected to be from collective projects (Reis et al., 2021).

Although there remains some debate in the literature about what constitutes an energy community, the EU policy as set out in RED II and the European Parliament's companion Internal Electricity Market Directive (IEMD) embraces the notion that energy communities should be citizen driven. The IEMD uses the moniker "citizen energy communities" (Reis et al., 2021).[5] This approach relies on community engagement in energy projects, recognizing that siting and operating renewable energy projects are often controversial and likely to divide communities. Projects often face substantial local resistance if the benefits are not adequately shared among the community, as happened with the Sharp Hills Wind Project.

Europe has a substantial history of renewable energy projects led by local cooperative enterprises (Candelise and Ruggieri, 2020).[6] It would seem that the business culture and social fabric of Europe lends itself to such an approach. This chapter considers whether North America—that is the United States and Canada—also exhibits characteristics that can promote cooperative community-centric approaches to developing clean energy projects. Are community energy projects already appearing? How might energy community concepts be adapted to the North American policy and regulatory context? Are there emerging policy initiatives to promote a more decentralized and distributed structure by facilitating neighborhood and small-scale community investment in renewable energy? Also, are there significant legal, legislative, or regulatory obstacles to energy communities in North America that must be considered?

The balance of this chapter is organized as follows:

- Section 2 examines the notion of an energy community;
- Section 3 provides a survey of community energy projects in North America using business models that are emerging among energy communities elsewhere as a template; and
- Section 4 provides an overview of the policy framework in North America, followed by the chapter's conclusions in Section 5.

[4] The chapter by Spasova and Braungardt provides an overview of the EU policy framework and its implementation in several member states.

[5] "Citizen Energy Communities" generally refer to virtual communities engaged in electricity generation that is not necessarily located near the community. "Renewable Energy Communities" generally refer to "citizen-driven" renewable energy projects more generally (e.g., electricity, biogas) where the community is located close to generation and engaged in consumption. In their chapter, Spasova and Braungardt speak to this distinction and provide an overview of energy communities in European member states.

[6] Other chapters in this book reference energy communities in the UK and European member states, Australia, and Brazil.

2. What is an energy community?

Walker and Devine-Wright's (2008) seminal paper proposed that an ideal energy community project is "entirely driven and carried through by a group of local people" and "brings collective benefits to the local community." Their research demonstrates that direct community involvement translates to greater acceptance and support for a project, and contributes to increased understanding and support for renewable energy projects overall. However, their work was heuristic and subsequent literature strives to more precisely define the concept of energy community and distinguish it from other renewable energy projects.[7] This has led to discussion on matters such as, among others: whether and to what degree the members of an energy community can be geographically dispersed; the extent to which projects are developed from the bottom up; models for decision-making; the level of citizen participation or local ownership; the degree of volunteer involvement; and the level of subsequent economic growth and job creation in the community. Consideration of these elements would seem to be a question of degree in determining where projects lie along the process and outcome parameters initially proposed by Walker and Devine-Wright (2008). Whether any of these elements alone or in combination are sufficient to qualify a project as an energy community project will not be debated here.

The principal objectives of an energy community are to encourage engagement in and support for a project or outcome, and the nature and level of community engagement among communities and community members will likely differ. A survey of energy communities in North America is not well served by a pedantic understanding of what constitutes an energy community. Too rigid a framework may eliminate consideration of models and approaches to community involvement that have not emerged elsewhere. Rather than relying on a precise definition, the analysis in this chapter reviews several emergent energy community business models as a means for identifying energy community activities in North America.

3. Energy communities in North America

Energy communities can take many forms and have adopted numerous business models, some of which are only now emerging. Reis et al. (2021) surveyed energy community business models in Europe and grouped them into eight broad archetypes: energy cooperatives, community collective generation, third-party sponsored communities, community energy service companies, community flexible aggregation, prosumer communities, local energy markets, and e-mobility cooperatives. Intentional communities, sometimes called ecovillages, also merit attention as a ninth business model that was not specifically cataloged by Reis et al. (2021).[8] These nine business

[7] The chapter by Rossetto et al. explores the characteristics of energy community.

[8] Although not mentioned in their rubric, arguably it may have been considered by Reis et al. (2021) as a subset of another business model archetype.

model archetypes provide a framework for the examination of energy communities in North America.

The survey of energy communities below sets out the fundamental characteristics of each business model archetype in the North American context: key participants; key activities; revenue streams; and key policy needs. The chapter by Benedetti and Stagnaro provides a table setting out similar characteristics for energy communities in Denmark and Germany. For each of the business model archetypes, an overview of the adoption of each model in North America is provided, along with some illustrative examples. The prosumer communities' model and the local energy market model are considered together in the analysis, recognizing that they are often closely related.[9] The objective is not to necessarily populate every business model archetype or to ensure that the North American energy community examples are perfectly aligned to any given model. The models are used as guideposts for this initial cataloging of North American energy communities.

3.1 Energy cooperatives

Energy Cooperatives, the most common form of energy community in Europe, are member-owned entities that invest in energy systems either for profit or as a nonprofit enterprise under a generally democratic structure with shared economic participation (see also Box 7.1).[10]

Box 7.1 Energy cooperatives

- Key participants: Residents, business owners, faith-based or social community members, or other members of a community of place. May be in the form of a member-owned large-scale regulated utility or small-scale neighborhood-based entity.
- Key activities: Invest in systems for the provision of energy to members in a defined geographic area, either for profit or as a nonprofit enterprise.
- Revenue stream(s): Where the cooperative is a regulated utility, revenue is usually from regulated tariffs for utility services. Generally, utility-scale cooperatives offer profit sharing among members. Small-scale neighborhood-based cooperatives may require a subscription on a pay-as-you-go basis or as a one-time contribution. In some cases, subscriptions are at no cost. Small-scale neighborhood-based cooperatives usually offer tariff rebates, credits on utility bills, or shared savings.
- Key policy need(s): When the cooperative is utility scale, regulatory and policy provisions are required to support investment in renewable energy. When the cooperative is a small-scale neighborhood-based entity, policies are required to enable neighborhood or community-scale energy systems. Net metering, net billing, or virtual net metering is also generally required.

[9] Reis et al. (2021, p. 9) note that local energy markets are typically developed as an extension of prosumer communities.

[10] The chapter by Spasova and Braungardt provides examples of cooperatives in several EU member states. The chapter by Chateau Terrisse et al. looks at an innovative multistakeholder cooperative.

There is a significant and lengthy history of utility-scale cooperative enterprises in North America. Electrification of much of rural western Canada is still provided by rural electrification associations,[11] and the United States has hundreds of member-owned utilities, together accounting for 42% of the country's distribution line miles and 5% of U.S. electricity sales.[12] Among the utility-scale electricity cooperatives in the United States are 63 generation and transmission cooperatives that provide wholesale power to 832 distribution cooperatives. By 2020, utility-scale cooperatives provided over 11.4 GW of renewable energy to their members.[13]

Small-scale neighborhood-based member-owned energy projects are emerging in North America; most commonly in the form of community collective generation or as intentional communities as discussed below.

3.2 Community collective generation

Community Collective Generation refers to communities of place engaged in shared self-generation and storage, usually by way of roof top solar installations, often on multitenancy buildings (see also Box 7.2).

Community collective generation is appearing in North America, most notably in jurisdictions with regulatory regimes and a policy structure that facilitate their integration into the traditional utility structures. In the United States, the majority of community solar installations are in jurisdictions with policies that provide opportunities for low- and mid-income utility customers to participate in the solar energy market. As of 2020, about a third of states have policies that facilitate community solar, generally through legislation that creates a third-party market for community solar. This

Box 7.2 Community collective generation

- Key participants: Residents, business owners, faith-based or social community members, or other members of a community of place. Usually in the form of a small-scale neighborhood-based entity.
- Key activities: Invest in shared self-generation and storage usually by way of roof top solar installations, often on multitenancy buildings, or on nearby land in a defined geographic area. Investment by cooperative members is usually subscription based.
- Revenue stream(s): Subscriptions on a pay-as-you-go basis or as a one-time contribution are common, but subscriptions may be at no cost. Subscribers are usually offered tariff rebates, credits on utility bills, or shared savings.
- Key policy need(s): Policies are required to enable neighborhood or community-scale energy systems. Net metering, net billing, or virtual net metering is also generally required.

[11] See, for example, Alberta Federation of Rural Electrification Associations http://www.afrea.ab.ca.
[12] Refer to https://www.cooperative.com/maps-facts-figures/Documents/Co-op-Facts-and-Figures.pdf.
[13] Refer to https://www.cooperative.com/maps-facts-figures/Documents/Co-op-Facts-and-Figures.pdf.

legislation requires solar project developers and utilities to follow certain regulations to enroll customers in community solar installations. However, there are community solar programs outside of these states that are typically developed and managed by local municipal or cooperative utilities.[14]

Cooperative Energy Futures (CEF) in Minnesota is a not-for-profit entity that finances and develops community-owned and operated low-income accessible community solar arrays.[15] CEF focuses on roof-top systems in the 200–250 kW range. However, they are also engaged in a number of 1.3 MW installations in collaborative joint ventures with municipalities.

CEF employs residential solar group purchasing, known as "Solarize" programs throughout the United States, loans, and leasing models to fund community developments and attract customers through both upfront and pay-as-you-go subscriptions. CEF targets residential, cultural, and faith communities but fills out additional subscriptions from local businesses and organizations as well as interested parties from other communities if needed to make the project viable. Although CEF prioritizes access for low-income subscribers and renters, some projects include large subscribers such as local institutions that serve as host sites and may use up to 40% of the energy generated. CEF serves as the project developer and long-term facilities manager for "community solar gardens" that are owned and controlled by the subscribers to the project. CEF claims to deliver 6.5% annual returns on investment for subscribers who invest up front and average savings equivalent to 8% of electric utility bills for pay-as-you-go subscribers.

TREC is an Ontario, Canada, member-based not-for-profit developer of community-owned renewable energy projects, similar to CEF.[16] TREC provides investment and support services to cooperatives, indigenous communities, and social enterprise initiatives to develop renewable energy projects. TREC finances projects through community bonds. To date, over $60M has been raised and invested in projects either directly or indirectly to finance arm's length ventures. Currently in Ontario cooperatives account for about 1 GW of renewable energy funded through community ownership or participation.[17]

3.3 *Third-party sponsored communities*

Third-Party Sponsored Communities are usually utility or technology company initiatives that sponsor energy projects in collaboration with local communities where the company retains the generation assets and investment risk, usually in exchange for long-term PPAs with the community (see also Box 7.3). However, these entities may also engage in crowdsourcing from communities of interest or affinity groups to fund renewable energy projects.

[14] Refer to https://www.energy.gov/eere/solar/community-solar-basics.

[15] Refer to https://www.cooperativeenergyfutures.com.

[16] Refer to http://www.trec.on.ca.

[17] Refer to http://www.trec.on.ca/community-power.

Box 7.3 Third-party sponsored communities

- Key participants: Utilities, technology companies, land developers, or cooperatives.
- Key activities: Sponsor energy projects in collaboration with local communities or nongeographically contiguous communities of interest, usually on a subscription basis. Sponsor retains ownership of the generation assets and retains investment risk and compensates community members with energy savings.
- Revenue stream(s): Revenues to the sponsor are generally in the form of long-term power purchase agreements (PPAs). Subscribers are usually offered tariff rebates, credits on utility bills, or shared savings. In the case of crowdsourcing, investors are generally compensated by way of interest-bearing bonds.
- Key policy need(s): Facility to sell power under PPAs. Net metering, virtual net metering, net billing, or feed in tariffs (FITs) may be required.

Sacramento Municipal Utility District (SMUD) in California has a "Neighborhood SolarShare" program that installs solar panels on individual homes and apartment buildings, where SMUD retains ownership of the PV system and maintains it on behalf of the customer.[18] Customers are compensated in the form of a bill credit less a monthly fixed operating and maintenance charge. SMUD also offers a "Community Solar" program for nonprofit organizations and low-income housing.[19]

Numerous organizations in North America are crowdsourcing investment dollars for renewable energy projects. RaiseGreen in the United States provides a listing platform for clean energy projects looking for investment dollars from corporate and individual investors.[20] In 2018, RaiseGreen helped New Haven Community Solar complete the first fully crowd funded solar project in the United States. The 9.24 kW solar project provides discounted electricity to Columbus House, a nonprofit organization.[21]

TREC, mentioned above, invests in its own cooperative renewable energy projects though "SolarShare," a program that crowdsources investment dollars from its members to fund solar projects throughout Ontario. Solar projects range in size from 10 kW rural systems to 600 kW arrays on industrial rooftops and on nonarable land. To date, TREC has over 14 MW of solar installations in over 30 locations.[22]

Another form of third-party sponsored communities are intentional communities and ecovillages, discussed below. These ecologically planned communities are gaining popularity in North America. Southwoods is a 19-acre seniors and affordable housing project in Edmonton, Alberta, where natural gas-fired combined heat and power generators power the community's homes. Excess heat is routed to a thermal energy

[18] Refer to https://www.smud.org/en/Corporate/Landing-Pages/Neighborhood-SolarShares.

[19] Refer to https://www.smud.org/en/Going-Green/Community-Solar.

[20] Refer to https://www.raisegreen.com.

[21] Refer to https://www.raisegreen.com.

[22] Refer to http://www.trec.on.ca/wp-content/uploads/2016/06/TREC_Primer_Jun28_Approved_Final-LR.pdf.

exchange and storage system to heat residential units.[23] Also in Edmonton, Blatchford's district energy sharing provides centralized energy to all the buildings within the community, tying in renewable energy sources like geo-exchange, sewer heat recovery, and solar panels.[24]

3.4 Community energy service companies

Community Energy Service Companies are collaborative ventures between companies and communities to jointly create and operate energy service or renewable energy supply entities, whereby the participants share in any savings (see also Box 7.4).

Private–public partnerships bring real estate developers and commercial property owners together with renewable energy service providers and community stakeholders, such as municipalities or large institutions like college or corporate campuses, to develop renewable energy projects. These may be as a retrofit to existing buildings or as an element of new construction.

2030 Districts Network[25] is a not-for-profit organization that promotes energy efficiency projects through public–private partnerships in major urban centers throughout the United States and Canada. The approach engages the private sector to adopt energy efficiency alternatives, in association with municipal districts or institutional interests. 2030 Districts Network now accounts for some 520 million square feet of commercial space with a low carbon footprint in 23 cities.

Smart Sustainable Resilient Infrastructure Association (SSRIA) is an Alberta-based member organization that supports building projects that reduce GHG emissions. SSRIA also provides funding for these projects.[26]

Box 7.4 Community energy service companies

- Key participants: Utilities, technology companies, land developers, municipalities
- Key activities: Collaborative ventures between companies, building owners, developers, or municipalities to jointly create and operate energy service or renewable energy supply entities, whereby the participants share in any savings that accrue. Often as private–public partnerships.
- Revenue stream(s): Equipment supply revenue, installation costs, and maintenance fees to the utility or technology company. Savings on utility costs.
- Key policy need(s): Changes to municipal zoning may be required and policies that allow the interconnection of microgrids. Net metering, virtual net metering, net billing, or FITs may also be required.

[23] Refer to https://www.gssintegratedenergy.com/gss-smart-microgrid.
[24] Refer to https://www.edmonton.ca/city_government/utilities/blatchford-renewable-energy.aspx.
[25] Refer to https://www.2030districts.org.
[26] Refer to https://www.ssria.ca.

3.5 Community flexible aggregation

Community Flexible Aggregation refers to collective demand-side management arrangements between aggregators and a community of geographically dispersed customers through bilateral contracts for the delivery of flexible energy consumption, both dispatchable and nondispatchable, in exchange for lower energy bills (see also Box 7.5).

Aggregation pools the energy efficiency capabilities of multiple market participants to produce an energy offset. It is frequently a component of demand response services used by utilities or independent system operators.

Aggregators such as Enernoc,[27] Itron,[28] and CPower[29] have been in North America for decades, mostly targeting industrial and commercial customers with large potentially interruptible loads. However, as technology has entered the market allowing residential and small commercial customers to better manage their energy usage, often remotely either by the customer or by the aggregator, residential customers are joining the aggregation community.

The value of aggregators to independent system operators in the management of load on their grids is growing, as a nonwires alternative to both generation and grid management. By 2018, Itron alone had some 2000 MW of curtailable load in its portfolio. ISO New England and the New York ISO both allow aggregated demand response to participate in their markets.[30] U.S. Federal Energy Regulatory Commission Order 2222, issued in September 2020, calls for enabling distributed energy resources aggregators to compete in all regional wholesale markets.[31]

Box 7.5 Community flexible aggregation

- Key participants: Third-party aggregation entities, utilities, ISO's, industrial, commercial, or individual utility customers.
- Key activities: Demand side management arrangements between community aggregators and customers. Customers are compensated with a percentage of the market value of the demand side offset achieved by the aggregator.
- Revenue stream(s): Aggregator sells reduced demand to the utility or ISO at market prices and shares the proceeds with its customers.
- Key policy need(s): Technology to curtail demand at customer's sight remotely or to monitor the customer's own curtailment. Net metering, virtual net metering, and net billing may be required.

[27] Enernoc was recently acquired by Itron.

[28] Refer to https://www.itron.com.

[29] Refer to https://cpowerenergymanagement.com. Note that CPower was acquired by LS Power Equity Advisors, LLC in 2018.

[30] Refer to Association of Power Producers of Ontario, https://magazine.appro.org/news/ontario-news/5616-1534211061-aggregators-harnessing-powerful-drivers-for-innovation-and-savings.html.

[31] Refer to https://www.ferc.gov/media/ferc-order-no-2222-fact-sheet.

Community choice aggregation (CCA),[32] also called municipal aggregation and government energy aggregation in some states, is a program that allows local municipalities to procure energy from a supplier other than their franchise utility or generate power on behalf of their residents, businesses, and municipal accounts while still receiving transmission and distribution service from their existing utility provider. There is debate as to whether CCAs are energy communities. However, where local governments procure and deliver renewable energy to a community, they exhibit some of the characteristics of a community energy service company. CCAs are nonprofit, unregulated, and arguably customer owned.

Some CCAs establish new revenue streams to support local renewable energy programs. CCAs may serve as a gateway to community-centric renewable energy business models within their member communities. MCE, one of the largest CCAs in California, is investing in community rooftop solar paired with storage and adopting a distributed energy resources management system.[33]

In the United States, CCA is available in 10 U.S. states: California, Illinois, Maryland, Massachusetts, New Hampshire, New Jersey, New York, Ohio, Rhode Island, and Virginia.[34] However, several other states are currently considering legislation that would sanction CCAs.[35] California has the greatest penetration of CCAs, with some 11M customers in over 200 communities served by 24 CCAs (McDonough et al., 2021).[36,37] Fig. 7.1 provides a map of CCAs as of 2021.

3.6 *Prosumer communities/local energy markets*

Prosumer Communities are prosumers who engage in peer-to-peer trading relationships that benefit from asset purchases and financing at scale, participate in collective energy efficiency initiatives, and expand their market reach, often through creation of a local energy market. Local Energy Markets are prosumer communities working together to increase peer-to-peer trading and reduce power trading with external entities, allowing community members to control to whom they sell and from who they buy (see also Box 7.6).

Prosumer communities require the facility for peer-to-peer trading. Blockchain developers such as ConsenSys are now developing peer to peer trading platforms, such as Grid+, which uses blockchain to give consumers direct access to wholesale energy markets (Bora, 2021). The Rocky Mountain Institute and Grid Singularity have joined forces to launch Energy Web Foundation and create open source applications for

32 Community Choice Aggregation is distinct from Community Flexible Aggregation which is explained above.

33 Refer to https://www.mcecleanenergy.org.

34 Refer to www.leanenergyus.org.

35 Arizona, Colorado, Connecticut, Michigan, New Mexico, Oregon, and Washington.

36 Refer to https://cal-cca.org.

37 Maps of active CCAs in the U.S., among other programs referred to in this chapter, can also be found at https://ilsr.org/community-power-map.

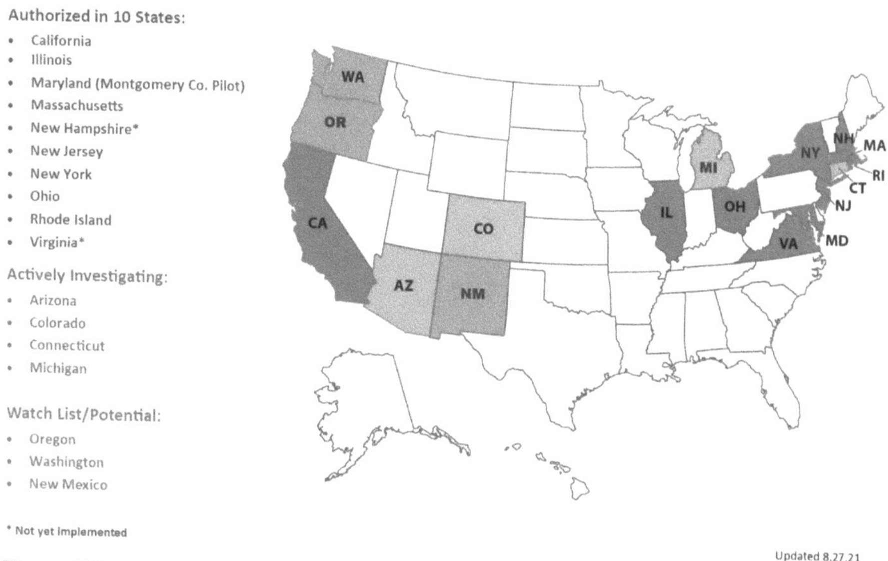

Figure 7.1 Map of CCAs as of 2021.
Source: www.leanenergyus.org.

> **Box 7.6 Prosumer communities/local energy markets**
>
> - Key participants: Residents, business owners, faith-based or social community members, or other members of a community of place.
> - Key activities: Peer-to-peer trading among members, asset purchases, and financing at a scale not available to the individual, collective energy efficiency initiatives, and potentially peer-to-peer trading or energy exports beyond the prosumer community.
> - Revenue stream(s): Savings on purchases of energy through peer-to-peer trading, bulk purchases of energy or energy-related equipment, the sale of aggregated energy, or aggregated demand side response. May form local energy markets to trade among themselves or with others.
> - Key policy need(s): The ability to engage in peer-to-peer trading behind the meter. Net metering, virtual net metering, and net billing may be required.

energy trading that allow "any energy asset owned by any customer to participate in any energy market."[38]

However, there are very few examples of prosumer communities or local energy markets appearing in North America, largely because the technology and regulatory structures for peer-to-peer trading are not yet well established (Deign, 2019). One of the first peer-to-peer local energy marketplace platforms in North America has been

[38] Refer to https://www.energyweb.org.

developed by LO3 Energy and Green Mountain Power (GMP) in Vermont.[39] The LO3 Energy trading platform, called Pando, is provided by GMP to its customers. The offering, called Vermont Green, connects local businesses such as ski resorts, craft breweries, and dairy farms with home solar owners who can market their excess energy production. LO3 has also developed the LO3 Brooklyn microgrid project, a marketplace that allows prosumers to sell their excess solar energy to New York City residents across the existing grid infrastructure.[40]

3.7 E-mobility cooperatives

E-Mobility Cooperatives are composed of members who share community public transportation, carpooling or ride sharing, usually with electric vehicles, often including vehicle to grid and grid to vehicle battery storage to take advantage of time of use rates. They may also include or connect to local renewable energy resources (see also Box 7.7).

At the time of writing, the author is not aware of any E-Mobility cooperatives in North America. However, the building blocks to support e-mobility cooperatives are beginning to appear. DCBEL has developed an EV charging system that uses solar power to charge an EV and allows the EV battery to be used as home back-up power.[41] Lion Electric, a developer of electric trucks and buses, recently announced a collaboration with Nuvve Corporation, a developer of vehicle-to-grid (V2G) technologies, to make all of its products V2G capable.[42] As a result, electric school buses in White Plains, New York, are now supplying electricity back to Con Edison utility customers as part of a V2G pilot project.[43]

Box 7.7 E-mobility cooperatives

- Key participants: Individuals, business owners, public entities, or other members of a community of place.
- Key activities: Use vehicles fueled by renewable energy to engage in ride sharing or carpooling to reduce GHG emissions. May include the use of battery storage and other means to benefit from time of use rates. May be used to optimize local renewable energy generation.
- Revenue stream(s): Savings on energy costs for transportation. Potential revenues are made available by using battery storage to produce and deliver ancillary services.
- Key policy need(s): The ability to engage in peer-to-peer exchanges and planning for carpooling and ride sharing. A pool of EVs and access to EV charging facilities.

[39] Refer to https://lo3energy.com.
[40] Refer to https://www.brooklyn.energy.
[41] Refer to https://www.dcbel.energy/blackout-power.
[42] Refer to https://thelionelectric.com/img/medias/Lion%20Nuvve%20PR_Final%2020201118.pdf.
[43] Refer to https://thelionelectric.com/img/medias/LION_Press_Release_White%20Plains%20EN%20FINAL.pdf.

3.8 *Intentional communities and ecovillages*

Intentional Communities or Ecovillages are communities with significant localized planning of buildings and local energy systems, focused on energy efficiency and renewable energy use for dwellings, transportation, food production and consumption, and waste management (see also Box 7.8).

One aspect of energy communities is local planning of energy systems, including integrating energy equipment micrositing and local distribution into the community. There are many variations in intentional community organization and design, and intentional communities exist in some form on every continent. They range from homeowner and condominium associations through coop-housing, retirement villages, and communities formed around philosophical, political, religious, or other world views.[44]

A common aspect of intentional communities is broadly participatory decision-making about aspects of architecture, engineering, building layout, construction techniques, and more (Barani et al., 2018). Many intentional communities focus on long term sustainability, equity, and inclusion by carefully integrating cultural, social, and economic aspects of community design and organization with the natural environment and conservation goals (Singh et al., 2019).

One study of North American ecovillages showed that residents use half as much total energy as other nearby residents, because of differences in energy use intensity for homes, transportation, food consumption, and waste disposal (Daly, 2017).

A variation of intentional communities is developer-led ecovillages, a form of third-party sponsored communities discussed above (Maguire, 2017). A current example from Ann Arbor, Michigan, is a planned neighborhood development project called

Box 7.8 Intentional communities/ecovillages

- Key participants: Residents, community planners and designers, builders, developers, lenders, local construction and energy code officials, and utilities.
- Key activities: Highly localized planning of buildings and local energy systems. Often focused on energy efficiency and renewable energy use for households, transportation, food production and consumption, and waste management. Any and all aspects of communal living can be included.
- Revenue stream(s): Highly variable depending on community organization and management styles and systems. Residents often buy into intentional communities.
- Key policy need(s): Ability for residents to engage in self-determination and design and implementation of what are often innovations in energy systems. Policy changes are often required to enable neighborhood or community-scale energy systems.

[44] See, for example, Foundation for International Community, https://www.ic.org/; and, Global Ecovillage Network North America, https://ecovillage.org/region/genna/.

"Veridian at County Farm."[45] Developers are building an all-electric net-zero community, applying comprehensive sustainable building design and construction practices developed by the International Living Future Institute, known as the Living Building Challenge.[46] The developer and electric utility company, DTE, are working together to treat this development as a pilot project to demonstrate microgrid and grid-interactive community capabilities and nonwires solutions (Dyson et al., 2018).[47] They are collaborating to integrate all-electric homes, electric vehicle infrastructure, demand response, and on-site solar energy and battery storage.

3.9 Observations on energy communities in North America

As the foregoing illustrates, many forms and types of energy communities are emerging in North America, mostly where legislative and regulatory frameworks promote, or at least allow, their development. Although enabling legislation is not a prerequisite for energy communities,[48] 84% of community solar projects in 2020 were in states with legislation promoting such projects (Heeter and O'Shaughnessy, 2021). Clearly, policy support and a regulatory regime that facilitate the adoption of energy communities make a difference. These elements are discussed in the next section.

The dominant policy approach in North America relies largely on engaging existing institutional players and new entrants through economic incentives, often by creating a third-party market for the development of community projects or by engaging utilities directly in community generation. This market-based approach tends to reduce the tensions between utilities and energy communities because utilities are encouraged to profitably engage with communities and other market participants in local energy projects. As referenced above, some utilities like SMUD and GMP are on the forefront of local generation and energy trading. However, there is still significant resistance among utilities to policies that are seen as encroaching on the monopoly franchise, as discussed below.

Nonetheless, community solar in the form of neighborhood-based shared solar installations is gaining popularity. Community members invest in these off-sight PV systems, usually on a subscription basis, and receive benefits most frequently in the form of bill rebates or credits. In 2020, there were at least 1184 MW of community solar capacity distributed across 811 projects in 39 states and Washington, D.C. Over 90% of community solar, measured in MW, is clustered in 10 states,[49] with Minnesota and Massachusetts accounting for 66% (Heeter and O'Shaughnessy, 2021).

45 Refer to Veridian at County Farm, https://thrive-collaborative.com/veridianatcountyfarm.

46 Refer to International Living Future Institute, *Living Building Challenge*, and *Living Community Challenge*, https://living-future.org/lbc/ and https://living-future.org/lcc/.

47 See also U.S. Department of Energy, Office of Energy Efficiency and Renewable Energy, *Grid-Interactive Efficient Buildings*, https://www.energy.gov/eere/buildings/grid-interactive-efficient-buildings.

48 There are about 279 MW in 129 projects located in 23 states without enabling legislation.

49 Minnesota, Massachusetts, Georgia, Colorado, Texas, Florida, Arizona, Utah, Oklahoma, and Missouri.

A market for crowd-sourced investments, often by not-for-profit entities that also provide expertise and management of community projects, has also developed, offering investment returns to affinity groups interested in promoting renewables. This approach has also manifested itself in public–private renewable energy partnerships among business entities and community stakeholders, such as municipalities or large institutions. And the market for nonwires alternatives[50] through aggregation and other resources is expanding in North America.

Although there has been much discussion in the literature and media about the emergence of the prosumer, there is little evidence of prosumer communities developing in North America, largely because the policy and regulatory landscape has not evolved to provide for peer to peer trading.

The next section provides an overview of the policy landscape in North America.

4. The North American policy landscape

4.1 Overview of policy initiatives

Enabling legislation promotes the development of energy communities. 21 U.S. states and the District of Columbia have adopted policies that lend themselves to the development of such communities. Many of the policy initiatives in the United States are aimed at providing opportunities for low- and mid-income utility customers to participate in the solar energy market by fostering a community-based approach. Many of these policies achieve this objective by creating a third-party market for the installation of community solar facilities. Fig. 7.2 provides a snapshot overview of the states with enabling policies and their success in promoting community energy.

A few jurisdictions in Canada, most notably Alberta and Ontario, have also adopted legislative and regulatory policies to facilitate community engagement. For example, the Alberta Small Scale Generation Regulation allows distribution connected community generation to sell power into the provincial power pool.[51] Ontario has approved a feed-in tariff to promote cooperative renewable energy projects.[52]

Table 7.1 provides an overview of the major policy initiatives currently in North American jurisdictions, including the recipients of each program and the type of incentives provided.

[50] Nonwires alternatives are distributed energy resources, microgrids, demand side management, grid management technologies, and other alternatives that defer or replace the installation of transmission and distribution infrastructure.

[51] Refer to https://mccac.ca/2020/04/03/connecting-to-albertas-grid-with-community-generation, https://www.qp.alberta.ca/documents/Regs/2018_194.pdf.

[52] Refer to https://www.ontario.ca/document/renewable-energy-development-ontario-guide-municipalities/40-feed-tariff-program.

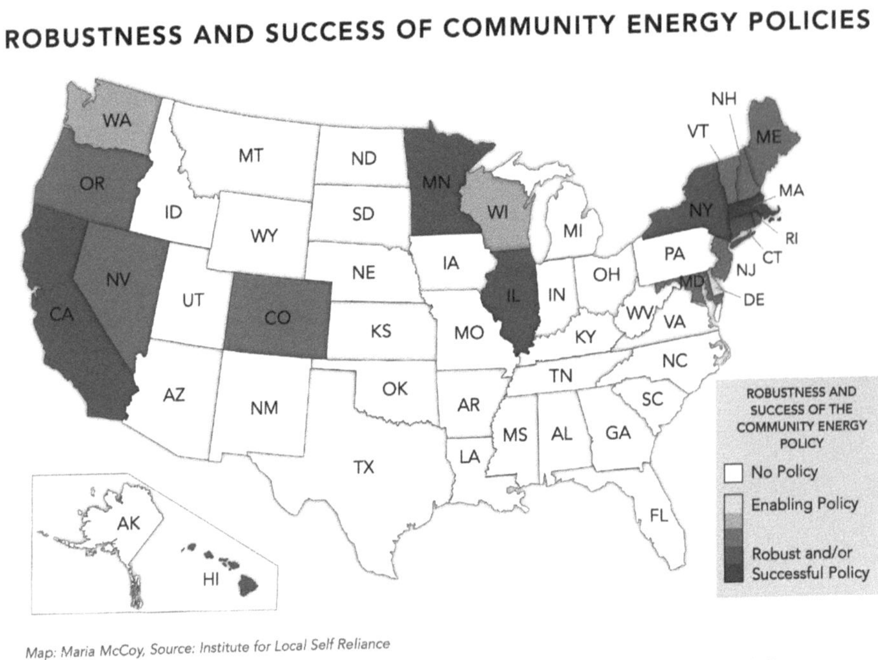

Figure 7.2 Robustness and success of community energy policies in the United States.

Table 7.1 North American renewable energy policies.

Jurisdiction	Policy initiatives	Recipients	Incentives
Alberta	• Micro generation regulation/load displacement installations under 5 MW/distribution connected. (see above) • Small scale generation regulation (see above) • Aggregation	• Businesses • Homeowners • Municipalities	
Arizona	• Solar community plans for residential, multifamily and commercial applications for low- and mid-income customers	• Low/mid-income • Multifamily	• Bill credits
California	• Solar on multifamily affordable housing program (SOMAH) • Disadvantaged communities—single-family affordable solar housing program (DAC-SASH) • Community solar green tariff program • Disadvantaged communities—green tariff program • Community choice aggregation	• Low/mid-income • Homeowners • Mutlifamily • Renters • Businesses • Organizations • Municipalities	• Rebates • Grants • Bill credits

Table 7.1 North American renewable energy policies.—cont'd

Jurisdiction	Policy initiatives	Recipients	Incentives
Colorado	• Colorado rooftop low-income program • Colorado community solar gardens	• Low/mid-income • Homeowners • Mutlifamily • Renters	• Grants • Bill credits
Connecticut	• Connecticut solar for all • Smart-E loans • Shared clean energy facilities program • EnergizeCT health and safety revolving loan fund • Low-income multifamily energy	• Low/mid-income • Homeowners • Mutlifamily • Renters • Businesses • Organizations	• Rebates • Financing
District of Columbia	• DC solar for all	• Low/mid-income • Homeowners • Mutlifamily • Renters	• Grants
Hawaii	• Green energy market securitization program (GEMS)	• Homeowners • Mutlifamily • Renters • Organizations	• Financing
Illinois	• Solar for all distributed generation • Solar for all community generation • Solar for all nonprofit organizations and public facilities • Community choice aggregation	• Low/mid-income • Homeowners • Renters • Organizations • Municipalities	• Rebates
Maine	• Low-income heat pump water heater rebate • Heat pump rebate	• Low/mid-income • Homeowners	• Rebates
Massachusetts	• Whole-home air–source heat pump pilot • Mass solar loan • Solar massachusetts renewable target (SMART) program • Community choice aggregation	• Low/mid-income • Homeowners • Mutlifamily • Renters • Businesses • Organizations • Municipalities	• Rebates • Financing
Maryland	• Community solar pilot project • The community solar LMI-PPA grant program • LMI energy efficiency grant program • Community choice aggregation	• Low/mid-income • Homeowners • Renters • Municipalities	• Rebates • Grants

Continued

Table 7.1 North American renewable energy policies.—cont'd

Jurisdiction	Policy initiatives	Recipients	Incentives
Michigan	• Michigan saves	• Households • Businesses	• Financing
Minnesota	• Connecting low-income communities through efficiency and renewable resources (CLICERSs) • Xcel Energy's solar × rewards program	• Low/mid-income • Homeowners • Mutlifamily • Renters • Businesses • Organizations	• Rebates • Financing
New Hampshire	• Low-moderate income solar program • Community choice aggregation	• Low/mid-income • Homeowners • Mutlifamily • Renters • Organizations • Municipalities	• Bill credits
New Jersey	• Community solar pilot program • COOL and WARM advantage programs • Community choice aggregation	• Low/mid-income • Homeowners • Mutlifamily • Businesses • Organizations • Municipalities	• Rebates • Grants
Nevada	• Lower income solar energy program (LISEP)	• Low/mid-income • Businesses • Organizations • Homeowners • Multifamily	• Rebates • Grants
Ohio	• Community choice aggregation	• Municipalities	
Ontario	• Feed in tariff • Aggregation	• Businesses • Homeowners • Municipalities	
Oregon	• Solar innovation grants • Solar within reach • Oregon community solar program	• Low/mid-income • Homeowners • Renters • Organizations	• Rebates • Bill credits
Rhode Island	• Community solar • Community renewables program • Community choice aggregation	• Low/mid-income • Homeowners • Renters • Municipalities	• Rebates • Financing

Table 7.1 North American renewable energy policies.—cont'd

Jurisdiction	Policy initiatives	Recipients	Incentives
Vermont	• Community solar for community action • Home energy loan	• Low/mid-income • Homeowners • Multifamily • Renters	• Rebates • Financing
Virginia	• Norfolk solar qualified opportunity zone fund • Community choice aggregation	• Low/mid-income • Businesses • Organizations • Municipalities	• Financing
Washington	• Evergreen sustainable development standard (ESDS)	• Low/mid-income • Homeowners • Renters	• Grants

Note: Table developed by the author.
Sources: https://www.cesa.org/resource-library/resource/directory-of-state-low-and-moderate-clean-energy-programs, https://www.aps.com/en/About/Sustainability-and-Innovation/Technology-and-Innovation/Solar-Communities, https://www.illinoissfa.com/programs/distributed-generation, https://energy.nv.gov/Programs/Lower_Income_Solar_Energy_Program, http://www.energy.ri.gov/community-solar, https://www.sevca.org/news/236-community-solar-for-community-action, https://www.norfolksolar.org/programs.

4.2 Metering and tariffs

Metering and billing practices have implications for the establishment of energy communities. If the members of a community or the community as a whole—e.g., microgrids—are to be compensated for generation that is exported to the utility grid, then the necessary metering and billing arrangements must be in place. The common approaches to metering and billing are net metering, net billing, and gross metering, also called buy all/sell all. An additional approach known as virtual net metering has also been developed in North America, for shared generation facilities not located on the customer's site. Virtual net metering prorates the net metered amounts to the individual subscribers to a shared generating facility.

Every state and the District of Columbia have metering policies intended to compensate distributed generators (Zientara, 2021). The jurisdictions set out in Table 7.1 with policies that enable community generation have generally adopted complementary metering practices. However, net metering and the tariffs paid for energy are controversial (Szmolyan, 2020).

Some jurisdictions have approved a FIT that allows distributed generators to sell their energy into the grid at established rates. Ontario, for example, has had a FIT since

2009, in addition to net metering.[53] In Canada, every province has a net metering policy, except Alberta which has adopted net billing.[54]

Net metering regimes will remain controversial for the foreseeable future. Net metering rates are generally intended to keep the utility whole with respect to transmission and distribution costs recovery. However, issues arise with respect to the amount of costs to be recovered in the net metering rates, and in some jurisdictions—e.g., California—rates intended to recover the utility's above-market costs associated with their power portfolios (PICA) are being challenged by consumer interests.

5. Conclusions

Energy communities are emerging in North America, mostly in the form of neighborhood-based shared solar installations, and the majority of these are in jurisdictions with policies and a regulatory framework that facilitate their establishment. It is evident that North Americans are motivated to engage in neighborhood-based renewable energy initiatives if given the opportunity, and a market-based approach that provides incentives for utilities and entrepreneurs to engage the public in these initiatives appears to be effective. Indeed, this might be the only way to achieve widespread adoption in the North American context. Unfortunately, in some jurisdictions, the advent of even small distributed solar PV systems has been greeted by regulatory backlash (Baker, 2019).

Utilities are unlikely to voluntarily seek alternatives to the current regime until their markets are significantly threatened, or they are offered an economic incentive to engage customers in renewable energy initiatives, as demonstrated in many of the examples in this chapter. Regardless, utility customers are being increasingly afforded the opportunity to bypass the utility's services altogether or to manage their consumption so as to circumvent the utility's tariffs; all of which threatens the utility's revenue streams. Policy makers and regulators will be urged to respond, hopefully in ways that encourage the further adoption of renewable energy, including energy communities. However, one of the challenges to widespread adoption in North America is the patchwork of jurisdictions. Every state in the United States and every province and territory in Canada has its own legislative and regulatory regimes. There is no overarching jurisdiction capable of promoting the adoption of polices that would encourage energy communities.

An issue policy makers and regulators will need to consider is whether energy communities are, or can be, welfare enhancing. In other words, how can energy communities improve welfare for society at large by lowering overall economic costs. This question is considered briefly in the introduction to this book. It is self-evident that community members see their energy community as welfare enhancing, otherwise

[53] Refer to https://www.ontario.ca/document/renewable-energy-development-ontario-guide-municipalities/40-feed-tariff-program.

[54] Refer to https://www.energyhub.org/solar.

they would not participate. But in this context, welfare may not be measured solely in economic terms. There may be nonmarket (values-based) incentives at play here. Whether broader societal welfare is enhanced by energy communities is a more complex question, particularly when the energy community is connected to the utility grid. Achieving broader societal welfare is largely a matter of pricing the grid so as to avoid subsidies or other market distortions. Costing and pricing the effects of energy communities on the grid will be complex and controversial. The controversy around net metering mentioned above is but one example. In addition, policy makers may also want to consider nonmarket valuations of the societal benefits accruing from energy communities. All of these matters are beyond the scope of this chapter and this book, but are important areas for further research.

Bibliography

Baker, S.H., 2019. The energy justice stakes embedded in the net energy metering policy debates. In: Krakoff, S., et al. (Eds.), Beyond Zero-Sum Environmentalism, Northeastern Public Law and Theory Faculty Research Papers Series No. 350-2019.

Barani, S., Alibeygi, A., Papzan, A., 2018. A framework to identify and develop potential ecovillages: meta-analysis from the studies of world's ecovillages. Sustain. Cities Soc. 43, 275–289.

Bora, I., 2021. [online] Available at: https://medium.com/digital-notice-media-labs/how-does-blockchain-technology-disrupt-the-energy-industry-7ffa44c8b05f. (Accessed 2 March 2022).

Candelise, Ruggieri, 2020. Available at: https://www.osservatorio-economie-emergenti-torino.it/images/slides_and_videos/aissec_2020/Candelise.Ott_2020.pdf. (Accessed 7 June 2021).

Daly, M., 2017. Quantifying the environmental impact of ecovillages and co-housing communities: a systematic literature review. Local Environ. 22 (11), 1358–1377.

Deign, J., 2019. [online] Available at: https://www.greentechmedia.com/articles/read/peer-to-peer-energy-trading-still-looks-like-distant-prospect. (Accessed 2 March 2022).

Dyson, M., Prince, J., Shwisberg, L., 2018. The Non-wires Solutions Implementation Playbook—A Practical Guide for Regulators, Utilities, and Developers. Rocky Mountain Institute. https://rmi.org/insight/non-wires-solutions-playbook/.

Heeter, J., O'Shaughnessy, E., 2021. Sharing the Sun: Understanding Community Solar Deployment and Subscriptions, NREL Webinar, Jenny Heeter, Eric O'Shaughnessy. Clean Kilowatts, LLC Gabriel Chan, University of Minnesota. April 28, 2020.

Maguire, M., 2017. Affordable, developer-driven ecovillages: meeting an unmet need. Communities Life Cooperative Cult. 174, 54–57. https://www.proquest.com/docview/1883487969.

McDonough, et al., 2021. Available at: https://cal-cca.org. (Accessed 14 June 2021).

Reis, I.F.G., Gonçalves, I., Lopes, M.A.R., Henggeler Antunes, C., 2021. Business models for energy communities: a review of key issues and trends. Renew. Sustain. Energy Rev. 144, 111013.

Singh, B., Keitsch, M., Shrestha, M., 2019. Scaling up sustainability: concepts and practices of the ecovillage approach. Sustain. Dev. 27 (2), 237–244.

Szmolyan, A., 2020. [online] Available at: Disunity Among the United States: Navigating Net-Metering Without Getting Electrocuted (pepperdine.edu). (Accessed 2 March 2022).

Walker, G., Devine-Wright, P., 2008. Community renewable energy: what should it mean? Energy Pol. 36 (2), 497−500.

Zientara, B., 2021. [online] Available at: https://www.solarreviews.com/blog/the-state-of-net-metering-usa-2021. (Accessed 2 March 2022).

Further reading

Afrea.ab.ca, 2021. REA NEWS | AFREA. Available at: http://www.afrea.ab.ca. (Accessed 7 June 2021).

APPrO, 2021. Aggregators Harnessing Powerful Drivers for Innovation and Savings. Available at: https://magazine.appro.org/news/ontario-news/5616-1534211061-aggregators-harnessing-powerful-drivers-for-innovation-and-savings.html. (Accessed 7 June 2021).

aps, 2021. aps—Arizona Public Service Electric. Available at: https://www.aps.com/en/About/Sustainability-and-Innovation/Technology-and-Innovation/Solar-Communities. (Accessed 7 June 2021).

Auc.ab.ca, 2021. Available at: https://www.auc.ab.ca/regulatory_documents/ProceedingDocuments/2018/23066-D01-2018.pdf#search=Proceeding%2023066. (Accessed 7 June 2021).

Brooklyn.energy, 2021. Available at: https://www.brooklyn.energy. (Accessed 14 June 2021).

Clean Energy States Alliance, 2021. Directory of State Low- and Moderate-Income Clean Energy Programs. Clean Energy States Alliance. Available at: https://www.cesa.org/resource-library/resource/directory-of-state-low-and-moderate-clean-energy-programs. (Accessed 7 June 2021).

COOPERATIVE ENERGY FUTURES, 2021. COOPERATIVE ENERGY FUTURES. Available at: https://www.cooperativeenergyfutures.com. (Accessed 7 June 2021).

Cooperative.com, 2021. Available at: https://www.cooperative.com/maps-facts-figures/Documents/Co-op-Facts-and-Figures.pdf. (Accessed 7 June 2021).

CPower Energy Management, 2021. The Bridge to Energy's Future. CPower Energy Management. Available at: https://cpowerenergymanagement.com. (Accessed 7 June 2021).

Crypto News Point, 2021. How Does Blockchain Technology Disrupt the Energy Industry? Available at: https://www.cryptonewspoint.com/how-does-blockchain-technology-disrupt-the-energy-industry/. (Accessed 7 June 2021).

dcbel, 2021. Turn Your EV into an Emergency Backup Power Source. dcbel. Available at: https://www.dcbel.energy/blackout-power. (Accessed 7 June 2021).

Digitalcommons.pepperdine.edu, 2021. Available at: https://digitalcommons.pepperdine.edu/cgi/viewcontent.cgi?article=1207&context=jbel. (Accessed 7 June 2021).

districts.org, 2021. Home | 2030 Districts Project Portal. Available at: https://www.2030districts.org. (Accessed 7 June 2021).

Edmonton.ca, 2021. Blatchford Renewable Energy. Available at: https://www.edmonton.ca/city_government/utilities/blatchford-renewable-energy.aspx. (Accessed 7 June 2021).

Energy - European Commission, 2021. Renewable Energy Directive. Energy European Commission. Available at: https://ec.europa.eu/energy/topics/renewable-energy/renewable-energy-directive/overview_en. (Accessed 7 June 2021).

Energy Web, 2021. Energy Web—Energy Web. Available at: https://www.energyweb.org/. (Accessed 7 June 2021).

Energygov, 2021a. Community Solar Basics. Available at: https://www.energy.gov/eere/solar/community-solar-basics. (Accessed 7 June 2021).
Energygov, 2021b. Grid-Interactive Efficient Buildings. Available at: https://www.energy.gov/eere/buildings/grid-interactive-efficient-buildings. (Accessed 7 June 2021).
Energy.nv.gov, 2021. Lower Income Solar Energy Program. Available at: https://energy.nv.gov/Programs/Lower_Income_Solar_Energy_Program. (Accessed 7 June 2021).
Energy.ri.gov, 2021. Community Solar. Office of Energy Resources, Rhode Island. Available at: http://www.energy.ri.gov/community-solar. (Accessed 7 June 2021).
Enernoccom, 2021. Technologies and Innovation for Smart Energy Solutions. Enel X. Available at: https://www.enernoc.com. (Accessed 7 June 2021).
FERC, 2021. Available at: https://www.ferc.gov/media/ferc-order-no-2222-fact-sheet. (Accessed 7 June 2021).
Foundation for Intentional Community, 2021a. Communities Magazine Index. Available at: https://www.ic.org/communities-index/. (Accessed 7 June 2021).
Foundation for Intentional Community, 2021b. Intentional Communities—Find, Join, and Learn about Intentional Community. Available at: https://www.ic.org/. (Accessed 7 June 2021).
Global Ecovillage Network, 2021. GENNA Alliance | Regenerative Communities in North America. Available at: https://ecovillage.org/region/genna. (Accessed 7 June 2021).
Greentechmedia.com, 2021. Peer-to-Peer Energy Trading Still Looks like a Distant Prospect. Available at: https://www.greentechmedia.com/articles/read/peer-to-peer-energy-trading-still-looks-like-distant-prospect. (Accessed 7 June 2021).
Grocoff, M., 2021. Available at: https://thrive-collaborative.com/veridianatcountyfarm. (Accessed 7 June 2021).
Gssintegratedenergy.com, 2021. Available at: https://www.gssintegratedenergy.com/gss-smart-microgrid. (Accessed 7 June 2021).
Illinois Power Agency, 2021. Available at: https://www.illinoissfa.com/programs/distributed-generation. (Accessed 7 June 2021).
International Living Future Institute, 2021a. Living Building Challenge. Living-Future.org. Available at: https://living-future.org/lbc. (Accessed 7 June 2021).
International Living Future Institute, 2021b. Living Community Challenge. Living-Future.org. Available at: https://living-future.org/lcc. (Accessed 7 June 2021).
Leanenergyus.org, 2021. Available at: https://www.leanenergyus.org/. (Accessed 7 June 2021).
LO3 Energy, 2021. The Future of Energy | LO3 Pando | Blockchain, Transactive Grids, Microgrids, Energy Trading | LO3 Tokens and Information | LO3 Energy. Available at: https://lo3energy.com. (Accessed 7 June 2021).
Municipal Climate Change Action Centre, 2021. Large Solar Projects in Alberta Connected to Energy Grid. Municipal Climate Change Action Centre. Available at: https://mccac.ca/2020/04/03/connecting-to-albertas-grid-with-community-generation. (Accessed 7 June 2021).
Norfolk Solar QOZ Fund, 2021. Programs—Norfolk Solar QOZ Fund. Available at: https://www.norfolksolar.org/programs. (Accessed 7 June 2021).
Ontario.ca, 2021. Available at: https://www.ontario.ca/document/renewable-energy-development-ontario-guide-municipalities/40-feed-tariff-program. (Accessed 7 June 2021).
Qp.alberta.ca, 2021. Available at: https://www.qp.alberta.ca/documents/Regs/2018_194.pdf. (Accessed 7 June 2021).
Rahimi, M., energyhub.org, R., 2021. Solar Power in Canada (Complete Guides 2021). energyhub.org. Available at: https://www.energyhub.org/solar. (Accessed 7 June 2021).
Raise Green, 2021. Raise Green. Available at: https://www.raisegreen.com. (Accessed 7 June 2021).

RMI, 2021. The Non-wires Solutions Implementation Playbook. RMI. Available at: https://rmi.org/insight/non-wires-solutions-playbook/. (Accessed 7 June 2021).

Sevca.org, 2021. Community Solar for Community Action. SEVCA. Available at: https://www.sevca.org/news/236-community-solar-for-community-action. (Accessed 7 June 2021).

Sherry, J., 2019. The impact of community sustainability: a life cycle assessment of three ecovillages. J. Clean. Prod. 237, 117830.

SMUD, 2021a. Available at: https://www.smud.org/en/Corporate/Landing-Pages/Neighborhood-SolarShares. (Accessed 7 June 2021).

SMUD, 2021b. Available at: https://www.smud.org/en/Going-Green/Community-Solar. (Accessed 7 June 2021).

Solar Reviews, 2021. The State of Net Metering in the United States in 2021. Available at: https://www.solarreviews.com/blog/the-state-of-net-metering-usa-2021. (Accessed 7 June 2021).

SSRIA, 2021. SSRIA. Available at: https://www.ssria.ca. (Accessed 7 June 2021).

Thelionelectric.com, 2021a. Available at: https://thelionelectric.com/img/medias/Lion%20Nuvve%20PR_Final%2020201118.pdf. (Accessed 7 June 2021).

Thelionelectric.com, 2021b. Available at: https://thelionelectric.com/img/medias/LION_Press_Release_White%20Plains%20EN%20FINAL.pdf. (Accessed 7 June 2021).

Trec.on.ca, 2021a. Available at: http://www.trec.on.ca/wp-content/uploads/2016/06/TREC_Primer_Jun28_Approved_Final-LR.pdf. (Accessed 7 June 2021).

Trec.on.ca, 2021b. TREC | Renewable Energy. Available at: http://www.trec.on.ca. (Accessed 7 June 2021).

Trec.on.ca, 2021c. What Is Community Power? TREC | Renewable Energy. Available at: http://www.trec.on.ca/community-power. (Accessed 7 June 2021).

wwwitroncom, 2021. Home | Itron. Available at: https://www.itron.com. (Accessed 7 June 2021).

Energy communities: challenges for regulators and policymakers

8

Darryl Biggar[1,*] *and Mohammad Reza Hesamzadeh*[2]
[1]Australian Competition and Consumer Commission, Melbourne, VIC, Australia; [2]KTH Royal Institute of Technology, Stockholm, Sweden

1. Introduction

The June 2019 EU "Clean Energy Package" introduced into EU law the concept of "Energy Communities."[1] The term "energy community" does not have a precise definition. Different authors use the term in different ways.[2]

We define an energy community as a group of electricity producers or consumers (potentially with their own generation or storage), each separately metered, which are granted the ability to reallocate, for billing purposes, their metered consumption or production within the group at a given point in time. Under this definition, the production observed at any one meter within the group can be used to offset the consumption at another meter within the group at the same point in time.

This reallocation is sometimes described as *virtual trade* or *virtual exchange* of electricity within the group. The word "virtual" refers to the fact that the change in production/consumption is only for billing purposes; underlying flows on the network remain the same. This definition captures a variety of arrangements, including many forms of community-owned generation and storage, so-called "virtual net metering," "virtual storage," and peer-to-peer trading.

As an example, under Brazil's "net metering" arrangements, generation measured at one meter creates "credits" which can be used to offset the consumption at another meter, whether that second meter is owned by the same customer (referred to as "remote self-consumption"), on the same site as the first meter ("aggregate net metering"), or in a completely different location ("community net metering").[3] In the EU, the same concept is referred to by CEER as "collective self-consumption" or "virtual

* The views expressed here are those of the authors and do not reflect the views of Australian Competition and Consumer Commission or the Australian Energy Regulator.

[1] Strictly, the legislation introduces two new terms: Renewable Energy Communities and Citizen Energy Communities. These terms can be found in the revised Renewable energy directive (2018/2001/EU) and the Directive on common rules for the internal electricity market (EU, 2019/944). See the summary of the EC and Interreg (2018), Caramizaru and Uihlein (2020).

[2] The chapter by Rossetto et al. sets out a typology of uses of the term "energy community." Many arrangements, which have been referred to as an "energy community," such as the joint, cooperative, or club ownership of a wholesale-connected generation asset, raise no particular public policy concerns. CEER (2019) refers to these as "unproblematic from a regulatory point of view."

[3] See Vieira et al. (2016) and the chapter by Hochstetler and Born.

Energy Communities. https://doi.org/10.1016/B978-0-323-91135-1.00002-X

sharing over the grid."[4] Some European states have enacted legislation to enable collective self-consumption.[5]

Using this definition, an energy community is financially beneficial to the members if and only if there are differences in the price faced at the margin by different members of the group—in particular, differences in the retail price paid by members of the group who are net exporters and members who are net importers.[6] Where there are differences in retail prices across members of the group, there are two possible sources of gains to the members of the group:

1. Reallocation of consumption from customers facing high prices to customers facing low prices ("arbitrage" across retail prices differences) resulting in a reduction in the total retail payments and
2. Improved allocation of generation and load within the members of the group (that is, improved efficiency of "dispatch" within the group).

Advocates of energy communities also point to other potential benefits from the establishment of an energy community:

- Allowing retail customers without access to generation sites (such as access to roof space or land area) the potential to obtain the same benefits of embedded generation that are enjoyed by other local residents, promoting a sense of fairness and equity, and building a sense of community and social cohesion.
- Promoting efficiency by encouraging investment in embedded generation at a larger (more efficient) scale.
- Promoting the take-up of renewable generation by enhancing the benefits received by owners of embedded generation.

These are valid benefits to specific customers, but good public policy practice focuses on the impact on economic welfare for the society as a whole. Do energy communities give rise to net social benefits from the perspective of the sector as a whole? This chapter seeks to address the following questions:

1. Under what circumstances does creating an energy community improve overall economic welfare?

[4] See CEER (2019) and the chapter by Lo Schiavo et al.

[5] CEER (2019): "Some Member States, such as France and Austria, have developed a framework for collective self-consumption, where energy can be shared within a group of customers, without requiring the direct involvement of a supplier. With the new provisions from the [Clean Energy Package], this kind of direct sharing of electricity will become a right, without necessarily requiring active involvement from the supplier of the remaining electricity."

[6] Here, the retail price refers to the change in revenue associated with a small change in consumption. As explained further below, this can be different to the nominal retail (per kWh) charge the customer faces. To be clear, we are not saying that there is no benefit from collective action by members of community in the case where members face the same price at the margin at all times. However, in that case, collective action does not take the form of an energy community. For example, a remote community may choose to install a large generator to take advantage of economies of scale, but, as long as this generator faces the same price as other members of the community, there is no additional financial benefit from allowing for the virtual exchange of power, which is the foundation of the energy community concept as defined here.

2. Does creating an energy community improve the efficiency of the incentives to invest in local generation?
3. Is the concept of an energy community compatible with the possibility of congestion on the local distribution network?
4. Is allowing the creation of an energy community compatible with historic retail tariff structures?

As explained further below, in most jurisdictions, retail tariff structures are not efficient.[7] Consequently, individual retail customers make inefficient decisions regarding when to use their own (on-site) generation and consumption assets, and when to invest in new generation and consumption assets.

Allowing the creation of energy communities amplifies and extends the consequences of inefficient retail tariffs. This generally reduces overall economic welfare. In addition, allowing the creation of an energy community may reduce the ability of the distribution network to charge different tariffs to different customers at different network locations, which may undermine the ability of the system to manage network congestion. Energy communities may also undermine the ability of retailers to offer certain tariff structures. This analysis suggests that rather than creating energy communities, policymakers should focus on reform of retail tariff structures. With efficient retail tariff structures in place, energy communities would offer no economic benefit.[8]

This chapter has three main sections:

- Section 2 introduces some key ideas, such as the definition of an energy community and related concepts such as "virtual net metering" and "peer-to-peer trading." We illustrate the operation of an energy community using simple worked examples.
- Section 3 highlights the regulatory problems and challenges arising from energy communities.
- Section 4 discusses various policy responses and concludes.

2. Preliminaries

2.1 The definition of an energy community

What, exactly, is an energy community? The term "energy community" is defined relatively broadly in the European legislation[9], and historically has been used to refer to a wide range of arrangements. For the purpose of this chapter, the key defining feature of an energy community is the ability to virtually share power within the community as though that shared power is "behind the meter," with the effect of changing the retail payments made by each member of the community.

[7] See Next10 (2021), Joskow and Wolfram (2012), Borenstein (2009).

[8] There could be other noneconomic benefits, such as promoting community cohesion.

[9] See Directive (EU) 2019/944 of the European Parliament and of the Council of 5 June 2019 on common rules for the internal market for electricity and amending Directive 2012/27/EU (Text with EEA relevance.), article 2.

A typical example is the concept of "net metering" as it is applied in Brazil.[10] As noted above, in Brazil, generation (exports) at one metered location creates "credits" which can be used to offset consumption (imports) measured at another meter, for billing purposes.[11] This concept can be generalized slightly to allow *any* customer (whether a producer, consumer, or storage facility) to increase their net consumption at a point on the network, thereby creating "credits" which can be used to reduce the net consumption at another metered point on the network, for billing purposes.

This chapter uses the following definition:

> *An* energy community *is a group of electricity retail customers (whether net producers or net consumers) who are each separately metered, and who operate under a separate retail contract, but who are permitted to virtually reallocate their total metered consumption or production between the members of the group for retail billing purposes, at each point in time, and to make internal financial transfers within the group.*

This definition is a somewhat narrower definition of energy communities than the broad taxonomy proposed in the chapter by Rossetto et al. Using that taxonomy we focus on just one type of energy community—an energy community that allows the virtual sharing of power among the members of the community.

This definition covers a wide variety of arrangements, including the following scenarios:

- An apartment building with separately metered apartments installs a solar PV facility on the roof, the output of which is allocated to offset the consumption of each apartment individually (as though that output is behind-the-meter for each apartment).[12]
- A large business, operating over several geographically separated sites, installs solar PV on one of its sites and uses the output of that solar PV to offset its metered consumption at another site.[13]
- A city council installs a large battery storage facility; residents of the local community are allowed to store their net exports (from rooftop solar PV generation) in that facility and to draw down that storage in a way that offsets their consumption (as though that output is behind-the-meter) at a later time.

[10] See Vieira et al. (2016) and the chapter by Hochstetler and Born.

[11] Kenning (2015): "Under the revision, Brazil now has 'virtual net metering', which means any company or consumer can install an energy system at different points of electricity use and still get credits, which can be used to abate consumption costs on another unit. [T]his 'landmark' revision from ANEEL creates a new business opportunity; now any group of clients can invest together in a single PV system for example and receive a share of the electricity generated and reduce their consumption from the grid, at a level proportional to the financial resources each entity has invested in the system."

[12] This is described as "aggregate net metering" in Vieira et al. (2016) and "shared generation" in the chapter by Hochstetler and Born.

[13] This is described as "remote self-generation" in the chapter by Hochstetler and Born and as "remote self-consumption" in Vieira et al. (2016).

- A local community establishes a "peer to peer" trading mechanism for electricity which allows for virtual transactions among members of the community, where some customers sell their surplus power (and are compensated) and other customers purchase power and use that purchased electricity to offset their own consumption (as though that purchased electricity is behind-the-meter).

For concreteness, this definition does not cover (a) local generation or storage which is located behind the meter for an individual customer (even if that generation or storage is community owned) and (b) local generation or storage which is compensated at its local metered rate (even if that generation or storage is community owned).[14]

This chapter focuses on the case where the energy community allows for the reallocation of metered consumption or production *at a specific point in time*. But it is possible to envisage allowing for reallocation of metered consumption or production *across time*. Such an energy community would allow for intertemporal arbitrage of differences in the retail price at the margin.[15] For the purposes of this chapter, this possibility is put aside.

2.2 Price differences as the driver of energy communities

At each point in time, each member of the community faces a price for producing or consuming an additional unit of electricity. This price is the change in the total retail bill of the customer in response to a small change in production or consumption. This will typically be the retail price ($/kWh) for a net consumer, or the feed-in tariff for a net exporter.[16]

Under the definition set out above, the primary effect of the energy community is to allow production or consumption to be shifted among members for billing purposes. As shown in the Appendix, this yields financial benefits for the members of the community if and only if members of the community face a different effective price at the margin.

[14] Many of the projects undertaken by community solar groups in Australia appear to fall into this category; see Community Solar, Coalition for Clean Energy, and Local Power Plan. This includes community initiatives to construct a centralized generation or storage facility, as well as "group purchasing" arrangements where members of a community might jointly agree on a supplier to provide rooftop solar to each of them individually. Many renewable generation projects offer investment opportunities for the local community (e.g., selling a proportion of the shares to the local community) or pay dividends or compensation to the local community, but such initiatives do not constitute energy communities under the definition above. As an example, see the Victorian government booklet on Community Energy Projects.

[15] Mejdalani et al. (2018) suggest that this is possible under net metering in Brazil.

[16] In some cases, the relevant price is very different, as in the case where the customer's retail bill depends in part on peak demand—in this case, the effective retail price at times approaching peak demand will typically exceed the standard retail price by a large margin. The same effect may arise where the customer purchases a fixed level of capacity and pays a penalty for exceeding that capacity.

Differences in retail prices across neighboring customers could arise from several different sources:

- Differences in the retail price paid for net exports versus net imports.[17] This difference in the retail price for imports and exports potentially gives rise to differences in the retail price (at the margin) faced by different customers at the same point in time, especially in the following circumstances:
  - When there are differences in the opportunities to invest in embedded generation across the members of the community[18] and
  - When there are economies of scale in generation, so that it is more efficient to install a single, larger generation facility than an equivalent capacity in a number of smaller-scale facilities.
- Differences in the nature of retail contracts, giving rise to differences in retail prices faced by neighboring customers at the same point in time (e.g., fixed-rate vs. time-varying, or differences in the extent to which retail contracts pass through variations in network charges).
- Differences in consumption patterns for customers on tariffs where the prices vary with volume (such as so-called "block" tariffs).
- In the case of tariffs based on peak demand, differences in the timing of peak demand across retail customers.
- Differences in the cost of serving customers arising from real, physical differences in the level of losses, or congestion, on the distribution network.[19]

2.2.1 Example 1: simple two-member community

To illustrate the impact of an energy community, consider an energy community consisting of two members, a generator facing a price P_G and a load facing a price P_L. The variable cost of the generation is assumed to be below P_G and the marginal value of the load is assumed to be above P_L. Before the energy community is created, the generator is assumed to be producing at the rate G and the load is consuming at the rate L. The net payments to the retailer are therefore $P_L L$ for the load and $-P_G G$ for the generator.

With the energy community in place production L and generation G remains the same, but, for billing purposes, a volume Δ of the output of the generator is used to create "credits" which are applied to offset the consumption of the load. The net payments to the retailer are now $P_L(L-\Delta)$ for the load and $-P_G(G-\Delta)$ for the generator.

[17] For example, in Australia, retail customers who are net importers typically pay a retail price in the range 20–30 c/kWh (the precise amount depends on the nature of the retail contract), but retail customers who are net exporters are typically paid the feed-in tariff (roughly 7–9 c/kWh) or the wholesale spot price. This is primarily due to the structure of distribution network charges. In jurisdictions with "net metering" rules, the price paid for export (the feed-in tariff) is the same as the price paid for import (the normal consumption tariff).

[18] This might arise, for example, due to differing access to roof space, or the different incentives of landlords and tenants.

[19] For example, at times of peak load on the distribution network, the system operator may increase the network tariff in the congested region in order to curtail load or elicit further supply, in order to balance supply and demand without the need for a network augmentation. This may lead to differences in the retail price faced at the margin by different members of the community.

Overall the payments to the retailer drop by the amount $(P_L - P_G)\Delta$. This is also the financial benefit (the gain) to the members of the community:

$$S = (P_L - P_G)\Delta$$

In other words, provided the price paid by load is higher than the price paid to generation, there is a potential financial benefit from reducing the output of the generation (for billing purposes) and using the difference to reduce the consumption of the load. The greater the volume that can be shifted in this way, the greater the financial gain to the members of the community. In this example, this gain comes at the expense of the retailer(s).

The distribution of this surplus depends on the level of any side-payments or transfer payments. Let us define the transfer payment to the load as $T_L = P'_L\Delta$ (this implicitly defines a price $P_L{}'$). Similarly, let us define the transfer payment to the generator as $T_G = -P'_G\Delta$.

It is easy to check that the generator and load are better off and the energy community at least breaks even if and only if the generator and the load face different retail prices at the margin, and the community increases the effective price to the generator and/or lowers the effective price to the load. These conditions imply that the four prices in this simple example must satisfy the following relationship:

$$P_G \leq P'_G \leq P'_L \leq P_L$$

Different arrangements for the energy community allocate the surplus S between the parties in different ways. For example, in the case where $P'_L = P_L$ and $P_G = P_G{}'$, all of the surplus accrues to the community-operator (and none to the members of the community—the generator and the load). The opposite case is the case where the community-operator earns no revenue $P'_G = P_L{}'$. In this case, all of the surplus accrues to the members of the community. But the allocation of the surplus to the generator versus the load is arbitrary. One extreme is the case where the generator is implicitly paid the original generator price (P_G, the feed in tariff) so that $P_G = P'_G = P_L{}'$. In this case, all of the surplus accrues to the load. The opposite case is where the generator is implicitly paid the customer's original retail price (P_L), in which case all of the surplus accrues to the generator.

This example can be extended to illustrate how an energy community may lead to an improvement in the efficiency of "dispatch." Specifically, let us assume that the load also has the ability to generate electricity on-site. The variable cost of the generation belonging to the generator is assumed to be C_G and the variable cost of the generation belonging to the load is assumed to be C_L. We will assume that the initial retail prices (before the creation of the energy community) and the variable costs satisfy the following relationship.

$$P_G \leq C_G \leq C_L \leq P_L$$

With this assumption, before the creation of the energy community, the load has an incentive to use its (more expensive) on-site generation, while the cheaper off-site generator will not be turned on. This is inefficient and gives rise to a potential welfare gain from the creation of the energy community.

Let us suppose an energy community is created, shifting an amount Δ of production from the load generator to the off-site generator and, at the same time, offsets that shift by moving the same amount of generation, for retail billing purposes, in the other direction. This leaves the retail payment for the generator and the load the same. However, this action reduces the production cost of the load by $C_L\Delta$ and increases the production cost of the off-site generator by $C_G\Delta$. The total increase in surplus is $(C_L - C_G)\Delta$.

As before this surplus can be shared between the parties according to the definition of the transfer payments. If we define the transfer payments as $T_L = P'_L\Delta$ and $T_G = -P'_G\Delta$, the surplus can be fully captured by the load if $P'_L = P'_G = C_G$, fully captured by the off-site generator if $P'_L = P'_G = C_L$ and captured by the community itself (as a separate entity) when $P'_L = C_L$ and $P'_G = C_G$.

2.2.2 Example 2: simple three-member community

To gain more intuition, let us consider a slightly more concrete scenario. In 2015, a local council in Australia sought to share power between "a large rooftop installation going on top of its sporting center, which has ample roof space and not a lot of consumption" and "the neighboring sewage plant, which has little roof space for a solar array, but has heavy energy needs." The argument of the council in support of the arrangement focused on the differences in the retail prices paid by the two facilities:

> *"Under current rules, the excess solar output from an array on the sporting center has to be sold back to the grid, for a payment of just 6c/kWh in NSW, if you are lucky (the payment is actually voluntary). But that output might be sold back by the retailer to the sewage plant at full commercial rates, which are likely more than 5 times that amount."*[20]

Drawing on this scenario, let us suppose there is a small utility-scale solar farm, producing 1 MW of energy, connected to the distribution network. It has two neighbors, a factory which consumes 900 kW and a municipal waste facility, consuming 300 kW. In the status quo, the solar farm is able to export to the distribution network, receiving the feed-in tariff of 5 c/kWh. The factory and the waste facility each pays the retail tariff of 20 c/kWh for the energy they consume. This scenario is illustrated in Table 8.1.

The difference in the retail prices faced by the different consumers gives rise to a motivation to create an energy community. The energy community allows 1000 kW of output of the generator to be virtually moved (perhaps in the form of "credits") from the 5 c/kWh feed-in tariff to the 20 c/kWh retail tariff, creating $150/h in private surplus to the members of the community.

[20] See Byron Shire Council virtual net metering pilot.

Table 8.1 A simple energy community with the surplus allocated in a way that leaves the loads better off.

Customer	Consumption (kWh)	Retail bill ($/h)	Virtual realloc. (kWh)	Consumption (after realloc.)	Retail bill (after realloc. Before transfers)	Final retail bill (after transfers)
Solar farm	−1000	$−50	1000	0	$0	$−50
Factory	900	$180	−800	100	$20	$60
Waste fac.	300	$60	−200	100	$20	$30
Total	200	$190	0	200	$40	$40

As noted above, how this surplus is allocated between the members of the community is a decision for the community. For example, we might allocate 800 kW of the output of the generator to the factory and 200 kW to the load. This might be associated with payments of $40/h from the factory and $10/h from the waste facility back to the generator so that it is left as well off as before. This results in $120/h of surplus being paid to the factory (reducing its retail bill from $180/h to $60/h) and $30/h being paid to the waste facility (reducing its bill from $60/h to $30/h), as illustrated in Table 8.1.

The activity of the energy community is best viewed as a form of arbitrage—shifting load from retail customers (producers or consumers) facing high retail prices to retail customers facing low retail prices. This arbitrage will continue as long as differences in the retail prices exist—to the point where all members of the community face the same retail price at the margin. At this point, the total surplus from the energy community is maximized.[21]

In this example, there is no benefit from changing the real production and consumption of the members of the community. The financial benefit to the members of the energy community comes exclusively from the reduced revenue for either the retailer or the distribution business.[22] As long as distribution network charges differ across customers at the same location on the network, distribution businesses stand to lose revenue from the establishment of an energy community and therefore could be expected to oppose such developments and/or seek changes in their tariff structure.

3. Regulatory challenges arising from energy communities

We have seen that energy communities can be privately beneficial for the members of the community, by allowing a form of arbitrage across price differences faced by different retail prices at the margin. But are energy communities beneficial for the economy as a whole? From a public policy perspective, does the creation of an energy community promote overall efficiency and equity?

3.1 Energy communities tend to amplify the impact of inefficient retail tariffs

It is widely recognized that retail prices for electricity typically differ widely from the theoretical ideal of the dynamic short-run marginal cost (SRMC), for several reasons:

21 This virtual arbitrage can be compared with the physical arbitrage that arises from expansion of the underlying network. To the extent that price differences between retail customers reflect congestion and losses on the underlying network, actions which reduce those price differences may improve welfare. Investment to physically improve connections between and within communities is a form of collective action which may lead to welfare improvements (where the network is not too costly).

22 Transmission connected generators and loads at a specific connection point face a price which is the same for net injections as for net withdrawals and therefore are not affected by virtual sharing arrangements within each connection point.

1. A typical retail electricity contract insulates customers from dynamic variation in wholesale spot prices. Instead retail customers typically face a time-averaged and geographically averaged retail price which may be above or below the wholesale spot price at any given point in time.[23]
2. Often, retail customers that are net exporters are paid a lower "feed-in tariff" which itself is often highly averaged over time and across locations.[24]
3. Distribution network charges typically do not reflect the SRMC of delivering electricity power over the distribution network. In particular, it is common for the variable component of distribution network charges to be time-averaged and above SRMC most of the time, providing a contribution to the fixed and common costs. It is rare for distribution charges to reflect distribution network congestion.[25]
4. Retail costs (which are also largely independent of consumption) are also often recovered through variable (per kWh) charges—further contributing to a divergence between the retail price and SRMC.[26]

These problems in the retail tariffs have consequences for end-customers' electricity usage and electricity investment decisions. Specifically, retail customers do not typically face the correct incentives regarding *usage* decisions (such as decisions as to when to charge an electric vehicle, or when to switch off an air-conditioning unit, or when to draw on electricity from a local storage). Neither do retail customers typically face the correct incentives regarding *investment* decisions in electricity consuming appliances. For example, retail customers typically have little or no incentive to invest in devices that increase consumption when wholesale prices are negative, but have overly strong incentives to invest in devices that consume electricity at peak times (such as large air conditioners).[27]

In particular, when retail prices are time-averaged and inflated above SRMC, retail customers have an incentive to invest in on-site embedded generation that is stronger than is socially desirable. This leads to overinvestment in on-site generation, an inefficient mix of generation in the power system overall, and inequities between those customers who have the facilities to install local generation and those that do not.

There are also other concerns. As noted above, distribution network tariffs are often structured in such a way that variable charges are above variable costs. In this context, a reduction in the volume of sales through the distribution network (due, for example,

[23] The retail tariff may also involve "block tariffs"; customers on higher block charges will face a price which is further removed from the SRMC.

[24] As noted earlier, this arises in Australia because of the asymmetry in distribution network charges. Distribution networks charge a positive amount for net imports, but do not make an equal-and-opposite payment for net exports at the same location.

[25] In principle, fixed costs should be recovered through fixed charges. It is common for electricity tariffs to have both a fixed component and a variable component. The fixed component should, in principle, be chosen to be high enough as to allow the variable component to only reflect variable costs.

[26] A recent study in California estimated that the retail price for electricity was double the estimated marginal cost for customers of Southern California Edison, and more than triple for PG&E and SDG&E. Next10 (2021).

[27] Next10 (2021) observe that "In California, overpricing electricity will inefficiently discourage some households from considering electrification of space heating, water heating, clothes drying, vehicle transportation, and other services that can switch between energy sources."

to the investment in local generation) reduces the revenue received by the distribution business more than the reduction in its costs. If the distribution network is to continue to break even, it must increase its charges, shifting costs onto customers who are unable to install local generation. In addition, by further increasing the retail charges relative to SRMC, it increases the incentive to invest in local generation, potentially contributing to a "death spiral."[28]

Although these inefficiencies of retail tariffs are well-recognized, there have historically been natural limits on the extent to which these inefficiencies could affect market outcomes. Many small customers have not had the willingness or ability to actively control their electricity production or consumption, or to invest in local production or consumption assets. Many customers cannot access rooftop solar at all (perhaps because they live in multi-unit buildings, or are tenants). Even for customers who have access to roof space, the available roof space is typically limited. Few retail customers can install hectares of solar panels, or large numbers of wind turbines. Most retail customers face a feed-in tariff which is much lower than the typical retail price paid for net imports, discouraging investment in embedded generation which exceeds the size of the on-site load. All of these factors limit the incentive to invest in, and the optimal size of, an on-site, embedded generator.

But, the creation of an energy community reduces these barriers and therefore greatly increases the incentives to invest in local, embedded generation, exacerbating the problems discussed above.

For example, consider the simple case in which each local consumer of electricity faces a flat retail price of, say, 30 c/kWh, and is paid a feed-in tariff of 10 c/kWh for each unit exported. Let us assume that each consumer has the option of installing an embedded generator which is, for the sake of argument, not socially efficient (that is, it does not form part of an overall optimal mix of generation plant in an ideal electricity system).

Let us assume that, as a stand-alone embedded generator, this generator would not be viable at the feed-in tariff of 10 c/kWh, but would be highly profitable if it were paid the retail price of 30 c/kWh. Let us also assume that, each customer is so small that, due to economies of scale (or lack of space), it is not viable for each individual customer to install their own embedded generator. In the absence of the energy community, no customers would choose to install on-site embedded generators. The inefficiency in the retail tariffs although present would have no adverse impact.

However, with the creation of an energy community, the embedded generator can be constructed on a much larger scale, overcoming the scale efficiency problems. Such a generator could be, in effect, paid the retail price (30 c/kWh) and could be highly profitable. Energy communities all over the country could seek to emulate this approach and set up their own embedded generators, even though such generators are socially inefficient.

[28] This point is well recognized in the literature: "[A]s more and more households adopt behind-the-meter (BTM) solar photovoltaic (PV) panels, cost recovery is disproportionately shifted onto the bills of solar nonadopters. … [A]s residential solar adoption increases, system costs are being recovered from a shrinking base." Next10 (2021), page 7. See also Vieira et al. (2016).

As long as retail tariffs are inefficient, allowing groups to form an energy community makes viable inefficient generation investments that would not be viable in the absence of the energy community. This exacerbates the problems of inefficient retail tariffs. A solution is tariff reform, as discussed further below.

3.2 Energy communities undermine typical distribution network pricing schemes

One of the primary motivations for the creation of an energy community is the arbitrage of retail price differences between the members of the community. One of the key sources of such price differences is differences in the distribution network tariff—especially differences between the distribution tariff charged when the customer is a net importer and when the customer is net exporter.

The creation of an energy community allows the arbitrage of such price differences, resulting in a lower total payment to the distribution network. If the distribution network must break even, this distribution network is forced to increase its prices, to restore the drop in revenue. Depending on how this increase is implemented, this can further exacerbate the problem of inefficient tariffs.

To illustrate, in Australia, distribution networks normally charge a flat time-averaged charge of approximately 8 c/kWh for net imports of power.[29] On the other hand, distribution networks neither charge nor pay compensation for net exports (i.e., the distribution charge for the export service is zero[30]). This difference between the distribution tariff for net exports and net imports can give rise to strong incentives to create an energy community and that energy community may result in a significant drop in revenue for the distribution network.

As an example, suppose that a group of consumers is consuming at the rate of 110 kW, and paying a distribution network charge of 8 c/kWh, resulting in a payment of $8.8 per hour. Now suppose that these consumers install a local generator producing 100 kW of output and, at the same time, create an energy community to virtually redistribute the output of this generator across the consumers. The total consumption of the consumers (after virtual redistribution) is 10 kW and the total payment to the distribution network is now $0.8/hour. The energy community has resulted in a substantial drop in the revenue received by the distribution network.

If the distribution business is to break even it will have to increase at least some of its tariffs. This increase in tariffs may further enhance the incentive to create an energy community and to invest in embedded generation, exacerbating the effect of inefficient tariff structures.[31]

[29] See https://www.energyaustralia.com.au/home/electricity-and-gas/solar-power/feed-in-tariffs.

[30] Rather than −8 c/kWh as theory suggests.

[31] The problem here is the inefficient structure of distribution network charges. Ideally distribution network charges would be symmetric for generators and loads (i.e., if loads are charged 15 c/kWh for extracting power from the network at a specific location, generators should be paid 15 c/kWh for injecting power at the same location).

This problem has been recognized in jurisdictions that have implemented virtual net metering. For example, in Brazil, Hochstetler and Born observe in their chapter that virtual net metering enhances the incentive to install embedded generation, thereby reducing the revenue of the distribution network: "The revenue shortfall is eventually picked up by the customers that do not participate [in] the net metering program (including the poorest consumers)." They also note the importance of more sophisticated tariffs in the form of locational and dynamic pricing. In 2019, the Brazilian regulator ANEEL proposed to reduce the size of the credit received for exported generation and to introduce a new distribution network fee for exported power.[32]

3.3 Energy communities undermine efficient handling of local network congestion

Energy communities exist to arbitrage differences in retail prices faced by different members of the community. This arbitrage tends to equalize the retail prices faced by retail customers at the margin, across the community.

But it is well known that circumstances can arise where it is efficient for local prices for electricity to differ across locations. This arises, for example, where the two locations experience different costs due to differences in losses, or differences in local network congestion.[33] In this context, the tendency of an energy community to equalize retail prices between members will actively undermine the achievement of efficient outcomes in the power system. It may also undermine the ability of the system operator to maintain the system in a secure operating state.

For example, where the local network is experiencing congestion in the export direction, the efficient local electricity price may be low (or even negative). But, in an energy community, if some customers face a lower electricity price than others, there is a strong incentive to virtually share electricity across the community. As we have noted above, this tends to increase the effective price facing customers at the low-priced location, and tends to reduce the effective price facing customers at the high-priced location. Customers at the low-price location will likely respond by increasing their net exports and customers facing a higher price will likely respond by reducing their net exports. But this is exactly the opposite of what is required to efficiently manage the congestion. The system operator will likely be forced to make an ad hoc intervention in the market to preserve system security.

The key problem here is that the virtual sharing implicit in an energy community is incompatible with the price differences that are necessary to reflect losses and congestion. If the losses and congestion are to be efficiently priced while retaining the energy community, the geographic scope of the energy community must be limited to those customers which face the same SRMC (that is, customers which are at the same pricing

[32] Bellini (2019).

[33] In order for this to arise, the two locations need to be far enough apart electrically in the distribution network. This allows us to define what we mean by "local" communities.

node in an efficiently priced market). Conversely, if the energy community is to be retained, it may be necessary to overbuild the local network to eliminate local congestion.[34]

4. Conclusions

The EU concept of "energy communities" has parallels in other jurisdictions in the form of "virtual net metering," "virtual power plants," and "peer-to-peer trading." The push to allow energy communities is driven, in part, by the perceived financial benefits to the members of the community. As long as energy consumers or producers in a local area face different retail prices at the margin, there arises an incentive to engage in "virtual" sharing of behind-the-meter production or consumption.

Energy communities may offer some overall social benefit. Energy communities may (in certain circumstances) allow for enhanced dispatch of production and consumption within the community itself. But this effect is likely to be small. Energy communities may also allow for greater exploitation of economies of scale in generation.

However, at the same time, a primary motivation for the creation of an energy community arises from the reduction in revenue paid to the retailer (and therefore also the distributor). This motivation will often arise even if there is no social benefit from the creation of the energy community at all.

From the public policy perspective, the analysis set out here suggests that when considering whether to establish an energy community, some caution is required. Doing so risks exacerbating and amplifying existing problems arising from the inefficiency of current retail tariff structures. In particular:

(a) In many cases, retail tariff structures give rise to inefficient incentives to install local (on-site) generation to offset local consumption. Energy communities, by allowing virtual sharing of power over a wide area, greatly expand the incentive to install embedded generation, potentially leading to an inefficient mix of generation investment.

(b) Energy communities, by allowing members of the community to arbitrage price differences across the community, reduce the revenue received by the distribution network, which likely forces the distribution network to raise its charges, exacerbating inequities and increasing incentives to form an energy community.

(c) Energy communities have the effect of reducing or eliminating retail price differences faced by different retail customers. But such price differences can be both efficient and essential for the secure operation of the power system. The elimination of such price differences can

[34] This concern about the incompatibility of virtual sharing and peer-to-peer trading with local network congestion was raised by CEER: "virtual energy sharing will only have a positive technical impact on network costs if it incentivizes its participants to change their consumption or production pattern in a way that is consistent with the needs of the system. This will only happen if consumption is actively managed by the participants of the sharing scheme and the physical limitations of the network or the power system is taken into consideration. This is not a trivial task and requires deep knowledge of the grid, as it implies accounting for the grid's technical current and voltage limits adjusted for real-time losses, in order to avoid network constraints." CEER (2019).

undermine the ability of the power system operator to efficiently and reliably manage the power system in the face of network constraints.

This analysis does not suggest that energy communities yield no benefits at all. As we have seen, energy communities can enhance the incentives for local (embedded) generation. Such generation can provide a range of benefits, such as reducing the need for network augmentation or reducing greenhouse emissions. The claim here is not that energy communities yield no benefits, but that they exacerbate the existing tendency toward inefficient investment, raising the overall costs of the power system, and shifting some of those costs to consumers outside an energy community.

The solution to these problems—as with nearly all the problems that arise in the retail electricity market—is to move toward more efficient retail tariff structures. This implies a move toward distribution-network tariffs that (at least at the margin) are dynamic and vary with time and with location on the network. Ideally, all retail customers would face the efficient (SRMC) retail price at the margin. This would be coupled with fixed charges sufficient to recover the fixed costs of the network. Such reforms are not likely to be easy, but are almost certainly in the long-term interests of electricity users and consumers.[35]

If retail customers faced efficient prices, there would be no need to establish mechanisms for sharing or trade of power across different locations—instead, each producer or consumer would simply trade as much power as he/she would like to at each network location. Energy communities, as defined here, would be redundant.

This chapter focused on a particular definition of energy communities—sometimes referred to as "collective self-consumption"—involving virtual sharing of power within the community. Other forms of community energy initiatives, such as community ownership of local generation or storage, do not suffer from the problems set out here. As noted above, there may be benefits from such initiatives in the form of community cohesion and building awareness of the environmental impacts of energy generation.

The analysis set out here suggests that if energy communities are to be permitted they should be strictly limited in geographic scope (e.g., to the residents of the same apartment building) and should be accompanied by a move to more efficient tariff structures. Potentially, the "carrot" of an energy community could be used as an incentive to shift retail customers to more cost-reflective tariffs.

The analysis here is a reminder of the importance of retail tariff reform—in particular, the importance of moving toward dynamic cost-reflective retail tariffs. How this can be achieved while retaining the insurance function of existing retail contracts is the subject for further research.

[35] This would almost certainly require the establishment of a Distribution System Operator (DSO) or Distribution Market Operator (DMO) to determine efficient locational marginal prices on the distribution network through a market mechanism. In addition, mechanisms would need to be developed to ensure that retail customers are largely insulated from these price variations (only facing these prices at the margin).

Appendix

Let us suppose we have a set of prosumers, each of which may have either load and/or generation. At a given time, the rate of on-site consumption of prosumer i is given by L_i (kW) and the rate of on-site generation is given by G_i (kW). We assume that the on-site consumption yields benefits to the prosumer at the rate $V_i(L_i)$ (\$/h) and the on-site generation incurs costs to the prosumer at the rate $C_i(G_i)$.

The prosumer is also assumed to be able to purchase power from (or sell power to) the local grid. The net, metered consumption of the prosumer (the rate at which power is extracted from the local grid) is $Z_i = L_i - G_i$ (kW). When the net consumption is Z_i, the retail payment of the customer is $R_i(Z_i)$ (\$/h). The total retail payment of all the members of the energy community is therefore $\sum_i R_i(Z_i)$.

In the absence of the energy community, at each point in time, the task of the prosumer is to find the combination of consumption and generation which maximizes the net benefits less costs. This can be described as follows:

$$max_{G_i,L_i} V_i(L_i) - C_i(G_i) - R_i(L_i - G_i)$$

It is convenient, for our purposes, to express this task in two parts: finding the optimal combination of generation and load for a given level of net consumption and finding the optimal net consumption. Specifically, let us define the net benefit to the prosumer as follows:

$$B_i(Z_i) = max_{G_i,L_i} V_i(L_i) - C_i(G_i)$$

subject to: $Z_i = L_i - G_i$

Then the task of each prosumer can be expressed as follows:

$$max_{Z_i} B_i(Z_i) - R_i(Z_i)$$

Now let us consider the effect of the energy community. An energy community is defined to be (a) a rate of net consumption Z_i' for each prosumer, (b) a redistribution or reallocation of power ΔZ_i for billing purposes, so that the total retail payment for customer i is $R_i(Z_i' + \Delta Z_i)$ where $\sum_i \Delta Z_i = 0$,[36] and (c) a set of financial payments from the members of the community to the central community-operator T_i.

With the energy community in place, each customer's net payoff is as follows:

$$B_i(Z_i') - R_i(Z_i' + \Delta Z_i) - T_i$$

[36] ΔZ_i represents the amount of "credits" created (when positive) or consumed (when negative).

The total net payment to the energy community (that is the total payments by the members to the community-operator) is $\sum_i T_i$.

There are two requirements for the energy community:

- First, membership of the community is voluntary, so each member must be individually better off within the community than outside it (more precisely, each member must be no worse off with at least one member strictly better off):

$$\forall i, B_i(Z_i') - R_i(Z_i' + \Delta Z_i) - T_i \geq B_i(Z_i) - R_i(Z_i)$$

- Second, the community-operator must at least break even:

$$\sum_i T_i \geq 0$$

The total gain to the members of the community from the creation of the community is the following:

$$S = \sum_i \left(B_i(Z_i') - R_i(Z_i' + \Delta Z_i)\right) - \sum_i \left(B_i(Z_i) - R_i(Z_i)\right)$$

It follows immediately that the energy community can only exist if the total gain S is positive. From this last equation, it is clear that we can break the total gain from the energy community down into two parts:

- The reduction in retail payments:

$$\sum_i \left(R_i(Z_i) - R_i(Z_i' + \Delta Z_i)\right)$$

- An increase in the efficiency of dispatch:

$$S = \sum_i \left(B_i(Z_i') - B_i(Z_i)\right)$$

In the case where the energy community leads to no change in the physical production and consumption (i.e., where $\forall i,\ Z_i' = Z_i$), the only benefit from the energy community comes from the reduction in retail payments from the virtual reallocation: $\sum_i \left(R_i(Z_i) - R_i(Z_i' + \Delta Z_i)\right)$.

The price faced by prosumer i at the margin is defined as follows:

$$P_i = \frac{dR_i}{dZ_i}$$

It follows that, in order for the energy community to exist, there must exist differences in the prices faced by prosumers at the margin. This follows because, before the energy community exists, Z_i is assumed to maximize $B_i(Z_i) - R_i(Z_i)$. If there are no differences in the prices faced at the margin, then no virtual reallocation of net consumption can change the total retail payment $\sum_i R_i(Z_i)$ and no change in the dispatch can improve the total net benefit $\sum_i B_i(Z_i)$, so the energy community cannot lead to an improvement in welfare for the members of the community.

If there are differences in the prices faced by prosumers at the margin, there is a potential for an energy community to exist, yielding total surplus S as defined above. The distribution of this surplus between the members of the community is arbitrary and depends on the financial payments T_i.

References

Bellini, E., November 2019. Brazilian regulator wants to reduce net metering credit and introduce grid fee. PV Mag. URL. https://www.pv-magazine.com/2019/11/06/brazilian-regulator-wants-to-reduce-net-metering-credit-and-introduce-grid-fee/.

Borenstein, S., 2009. Time-varying retail electricity prices: theory and practice. In: Electricity Deregulation. University of Chicago Press, pp. 317–357.

Caramizaru, A., Uihlein, A., 2020. Energy Communities: An Overview of Energy and Social Innovation. Technical report. URL. https://www.interregeurope.eu/fileadmin/user_upload/plp_uploads/policy_briefs/2018-08-30_Policy_brief_Renewable_Energy_Communities_PB_TO4_final.pdf.

CEER, June 2019. Regulatory Aspects of Self-Consumption and Energy Communities. Technical report. URL. https://www.ceer.eu/documents/104400/-/-/8ee38e61-a802-bd6f-db27-4fb61aa6eb6a.

Interreg, 2018. Renewable Energy Communities: A Policy Brief from the Policy Learning Platform on Low-Carbon Economy. Technical report. URL. https://www.interregeurope.eu/fileadmin/user_upload/plp_ uploads/policy_briefs/2018-08-30_Policy_brief_Renewable_Energy_ Communities_PB_TO4_final.pdf.

Joskow, P.L., Wolfram, C.D., 2012. Dynamic pricing of electricity. Am. Econ. Rev. 102 (3), 381–385.

Kenning, T., November 2015. Brazil Approves 'historic' Net Metering Revision. PV Tech. URL. https://www.pv-tech.org/brazil-approves-historic-net-metering-revision/.

Mejdalani, A., Lopez, D., Hallack, M., December 2018. The Impact of Net Metering Policy Design on the Adoption Rate of Solar Photovoltaic Systems: A Simulation Using Calibrated Data from brazil. Technical report. URL. https://publications.iadb.org/publications/english/document/Net_Metering_Technical_Note_Simulation_PV_Adoption_Brazil_ Covered_final.pdf.

Next10, 2021. Designing Electricity Rates for an Equitable Energy Transition. Technical report. URL. https://www.next10.org/sites/default/files/2021-02/Next10-electricity-rates-v2.pdf.

Vieira, D., Shayani, R.A., De Oliveira, M.A.G., August 2016. Net metering in Brazil: regulation, opportunities and challenges. IEEE Latin Am Trans 14 (8), 3687–3694.

Part Two

The appeal of energy communities to customers and citizens

What motivates private households to participate in energy communities? A literature review and German case study

9

André Hackbarth and Sabine Löbbe
Reutlingen University, Reutlingen, Germany

1. Introduction

Energy communities can contribute to the decarbonization, decentralization, and democratization of the energy system, because they can expand the scope of action of private individuals. In this vein, EU legislation, which is described in several chapters in this volume, such as the chapter by Spasova and Braungardt, supports citizen and renewable energy communities. It aims at lowering the barriers for citizens and smaller companies to share or trade electricity among themselves and potentially become collectively self-sufficient (see also Roberts et al., 2019; Caramizaru and Uihlein, 2020).

Energy communities can be set up in many different ways depending on their:

- Organization, e.g., decentralized peer-to-peer trading versus centralized management and optimization by a service provider;
- Proximity, e.g., collective self-consumption within a building or on a local, regional, or national scale;
- Reliance on the public grid;
- Scope of services provided, e.g., electricity or heat, and included energy sources, for instance, exclusively renewable energy; and
- Goals and objectives, e.g., self-sufficiency, mutual supply of electricity, provision of flexibility or grid services, or economic benefits for the group.

Depending on the chosen characteristics and goals (see also the chapter by Rossetto et al.), the potential individual and collective benefits of these energy communities can be manifold (see, e.g., the chapters by Koltunov et al., Robinson and Del Guayo, as well as Biggar and Hesamzadeh). Nevertheless, participation of one of the most important stakeholders—household customers—is not a given. Furthermore, these benefits may not reach all potential participants, as participation may require investments for devices, such as metering and control technology, that some customers may not have the financial resources to buy.

This chapter examines individuals' heterogeneous preferences for energy communities and their attributes to identify the most promising customer segments and the

Energy Communities. https://doi.org/10.1016/B978-0-323-91135-1.00026-2

potentially most appealing energy community models and offerings. The remainder of the chapter is organized as follows:

- Section 2 reviews the scientific research on motives, preferences, and participation of private households concerning energy communities.
- Section 3 introduces the empirical results of a Germany-wide survey among 1000 individuals concerning their preferences for energy community attributes and describes different customer segments.
- Section 4 discusses and compares the results of prior sections, followed by the chapter's conclusions in Section 5.

2. Preferences for energy communities identified in the scientific literature

In recent years, a growing number of studies have investigated the motivation of individuals to participate in energy communities in general and in peer-to-peer (P2P) trading in particular, and have assessed individual preferences for specific attributes of such communities. The following summary[1] is based on a detailed review of this international research (see Adams et al. (2021)), analyzing 36 empirical peer-reviewed studies mainly conducted in Europe, the United States of America, and Australia. It finds that the main motives for participating in energy communities can be summarized as follows:

- *Economic benefits* are one of the most important factors that increase individuals' willingness to participate in energy communities (Mengelkamp et al., 2019; Kirchhoff and Strunz, 2019; Löbbe et al., 2020; Ableitner et al., 2020; Hahnel et al., 2020). There are different expressions of economic benefits from the customer's perspective, encompassing increased self-consumption of own electricity generation, reduced costs of energy purchased, or increased revenues from selling surplus electricity. As the lowest common denominator, Plewnia and Guenther (2020) find that the avoidance of financial losses for the individual can be considered as a necessary condition for participation in energy communities. This, however, might not be sufficient to trigger actual participation (Singh et al., 2018; Mengelkamp et al., 2018; Scuri et al., 2019).
- *Autonomy* was found to increase the willingness to participate in energy communities. This includes the preference for being independent from an energy provider (Hackbarth and Löbbe, 2020), the possibility of a self-determined decision about how to obtain the energy needed (Ableitner et al., 2019), or codetermination of that decision in an energy community (Löbbe et al., 2020). Greater autonomy is linked to the desire to play a greater part in the energy transition, to cocreate, or to develop deeper knowledge about one's own energy consumption (Mengelkamp et al., 2018; Wilkinson et al., 2020; Klein et al., 2020). However, individuals are also willing to accept some loss of control to receive price cuts (Kubli et al., 2018).
- *Self-sufficiency,* or *energy autarky,* is an important driver for some individuals to participate in new energy models (Ecker et al., 2018; Hahnel et al., 2020). For instance, Hahnel et al. (2020) found that about 1/3 of the participants in their energy community study were making

[1] Contributions from Jäger (2021) are particularly acknowledged.

individual autarky-increasing energy trading decisions. Such a focus on individual autarky of a large share of prosumers can reduce the amounts of traded energy within an energy community. Consequently, it may decrease its cost-effectiveness or collective self-sufficiency (Wörner et al., 2019). On the other hand, householders also recognized the need to include grid stability in the design of energy communities and their trading or incentive schemes to prevent potentially negative effects of such communities on grid balancing (Smale and Kloppenburg, 2020).

- *Environmental benefits* of energy communities are found to be major drivers for individuals' willingness to participate (Mengelkamp et al., 2018; Hackbarth and Löbbe, 2020; Smale and Kloppenburg, 2020; Klein et al., 2020; Ableitner et al., 2020), especially for those individuals who are committed to a sustainable lifestyle (Hackbarth and Löbbe, 2020; Wilkins et al., 2020).
- *Community spirit*—the notion of solidarity, cooperation, or the possibility to share generation and consumption within an energy community—is an essential reason for participation (Hackbarth and Löbbe, 2020; Singh et al., 2018). That is, some individuals not only attach greater importance to social relationships with other community members than to material gain, but they are also more willing to share energy with socially and physically close participants (Singh et al., 2018). In such energy communities where social relationships are central, e.g., energy communities among family or friends, anonymity is counter-productive (Singh et al., 2017). However, in other cases, potential participants may prefer anonymity. Hence, the importance of community spirit or anonymity depends on the attributes of the respective community and its specific members (Fell et al., 2019).
- *Regionality*, i.e., the regional provenance of the energy community, seems to increase the willingness to participate in an energy community or to pay more for such locally generated electricity (Löbbe et al., 2020; Wörner et al., 2019; Mengelkamp et al., 2019). The more people are regionally aware or rooted and the more importance they ascribe to the social value of communities, the more they cherish the regionality of energy communities (Hackbarth and Löbbe, 2020; Mengelkamp et al., 2018; Klein et al., 2020). The relevance of regionality varies among respondents and also among regions (Mengelkamp et al., 2019). Furthermore, regionality and local benefits of energy communities are more important to people who prefer self-sufficiency (Fell et al., 2019; Ableitner et al., 2020). However, concerns about security of supply and preferences for anonymity can lead to potential members favoring energy communities that are not too restricted in terms of location (Fell et al., 2019; Mengelkamp et al., 2019).
- *Convenience* and *the simplicity of participation*—i.e., ease of use and a minimal additional complexity of energy communities and their underlying technological and data requirements or trading schemes—seem to be beneficial for a community's success (Kirchhoff and Strunz, 2019; Scuri et al., 2019; Hackbarth and Löbbe, 2020). However, mixed results are reported for the importance of automation and financial transparency, as well as data security and privacy (Ableitner et al., 2019; Mengelkamp et al., 2019; Scuri et al., 2019; Wilkins et al., 2020).

Altogether, two main *motive clusters* appear to be central to potential energy community participants: economic benefits and "social" benefits. Accordingly, energy communities can emphasize either the social exchange among members, the democratic decision-making system (Creamer et al., 2018), and other noneconomic benefits, or they can accentuate cost reductions and economic benefits (Morstyn et al., 2018) to attract participants. If, on the other hand, these dichotomous preferences are not adequately represented in the energy community's design, the proposal is unlikely to be welcome—for instance, if an energy community that is focused on economic gains or individual self-sufficiency goals is offered to a customer segment whose willingness to engage is driven by noneconomic motives and social values (Wilkinson et al., 2020).

Generally, the willingness to participate in energy communities appears to be relatively high, but with regional differences (Mengelkamp et al., 2018; Fell et al., 2019; Löbbe et al., 2020; Hahnel et al., 2020). Regarding *socio-economic characteristics*, younger or middle-aged (Mengelkamp et al., 2018; Fell et al., 2019; Hackbarth and Löbbe, 2020) and better educated (Hahnel et al., 2020; Hackbarth and Löbbe, 2020) individuals seem more likely to join an energy community. Larger households on average also have a greater willingness to participate (Hahnel et al., 2020; Mengelkamp et al., 2019), while income and home ownership only show mixed results (Hackbarth and Löbbe, 2020; Löbbe et al., 2020; Wilkinson et al., 2020).

Attitudes that positively affect the willingness to participate in energy communities are individuals' environmental awareness and their interest in participating in the energy transition (Fell et al., 2019; Wilkinson et al., 2020), as well as their technological interest. That is, willingness to participate is higher among participants who are early adopters of technological innovations (Mengelkamp et al., 2018; Fell et al., 2019; Hackbarth and Löbbe, 2020; Löbbe et al., 2020; Ableitner et al., 2020).

3. Preferences for energy communities identified in a German case study

To obtain a realistic view of private households' preferences for energy communities in Germany, an online survey was carried out in July 2019 among 1000 members of an online access panel.[2] The main goal of the survey was to analyze private households' preferences regarding centrally managed energy communities in general—a concept which is close to the one described in the chapter by Hadler—and their specific attributes in particular. Moreover, the task was to identify potential heterogeneity of preferences of different segments of respondents regarding these attributes.[3] The results were intended to support recommendations regarding target groups, value propositions, and communication strategy for energy community managers.

The survey was representative of the German population on gender, education, and regions. However, middle-aged individuals and those living in multiperson households were overrepresented, whereas home owners were underrepresented. 89% of

[2] For more detailed information on the survey, see Hackbarth and Löbbe (2022).

[3] According to Kolmogorov–Smirnov tests, the variables were not normally distributed, so that nonparametric tests—i.e., Mann–Whitney U test or Pearson Chi squared test, depending on the scale of the respective variables—were used to identify significant differences of preferences regarding energy communities and their attributes between different segments of respondents, e.g., interested versus uninterested individuals.

participants were pure consumers without own production or storage, obtaining electricity entirely from the public power grid. They are referred to as "consumers" in the following analysis. About 3% of participants in the survey were "prosumers" who already owned a renewable energy generation unit. The remaining roughly 8% of respondents were consumers or prosumers and prosumagers planning to install a (further) power generation asset in the coming 2 years, mainly a photovoltaic (PV) system with or without a battery. They are referred to as "planners." This differentiation is important because the three respondent categories have different options and can obtain different potential benefits when engaging in energy communities (Sioshansi, 2019). Furthermore, there is an increasing share of prosumers who are falling out of the federal support scheme in Germany in the coming years. They will therefore be searching for a solution to sell their surplus electricity and to economically operate their PV system without the fixed feed-in tariff in the near future.

Roughly 74% of the 1000 participants were willing to participate in an energy community; we refer to them as "interested" respondents:

1. More than 60% of these interested individuals would only participate in an energy community if they would not have to pay more for electricity than currently. They are referred to as the "cost-focused" target group.
2. Roughly 36% of the interested individuals would only participate if switching and participation were uncomplicated, and about 3% would even participate if energy communities were more expensive. These individuals comprise the so-called "convenience-focused" target group.

The remaining about 26% of respondents are referred to as "uninterested" or "unwilling" (see also Löbbe et al., 2020; Hackbarth and Löbbe, 2022).

The following differences in *socio-economic characteristics* between the groups were found:

- *Age:* Interested respondents are statistically significantly younger on average compared to uninterested respondents, who are about 46 years old. Of the interested respondents, the convenience-focused group has the lowest average age at about 43 years, while the cost-focused group lies in-between at about 45 years.
- *Gender:* 54% of respondents in the convenience-focused group are male, while women represent 51% of the cost-focused group and 55% of the uninterested group, respectively. The gender difference between the uninterested and the convenience-focused group is statistically significant.
- *Education:* Those willing to participate in energy communities (i.e., the interested group) on average have a statistically significantly higher educational level than unwilling individuals, with the convenience-focused group being educated best, while members of the cost-focused group are in-between.
- *Household size:* The number of persons living in a household is statistically significantly larger in both interested groups—even more so in the convenience-focused group than the cost-focused group—compared to the uninterested group.
- *Home ownership:* In all groups, a minority of respondents lives in their own house or apartment, although the ownership rate of 39% and 43% in both interested groups is statistically significantly larger than in the uninterested group (29%), with the convenience-focused group having the highest level of home ownership.

Furthermore, the three groups were *hardly familiar with the concept of energy communities*. That is, only 2% of the individuals in the cost-focused group, 3% of the uninterested respondents, and 4% of the members of the convenience-focused group already knew about currently existing energy community offers, with the difference between both interested groups being statistically significant.

The examination of individuals' preferences for specific *energy community attributes* in Fig. 9.1 shows that the majority of interested respondents would prefer an energy community that includes family and friends. This is true especially in the convenience-focused group, but also for those who are not interested in energy communities.[4] All respondents also generally prefer local or regional energy communities over national or supraregional energy communities. At the same time, many potential members, especially in the cost-focused group, as well as a large share of uninterested respondents, are unsure regarding their preferred energy community type. The preference differences between both interested groups and the uninterested group are statistically significant for all energy community type options.

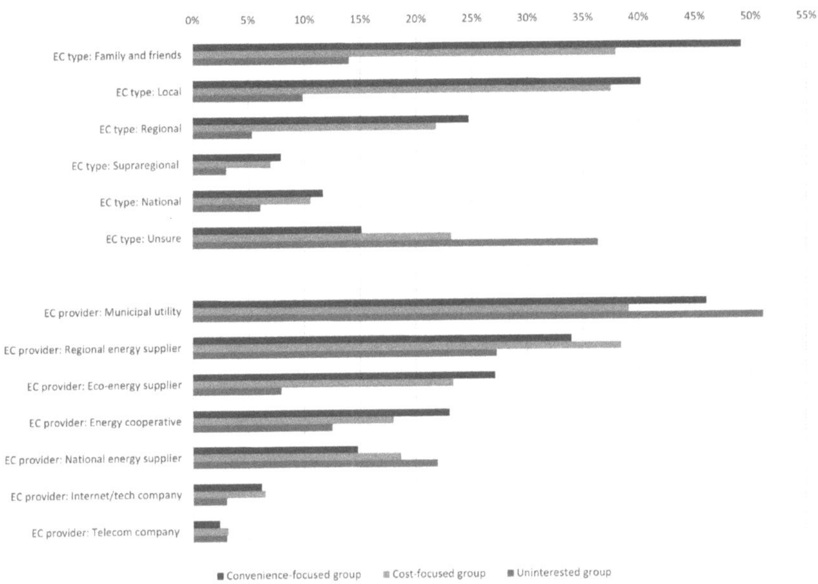

Figure 9.1 Preferred energy community type (*top*) and preferred community operator (*bottom*) by customer segment.
Notes: Figure is based on questions with multiple selection opportunity; *EC*, energy community.

[4] All survey participants were asked to indicate their preferences for energy communities and their specific attributes, even if they were not interested in participating in such communities at the moment. These results are instructive for identifying how this currently reluctant customer group can be attracted in the foreseeable future.

Moreover, survey participants would predominantly prefer municipal utilities as *managers of energy communities*. This holds true especially in the uninterested group (51%), followed by regional energy suppliers, eco-energy providers in both interested groups or the national energy supplier in the uninterested group, and cooperatives.[5] Internet or tech companies and telecom companies are the least preferred managers of energy communities in all groups. However, statistically significant differences can be observed between interested and uninterested respondent groups regarding their preferences for regional energy suppliers, eco-energy suppliers, energy cooperatives, and internet/tech companies as energy community managers, with all of these being more preferred by interested respondents. Between both interested groups, statistically significant differences can also be observed: municipal utilities and energy cooperatives are preferred more by respondents of the convenience-focused group.

To design attractive energy communities, it is important to know the attributes which are most appealing to potential participants and customers. Fig. 9.2 gives an overview of the importance ratings of the 3 respondent groups for 12 different *energy community features*. Overall, cost is the most important attribute across all groups, followed by service portal, regionality, utilization of revenues, self-sufficiency, and

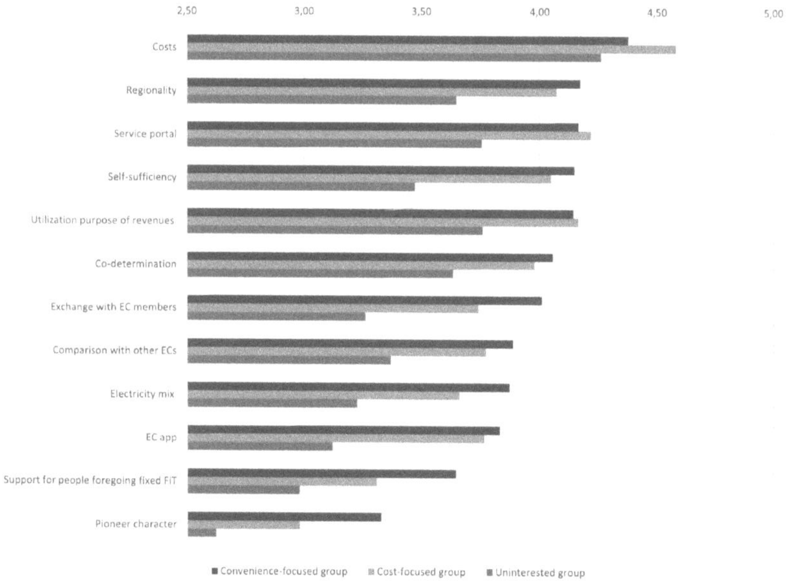

Figure 9.2 Importance of energy community attributes by customer segment.
Notes: Importance was measured on a five-point Likert scale ranging from "1 = not at all important" to "5 = very important"; *EC*, energy community.

[5] Regional and national energy suppliers are competitive electricity retail companies which provide energy to customers only on a regional, i.e., a geographically limited area (which can range from county to federal state), or national level, respectively.

codetermination, albeit in slightly different order depending on the respective group. These attributes are followed in importance by the possibility to compare oneself—and exchange—with other community members, the electricity mix of the energy community, and the availability of an energy community app. Supporting people foregoing fixed feed-in tariffs[6] and the pioneer character of energy communities are the least relevant attributes across all groups.

Respondents in the convenience-focused group attach a higher importance to almost all attributes compared to respondents in both other groups, with three exemptions: costs, the availability of a service portal, and the utilization purpose of community revenues are more important in the cost-focused group. These differences in importance ratings between interested respondents and uninterested respondents are statistically significant for almost all attributes but costs. The same holds true for differences between both interested groups, with the only exceptions being the importance ratings of the energy community app, service portal, utilization purpose of revenues, and codetermination of decisions by community members.

Most respondents in all three groups, especially in both interested groups in general and the cost-focused group in particular, would prefer an all-inclusive carefree package. An option with greater freedom of choice and personal responsibility with regard to the *technical components* appears to be less attractive. The latter option, however, is more preferred in the uninterested group, as Fig. 9.3 shows.

Another interesting characteristic of energy communities in the context of its social role (see also the chapter by Hanke et al.) is the *distribution of potential surplus revenues* within the community. Such additional revenues can be generated when the electricity supply exceeds the demand within the energy community and this surplus is sold on the wholesale market. Surplus revenues can also be generated by the delivery of grid services, e.g., enabled by individual battery storages. As can be seen in Fig. 9.3, respondents of all three groups have comparable preferences regarding the distribution

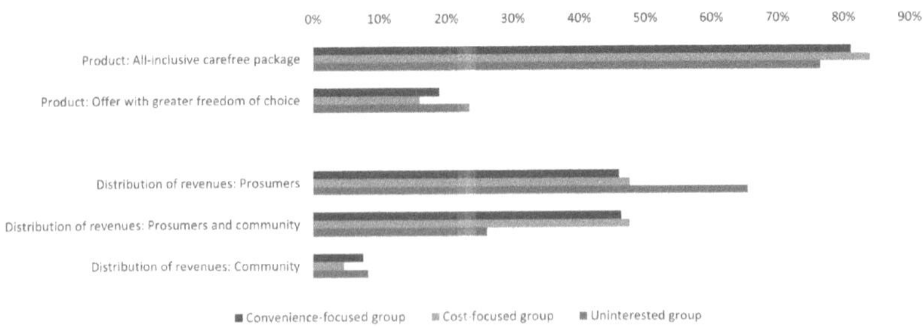

Figure 9.3 Preferred energy community product (*top*) and preferred distribution of revenues (*bottom*) by customer segment.

[6] Energy communities represent an opportunity for prosumers to continue operating their PV systems for which the fixed feed-in tariff ends after 20 years of operation. Furthermore, several of the energy community products currently on the market require prosumers who still receive the federal subsidy for their PV systems to assign the fixed federal feed-in tariff to the company that provides the energy community services.

of such revenues. This means that a full distribution to the owners of the respective energy generation units or batteries is equally or more appreciated than a partial distribution between the prosumers and the community. Distributing the revenues entirely to the community is the least preferred option in all three groups. Statistically significant differences between interested and uninterested respondents are found for almost all of the three revenue distribution options.

Finally, analyzing the *ownership of energy generation units*, in all three groups the vast majority of respondents are pure electricity consumers, followed by planners of PV systems, and prosumers. Consumers have the statistically significantly largest share in the uninterested group. However, since currently uninterested respondents most likely will not be energy community members in the foreseeable future, it is worthwhile to have a closer look at the two interested groups. Roughly 87% of these interested individuals are consumers, about 11% are planners, and only about 3% are prosumers. 62% of the consumers and 76% of the prosumers are in the cost-focused group, i.e., they are only willing to participate in an energy community if costs will fall or at least not rise. By contrast, 57% of planners are willing to switch to an energy community if participation is uncomplicated or even if it is more costly than their current tariff. Thus, they are more likely to be members of the convenience-focused group. These differences regarding the conditions of participation between the groups are statistically significant. Consequently, the preferences of interested consumers and prosumers for energy community attributes are more comparable to those of the cost-focused group, whereas preferences of planners are more similar to those of members of the convenience-focused group (for a detailed analysis, see Hackbarth and Löbbe, 2022).

4. Motivations to participate in energy communities

This section summarizes, compares, and discusses the main findings of the review of the international scientific literature in Section 2 and the survey results in Section 3 on motivations and preferences of potential members of energy communities.

In general, private households were found to be *open to the concept of energy communities*, with up to ¾ of respondents willing to participate (Fell et al., 2019; Löbbe et al., 2020; Hahnel et al., 2020). However, energy communities currently are also rather unknown to energy customers, as indicated by the case study results in Section 3. Moreover, individuals who are interested in energy communities on average seem be to be younger, live in larger households, are better educated, and own their homes. The first two characteristics apply in particular to planners, while the latter two are especially true for prosumers (Hackbarth and Löbbe, 2022).

However, private households that are willing to participate in an energy community not only differ in many ways from their uninterested counterparts, but also among themselves. The interested participants can be divided into two main segments with regard to the dominance of economic versus noneconomic attributes in the decision to participate. That is, while costs of participating in an energy community on average can be the most important factor in both segments, it is the decisive factor in the first and much larger more *"economically focused" segment*—labeled cost-focused group in the German case study. By contrast, in the more *"socially focused" segment*—the so-called convenience-focused group in the German case study—other noneconomic or community-oriented value propositions of energy communities prevail, although costs must not be neglected. These propositions cover the community spirit or the possibility to exchange with other community members, regionality, and environmental benefits. These other attributes can mitigate the importance of economic factors and are strong reasons for all individuals in this segment to participate, and in some cases to pay a premium compared to their current electricity tariff. This general dichotomy in motives to participate and the differences in preferences for energy community attributes between both interested groups is depicted in Fig. 9.4 (see also Jäger, 2021). On the household level, the importance of specific energy community attributes might differ from this aggregated ranking.

Furthermore, as the results of the German case study indicate, the current or future *ownership of an energy generation unit* seems to be linked to the importance interested individuals ascribe to the different energy community attributes. Hence, it is also related to their classification into the two main segments:

- The majority of *consumers* and especially *prosumers* belong to the cost-oriented group which attaches greater importance to the costs of the energy community and the self-utilization of community revenues—concerning prosumers, this probably stems from the comparably high fixed feed-in tariff that they now receive in Germany.
- The majority of *planners* belong to the convenience-focused, socially oriented group which attaches greater importance to the exchange of experiences with other community members and community self-sufficiency, and is also more willing to share the revenues of the generation units within the community. Furthermore, almost all planners in the German case study would be willing to install a new PV system without a fixed federal feed-in-tariff, which often is a prerequisite for participating in an energy community.

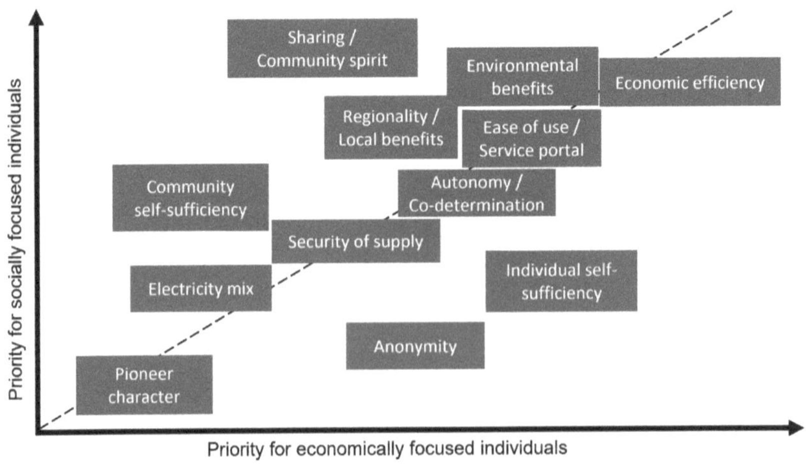

Figure 9.4 Priority of energy community attributes for the different household segments.

5. Conclusions

Energy communities have the potential to increase the codetermination and financial participation of private households in the energy system and are therefore supported by legislation on the European level. Hence, this chapter analyzed the motivations of private households to participate in an energy community and the importance they ascribe to various community features. Based on these findings, the following *recommendations* for a more customer-oriented design of energy communities can be derived for those initiating, developing, and implementing such communities—be it individuals or cooperatives, corporations, incumbents, or start-ups alike.

First, energy community managers should introduce *tailored engagement strategies and marketing measures* to increase the awareness and knowledge about energy communities in the population, to thereby raise their acceptance and the willingness to actively participate in such communities (see chapters by Li as well as by Menke et al. for examples of actual implementations of participation measures). Furthermore, participants of the first energy community projects should be actively involved in the decision-making processes of defining the characteristics of the energy community offers. They should also be accompanied by extensive *customer service and technical support* to manage expectations and ensure success, and that members are satisfied, wish to remain, and spread the positive experience (see also Löbbe et al., 2020).

Second, energy community managers should develop services that take the *importance of regionality* into account, since individuals can best be attracted by energy communities with a local or regional character. For many individuals, socially close energy communities with family and friends are interesting as well. In the same vein, locally or regionally rooted managers of such energy communities are the most favored ones. However, the results indicate that this preference for regionality should not be equated with a high willingness of potential energy community members to be active participants in the community, or to take on positions of responsibility. Hence, further *community design features*, such as all-inclusive carefree packages, web-based services, as well as the policy on distribution and utilization of revenues within the community are also crucial to attract and integrate members.

Third, based on the results, shaping *cost-competitive energy communities* seems to be particularly necessary. Cost on average is the most decisive attribute for participation among the majority of individuals. However, cost-competitiveness is a challenging task for energy community managers, because additional metering and control technology, the need for customized solutions, and an initially small number of customers lead to high up-front capital costs, and high average costs, in a mainly price-driven electricity market. Moreover, in the energy sector, electricity costs are to a large extent shaped by national taxes and charges and, thus, can hardly be influenced by energy community managers. Therefore, to reach break-even with this business model, governmental support and a large number of community members is ultimately needed. Consequently, national policymakers should introduce *supportive regulation* as well as *incentive or funding schemes* to increase the general economic attractiveness of energy communities. For instance, Austria plans to reduce the tariffs for using local and regional electricity grids by up to 64% for members of energy communities and their electricity transactions. With a growing number of participants and economies of scale, more cost-efficient energy community offers will be possible, probably increasing the demand for energy community services from more cost-conscious consumers.

Fourth, to get this process started, a first step for energy community managers could be to aim their product development and marketing strategies at *planners*, as they are less focused on costs and more receptive to other, especially socially and sustainability oriented features of energy communities. Furthermore, as indicated by the German case study, planners seem to be more knowledgeable about energy community offers available on the market than consumers and prosumers. Consequently, planners are the most promising initial target group for energy communities. As mentioned above, this holds true especially in a market such as Germany, where prosumers can still get comparatively high fixed feed-in tariffs. Given planners' potential acceptance of price premiums, priority should be given to highlighting the noncost benefits of energy communities. The preference of planners for regional solutions and the importance they attach to community spirit as well as to other social values should be transferred into specific offers. For instance, planners can be addressed particularly well with measures that support social exchange and try to decrease anonymity, as they are particularly interested in energy communities with socially close members. They have greater potential for service bundles, because they do not already own a renewable generation unit, battery storage, or energy management system. They may be further attracted by charging infrastructure solutions and specific electric vehicle charging tariffs managed by the community. This offers additional value creation potentials.

Fifth, energy community managers could *offer different community designs and goals* to appeal to the different customer groups and their heterogeneous underlying preferences from the start. That is, in addition to developing a socially focused energy community for some groups, especially planners, developers could offer a slimmed down low-cost energy community with platform-based self-services, without the social overlay, to address the economically focused majority. In the case of Germany, this approach would be particularly interesting to address the increasing group of prosumers that are falling out of the federal support scheme after 20 years of subsidization. This would entail offering only a limited number of standardized hardware and software options, e.g., PV systems, batteries, inverters, or energy management tools. On the contrary, the energy community offer for the socially focused target group should be service oriented and include a high level of community-supporting activities.

Finally, since *consumers* currently are the group most reluctant to join an energy community and are also unable to directly participate financially in an energy community through an own energy generation unit, *specific measures are needed* to attract their interest. These could include distributing parts of the revenues in the community so that not only prosumers but all community members would benefit. Through this and the resulting lower tariffs, energy communities may not only become more attractive to economically focused nonprosumers but also to poorer households and, consequently, a means to increase energy justice and a fair energy transition (see also the chapter by Hanke et al.).

References

Ableitner, L., Meeuw, A., Schopfer, S., Tiefenbeck, V., Wortmann, F., Wörner, A., 2019. Quartierstrom—Implementation of a Real-World Prosumer Centric Local Energy Market in Walenstadt. Switzerland. http://arxiv.org/abs/1905.07242.

Ableitner, L., Tiefenbeck, V., Meeuw, A., Wörner, A., Fleisch, E., Wortmann, F., 2020. User behavior in a real-world peer-to-peer electricity market. Appl. Energy 270. https://doi.org/10.1016/j.apenergy.2020.115061.

Adams, S., Brown, D., Cárdenas Álvarez, J.P., Chitchyan, R., Fell, M., Hahnel, U.J.J., Hojckova, K., Johnson, C., Klein, L., Montakhabi, M., Say, K., Singh, A., Watson, N., 2021. Defining the Social and Economic Value Propositions and Success Factors of Peer-To-Peer (P2P), Community Self-Consumption (CSC), and Transactive Energy (TE) Models (Manuscript under peer-review, forthcoming).

Caramizaru, A., Uihlein, A., 2020. Energy Communities: An Overview of Energy and Social Innovation. Publications Office of the European Union, Luxembourg. https://doi.org/10.2760/180576. EUR 30083 EN, JRC119433.

Creamer, E., Eadson, W., van Veelen, B., Pinker, A., Tingey, M., Braunholtz-Speight, T., Markantoni, M., Foden, M., Lacey-Barnacle, M., 2018. Community energy: entanglements of community, state, and private sector. Geography Compass 12, e12378. https://doi.org/10.1111/gec3.12378.

Ecker, F., Spada, H., Hahnel, U.J.J., 2018. Independence without control: autarky outperforms autonomy benefits in the adoption of private energy storage systems. Energy Pol. 122, 214–228.

Fell, M.J., Schneiders, A., Shipworth, D., 2019. Consumer demand for blockchain-enabled peer-to-peer electricity trading in the United Kingdom: an online survey experiment. Energies 12 (20), 3913.

Hackbarth, A., Löbbe, S., 2020. Attitudes, preferences, and intentions of German households concerning participation in peer-to-peer electricity trading. Energy Pol. 138. https://doi.org/10.1016/j.enpol.2020.111238.

Hackbarth, A., Löbbe, S., 2022. Analyzing motivations to participate in energy communities—results of a survey among German private households. In: Reutlinger Diskussionsbeiträge zu Marketing and Management. Reutlingen University, Reutlingen, Germany, forthcoming.

Hahnel, U.J.J., Herberz, M., Pena-Bello, A., Parra, D., Brosch, T., 2020. Becoming prosumer: revealing trading preferences and decision-making strategies in peer-to-peer energy communities. Energy Pol. 137. https://doi.org/10.1016/j.enpol.2019.111098.

Jäger, L., 2021. Kundenbedürfnisse in Energiegemeinschaften – was sind die wichtigsten Verbraucherpräferenzen, die das Design und die Akzeptanz von Energiegemeinschaftsmodellen in Europa einschränken oder prägen?. In: Student Research and Development Study. Reutlingen University, Reutlingen, Germany.

Kirchhoff, H., Strunz, K., 2019. Key drivers for successful development of peer-to-peer microgrids for swarm electrification. Appl. Energy 244, 46–62.

Klein, L.P., Matos, L.M., Allegretti, G., 2020. A pragmatic approach towards end-user engagement in the context of peer-to-peer energy sharing. Energy 205. https://doi.org/10.1016/j.energy.2020.118001.

Kubli, M., Loock, M., Wüstenhagen, R., 2018. The flexible prosumer: measuring the willingness to co-create distributed flexibility. Energy Pol. 114, 540–548.

Löbbe, S., Hackbarth, A., Stillahn, T., Pfeiffer, L., Rohbogner, G., 2020. Customer participation in P2P trading: a German energy community case study. In: Sioshansi, F. (Ed.), Behind and beyond the Meter—Digitalization, Aggregation, Optimization, Monetization. Academic Press, London, UK, pp. 83–104.

Mengelkamp, E., Schönland, T., Huber, J., Weinhardt, C., 2019. The value of local electricity—a choice experiment among German residential customers. Energy Pol. 130, 294–303.

Mengelkamp, E., Staudt, P., Gärttner, J., Weinhardt, C., Huber, J., 2018. Quantifying factors for participation in local electricity markets. In: 15th International Conference on the European Energy Market (EEM), pp. 1–5. https://doi.org/10.1109/EEM.2018.8469969.

Morstyn, T., Farrell, N., Darby, S.J., McCulloch, M.D., 2018. Using peer-to-peer energy-trading platforms to incentivize prosumers to form federated power plants. Nat. Energy 3, 94–101.

Plewnia, F., Guenther, E., 2020. The transition value of business models for a sustainable energy system: the case of virtual peer-to-peer energy communities. Organ. Environ. 34 (3), 479–503. https://doi.org/10.1177/1086026620932630.

Roberts, J., Frieden, D., Gubina, A., 2019. Energy Community Definitions. Explanatory Note, Compile Project: Integrating Community Power in Energy Islands, May. https://www.compile-project.eu/wp-content/uploads/Explanatory-note-on-energy-community-definitions.pdf.

Scuri, S., Tasheva, G., Barros, L., Nunes, N.J., 2019. An HCI perspective on distributed ledger technologies for peer-to-peer energy trading. In: Lamas, D., Loizides, F., Nacke, L., Petrie, H., Winckler, M., Zaphiris, P. (Eds.), Human-Computer Interaction—INTERACT 2019, Lecture Notes in Computer Science 11748. Springer, Cham, Switzerland, pp. 91–111.

Singh, A., Strating, A.T., Romero Herrera, N.A., Mahato, D., Keyson, D.V., van Dijk, H.W., 2018. Exploring peer-to-peer returns in off-grid renewable energy systems in rural India: an anthropological perspective on local energy sharing and trading. Energy Res. Social Sci. 46, 194–213.

Singh, A., Strating, A.T., Romero Herrera, N.A., van Dijk, H.W., Keyson, D.V., 2017. Towards an ethnography of electrification in rural India: social relations and values in household energy exchanges. Energy Res. Social Sci. 30, 103–115.

Sioshansi, F. (Ed.), 2019. Consumer, Prosumer, Prosumager—How Service Innovations Will Disrupt the Utility Business Model. Academic Press, London, UK.

Smale, R., Kloppenburg, S., 2020. Platforms in power: householder perspectives on the social, environmental and economic challenges of energy platforms. Sustainability 12 (2), 692. https://doi.org/10.3390/su12020692.

Wilkins, D.J., Chitchyan, R., Levine, M., 2020. Peer-to-peer energy markets: understanding the values of collective and community trading. In: Proceedings of the 2020 CHI Conference on Human Factors in Computing Systems. Association for Computing Machinery, New York, NY, USA, pp. 1–14. https://doi.org/10.1145/3313831.3376135.

Wilkinson, S., Hojckova, K., Eon, C., Morrison, G.M., Sandén, B., 2020. Is peer-to-peer electricity trading empowering users? Evidence on motivations and roles in a prosumer business model trial in Australia. Energy Res. Social Sci. 66, 101500. https://doi.org/10.1016/j.erss.2020.101500.

Wörner, A., Ableitner, L., Meeuw, A., Wortmann, F., Tiefenbeck, V., 2019. Peer-to-peer energy trading in the real world: market design and evaluation of the user value proposition. In: 40th International Conference on Information Systems, ICIS 2019. https://aisel.aisnet.org/icis2019/sustainable_is/sustainable_is/2.

Community energy initiatives as a space for emerging imaginaries? Experiences from Switzerland

10

Benjamin Schmid[1,4], *Mònica Serlavós*[2] *and Léon Francis Hirt*[3]
[1]Swiss Federal Research Institute WSL, Birmensdorf, ZH, Switzerland; [2]Association Suisse pour l'énergie citoyenne, Genève, Switzerland; [3]University of Geneva, Genève, Switzerland; [4]Zurich University of Applied Sciences ZHAW, Zürich, Switzerland

1. Introduction

Community energy has emerged in recent years as a viable option to enact energy transitions, notably by moving toward more decentralized forms of energy production and promoting new forms of social organization (Becker al., 2017). As illustrated by this edited volume, community energy is a heterogeneous phenomenon (see, e.g., the chapters by Rossetto et al. and Biggar and Hesamzadeh; Seyfang et al., 2014) as it takes various forms and is linked to numerous business models (see, e.g., the chapters by Kolesar as well as Barnes and Hansen). In addition, there is often an implicit understanding of what the future of energy should look like. This particular future is generally framed along the lines of a renewable future that will lead to a more sustainable society. Thus, this shared vision of a desirable future, a so-called *imaginary,* tends to be merely framed in techno-economic terms, conveying the idea that the central aim and focus of community energy, but also broadly energy transitions, is to increase the share of renewable energy technologies or economic efficiency in the energy market.

Although this view of the future appears straightforward, it may misrepresent or fail to capture other framings or conceptions of the future, both among community energy initiatives themselves and in society in general. Indeed, it remains elusive and understudied what types of renewable futures current actors taking part in community energy initiatives are in effect trying to achieve, and what these visions of the future might entail for these actors in the present. There is a risk that a potentially key transformative contribution of community energy is neglected if such alternative futures are overlooked. Indeed, community energy may provide a setting for various actors to come together and build new imaginaries of the future (e.g., by providing a space to consider what a just energy transition might look like; see the chapter by Hanke et al.). This tendency to omit the fact that there may be diverging visions of the future may prevent necessary renegotiations of existing social institutions. In addition, uncovering diverging visions of the future may provide insights into societal contestation or support of ongoing processes that could affect energy transitions as a whole. Notably, how a particular future is envisioned and what role community energy

Energy Communities. https://doi.org/10.1016/B978-0-323-91135-1.00006-7

initiatives are expected to play in it directly determines how these initiatives will be supported by public policy. There is thus a pressing need to investigate what imaginaries of the future are emerging to provide additional tools to understand ongoing community energy developments as well as to lead meaningful and purposeful energy transitions.

This chapter suggests a complementary way to explore community energy initiatives. It does so by bringing together community energy initiatives and the concept of sociotechnical imaginaries, a strand of research of science and technological studies. The concept of sociotechnical imaginaries describes how collective and shared visions of desirable futures can be achieved through technological and scientific developments (Jasanoff, 2015). Specifically, this chapter explores sociotechnical imaginaries in community energy initiatives in Switzerland, focusing on their aspirations and how they try to perform them in the present. The Swiss context has seen an increase in recent years of several community energy initiatives—mostly in the form of cooperatives and municipal projects—which makes Switzerland an interesting setting for empirical work. Overall, this chapter contributes to the ongoing developments and debates surrounding community energy by addressing the need to investigate the plurality of imaginaries that emerge within community energy settings and the way these imaginaries are enacted in the present. It thus provides new insights into how individuals and collectives engage in community energy initiatives and how community energy might be better researched and governed.

The chapter is organized as follows:

- After this brief introduction, Section 2 presents the Swiss context regarding its electricity system and the development of community energy initiatives in recent years;
- Section 3 introduces the analytical approach, that is how the concept of sociotechnical imaginaries is operationalized as well as the data sources;
- Section 4 depicts emerging imaginaries in Swiss energy community initiatives; and
- Section 5 offers a reflection on the emerging imaginaries, highlighting their plurality and looming tensions, followed by the chapter's conclusions in Section 6.

2. The Swiss context

2.1 The Swiss electricity system

The Swiss electricity system features low-carbon production, largely composed of hydropower (56% of the total production) and nuclear power (35%). In contrast, new renewables (wood-burning, biogas plants, photovoltaic plants, wind turbines) account for only a small share (4%) (SFOE, 2020, 2019 figures). While electricity production capacities are largely owned and operated by large companies (Verhoog & Finger, 2016), electricity supply is highly decentralized. In 2019, there were 632 mostly not unbundled[1] distribution companies supplying electricity (ElCom, 2020). Although

[1] No organizational unbundling of distribution grid operation and electricity supply.

there are various degrees of privatization and legal forms, these distribution companies are largely publicly owned by cantons—Swiss member states—and municipalities (89% of total share capital, SFOE, 2020).

This structure is mainly due to the regulation of the electricity market. Switzerland's electricity market has not yet been fully liberalized, meaning that supply to small end consumers (annual electricity consumption of <100 MWh) falls under the territorial monopolies of individual distribution companies. Small end-users are therefore prevented from freely choosing their own supplier. This has major implications for producers of renewable electricity that do not have their own grid, such as community energy initiatives, since this regulation also limits the sales options for generated renewable electricity. If the electricity is not consumed on site, the only remaining sales channel is to the distribution company responsible for the respective supply area. The distribution companies are obligated to purchase the generated electricity at a regulated price. However, the level of remuneration varies greatly between individual distribution companies (VESE, 2021), which leads to high uncertainty for the producers and in some cases severely impairs cost-covering operation of PV plants.

Swiss energy (transition) policy has strong federalist characteristics. At the national level, the Energy Strategy 2050 provides the overall framework and sets the goals of phasing out nuclear energy and increasing annual electricity generation from renewables (except hydropower) to 11,400 GWh by 2035 (increase by 142%, reference year 2019, SFOE, 2020; EnG, 2016). As part of this energy strategy, a fixed feed-in tariff was introduced in 2009 to promote renewables. This feed-in tariff replaced remuneration by distribution companies and provided a nationally uniform compensation scheme. But since the subsidy volume was limited, there was no prospect of receiving this feed-in tariff for new plants as early as 2012. In 2014, a one-time investment grant for small PV systems (<30 kW) was introduced as an alternative (increased to 100 kW from 2018).

In 2018, a new energy act was implemented that allowed collective self-consumption among multiple consumer parties and across adjacent properties, opening up sales options for energy communities. Apart from this collective self-consumption regulation, however, there has been no recognition of community energy in Swiss energy legislation to date. So far, no such recognition is envisaged contrary to the EU legislation.

In accordance with the principle of subsidiarity governing Swiss federalism, the subnational levels—i.e., the cantons and municipalities—have key roles in implementing and shaping energy policy. For a long time, energy supply was mainly a task of municipalities, and until today, they retain significant competencies (Sager, 2014). They can also have major influence on local energy systems through ownership and control of local distribution companies (Hofmann & Richert, 2017; see also Hirt et al., 2021 for the role of municipalities in PV development).

2.2 Community energy in Switzerland

Community energy in Switzerland takes many forms, especially given the proximity between municipalities and local distribution companies. Citizens can often directly participate in decision-making processes related to the strategic direction of municipality-owned distribution companies through direct democratic institutions.

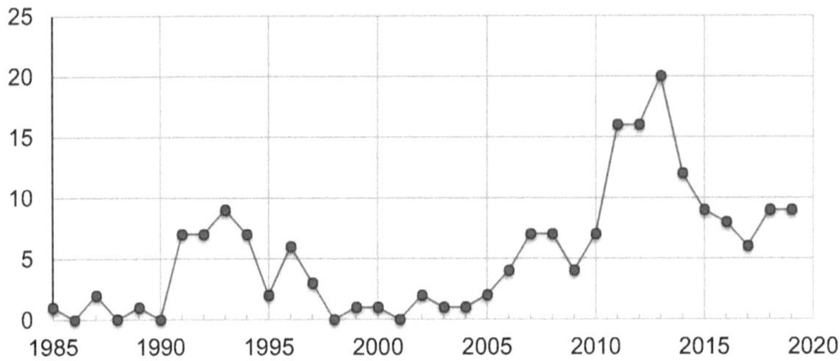

Figure 10.1 Number of annual new foundations of community energy initiatives (energy cooperatives active in energy generation) in Switzerland, 1985–2019.
Sources: Based on own data, see also Schmid, B., Musiolik, J., 2021. Research Report on Cooperative Organisational Models for Renewable Energy in Switzerland. SONNET: EU Horizon 2020 Grant agreements no: 837498. https://zenodo.org/record/4727581. Accessed 6 May 2021.

In a typology of community or citizen energy in Switzerland, Serlavós (2018) therefore includes not only new or emerging civil society actors (such as associations and cooperatives) but also municipal projects that emerge in collaboration with local distribution companies. Moreover, a significant proportion (about 150 in 2015) of the 632 distribution companies are themselves cooperatives that emerged in the wake of electrification in the early 20th century and continue to exist today (Rivas et al., 2018). Against the background of the anti-nuclear movement and later the climate change debate, nearly 200 new energy cooperatives have been founded since 1985 (in two founding waves, see Fig. 10.1). The spike of new foundations between 2009 and 2012 was largely due to the available feed-in tariff. Despite this substantial growth, in 2016, these energy cooperatives only accounted for approximately 1%–2% of total installed capacity of photovoltaics (Rivas et al., 2018).

Overall, community energy in Switzerland is to be understood as a broad phenomenon, both in terms of its extent and its manifested forms, which echoes other contributions and observations in this volume (see, e.g., chapters by Rossetto et al., Biggar and Hesamzadeh). At the core of the phenomenon are energy cooperatives, while other forms, such as municipal projects, can also be included on the periphery.

3. Applying the concept of sociotechnical imaginary to Swiss community energy initiatives

To investigate the plurality of envisioned futures that emerge in community energy initiatives, the concept of sociotechnical imaginaries is applied. According to Jasanoff (2015, p. 6), sociotechnical imaginaries are defined as "*collectively held, institutionally stabilized, and publicly performed visions of desirable futures, animated by shared understandings of forms of social life and social order attainable through, and supportive of, advances in science and technology."* Three aspects of this definition are worth highlighting. Firstly, it emphasizes that imaginaries may emerge from either individuals or small groups of individuals (see, e.g., Kim, 2015; Mutter, 2019) and potentially gain traction overtime to establish themselves as dominant sociotechnical imaginaries. Secondly, the definition highlights the link between visions of desirable futures and how they are performed in present-day practices. This means, in the context of community energy initiatives, that practices that are currently being performed by these initiatives are related to imagined desirable futures. For instance, cooperatives provide their members with broad participation opportunities, which is a practical component of a vision of the future that consists in assuring codetermination rights for a broad public in energy-related matters. Thus, by investigating the practices and visions of community energy initiatives, it is possible to capture the potential plurality of imaginaries that emerge within these initiatives. Thirdly, a key advantage of the concept of sociotechnical imaginaries is that it provides a more holistic depiction of community energy initiatives. Applying this concept contributes to preventing the risk of succumbing to single broad narratives of technological and scientific projects and therefore ignoring other and parallel social dynamics that also characterize energy transitions, in which the social, technology, and science coevolve together. Also, it provides an alternative way of understanding the engagement of individuals or groups of individuals in community energy initiatives.

To operationalize the concept of sociotechnical imaginaries in this article, two dimensions are used that capture essential ideas of the definition of imaginaries as explained above: The first dimension captures the idea of visions of desirable futures present in community energy initiatives, which are termed *aspirations*. The second dimension captures the idea that visions of desirable futures translate into present-day practices by community energy initiatives, which are termed *performances* (Table 10.1). For both dimensions—aspirations and performances—indicators were

Table 10.1 Synthetic overview of operationalization and data.

Dimensions	Features of emerging new sociotechnical imaginaries (applied to CE initiatives)	Sources of data
Aspirations – Linked to a desirable future	• Vision of a future energy system • Operational goals at an organizational level • Motivations of members • Envisioned role of technology in the transition	• **Case Studies 1:** Case studies of 20 Swiss initiatives including cooperatives, associations, and municipal projects (Serlavós, 2020) • **Case Studies 2:** Case studies of four energy cooperatives (Schmid, 2019) • **Survey 1:** Survey of 136 Swiss energy cooperatives (Schmid, 2019)
Performances – Enactment of the future as of today	• Main activities • Organizational governance • Relationship of community energy initiatives with other local actors	

developed by iterating between the theoretical concept of sociotechnical imaginaries and the data collected on community energy in Switzerland as part of two different doctoral studies (Schmid, 2019; Serlavós, 2020, see Table 10.1).

4. Evidence of emerging sociotechnical imaginaries

4.1 Aspirations of Swiss community energy initiatives

The aspiration dimension of socio-technical imaginaries of Swiss community energy initiatives is examined using the following four indicators:

- Vision of a future energy system;
- Operational goals at an organizational level;
- Motivations of members; and
- Envisioned role of technology in the transition.

The first indicator used to analyze the aspiration dimension is the *vision of a future energy system*. All studied initiatives (see Table 10.1) called for an indigenous production of renewable energy—whether at a local, regional, or national level—to improve the energy autonomy of Swiss territories. Although community energy initiatives rarely envisioned meeting the total energy needs of a region single-handedly, they nevertheless aimed to help reduce the amount of imported energy, mainly through electricity production, as part of their aim to increase energy autonomy. Furthermore, there was disagreement among Swiss energy cooperatives when it came to actor diversity in the future energy system, specifically regarding the anticipated scale of activities and the size of the initiatives. When asked whether many energy cooperatives will be developing new projects at a supraregional level in the future, only around half of the energy cooperatives agreed. And when asked whether there will be an increasing number of large energy cooperatives in the future, only around one third agreed.

There is more agreement on the vision of a future energy system when it comes to the outlook for professionalization and organizational structures. More than two thirds of energy cooperatives expected to remain heavily dependent on volunteer work and did not anticipate moving toward a hierarchical form of organizational structure. Further, these energy cooperatives believed they will continue to provide members with extended opportunities for codetermination through decision processes. In that sense, more than 90% of the surveyed Swiss energy cooperatives (Survey 1, see Table 10.1) rated the possibility for many members to participate in energy production as relevant when choosing the cooperative's legal form. The same applied to democratic codetermination, which had similarly high values. Both aspects represent key characteristics or ideals of the cooperative model. These measures indicate that democratic participation going beyond simple financial investment constitutes a widely shared aspiration among Swiss energy community initiatives. These results reaffirm the strong emphasis placed on citizen participation in Swiss community energy initiatives; and that energy cooperatives do not expect to replace incumbent energy supply infrastructures but complement the current energy system.

The second indicator looks at community energy initiatives' *operational goals at the organizational level.* Survey results (Survey 1, see Table 10.1) showed that such operational goals went beyond a techno-centered understanding of energy transitions and highlighted social and environmental aspects. Fig. 10.2 precisely shows the relevance of different goals for Swiss cooperatives (Survey 1, see Table 10.1). Among others, these included tackling global warming, promoting social acceptance of renewable energies, raising awareness, knowledge transfer, reducing energy consumption, or contributing to building better conditions for political engagement. Whereas goals linked to energy transitions were widely shared, there was greater heterogeneity among initiatives regarding localism, autonomy, or even financial goals.

The third indicator focuses on *motivations of community energy members,* with particular attention to how profitability is understood. Case studies (Case Studies 1, see Table 10.1) show[2] that these motivations were, in order of importance, environmental, infrastructural, personal, political, social, and economic. As Fig. 10.3 shows, profitability did not appear as a priority for most of the initiatives' members, whereas environmental

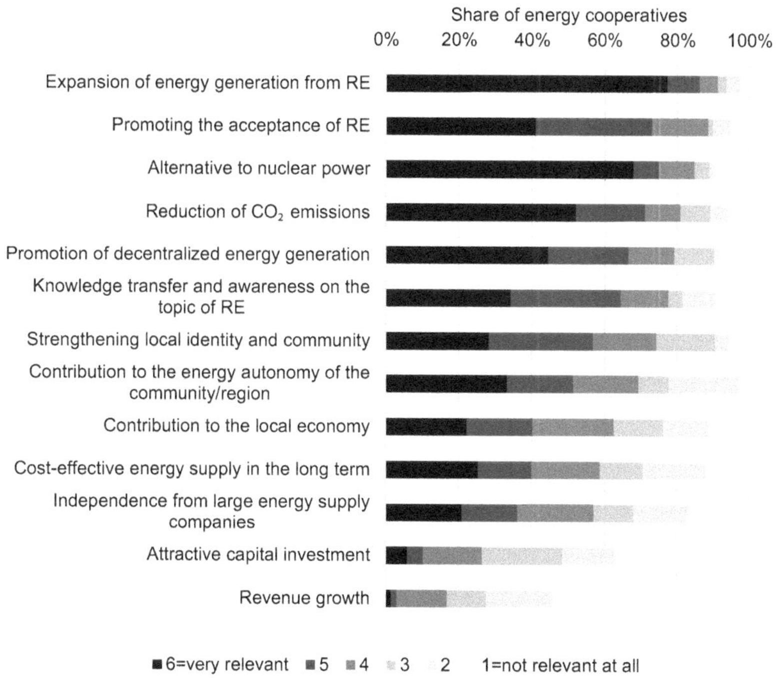

Figure 10.2 Relevance of goals for community energy initiatives in Switzerland (energy cooperatives founded after 1985, $n = 65-79$).
Source: Based on own data (Survey 1, see Table 10.1).

[2] These different categories include, respectively, reduction of CO_2 emissions; improvement of energy autonomy and energy security; change of individual values and link to nature; change of the energy system; contribution to the common well-being and reinforcing social link; economic development; and return on investment.

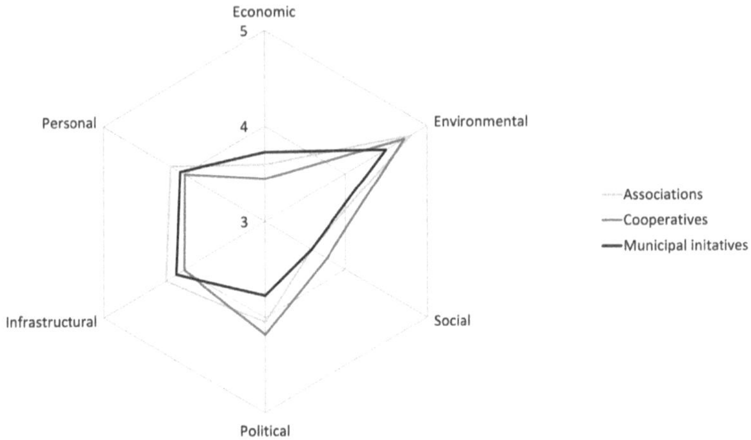

Figure 10.3 Comparison of motivations of community members, according to the juridical form. Notes: The vertical axis values correspond to: 3 (moderately important), 4 (very important), 5 (extremely important).
Source: Based on own data (Case Studies 1, see Table 10.1).

goals were central. Interestingly, motivations differed according to the juridical status and the level of participation of the members. In the case of cooperatives and associations, where citizen engagement is generally higher, political and social motivations came right after environmental ones. On the contrary, in the case of municipal initiatives, where citizens may only contribute through financial participation, environmental motivations still had precedence, but economic motivations came in second place.

Overall, what most initiatives agreed on is the need to empower citizens to shape and implement their own energy policy. According to them, current incumbent actors did not do enough to advance the energy transition in Switzerland. This aim prevailed over the traditional purpose of cooperatives in meeting the needs of their members.

The fourth indicator captures the *envisioned role of technology in the transition.* Indeed, case studies (Case Studies 1, see Table 10.1) showed a wide range of divergent approaches regarding the role of technology in energy transitions, ranging from high-tech to low-tech scenarios. On the one hand, some members strongly believed that technology would play a core role in finding ways to change the energy system toward a more sustainable one. On the other hand, some members considered that the main lever for change is to transform individual and social behavior.

4.2 Performances of Swiss community energy initiatives

The performance dimension of socio-technical imaginaries of Swiss community energy initiatives is examined using the following three indicators:

- Main activities;
- Organizational governance; and
- Relationship of community energy initiatives with other local actors.

The first indicator of the performance dimension relates to the *main activities* of community energy initiatives. These are all the concrete actions through which the initiatives attempt to fulfill their aspirations. A central aspiration of Swiss community energy initiatives was the expansion of renewable energies. Accordingly, the most commonly used technologies were small to medium photovoltaic systems to generate electricity. Among the energy cooperatives, 77% were active in renewable electricity generation. Of these, 93% used photovoltaics with installations averaging 60 kWp. Compared to that, only 28% of the energy cooperatives were active in heat generation and distribution (Survey 1, see Table 10.1). Similar figures apply when including not only energy cooperatives but also other forms of community energy initiatives (Case Studies 1, see Table 10.1).

The second indicator draws attention to the *organizational governance* of the studied initiatives. This includes the rules and processes governing how decisions are made and implemented within the initiatives in order to pursue their goals. In their practical expression, aspirations about this were first of all manifested in the choice of a specific juridical form, which determines key features of organizational governance. In community energy projects led by distribution companies or municipalities, rights of the participants were usually limited, and participation was restricted to a financial nature (Case Studies 1, see Table 10.1). In the case of energy cooperatives, however, Swiss cooperative law stipulates mandatory rules on the purpose—support the interests of members—and democratic decision-making—one-person-one-vote principle—of cooperatives (Swiss Civil Code OR 828–926). Similar rules apply to initiatives constituted in the juridical form of associations. By ensuring broad participation *opportunities* for citizens, energy cooperatives thus realized their aspiration of citizen participation including codetermination rights in their performances.

However, the situation is more nuanced when looking at actually *realized* broad participation, where differences between municipality-led projects and cooperatives become apparent. On the one hand, some results indicated extensive participation, especially of cooperative members. As Table 10.2 shows, for example, more than

Table 10.2 Rate of participation of members in different activities and tasks according to the juridical form.

	Associations (%)	**Cooperatives** (%)	**Municipal initiatives** (%)
General assemblies	28.2	56.3	2.6
Work meetings	15.4	30.3	7.9
Raising awareness	12.8	28.8	10.5
Internal communications	17.9	31.3	7.9

Source: Based on own data (Case Studies 1, see Table 10.1).

half of energy cooperative members reported to have participated in general assemblies and almost one third to have attended working meetings or awareness-raising activities (Case Studies 1, see Table 10.1). These figures are much lower among associations and municipal initiatives.

Moreover, about 70% of energy cooperatives operated exclusively on a volunteer basis, which might indicate a culture of far-reaching participation (Survey 1, see Table 10.1). On the other hand, even among energy cooperatives, operational participation was far from being evenly distributed. Often, it was only a few individuals who actively shaped the business, whereas most members kept to passive participation. Occasionally, "easy" participation without the need for additional efforts was even advertised (Case Studies 1, see Table 10.1) as an inducement to join the community. Although this can help to acquire additional capital for the construction of renewable energy facilities, there is a risk of being at odds with actual realized broad participation, which is supposed to be characteristic of citizen energy initiatives (Serlavós, 2020).

Finally, community energy initiatives do not operate in a vacuum. Hence, the third indicator concerns the *relationship of community energy initiatives with other local actors* regarding energy supply and governance. Given the high degree of decentralization of the political and energy system in Switzerland, the performance of Swiss community energy initiatives was often highly intertwined with local structures (see also Schmid et al., 2020). Particularly, this included their organizational linkages with and dependence on incumbent actors, municipalities, and local distribution companies. In fact, half of all energy cooperatives had at least one municipality as a member. As shown in Fig. 10.4, up to 70% of these cooperatives received substantial support from municipalities—most often by providing roof space (44%), by offering direct financial aid in the form of loans or guarantees (32%), or by purchasing the generated electricity on generous terms (24%) (Survey 1).

In some cases, municipalities even made specific adjustments in their energy policies to better support local energy community initiatives (Serlavós, 2020). As case studies showed, even without formal municipal membership, many individuals involved in community energy initiatives had close ties to municipal actors or engaged

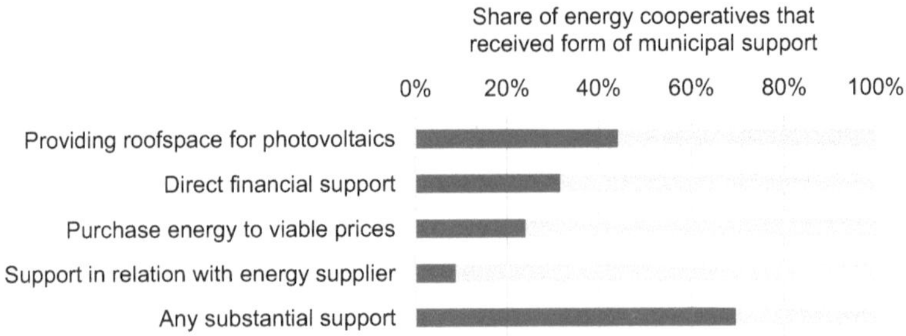

Figure 10.4 Frequency of different forms of municipal support among Swiss community energy initiatives (energy cooperatives, $n = 79$).
Source: Based on own data (Survey 1, see Table 10.1).

themselves in municipal legislative or executive bodies (Case Studies 1 and 2, see Table 10.1). In some cases, municipalities themselves initiated projects allowing citizens to finance renewable installations (Case Studies 1, see Table 10.1).

When it comes to the relationship between energy cooperatives and distribution companies, organizational linkages were often less pronounced than with the municipalities. Yet, the cooperatives were often highly dependent on them. For instance, only 15% of energy cooperatives had distribution companies as members. Moreover, energy cooperatives perceived their support as less substantive or even absent (Case Studies 2, Survey 1). Nevertheless, Swiss community energy initiatives were exceptionally dependent on distribution companies. This is because they retain territorial supply monopolies as Switzerland's electricity market is not yet fully liberalized (see Section 2.1). Typically, they are the only valid purchasers of generated energy that is not self-consumed on site. Through their feed-in price policy, they thus have a decisive influence on the development of community energy initiatives. As case studies showed, this dependency can be advantageous for community energy initiatives and facilitate their rapid development if the local distribution company offers generous and long-term predictable purchase conditions. However, it can render a successful development virtually impossible if distribution companies only pay the minimum amount required by law (Case Studies 1 and 2, see Table 10.1).

5. Reflecting on community energy from a sociotechnical imaginaries perspective

Building on the empirical evidence presented in the previous section, the three following points are further discussed:

- Community energy initiatives reveal alternative and plural sociotechnical imaginaries;
- Tensions arise between aspirations and performances in internal and external relations of community energy initiatives; and
- The democratic characteristics of community energy initiatives enables them to deal with these tensions and therefore to develop new feasible imaginaries.

Community energy initiatives exhibit many common features that point toward alternative sociotechnical imaginarie*s*, including indigenous and decentralized production of renewable energies, democratic governance, and limited relevance of profitability. By embodying these features, community energy organizations contribute to energy transitions not only from a technological perspective but also from a social and political one. All these elements suggest that, through their present-day performances, community energy initiatives enact visions of a desirable future that challenge dominant discourses and practices that currently prevail in energy system organisation and research. However, plurality among community energy initiatives is very common, as illustrated by the wide range of goals, motivations, or business models, among others. This diversity contributes to making community energy initiatives a credible strategy for a more democratic and ecological energy transition. In that sense, it can be argued that plurality in community energy initiatives constitutes a strong feature to transform the current energy system as well as the underlying imaginary.

By analyzing community energy initiatives with the concept of sociotechnical imaginaries, it is possible to juxtapose their aspirations and performances and to identify tensions that arise between these two dimensions, i.e., aspects in which the present-day practices do not fully correspond to the aspired futures. Interestingly, such tensions occur both in internal and external relations of community energy initiatives. In internal relations, i.e., within the organization of community energy initiatives, a key tension arises as initiatives envisage broad participation and effective codetermination of many members in their aspirations, especially in their organizational goals. However, in their daily performances, the actual extent of participation in operational aspects is often limited to a few key actors. Similarly, the choice of the juridical form of the cooperative or the association as well as the stated organizational goals usually restrict profit-seeking practices. Yet, this does not exclude the fact that members are certainly motivated by financial prospects, even if it is not the main priority (see also Fleiβ et al., 2017). Finally, there is an internal tension when it comes to the role of technology. In their aspirations, on the one hand, members of community energy initiatives embrace broad pluralistic ideas about the role of technology in the energy transition, including low-tech models or sufficiency as central pillars of energy transitions (see also Salomon et al., 2015). On the other hand, at least in the case of Switzerland, the initiatives have so far been mainly active in the expansion of photovoltaics, while other activities are rather rare. In external relations, i.e., the interactions of the initiatives with their environment, there is a major tension when it comes to the role of community energy in the energy system. In the stated aspirations, the development of community energy is often envisioned as the emergence of new actors and institutions representing an independent alternative to the incumbent system. However, in current performances, such initiatives are often strongly dependent on the support and cooperation of incumbent actors. This is reinforced by the fact that they are also often closely intertwined with them on a personal level, which makes it complicated to establish a clear distinction from the incumbent system. Regarding the need to meet the claim of effective citizen participation and codetermination, community energy initiatives are thus faced with the challenge of maintaining sufficient independence without jeopardizing the relationships with incumbent actors, which are necessary for the successful implementation of their energy-related goals.

Such tensions between aspirations and performances, however, should not simply be dismissed as inconsistencies. Rather, the need to deal with them underscores how central the democratic characteristics of community energy are for new imaginaries to emerge and develop. Only if new visions of the future are continuously confronted with their practical implementation, then the conditions are enabled for such emerging imaginaries to scale up. This confrontation, however, does not imply the subordination of aspirations to the practical dimension, but rather the necessity of a constant process of deliberation between the plurality of aspired futures and present realities and scope of action of the initiatives. It can be assumed that extensive opportunities for effective participation are essential for these deliberative processes to be fruitful. For these deliberation processes, extensive opportunities for effective participation, going beyond mere financial part-taking, are essential. Hence, this indicates that the democratic claims of community energy initiatives are not to be seen only as a value in themselves or as a means of educating members. Rather, these claims should be deemed to be a central prerequisite for linking and updating aspirations in emerging imaginaries to a feasible implementation, thus creating the potential for imaginaries to develop and a central contribution of community energy to unfold.

6. Conclusions

This chapter investigates alternative sociotechnical imaginaries in Swiss community energy initiatives in order to better understand how they can contribute to the energy transition. The study reaffirms the plurality of community energy initiatives, both in terms of their views of the future—aspirations—and the enactment of those visions as of today—performances. With this plurality of visions and performances, community energy can be considered as a space for new imaginaries to emerge and to develop, thus making a pivotal contribution to a sustainable energy transition. Energy community initiatives provide experimental spaces in which future energy systems can be deliberatively contemplated and shaped, going beyond mere technical expansion of renewables or economic efficiency. A key advantage is that such aspirations do not remain mere mental constructs, but that they are put directly into practice and thus are constantly challenged. In this way, dealing with the tensions that arise between aspirations and implementation is a fundamental process for mainstreaming new and viable imaginaries. The democratic characteristics that are inherent to community energy initiatives in most cases provide a potentially fruitful setting to structure and sustain the necessary deliberation of these processes and thus to achieve a sustainable energy transition.

It follows that for community energy, democratic design with broad and effective participation and codetermination of many actors should be a core value if the genesis of new imaginaries is to remain an important contribution. In turn, policymakers and researchers are required to consider such qualitative contributions of community energy more thoroughly and not only to focus on economic and technological efficiency in the design of policy instruments. At the same time, it is important to refrain from establishing overly protected niches for community energy projects, since it is precisely the confrontation between their aspirations and their implementation in the given system that allows them to make an essential contribution to a sustainable energy transition through the development of new viable socio-technical imaginaries.

Bibliography

Becker, S.L., Franke, F., Gläsel, A., 2017. Regime pressures and organizational forms of community-based sustainability initiatives. In: Environmental Innovation and Societal Transitions.

ElCom – Federal Electricity Commission, 2020. Report on the Activities of ElCom 2019. https://www.elcom.admin.ch/elcom/de/home/dokumentation/berichte-und-studien/taetigkeitsberichte.html. (Accessed 6 May 2021).

EnG, 2016. Federal Energy Act of 30 September 2016 (Status as of 1 January 2021) (SR 730). https://www.fedlex.admin.ch/eli/cc/2017/762/de. (Accessed 6 May 2021).

Fleiß, E., Hatzl, S., Seebauer, S., Posch, A., 2017. Money, not morale. The impact of desires and beliefs on private investment in photovoltaic citizen participation initiatives. J. Clean. Prod. 141, 920–927.

Hirt, L.F., Sahakian, M., Trutnevyte, E., 2021. What socio-technical regimes foster solar energy champions? Analysing uneven photovoltaic diffusion at a subnational level in Switzerland. Energy Res. Social Sci. 74, 101976.

Hofmann, B., Richert, J., 2017. Effektivere Mehrebenenpolitik im Bereich Strom. EGI Working Papers Series. University of St. Gallen. https://www.alexandria.unisg.ch/250717/. (Accessed 6 May 2021).

Jasanoff, S., 2015. Future imperfect: science, technology, and the imaginations of modernity. In: Jasanoff, S., Kim, S.-H. (Eds.), Dreamscapes of Modernity: Sociotechnical Imaginaries and the Fabrication of Power. University of Chicago Press, Chicago and London, pp. 1–33.

Kim, S.-H., 2015. Social movements and contested sociotechnical imaginaries in South Korea. In: Jasanoff, S., Kim, S.-H. (Eds.), Dreamscapes of Modernity: Sociotechnical Imaginaries and the Fabrication of Power. University of Chicago Press, Chicago and London.

Mutter, A., 2019. Mobilizing sociotechnical imaginaries of fossil-free futures – electricity and biogas in public transport in Linköping, Sweden. Energy Res. Social Sci. 49, 1–9.

Rivas, J., Schmid, B., Seidl, I., 2018. Energiegenossenschaften in der Schweiz: Ergebnisse einer Befragung, vol. 71. WSL Berichte. https://www.dora.lib4ri.ch/wsl/islandora/object/wsl%3A18943. (Accessed 6 May 2021).

Sager, F., 2014. Infrastrukturpolitik: verkehr, Energie und Telekommunikation. In: Knoepfel, P., Papadopoulos, Y., Sciarini, P., Vatter, A., Häusermann, S. (Eds.), Handbuch der Schweizer Politik. Manuel de la politique Suisse. Verlag Neue Zürcher Zeitung, Zürich, pp. 721–748.

Salomon, T., Jedliczka, M., Marignac, Y., 2015. Manifeste NégaWatt : En route pour la transition énergétique ! Actes Sud Editions, Arles.

Schmid, B., 2019. Energy Cooperatives in Switzerland – A Study of Energy Cooperatives and Their Interrelations with Local Governments in the Swiss Federalist System. Dissertation. University of Bern.

Schmid, B., Meister, T., Klagge, B., Seidl, I., 2020. Energy cooperatives and municipalities in local energy governance arrangements in Switzerland and Germany. J. Environ. Dev. 29 (1), 123–146.

Schmid, B., Musiolik, J., 2021. Research Report on Cooperative Organisational Models for Renewable Energy in Switzerland. SONNET: EU Horizon 2020 Grant agreements no: 837498. https://zenodo.org/record/4727581. (Accessed 6 May 2021).

Serlavós, M., 2018. Les citoyens deviennent acteurs de la transition énergétique - l'exemple de l'énergie citoyenne en Suisse romande. In: Niwa, N., Frund, B. (Eds.), Volteface. La transition énergétique: un projet de société. Lausanne, Paris: Editions d'en bas. Editions Charles Léopold Mayer.

Serlavós, M., 2020. L'énergie citoyenne : levier pour une société autonome et durable ? L'exemple d'une transition énergétique et des imaginaires en Suisse romande. Dissertation. University of Lausanne.

Seyfang, G., Hielscher, S., Hargreaves, T., Martiskainen, M., Smith, A., 2014. A grassroots sustainable energy niche? Reflections on community energy in the UK. Environ. Innov. Societal Trans. 13, 21–44.

SFOE – Swiss Federal Office of Energy, 2020. Schweizerische Elektrizitätsstatistik 2019. https://www.bfe.admin.ch/bfe/de/home/versorgung/statistik-und-geodaten/energiestatistiken/elektrizitaetsstatistik.html. (Accessed 6 May 2021).

Verhoog, R., Finger, M., 2016. Governing energy transitions: transition goals in the Swiss energy sector. In: Dorsman, A., Arslan-Ayaydin, Ö., Baha Karan, M. (Eds.), Energy and Finance. Springer International Publishing, Cham, pp. 107–121.

VESE – Association of independent power producers, 2021. pvtarif.ch. https://www.vese.ch/pvtarif/. (Accessed 6 May 2021).

The construction of a citizen-centered ecosystem for renewable energies in France

11

Pascale Château Terrisse [1], *Hajar El Karmouni* [2] *and Marion Maignan* [3]
[1]Toulouse INP ENSAT, Toulouse, France; [2]Université Paris Est Créteil, Paris, France;
[3]Université Picardie Jules Verne, Amiens, France

1. Introduction

This chapter focuses on an innovative multistakeholder regional cooperative, Enercoop Languedoc Roussillon (ELR). ELR brings together a wide variety of members, including public and private actors, as well as citizens and energy communities. This cooperative is a member of the Enercoop network, which produces 100% renewable energy and promotes a sustainable energy transition[1] based on the NegaWatt 2050 energy scenario in France. That scenario stresses the need for a decrease in energy consumption, «NegaWatt» being a term conceptualized by Amory Lovins, designating the energy saved by the change of practices.[2] It illustrates how France could shift to a climate-friendly, nuclear-free, and sustainable energy future through an approach based on three dimensions: energy sufficiency, energy efficiency, and the production of renewable energy.

In line with this scenario and according to the French Agency for ecological transition (ADEME being the French acronym), the energy transition implies both the adoption of behaviors favoring energy saving and the replacement of fossil fuels and radioactive materials by renewable energy sources. The French law on "Energy transition for Green Growth," passed in 2012, aims to reduce French final energy consumption by 50% in 2050 compared to 2012. It also plans to reduce nuclear power and to increase the share of renewable energy to 40% by 2030. However, renewable energies currently represent only about 10% of final energy consumption and 19.1% of French electricity generation.[3] Renewable energies in France integrate approximately 10 different subsectors. Wood energy and hydraulic energy are still the most developed, but wind energy and electric heat pumps are among those that have made significant progress in recent years. At the same time, the importance of renewable energies within the French economy has

[1] The energy transition refers to the transition from an energy system based mainly on carbon resources to an energy mix based on renewable resources (see also ADEME).
[2] https://negawatt.org/La-demarche-negaWatt.
[3] The final consumption of primary energy was 249 Mtoe in 2019, compared to 259 Mtoe in 2012, a decrease of 3.8% (the objective announced by the law is −7% by 2023).

Energy Communities. https://doi.org/10.1016/B978-0-323-91135-1.00021-3

increased: in 2017, eight billion euros were invested, and more than 60,000 equivalent full-time jobs were created in this sector.

For French consumers, many offers are available from renewable energy suppliers. Current regulations do not require a supplier to produce renewable electricity in order to advertise their offer as a "green solution,"[4] so several energy suppliers advertise an environmentally conscious option without actively contributing to the development of additional renewable energy production. Greenpeace ranks companies according to the percentage of their revenue invested in renewable energy. Enercoop is ranked first by Greenpeace as the cooperative is a 100% renewable energy supplier and supports small independent producers and energy communities with newly built renewable energy capacity. Enercoop is structured through a national cooperative, Enercoop Nationale, and through 11 regional cooperatives mapping the regions of France. Languedoc Roussillon (ELR) is one of the 11 regional cooperatives of Enercoop and is located in the region of Occitanie in the South of France.

The case of ELR reflects a way of structuring and linking energy communities and deepens the understanding of a special form of legal entity, the multistakeholder cooperative, that makes it easier for citizens to team up and jointly act for energy transition. Indeed, in France, multistakeholder cooperatives are used as a tool to organize energy communities because they are the first legal form of cooperative that supports citizen-driven energy actions. In 2019, all energy cooperatives created were multistakeholder cooperatives.

Based on the case study of ELR and its ecosystem, this chapter investigates the following question:

How do multistakeholder renewable energy cooperatives, as part of an ecosystem supporting energy communities, contribute to the ecological, economic, and social transition?

This study analyzes the data collected through the Socio-Anthropological Study of Collective Frugality research project, following a comprehensive and inductive methodological approach. From 2017 to 2019, the research team conducted a series of 29 interviews and observations with ELR and its members in the region of Occitanie.

The chapter is organized as follows:

- Section 2 provides background to Enercoop and ELR;
- Section 3 presents the ecosystem;
- Section 4 examines the main alternative characteristics of the ecosystem;
- Section 5 identifies recommendations for building an alternative renewable energy ecosystem; and
- Section 6 offers concluding comments.

[4] However, they have to purchase a guarantee of origin for the energy they sell, but the financial support generated is not sufficient to develop new renewable energy production devices (https://cdn.greenpeace.fr/site/uploads/2017/03/ecolo-watt-rapport-complet.pdf).

2. Background to Enercoop and Enercoop Languedoc Roussillon

Created in 2005, Enercoop became a network leader and a major player in the French energy transition. At the end of 2020, Enercoop had 99,000 customers at the national level, 54,000 members, and 231 employees. Enercoop chose a status of multistakeholder social cooperative (SCIC is the French acronym).[5] More than half of its profits are reinvested in the company, in particular toward developing energy communities, i.e., citizen-owned renewable energy projects. In addition, both producers and consumers participate collectively in the decision-making process of the cooperative. Enercoop encourages the spin-off of regional cooperatives—11 cooperatives exist as of 2020.

ELR is one of these 11 regional cooperatives and is located in the region of Occitanie in the South of France. ELR is the result of decisions by working group of 20 stakeholders, including the following:

- Private renewable energy producers;
- Environmental organizations; and
- Members of the social and solidarity economy such as cooperative banks and organic supermarkets.

ELR is a supplier of renewable energy. It establishes contracts with its producers and consumers in order to build a supply channel of renewable energy anchored in the territory of Occitanie. ELR also supports citizens' communities through citizens' renewable energy production groups. ELR supports them and wants them to be a central element of the transition. Finally, ELR carries out lobbying activities to open up the energy market and develop renewable energy production. On the demand side, ELR provides training in energy saving and energy transition through courses at the university or through education tools such as Dr. Watt.[6] ELR also fights against energy insecurity with actions mainly carried out by an association called "Friends of Enercoop."

ELR has 11 employees and the same number of members on its Advisory Board. It has 3500 customers at the Occitanie regional level, including approximately 1400 shareholder customers. ELR is also a multistakeholder social cooperative that enables any citizen to become a shareholder. Every shareholder has a voice in the overall

[5] SCICs are multistakeholder cooperatives with a social vocation (Groupe Pap Scic, 2016). These organizational forms have emerged in several countries in recent years (Italy, Great Britain, and Canada) and propose to bring together multiple members around a project of collective interest. SCICs were created in France in 2001. This status responds to a need for collective innovation and multistakeholder management of cooperatives. They allow a diversity of actors, distinguished by their relationship to the activity (employee, volunteer, user, financier) and their nature (individuals or entities, public, or private organizations), to become partners. They are based on the implementation of a democratic governance among heterogeneous actors.

[6] Dr. Watt is a pedagogic training program proposed by Enercoop to its clients, which aims to reduce energy consumption in daily decisions. The service is free for Enercoop shareholders and fee based for others.

management of the Enercoop project, regardless of the number of shares they own. Cooperative governance improves organizational transparency, democracy, and openness (Bonnemaizon et al., 2019). Similarly, ELR allows producers to participate in the governance. The organization itself has six categories of cooperative shareholders including customers, producers, employees, partners, supporters, and local authorities. ELR enhances equity in decision-making as each group has a voice, thereby strengthening the bonds among a variety of actors. ELR therefore appears to be a key player in linking multiple and heterogeneous actors working for the ecological, economic, and social transition in Occitanie. Beyond the internal organizational perimeter of the cooperative, ELR is also in contact with, and relies on, a whole ecosystem to support the transition. In particular, it promotes the emergence and development of energy communities producing renewable energy. Enercoop defines the energy communities as citizens' collectives and "projects initiated, carried out, and financed by local actors like citizens, associations, and communities that aim to develop renewable energy installations through the creation of a participatory society, with local governance."[7]

3. The Enercoop Languedoc Roussillon ecosystem: linking and empowering energy communities based on citizens' collectives

ELR relies on an ecosystem to disseminate innovative social and environmental practices. The concept of ecosystem finds its origin in biology and refers to a dynamic partnership comprising a natural environment and the living beings that make it up. This approach highlights the complexity of systems that are linked by interdependent relationships. The notion of ecosystem applied to organizations also emphasizes the interactions, interdependencies, and coexistence among heterogeneous socio-economic initiatives that form a community of interest organized as a network (Stam, 2015). The concept of "ecosystem" is particularly relevant to analyze the system of actors supporting the energy transition and to consider its different scales of bottom-up actions: from citizens and energy communities to regional and national multistakeholders' cooperatives that support and organize them.

Therefore, the ELR ecosystem is composed of diverse yet complementary actors who can be classified into six categories according to their type of action in favor of the energy transition:

- *Organizations for advocacy of the ecological transition*: These actors are engaged in advocacy at the national level for the transition in different areas: renewable energy, short food chains and organic food, water, and biodiversity. Examples of such organizations are Biocoop, a cooperative network of organic supermarkets acting for the agricultural transition, and Greenpeace France, promoting an environmental transition.

[7] https://www.enercoop.fr/nos-cooperatives/languedoc-roussillon.

- *Organizations anchoring the transition in the territory:* These organizations are involved in projects of ecological, economic, and social transition anchored in Occitanie. In this category, several energy communities and private actors were identified that promote renewable energy projects, energy conservation, and recycling and regenerative agriculture. Energy communities are indeed one of the main actors through which Enercoop develops a decentralized model of renewable energy based on local citizens communities' projects.
- *Tailor-made organizations and financial institutions financing the transition:* These are financial players specializing in the energy transition. They realize forms of financing, mainly based on participative and alternative platforms, segmented by the size of the renewable energy production projects. The aim of these organizations is to support projects in line with the goals of the energy transition: energy investment in the production of renewable energy. Other funders in the ecosystem are well-known financial institutions. These are large banks or insurance companies involved in the development of the social and solidarity economy (such as MACIF and MAIF mutual insurance companies in France) that provide financial services. They also invest in the capital of Enercoop at a national level.
- *Dominant public and private actors of the centralized energy market:* These are the major public and private companies of the French centralized energy sector such as EDF, Total, ENGIE, and ENEDIS. In recent years, they have become involved in the renewable energy sector. In France, these stakeholders play a major role even in decentralized energy models. For example, EDF has an obligation to purchase the electricity produced by any player and these actors provide the distribution infrastructures to connect the electricity generated to the system. Their relationship to the transition is perceived as exclusively instrumental.
- *Governmental organizations supporting energy transition:* These organizations include actors such as the Government, the Ministry of Ecological and Solidarity Transition, as well as regional and departmental councils, supporting the transition. They provide information on the transition, carry out educational actions, and finance or support energy efficiency or renewable energy production. These public stakeholders are identified as important levers to develop the savings/efficiency themes of the transition pictured by the Negawatt Scenario, but they focus mainly on production in their financing choices and their discourse.
- *Engineering companies for renewable energy:* These enable the commercialization and production of renewable energy, in partnership or in competition with Enercoop, offering services such as energy audits and feasibility studies, functional and technical analysis, design of tailor-made energy efficiency solutions, energy performance contracting and monitoring, or training and delegation of skills. For example, Cayrol International works from the market research and design phases to the production launch of solar power energy community projects.

This description of the ecosystem shows that ELR developed a network that is both anchored locally, thanks to energy communities, and oriented toward the broader national goal of the transition. The multistakeholder social cooperative status appears as an open and flexible organizational structure that helps to identify joint solutions to favor energy transition, based on interdependence among multiple stakeholders and heterogeneous organizations. It offers the possibility of building an alternative decentralized energy system that connects local renewable energy generation with local consumption, as opposed to the centralized and top-down mainstream organization of the French energy sector.

Instead of relying on networks already established in the energy sector, the ELR connects energy, agriculture, and social issues, rather than treating them as being separate. It is an alternative ecosystem, which encourages solidarity, collaborative practices, and the sharing of skills among potentially competing organizations. For example, energy communities collaborate, exchange, and mutualize experiences, knowledge, and services. Engineering companies developing renewable energy projects work together with ELR to help citizens' renewable energy production groups. The aim of the ecosystem—contributing to the ecological transition—creates a bond among the participation organizations that enables them to overcome conventional barriers to collaboration. The links of interdependence between the members of the ecosystem appear as a strength that encourages exchanges among peers, the pooling of resources, and collective learning. However, the energy communities still have to rely on the existing electricity distribution network and balancing services provided by the dominant public and private actors of the centralized energy market.

4. The main organizing alternative characteristics of the ELR ecosystem

This mapping of the ecosystem demonstrates that ELR has thought out the development of its actions in favor of the energy transition. Faced with a challenge such as the transition, which goes beyond individual and organizational interests, this open and flexible form facilitates geographical and sectoral coherence.

4.1 A geographical coherence at national, regional, and local levels

The ecosystem of Enercoop reflects a multilevel geographic coherence, to implement a decentralized energy system and support energy communities. A multilevel structure can be observed from the national level with one national cooperative, to the regional one with the 11 regional cooperatives and the local one with 50 energy communities. In that way, it brings the producers and citizens closer together and enables decisions to be made at a local level. Local chains of production and distribution and close relationships at the communities' level are combined with a larger scale of action at the national level. This allows local coverage while, at the same time, aiming for national social change. Therefore, this ecosystem in Occitanie is much more than a regional variant of a national program with a top-down dynamic as it creates and relies on bonds of reciprocity top-down and bottom-up.

At the national level, ELP participates in the governance of Enercoop National, through the allocation of four seats for regional Enercoop cooperatives on the Board of Directors. Altogether, interests are organized according to seven categories of shareholders:

- Consumers, who want to be supplied with green energy;

- Renewable energy producers, who want to sell their production at a fair price;
- Employees, who want value-sharing, fair wages, and democratic working conditions;
- Founders, who ensure that the ethics of the project are respected;
- Governmental organizations, who aspire to a reasoned and green management of the territories;
- Regional cooperatives, which are looking for the support and expertise of Enercoop National in its business as a supplier of renewable energy; and
- Partners and supporters looking for the promotion and use of renewable energy.

The results of this multistakeholder social organization are many. For example, the tariffs of electricity are codeveloped by involving shareholders, directors, representatives of regional cooperatives, employees, and managers from focus groups. The construction of the tariff schedule follows a set of principles—fairness, transparency, and social and environmental commitment—and the prices are voted on in general assembly.

Each regional cooperative develops its own activities adapted to its territory and skills, including training, consulting, energy diagnostics, and support for project ownership or project management of energy communities. For example, Conflent Energie is an energy community created in December 2013 in Prades under a multistakeholder social cooperative status. Conflent Energie's objective is to achieve energy autonomy at local level through the production of renewable energy (to date, six photovoltaic power plants have been built), the reappropriation of energy control by local actors through support and advice, as well as raising awareness on these issues. With the support of Conflent Energies, another energy community, CatEnR, was created in May 2014. CatEnR is also a multistakeholder social cooperative and aims to finance and develop renewable energy in the Eastern Pyrenees. It has carried out several photovoltaic projects and accompanies local government in technical-economic studies to renovate buildings from an energy perspective.

4.2 A multisectoral coherence insured by cooperation values

Although energy communities structure the ecosystem around the production of renewable energies, the actors in the ecosystem adopt a multisectoral approach to the transition, integrating concerns about food, water, finance, and digital issues. Cooperation is emerging among the players involved in the transition in these different sectors. ELR, like other regional Enercoop cooperatives, is an electricity supplier to local organic food cooperatives (Biocoop) and is working to develop renewable energies on organic farms. The key issue of financing renewable energies and the search for autonomy from the "traditional" institutional financial sector is a recurring concern, which has led to the creation of tailor-made financing organizations such as Energies partagée and Energie Solidaire. This intersectoral dynamic can be found at the level of energy communities, whose members are involved in multisectoral initiatives, through the creation of associations and consumer cooperatives, participatory finance, popular education, and the use of free software.

The coherence of this multisectoral approached is insured by democratic values and practices adopted by what is referred to as the social and solidarity economy.[8] While taking financial risk and striving for an economic balance, the actors of the ecosystem wish to emerge from a "growth-based economy" that is overconsuming resources. Apart from the "historical" or "conventional" players in the energy and banking sector, the ecosystem is therefore mainly composed of social and solidarity economy organizations: associations, cooperatives, and mutual benefit societies of insurance. Moreover, the multistakeholder social cooperative status is the optimal social innovation of the social and solidarity economy, that is able to reunite multiple and heterogeneous actors. Finally, some organizations in the ecosystem, although they have "conventional" company status, have opted for democratic governance (one person, one vote).

The integration of these actors in the energy market, and their strong links to governmental organizations, places the ecosystem in a plural economy model. The transition is also an economic one: from a unique global model, based on economic growth, increased debt and the plundering of natural resources, to a decentralized federation of social and ecological preservation economies. Ecosystem development does not focus exclusively on economic growth. Instead its main goal is to preserve the resources and to encourage the adoption of this behavior and, in doing so, promoting a sustainable development. In contrast to the capitalist model, the ecosystem stakeholders support a "pluralistic economic model" (Polanyi, 1944) based on democratic values and promoting multiple interdependencies.

5. Recommendations for building an alternative renewable energy ecosystem

Finally, the case study of ELR's ecosystem allows us to identify three fundamental, interlinked pillars involved in the construction of ecosystems that promote economic, ecological, and social transition.

5.1 Pillar 1: a shared definition of the transition supported by a national scenario

The first pillar relates to the vision of the transition shared by the members of the ecosystem. This is an essential element in coordinating the players: it is necessary to identify converging values and objectives to promote the transition and the expected impact. In the case of ELR, the common reference for the energy transition is the

[8] "Social and solidarity economy is a sector designating enterprises and organizations, in particular co-operatives, mutual benefit societies, associations, foundations, and social enterprises, which have the specific feature of producing goods, services, and knowledge while pursuing both economic and social aims and fostering solidarity" (ILO Regional Conference on Social Economy, Africa's Response to the Global Crisis, October 2009).

Negawatt scenario, articulated around three major dimensions: efficiency, energy saving, and renewable energy production.

The ELR case indicates that a national scenario for the energy transition, whose assumptions and results are known and shared, makes it possible to build and disseminate this homogeneous vision, at least among the actors acting outside the classical capitalist models. It sets a blueprint that influences the cognitive maps of the actors, allowing them to find common goals. It can also be broken down into regional, local, or sectoral scenarios that allow each actor in the ecosystem to have shared references and objectives. These scenarios are used to structure the discourse and demonstrate the feasibility of the transition as well as to identify actions, variables, and measures to propose coherent solutions.

Nevertheless, the appropriation of the Negawatt scenario at a regional or local level often leads actors to privilege some of its dimensions and to hide others. In the case of ELR, a consensus is emerging around the need to produce renewable energy to move toward greater sustainability. This is the main dimension, mentioned by most of the ecosystem members. The production of renewable energies is then a way to reappropriate energy, in particular to preserve local resources. The two other dimensions of the Negawatt scenario—energy saving and efficiency—have less support from public and private actors. The region supports efficiency through housing renovation and an insulation program. The reference to energy saving remains marginal. The energy production dimension thus seems to be the one on which most actors can agree.

5.2 Pillar 2: public and private funding dedicated to renewable energy ecosystems

Renewable energy production projects have a very high initial investment, a long lifetime of 20–40 years, and relatively low operating costs. The ELR case shows us the relevance of financial instruments based on the complementarity between public and private actors of its ecosystem.

The government of the Occitanie Region appears to be an essential financial support for the production of renewable energies, particularly for energy communities. Indeed, a call for projects "Euro Citoyen"[9] carried out by the Region since 2014 has structured and facilitated several projects of energy communities producing renewable energies. Public support is important in managing financial risks and providing capital, but it must be complemented, often by private funding. Energie Partagée is a hybrid tool, both financial and associative, exclusively financing collective citizen projects in the production of renewable energies. In addition, a financial institution dedicated to the social and solidarity economy, La Nef, has very close and strong financial and governance links with Enercoop, being a cofounder of the organization and its bank. It sits on the Board of Directors of Enercoop and the regional Enercoops. It is a cofounder of

[9] A subsidy paid by the Region to citizen projects. For each citizen euro invested, the Region contributes 1€. The maximum amount paid by the Region for a project is 100.000€.

Energie Partagée and cofinances its projects. This bank also finances some energy communities, while others rely on the traditional banking system. So, private financial actors, as well as conventional banking actors with strong social and solidarity values, contribute to financing.

5.3 *Pillar 3: promoting knowledge transfer about renewable energies to citizens*

Finally, the plurality of actors of the ecosystems permits the transfer of knowledge and skills related to renewable energies. A final key factor of coherence and stability of the ecosystem is the reappropriation of energy issues and knowledge by all, and in particular by the citizens. Thus, as one of the results of the research, training activities and accompanying actors seem to be essential to support the transition. They are promoted by actors like engineering consultancies or governmental organizations. In the ELR ecosystem, the Association of Local and Renewable Citizen Energies incubates and accompanies citizen collectives in the Languedoc Roussillon. This association was created by ENERCOOP, the Occitanie Regional Government, and ADEME in 2015 with the goal of federating, accompanying, and training citizen collectives on governance, communication, and legal arrangements in the implementation of their renewable energies production projects.

In summary, a shared definition of the transition, the promotion of knowledge transfers on renewable energies, and private and public financing dedicated to renewable energies are key factors to build energy ecosystems that promote economic, ecological, and social transition.

6. Conclusion: the appeal of energy communities to citizens

Multistakeholder social cooperatives help energy communities created by citizens to produce renewable energies and to save energy. The ecosystems of the cooperatives are a great means to restructure the energy system to become more local, democratic, citizen centered, and decentralized. This new form of cooperatives contributes greatly to a more decarbonized and flexible energy system and can be thought of as "community energy enablers" that initiate and support citizens' energy projects capabilities by grouping, linking, and providing communities with technical and economic services, tools, and resources. More than just "community energy intermediaries" that have been depicted as nongovernmental organizations or think tanks (Hargreaves et al., 2013), multistakeholder social cooperatives can help to structure and organize energy communities in France in a way that emancipates citizens. Beyond the practical goals of producing renewable energy, communities could indeed be driven by cultural influences and cooperative arrangements and values (Wirth, 2014). Being shareholders of an energy multistakeholder cooperative, citizens can participate to renew and decentralize energy and organizational systems to support economic, ecological, and social transition. So, these cooperatives create communities that integrate citizens into the electricity system and into the broader economy, enabling them all to participate equally.

Bibliography

Bonnemaizon, A., El Karmouni, H., Maignan, M., 2019. Les visages de la démocratie dans les sociétés coopératives d'intérêt collectif (SCIC) au prisme du concept d'indétermination démocratique. Revue Interdisciplinaire Manag. Homme Entreprise (35), 50–76.

Groupe Picri PAP Scic, 2016. Les Scic, entreprises de demain. Le multisociétariat à l'épreuve de la gestion. RECMA 2 (340), 52–64.

Hargreaves, T., Hielscher, S., Seyfang, G., Smith, A., 2013. Grassroots innovations in community energy: the role of intermediaries in niche development. Glob. Environ. Change 23 (5), 868–880.

Polanyi, K., 1944. The Great Transformation. Beacon Press, Boston.

Stam, E., 2015. Entrepreneurial ecosystems and regional policy: a sympathetic critique. Eur. Plann. Stud. 23 (9), 1759–1769.

Wirth, S., 2014. Communities matter: institutional preconditions for community renewable energy. Energy Policy 70, 236–246.

Energy communities' social role in a just energy transition

12

Florian Hanke[1], *Rachel Guyet*[2] *and Mariëlle Feenstra*[3]
[1]European University Viadrina, Frankfurt, Germany; [2]Centre International de Formation Européenne, Nice, France; [3]University of Twente, Twente, Netherlands

1. Introduction

Energy justice envisions elements of a global energy system that fairly distributes both energy services' benefits and burdens and can be used as a framework to identify energy injustices (Sovacool et al., 2017). An example of energy injustice is that only homeowners owning a property with a roof benefit from financial incentives like tax reductions or subsidies to install photovoltaic (PV). Further, more than 50% of households in Germany are tenants. They do not own a roof for PV to produce clean energy or lack capital to cover upfront investment costs. Despite programs such as "Mieterstrom,"[1], tenants often remain excluded from energy policy promoting energy prosumption, that is producing a part of the energy consumed, with public subsidy programs. Not only are these households unable to prosume, but they also carry a considerable part of energy transition costs (Schneller et al., 2020). Simultaneously, up to 82 million European households struggle to pay their energy bills (Bouzarovski et al., 2020). Threatened by energy poverty, (energy) vulnerable groups[2] are often excluded from shaping energy transition. Energy justice looks at such dynamics to identify injustices in the energy system and ways to overcome them.

Over the past decades, energy communities have become major actors driving energy transition (see the chapter by Spasova and Braungardt; Caramizaru et al., 2020). In this light, academics and policymakers increasingly discuss energy communities as taking a central role in overcoming energy-related injustices with a democratic, equity enhancing approach (Baxter et al., 2020). For example, EU legislation in the form of the recast renewable energy directive (RED II) stipulates "opportunities for renewable energy communities to advance energy efficiency at household level and (...) fight energy poverty" (European Parliament and European Council, 2018 Recital 67). RED II further links an enabling framework "to promote and facilitate the development of renewable energy communities" (Art. 22) with the obligation to ensure the participation of all "consumers, including those in low-income or vulnerable households" (European Parliament and European Council, 2018 Art.22 (f)).

[1] Housing providers have the possibility to sell renewable energy produced on the premisses directly to tenants at an affordable cost, facilitated by a subsidy.

[2] Thomson et al. (2018) explore different facets of energy vulnerability among which low-income, high-energy costs, and low energy efficiency.

Energy Communities. https://doi.org/10.1016/B978-0-323-91135-1.00027-4

Regarding energy justice, some research analyzes the contribution of individual community energy initiatives to energy justice.[3] However, the extent to which these initiatives reach different social groups is still underresearched. Available data on energy communities in Germany reveal a relatively homogenous membership structure: members are mostly middle-aged men with a high education level and income. Women, young families, and low-income households remain the exception (see the chapter by Hackbarth and Löbbe). Furthermore, according to a study carried out by the authors, a vast majority of energy communities do not include vulnerable groups or lack specific measures to empower their participation (Hanke et al., 2021).

This chapter critically discusses energy communities' emerging social role in contributing to a democratic, welfare-enhancing, and just energy transition. It highlights different participatory prerequisites and their impact on the participation of vulnerable groups in energy communities.[4] The discussion mainly draws on data available for energy communities in Germany (see also chapters by Spasova and Braungardt as well as Benedettini and Stagnaro; Hanke et al., 2021).

The chapter is composed of four sections:

- Section 2 introduces energy communities' technical and social role and the link to energy democratisation;
- Section 3 introduces the energy justice framework to underline existing energy (in)justices in the energy system; and
- Section 4 discusses energy communities' contributions, challenges, and opportunities to contribute to energy justice followed by the chapter's conclusions.

2. Toward democratizing energy—energy communities' social role

Energy communities' role in energy transition entails a technical, ecological, and social component. To differentiate their role from other actors, energy communities should combine cost-effective and cost-competitive clean energy with greater equity in energy transition (European Parliament and European Council, 2018). Contributing to energy decentralization, they invest in new RE projects and energy efficiency measures. In the past 2 decades, they have become an essential factor in driving forward energy transition creating and investing in needed RE production capacities. By 2030, energy communities can produce up to 17% of installed wind

[3] See, for examples, Fuller and Bulkeley (2013) and Mundaca et al. (2018).

[4] As explored in greater detail in the chapter by Rossetto et al., defining energy community initiatives is difficult due to their diversity across different national contexts. In this chapter, we apply the definition of renewable energy communities in RED II. As a result, energy communities are (i) legal entities, (ii) based on open and voluntary participation, (iii) effectively controlled by members who live in the vicinity of the RE projects. In contrast to purely commercial actors, energy communities provide environmental or social benefits for members and the local community [25 Art.2 (16)].

capacity and 21% solar (Caramizaru et al., 2020). By 2050, energy communities could produce up to 37% of electricity.[5]

Energy communities' social role entails empowering all citizens to participate. Energy democracy reflects this social dimension and is closely linked to environmental and energy justice (Szulecki, 2018). Where energy justice investigates the moral implications of energy decisions (Sovacool and Dworkin, 2015), energy democracy propagates a political dimension. It stresses the importance of citizen control over an energy system controlled by private and profit-making energy suppliers (Szulecki, 2018). Repoliticizing the energy system connects individuals with decisions on energy production and consumption. In consequence, energy democracy promotes citizens' active involvement and participation in energy transition.

Instead of energy policy being a "technoscientific domain reserved for experts" (Szulecki, 2018), energy democratization shapes socio-political institutions to connect policymaking with the individual citizen. Energy democracy is thus a "political goal, in which citizens are the recipients, stakeholders as consumers or producers—and account holders of the entire energy sector policy" (Szulecki, 2018). However, this new form of energy governance demands high civic empowerment and broad participation of informed, aware, and responsible prosumers in an inclusive and transparent decision-making process. Therefore, coownership of energy infrastructures becomes necessary to create and maintain citizen empowerment and participation in the energy system (Szulecki, 2018). In this light, energy communities' social role is about enabling collective forms of energy coownership. Thus, energy communities pool individual resources to enable citizen participation (Holstenkamp et al., 2018).

To engage in a social role, energy communities should open their membership to citizens who are neither traditionally part of energy decision-making processes nor benefiting from it, e.g., low-income and vulnerable households. In a survey carried out by the authors, in Germany, out of 51 energy communities 18 report offering lower energy tariffs compared to market prices and energy efficiency measures (Hanke et al., 2021). Low-income, high energy prices, and low energy efficiency are the main drivers of energy poverty (Thomson et al., 2018). Thus, when energy communities empower these households to participate, their potential benefits, like lower tariffs and increased energy efficiency, could be shared to mitigate energy poverty (Hanke and Lowitzsch, 2020). Membership is the key to these benefits. In short: This chapter defines energy communities' social role as empowering all citizens, especially the most vulnerable, to participate in shaping energy transition.

[5] Access to the report via https://friendsoftheearth.eu/wp-content/uploads/2019/01/community_energy_booklet_final.pdf.

3. The energy justice framework

Injustices in the energy system can be linked to social class, race, gender or spatial, social, and economic inequalities.[6] The energy justice framework aims at identifying both ignored social groups affected by energy injustices, and processes to overcome such injustices (Jenkins et al., 2016). Over the past 30 years, economic inequality has steadily increased in almost all high-income countries, reflecting class society (Piketty, 2016). Moreover, economic inequality affects energy consumption and reduces opportunities to participate in shaping energy transition (Bouzarovski and Tirado Herrero, 2017). Thus, applying the energy justice framework to energy communities must entail questions of social and economic inequality and where they intersect. In the following, the three energy justice tenets, *procedural*, *distributional*, and *recognitional,* are applied to energy communities. This chapter discusses how these three tenets are interlinked.

3.1 Energy justice in energy communities

Procedural justice refers to equitable procedures that allow all local stakeholders to engage and participate in energy transition in a nondiscriminatory and inclusive way (Walker and Day, 2012). Regarding energy communities' social role, inclusive procedures offer different social groups access to energy communities' benefits and services and enable citizens to participate as members. Therefore, citizen participation enhances democratic legitimacy. However, without recognizing the inequality of opportunities for different citizen groups to participate, "citizen participation" fails to deliver its democratic and empowering promise and risks becoming empty rhetoric.

Thus, bringing about procedural justice requires inclusive procedures that respect the specific needs of different social groups to enable their participation in energy communities (Hanke and Lowitzsch, 2020). Procedures to facilitate membership of underrepresented groups, like low-income and energy-vulnerable households, need to be based on a thorough understanding of their living situations. In consequence, to establish inclusive procedures, energy communities need to start with recognizing such specific needs—the core of recognitional justice (Walker and Day, 2012).

Recognitional justice entails recognizing vulnerable groups' structural underrepresentation in energy communities and the unequal distribution of transition costs and benefits. It also entails recognizing the effect of economic and social inequality on opportunities for participation. Ideally, this leads to a discussion of prerequisites for participation and their impact on different social groups. As a result, procedures could emerge which respond to all citizens' needs and improve their opportunities for participation. Individuals belonging to different social groups have different capabilities to participate (Sen, 1999). In this light, *recognitional justice* researches into which sections of society are misunderstood or ignored or misrepresented (Walker and Day, 2012).

[6] See Feenstra and Özerol (2021), Healy et al. (2019), Lee and Byrne (2019).

Distributional justice looks at energy injustices linked to production and consumption (Jenkins et al., 2016). It highlights the question of who benefits from energy transition mechanisms. For example, subsidies for electronic vehicles are worth hundreds of bus tickets and could be used to finance free public transport, a measure available to vulnerable groups who cannot afford electric vehicles even with subsidies (Sovacool et al., 2019). The same applies to energy prosumption. Over the past decade, private households and energy communities received subsidies to install PV and wind energy. However, vulnerable households are not eligible for subsidies as they do not own a house or partake in energy communities, a circumstance raising distributional issues.

In most cases, access to energy communities' benefits, the basis for their democratic legitimacy, is linked to membership. Thus, inclusive procedures that allow vulnerable groups' participation are the basis for overcoming distributional injustices. Fig. 12.1 summarizes the main elements of procedural, recognitional, and distributional energy justice in energy communities.

3.2 Membership prerequisites and energy justice

To participate in an energy community, potential members need to know about it. They need to receive a membership invitation, pay for a share or membership fees, or have the time to volunteer (Hanke et al., 2021). Thus, conditions for membership influence opportunities for participation in energy communities. These, in turn, affect their equity-enhancing role in energy transition. For underrepresented, vulnerable, and energy-poor households, membership prerequisites need to be assessed from two different perspectives: the energy community and the household.

In many markets, as in Germany, energy communities are embedded in a highly competitive market environment; the discontinuation of state support, e.g., in

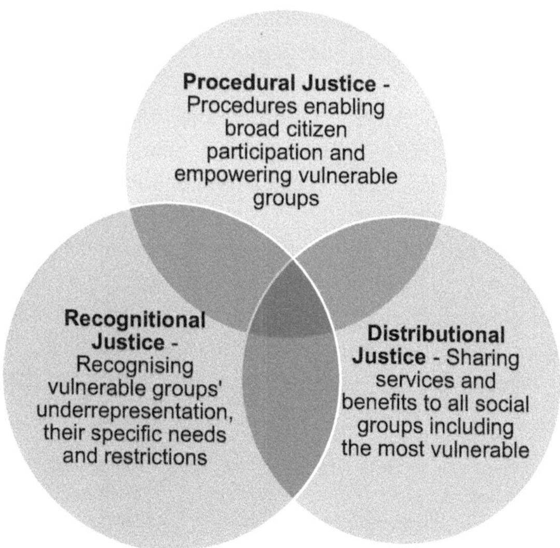

Figure 12.1 Energy justice in energy communities.

Germany, forces energy communities to focus on staying competitive and reduces the scope for inclusive action. Especially, smaller energy communities[7] have limited access to borrowing capital (Lowitzsch and Hanke, 2019b). Thus, they rely on their members' financial contributions to fund projects. Given that organizational resources such as financial and human capital are limited and used to produce clean energy and make profits (Drewing and Glanz, 2020), few resources are left to engage in a social role. In the end, these factors reduce energy communities' ability to open up to broader social groups (Hanke et al., 2021).

In Germany, motivations of energy communities' members are very diverse (Radtke, 2016). In general, smaller energy communities are motivated by contributing to energy transition and environmental protection. The larger the energy community in terms of members or investment volume, the more members might expect financial benefits (Radtke, 2016). Further, energy communities' initiators and founders often determine the organizational purpose and whether an organization engages in a social role. Often steering decision-making, they significantly shape the organizational culture, introduce rules, norms, and beliefs (Drewing and Glanz, 2020). The organizational culture, in turn, functions as a gatekeeper and often determines who becomes a member. Drewing and Glanz (2020) conclude that such decision-makers' biases often prevent the involvement of certain groups who are not perceived as sharing the same set of values, e.g., contributing to energy transition or who do not possess sufficient financial means to participate.

On a practical level, to gain new members, energy communities need to engage with different social groups by informing them about opportunities for participation. However, it is hard to reach vulnerable households; they often do not have access to social networks. In addition, the language used to describe projects often fails to consider the linguistic and cognitive barriers of some social groups. So, establishing effective communication channels is essential in reaching specific groups. The Electricity-Savings-Check (Stromsparcheck) provided by Caritas Germany has yearlong experiences in reaching out to energy-poor households and reports communication difficulties, highlighting the need for direct engagement, easily understandable language, and tangible benefits for the target group.

The next operational challenge is linked to financial participation. In Germany, the average minimum financial contribution—usually individuals buy shares to become members—amounts to 545 Euro (Dannemann, 2020). These amounts often determine who can benefit from energy communities' services.

Vulnerable households' perspectives and their ability to participate in and benefit from energy communities remains widely unknown—a logical consequence of their underrepresentation. Still, we can assume that energy literacy is not only required to switch energy providers, but also to understand energy communities' activities. In this light, receiving information about energy communities and applying that

7 Radtke (2016) distinguishes between energy communities with a focus on wind energy and solar energy. Both forms differ in terms of number of members and level of investment. Wind projects require higher investments but tend to have less citizens involved compared to solar projects with lower investments but higher membership numbers.

knowledge is an additional level of energy literacy. In addition, vulnerable groups usually do not have access to local social networks or individual players involved in local energy communities (Drewing and Glanz, 2020). Therefore, participation becomes difficult simply because households remain in the dark about opportunities to participate.

What is more, experiencing poverty affects vulnerable households' perception and choices.[8] Although empirical data on these dynamics in energy communities are not yet available, we assume that similar effects affect vulnerable groups' participation in energy communities (Lowitzsch and Hanke, 2019a). Also, potential members need to understand and mitigate economic risks linked to financial participation in energy communities (Holstenkamp and Kahla, 2016). In addition, vulnerable groups often rely on social welfare, e.g., subsidies for rent or electricity costs. Usually, the main prerequisite for receiving social welfare is the proof that a household does not have sufficient income or assets to sustain a minimum standard of living. Any extra income is automatically deducted from these payments. Therefore, any benefit from financial participation in an energy community, e.g., in the form of dividends, is an extra income that potentially reduces social welfare. Nonetheless, many vulnerable households rely on these payments. Thus, a reduction threatens these people's livelihood. In this case, they refrain from participating (Lowitzsch and Hanke, 2019a). Moreover, energy communities often rely on their members' time for volunteering operational tasks (Fischer et al., 2020). Vulnerable households can seldom commit time and skill to become members as they often deal with irregular working hours and other daily challenges (Haushofer and Fehr, 2014; Shafir, 2017). Once energy communities are aware of these difficulties, procedures can be developed to allow

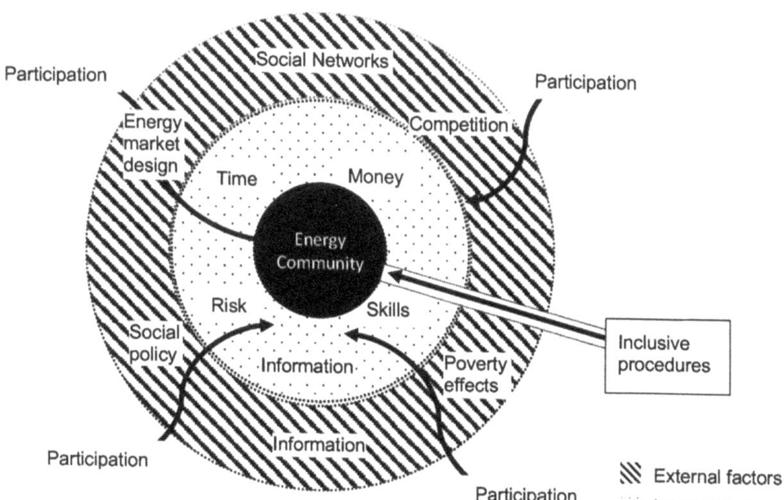

Figure 12.2 Membership prerequisites as barriers to participation.

[8] For examples, see (Haushofer and Fehr 2014).

wider participation of vulnerable households. Fig. 12.2 illustrates the most crucial membership prerequisites on an external and internal level and the inclusive function of procedural justice.

4. Toward energy justice in energy communities

How can these challenges for energy communities and their social role in contributing to energy justice be overcome? Energy justice framework is applied to describe prevailing challenges in this regard and point out some examples of good practice. Fig. 12.3 summarizes the challenges and policy options that could help energy communities address recognitional, procedural, and distributional justice. In contrast to the above, the discussion starts with *recognitional* challenges. Doing so, it is argued that inclusive procedures empowering vulnerable groups' participation only emerge when energy communities recognize vulnerable groups' specific needs and circumstances.

Recognitional challenges: Recognizing the need for a just distribution of costs, benefits, and opportunities for participating in energy transition is the first step toward energy justice. However, such a fair distribution across all members of society is far from being realized. Social and economic inequality, energy vulnerability, and poverty create a set of distinct and intersecting living conditions. Membership prerequisites

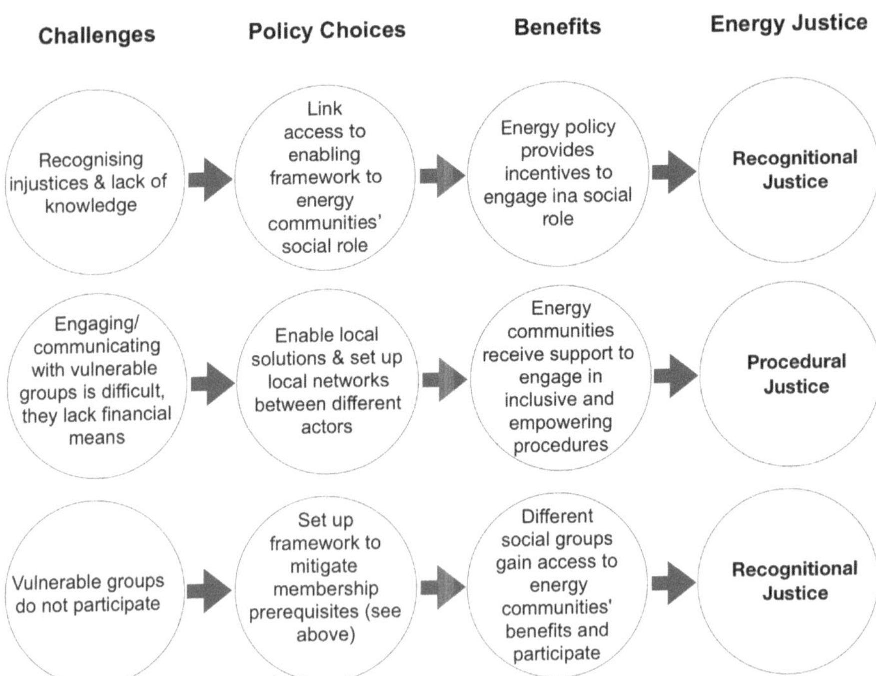

Figure 12.3 Applied energy justice in energy communities.

intersect with these conditions, preventing vulnerable households from participating in and benefiting from energy transition. As a result, vulnerable households remain without a voice in energy transition and are, in fact, often left behind. To develop empowering procedures, energy communities must recognize and understand vulnerable households' distinct living situations and preferences and how they intersect with membership prerequisites.

In practice, understanding vulnerable households is often challenging for energy communities that have never worked with these groups before. Moreover, even energy communities with some experiences still report that getting in contact with energy-poor or low-income households is difficult (Hanke et al., 2021). Local charities, NGOs, or municipalities already working with these groups have valuable experiences and knowledge to share with energy communities—such as Caritas (see above). Caritas trains so-called electricity-saving advisors who visit energy poor households and implement energy-saving measures onsite. Caritas has direct access to these households. Thus, in cooperation with a local energy community, actors like Caritas can help energy poor households access energy communities and maybe even lower electricity prices.

Procedural challenges: Once an energy community plans to engage with vulnerable groups, the question of inclusive and empowering procedures emerges. In general, inclusive procedures mean adjusting conditions for membership to the living conditions of vulnerable groups. For example, informing about the benefits of participation, e.g., lower energy tariffs and energy efficiency would attract attention. In Belgium, the energy community Pajopower reaches out to people in socially vulnerable neighbourhoods with their "Kyoto mobiel," a funny looking vehicle that they park in the community. People get curious and ask about this vehicle. In this way, Pajopower raises attention and gains trust. Like Caritas, local charities have often been working with vulnerable people for many decades. In their daily work, they could also start informing about the benefits of participating in energy communities and thus help energy communities engage with energy vulnerable households.

Further, inclusive procedures mitigate risks linked to participation. For example, the need to buy a share is a significant obstacle for low-income households. Offering special share prices for vulnerable groups below 50 Euros is one way of ensuring that low-income groups can participate. For example, in Belgium, the city of Eklo buys shares in the local energy community and hands them over to vulnerable households to enable their participation. Another example comes from the community-owned wind farm on the Danish Island of Ærø. In cooperation with a local bank, the energy community offers zero-interest loans to vulnerable households to finance their participation. However, in the end, social inclusion depends on whether or not the founders decide to promote procedures and initiatives to bring in vulnerable groups (Drewing and Glanz, 2020).

Distributional challenges: Energy communities can provide tangible benefits such as lower energy tariffs, access to clean energy and energy efficiency measures or dividends as an additional income, or access to social networks (Hanke et al., 2021). As mentioned above, such benefits depend on membership. Finally, through membership, publicly funded support for energy communities in the form of reduced system costs, e.g., levies and taxes, becomes available to different social groups. Thus, facilitating membership by establishing more favorable conditions of membership enhances distributional energy justice. Fig. 12.3 summarizes the challenges and policy options to support energy communities' impact on distributional justice.

Policy implications: Currently, energy communities' new social role receives considerable attention and support from EU bodies (European Commission, 2019). This creates political awareness of the need to merge technical feasibility and social equity in energy transition. In theory, energy communities could be more open to the participation of different social groups. Nonetheless, considering energy communities socially inclusive because they are local disregards the problematic conditions of membership and social inequality.

Therefore, energy policy should acknowledge at least three things:

- First, different social groups of different social classes, gender, and race face uneven capacities and power (Catney et al., 2014). Indeed, they have specific needs and can face difficulties in becoming full members. Yet, participation in energy communities would be of particular benefit for vulnerable groups affected by energy poverty.
- Second, energy communities have limited capacities to adjust their membership conditions to the needs of different social groups.
- Third, in a liberalized market environment, resources are subject to competition with limited scope for inclusive action.

Thus, as a basis for enabling conditions, energy policy needs to acknowledge energy communities' social value, which remains underrecognized in a market-oriented logic of efficiency and competition. For example, current energy policy reduces RE subsidies in Germany (Bundesministerium der Justiz und für Verbraucherschutz, 2021). The same policies continue to subsidize carbon-based energy production (Schmidt and Schumann, 2020). Such energy policy reduces energy communities' economic scope and their resources to address energy injustices. As long as fossil and nuclear subsidies continue to create an economic advantage for fossil and nuclear energy, energy communities producing clean energy must continue to receive subsidies to establish a level playing field on the energy market.

In this light, energy policies should distinguish between RE projects and communities engaging in a social role and those only serving financial interests. Social active energy communities who, for example, enable vulnerable groups' participation or similar activities should be supported. An enabling framework should safeguard financial remuneration on the energy market to cover energy communities' costs, e.g., through guaranteed feed-in tariffs or preferential treatments in public tenders. Energy communities should also be able to share electricity among their members without having to pay full grid tariffs and RE levies. In turn, energy communities would be able to provide affordable clean energy to their members. In general, an enabling framework should remove administrative and financial barriers. It would facilitate energy communities in pooling individual resources and reduce transaction costs for citizen participation in energy transition; it would further promote local networks within local communities and municipalities and consider the specific needs of vulnerable groups.

Similar frameworks recognizing organizations' social value exist already. For example, the concept *economy for the common good*[9] applies social and ecological in addition to financial indicators to account for and rank the social welfare added by an organization.[10] Organizations receive tax incentives when they engage in social and environmental business practices: corporations laying off people, despite increased operational profits, would carry a higher tax burden than a local business that just installed a wind turbine for clean energy supply. Consequently, engaging in social and ecological business practices becomes an economic advantage. This concept could be applied to energy communities and providers: Energy communities that score high on a social welfare ranking would gain access to an enabling framework that includes tax benefits and access to subsidies. Those energy providers that rank low on a social welfare ranking would carry the full tax burden. The extra tax revenues could finance subsidies for socially engaged energy production.

Another pathbreaking example comes from France. Here national legislation recognizes automatically as an energy community social housing providers who install PV and engage in RE production. It declares social housing residents as members by default. Thus, social housing residents, usually not able to participate, benefit from lower energy tariffs automatically. Another example in Spain shows how social welfare can empower low-income households: Instead of monthly payments, unemployed individuals capitalize (a part of) unemployment benefits as a lump-sum to set up or join Sociedades Laborales (a worker-owned company). A similar mechanism could channel social welfare payments in the form of energy bill subsidies to finance membership in an energy community and to help access more affordable clean energy (Hanke and Lowitzsch, 2020).

Such examples of aligning social with energy policy are groundbreaking. As a result, energy communities could contribute more than just clean energy. They could provide other services, e.g., energy advice, energy efficiency, e-mobility, and play a key role in realizing the sustainable development goals (SDGs) that increasingly frame EU strategy: SDG 10 addressing (energy) inequalities, SDG 5 empowering underrepresented groups, including women, SDG 7 promoting access to clean energy to all while contributing to action on climate change (SDG13).

Fig. 12.3 summarizes challenges and policy choices to support energy communities engaging in a social role.

5. Conclusion

Liberalizing the European energy system increased competition between different actors, and new market entrants such as energy communities appeared. To differentiate

[9] For more information, visit https://www.ecogood.org or https://bcorporation.net for similar ideas.

[10] Currently, economic success is measured by financial indicators. However, these financial indicators do not account for social and ecological costs and benefits. As a result, oil companies still rank among the most valuable companies on global stock markets although they destroy the ecological foundation of human life.

energy communities' role from other incumbents and legitimize public subsidies, energy communities combine cost-effective and cost-competitive clean energy with enhanced equity in energy transition. However, energy communities do not have the means to achieve the social dimension of energy transition on their own. Therefore, policymakers need to take ground-breaking political decisions to support energy communities' emerging social role.

While conflicts between a fair energy transition and the current liberalized energy market design prevail, some energy communities drive a democratic approach in energy transition; they create social and economic value and contribute to the SDGs. However, the current energy market does not account for energy communities' added social value. Membership prerequisites and social inequality limit their scope for further social action. Therefore, social and energy policy should recognize energy communities' social value, especially regarding the engagement and inclusion of energy poor and vulnerable households. In designing an enabling framework, policymakers could build on existing concepts that account for the social value brought in by these organizations. Such an enabling framework should only be accessible to energy communities engaging in a social role, e.g., by reaching out to energy-poor households. Only then would energy communities benefit from tax incentives, preferential treatments in public tenders, and financial schemes such as zero-interest loans or regulatory support. In this regard, energy communities could become frontrunners in linking and resolving prevailing challenges of social, economic, and energy policy.

Bibliography

Baxter, J., Walker, C., Ellis, G., Devine-Wright, P., Adams, M., Smith Fullerton, R., 2020. Scale, history and justice in community wind energy: an empirical review. Energy Res. Social Sci. 68, 101532. https://doi.org/10.1016/j.erss.2020.101532.

Bouzarovski, S., Thomson, H., Cornelis, M., Varo, A., Guyet, R., 2020. Towards an Inclusive Energy Transition in the European Union: Confronting Energy Poverty amidst a Global Crisis. Publications Office of the European Union. EU Energy Poverty Observatory, Luxembourg.

Bouzarovski, S., Herrero, S.T., 2017. The energy divide: integrating energy transitions, regional inequalities and poverty trends in the European union. Eur. Urban Reg. Stud. 24 (1), 69–86.

Bundesministerium der Justiz und für Verbraucherschutz, 2021. Gesetz Für Den Ausbau Erneuerbarer Energien (Erneuerbare-Energien-Gesetz - EEG 2021) Anlage 4 (Zu Den §§ 64, 103) Stromkosten- Oder Handelsintensive Branchen.

Caramizaru, A., Uihlein, A., European Commission, Joint Research Centre, 2020. Energy Communities: An Overview of Energy and Social Innovation. Aura Caramizaru Andreas Uihlein, LU.

Catney, P., MacGregor, S., Dobson, A., Hall, S.M., Royston, S., Robinson, Z., Ormerod, M., Ross, S., 2014. Big society, little justice? Community renewable energy and the politics of localism. Local Environ. 19 (7), 715–730.

Dannemann, B., 2020. Energiegenossenschaften 2020 - Jahresumfrage des DGRV.

Drewing, E., Glanz, S., 2020. Die Energiewende als Werk ausgewählter Gemeinschaften? Zur sozialen Exklusivität von Energiegenossenschaften. In: Energiewende und Megatrends - Wechselwirkungen von globaler Gesellschaftsentwicklung und Nachhaltigkeit, vol. 93. Transcript, pp. 275–302.

European Commission, 2019. The European Green Deal. COM(2019) 640 Final. European Union, Brussels.

European Parliament, European Council, 2018. Directive (EU) 2018/2001 of the European Parliament and of the Council of 11 December 2018 on the Promotion of the Use of Energy from Renewable Sources (Recast).

Feenstra, M., Özerol, G., 2021. Energy justice as a search light for gender-energy nexus: towards a conceptual framework. Renew. Sustain. Energy Rev. 138, 110668. https://doi.org/10.1016/j.rser.2020.110668.

Fischer, B., Gutsche, G., Wetzel, H., 2020. Who wants to get involved? Determinants of citizens' willingness to participate in German renewable energy cooperatives. Energy Res. Soc. Sci. 37.

Fuller, S., Bulkeley, H., 2013. 'Energy justice and the low-carbon transition: assessing low-carbon community programmes in the UK. In: Bickerstaff, K., Walker, G., Bulkeley, H. (Eds.), Energy Justice in a Changing Climate: Social Equity and Low-Carbon Energy. Zed Books, London, New York, pp. 61–78.

Hanke, F., Guyet, R., Feenstra, M., 2021. Do renewable energy communities deliver energy justice? Exploring insights from 71 European cases. Energy Res. Social Sci. 80, 102244.

Hanke, F., Lowitzsch, J., 2020. Empowering vulnerable consumers to join renewable energy communities—towards an inclusive design of the clean energy package. Energies 13 (7), 1615.

Haushofer, J., Fehr, E., 2014. On the psychology of poverty. Science 344 (6186), 862–867.

Healy, N., Stephens, J.C., Malin, S.A., 2019. Embodied energy injustices: unveiling and politicizing the transboundary harms of fossil fuel extractivism and fossil fuel supply chains. Energy Res. Social Sci. 48, 219–234.

Holstenkamp, L., Kahla, F., 2016. What are community energy companies trying to accomplish? An empirical investigation of investment motives in the German case. Energy Pol. 97, 112–122.

Holstenkamp, L., Kahla, F., Heinrich, D., 2018. Finanzwirtschaftliche Annäherungen an das Phänomen Bürgerbeteiligung. In: Holstenkamp, L., Radtke, J. (Eds.), Handbuch Energiewende und Partizipation. Springer Fachmedien Wiesbaden, Wiesbaden, pp. 281–301.

Jenkins, K., McCauley, D., Heffron, R., Stephan, H., Rehner, R., 2016. Energy justice: a conceptual review. Energy Res. Social Sci. 11, 174–182.

Lee, J., Byrne, J., 2019. Expanding the conceptual and analytical basis of energy justice: beyond the three-tenet framework. Front. Energy Res. 7.

Lowitzsch, J., Hanke, F., 2019a. Consumer (Co-)Ownership in renewables, energy efficiency and the fight against energy poverty – a dilemma of energy transitions. Claeys & Casteels Law Publishers BV 9 (3), 5–21.

Lowitzsch, J., Hanke, F., 2019b. Renewable energy cooperatives. In: Lowitzsch, J. (Ed.), Energy Transition: Financing Consumer Co-ownership in Renewables. Springer International Publishing, Cham, pp. 139–162.

Mundaca, L., Busch, H., Schwer, S., 2018. "Successful" low-carbon energy transitions at the community level? An energy justice perspective. Appl. Energy 218, 292–303.

Piketty, T., 2016. Das Kapital im 21. Jahrhundert. 1. Auflage in C.H. Beck Paperback. München: C.H. Beck.

Radtke, J., 2016. Bürgerenergie in Deutschland. Springer Fachmedien Wiesbaden, Wiesbaden.

Schmidt, N., Schumann, H., 2020. Wie Europas Staaten Ihre Eigenen Klimaziele Sabotieren. Investigate Europe.
Schneller, A., Kahlenborn, W., Töpfer, K., Thürmer, A., Wunderlich, C., Fiedler, S., Schrems, I., Ekardt, F., Lutz, C., Großmann, A., Schmidt-De Caluwe, R., 2020. Sozialverträglicher Klimaschutz – Sozialverträgliche Gestaltung von Klimaschutz Und Energiewende in Haushalten Mit Geringem Einkommen. TEXTE 66/2020. Umweltbundesamt, Dessau-Roßlau.
Sen, A., 1999. Development as Freedom, first ed. Knopf, New York.
Shafir, E., 2017. Decisions in poverty contexts. Current Opinion in Psychology 18, 131–136.
Sovacool, B.K., Burke, M., Baker, L., Kotikalapudi, C.K., Wlokas, H., 2017. New frontiers and conceptual frameworks for energy justice. Energy Pol. 105, 677–691.
Sovacool, B.K., Dworkin, M.H., 2015. Energy justice: conceptual insights and practical applications. Appl. Energy 142, 435–444.
Sovacool, B.K., Martiskainen, M., Hook, A., Baker, L., 2019. Decarbonization and its discontents: a critical energy justice perspective on four low-carbon transitions. Climatic Change 155 (4), 581–619.
Szulecki, K., 2018. Conceptualizing energy democracy. Environ. Polit. 27 (1), 21–41.
Thomson, H., Petrova, S., Bouzarovski, S., Simcock, N., 2018. Energy Poverty and Vulnerability: A Global Perspective. Routledge, Abingdon.
Walker, G., Day, R., 2012. Fuel poverty as injustice: integrating distribution, recognition and procedure in the struggle for affordable warmth. Energy Pol. 49, 69–75.

Part Three

Enabling technologies, community design, and business models

The digitalization of peer-to-peer electricity trading in energy communities

13

Emilio Ghiani[1], *Mario Mureddu*[1], *Marco Galici*[1], *Matteo Troncia*[2] and *Fabrizio Pilo*[1]
[1]University of Cagliari, Cagliari, Italy; [2]Universidad Pontificia Comillas, Madrid, Spain

1. Introduction

The achievement of decarbonization goals requires the reorganization of the energy system. The coupling of different energy systems, the increasing renewable electric energy production, and the electrification of final energy uses will characterize this reorganization. Green energy sources, typically nondispatchable, need advanced control and energy storage systems to exploit the flexibility of the Distributed Energy Resources (DERs). Integrating electrical power systems with gas and transportation networks is crucial to increase flexibility and operational options in this context. In Local Energy Communities (LECs), the integration of different energy systems, the operation of small-scale photovoltaic or nonutility scale RES, and the engagement of customers and prosumers are easier for both technical (e.g., the size of the community, the presence of different systems and energy vectors, etc.) and socio-economic reasons (e.g., energy price reduction, community engagement, attention to the local environment, etc.). For both energy and the power system, LECs can reduce operational issues thanks to the controllable behavior and the self-subsistence and allow the provision of ancillary services to the system.

Local energy markets involving users, local power producers, system operators, and market operators (e.g., aggregators) can maximize the local use of energy, reduce the RES and high electrical demand burden on both transmission and distribution systems, and favor the development of LECs. Furthermore, local distribution system operators (DSOs) and transmission system operators (TSOs) can exploit LEC flexibility for network operation, contributing to energy balancing, as shown in Fig. 13.1.

For this to work, LEC energy networks must rely on smart grids to operate DERs and flexible demand. The smart grid gives new business opportunities to all stakeholders in an LEC, such as power producers, energy suppliers, aggregators, and consumers (European Union, 2011). Data, energy, and services can be exchanged among LEC members or offered to DSO and TSO for system operation (Galici et al., 2019; De Zotti et al., 2018).

The Renewables Directive (RED II) (Directive, 2018) and the Electricity Market Directive (IEM) (Directive, 2019) introduce various innovations to foster LEC

Energy Communities. https://doi.org/10.1016/B978-0-323-91135-1.00018-3

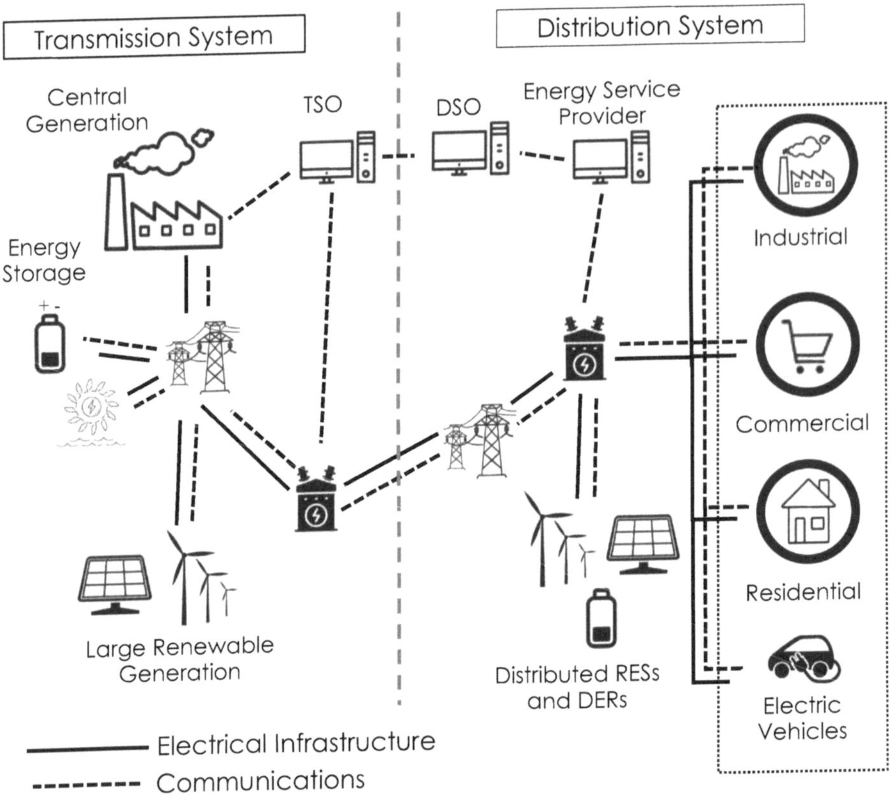

Figure 13.1 Interaction of LECs with TSO and DSO.

creation by promoting the citizens' partnership in projects concerning the exploitation of renewable sources and the active participation of consumers or prosumers (Sioshansi, 2017). According to the EU directives, member states must promulgate laws and rules to entitle energy communities to produce, use, store, and sell renewable energy with renewables power purchase agreements. DERs still face barriers to offering flexibility services in competitive ancillary markets, and the regulation has not made that participation worthwhile for most consumers yet (Sharifi et al., 2017). In Italy, the regulation for the ancillary service market allows all consumers and producers, possibly aggregated, to offer system services into the Ancillary Service Market (Decision 300/2017/R/eel, 2017). More recently, with the Decision 352/2021/R/eel, the Italian regulator launched new pilot projects to enable DSOs to procure with local markets or bilateral contracts local ancillary services offered by DERs for the distribution network operational needs. Regulatory sandboxes on the use of the flexibility provided by DERs are ongoing in most European Union (EU) countries (CEDEC et al., 2021). The regulation on local and system ancillary services, still at the early stage in most EU countries, will directly impact the development of local or renewable energy communities since they natively operate as an aggregate of diverse DERs capable of offering local and global services to DSOs and TSO. From a technical point of view, the main challenge is enabling the free aggregation of DERs and consumers without the need for centralized control systems that are expensive, not suited to operate many small consumers, and not flexible enough to accept bottom-up aggregation processes. The technology advancement on smart meters (SMs), Internet of Things (IoT), digital communication, cloud-based platforms for data storing and visualization services, artificial intelligence, and distributed control systems are paving the way to creating LECs with decentralized coordination that delivers economic and social benefits.

Describing the basic requirements, functionalities, and technologies for energy communities, this chapter is organized as follows:

- Section 2 introduces the technology for implementing an LEC;
- Section 3 describes distributed ledger technologies and blockchain for enabling energy and services transactions without any external authority;
- Section 4 discusses the concept of the local energy market; and
- Section 5 shows an example of the P2P market at laboratory scale followed by the chapter's conclusions

2. Digital enablers for energy communities

Measurement sensors, communication devices, data storage and processing, smart grid automation equipment (e.g., protection and network reconfiguration switchboards), and distributed generation interfaces, including intelligent appliances and smart home controllers, are the devices needed for an LEC.

Buildings and smart homes are LEC energy resources that act as nano grids capable of offering services and following a predefined energy pattern at the point of coupling with the network. The communication with the system operator and other LEC members uses a standardized communication infrastructure (Ghiani et al., 2017). The communication among home appliances, energy storage devices, and green generation is via a home area network (Fig. 13.2).

In this context, SMs are employed to work with an embedded energy management unit (EMU) that receives signals from market and system operators and offers services to optimize the user's techno-economic strategies. For instance, the SM/EMU enables participation in demand-side and energy management strategies to follow the electricity prices and maximizes the benefit of bidding in local and global ancillary markets. A typical example for peak shaving or shifting power usage to off-peak hours is illustrated in Fig. 13.2. The SM receives from the market or the system operator a price signal or command which is translated into the optimal strategy based on the SM input, contractual obligations, current and forecast demand and production, the state of charge of electric vehicles and storage devices, the time of the day, and the expected energy usage.

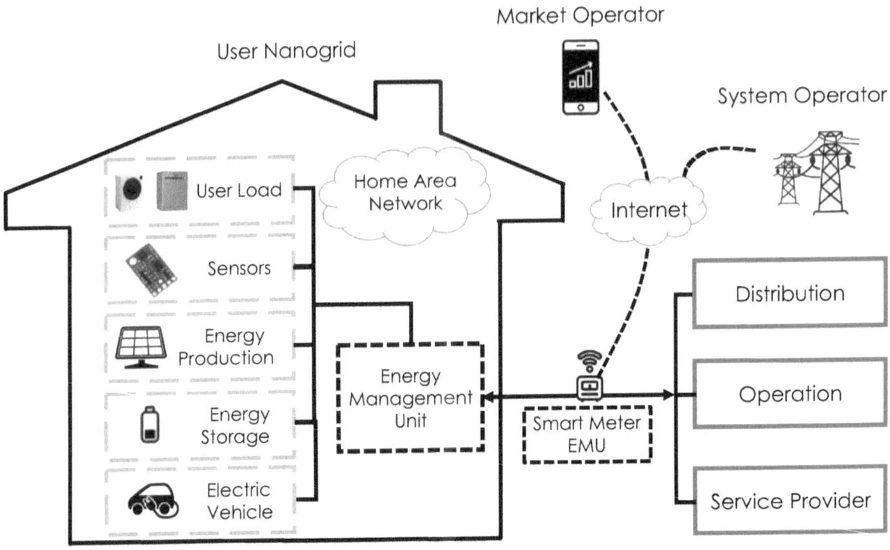

Figure 13.2 Home/building nano grid.

2.1 Local energy community

An LEC control center schematics is depicted in Fig. 13.3. In this architecture, the LEC members produce energy with renewable generators, consume energy, or do both. In addition, each LEC member has an SM/EMU that manages the real-time consumption and production. Stationary and mobile energy storage can be included to increase flexibility.

The Community Energy Management System optimizes the RES energy production and enables demand response with home automation systems.

The EU's Clean Energy Package requires that LECs give environmental and economic benefits to the community members (Energy European Commission, 2017). Therefore, Community Energy Management System enhances energy efficiency and maximizes LEC members' profits. The CEMS coordinates the energy management system at the user's location, compensating for the power fluctuations caused by renewable energy production variability and retrieving the energy that cannot be used instantaneously to maximize the LEC self-consumption.

As recommended by the EU association of regulators, the scheme adopted for LEC by the Italian regulation is virtual, with the distribution network owned and operated by a regulated distribution system operator. Each user is connected to the public network at the point of common coupling and is free to choose to be or not be an LEC member. Under rare circumstances, due to legislative and regulatory limits, the LEC may be physical, which means the community owns the distribution network, as in Ghiani et al. (2019).

In both cases, the CEMS optimizes the DERs operation according to system operators' and market operators' needs. For instance, CEMS may optimize to offer the

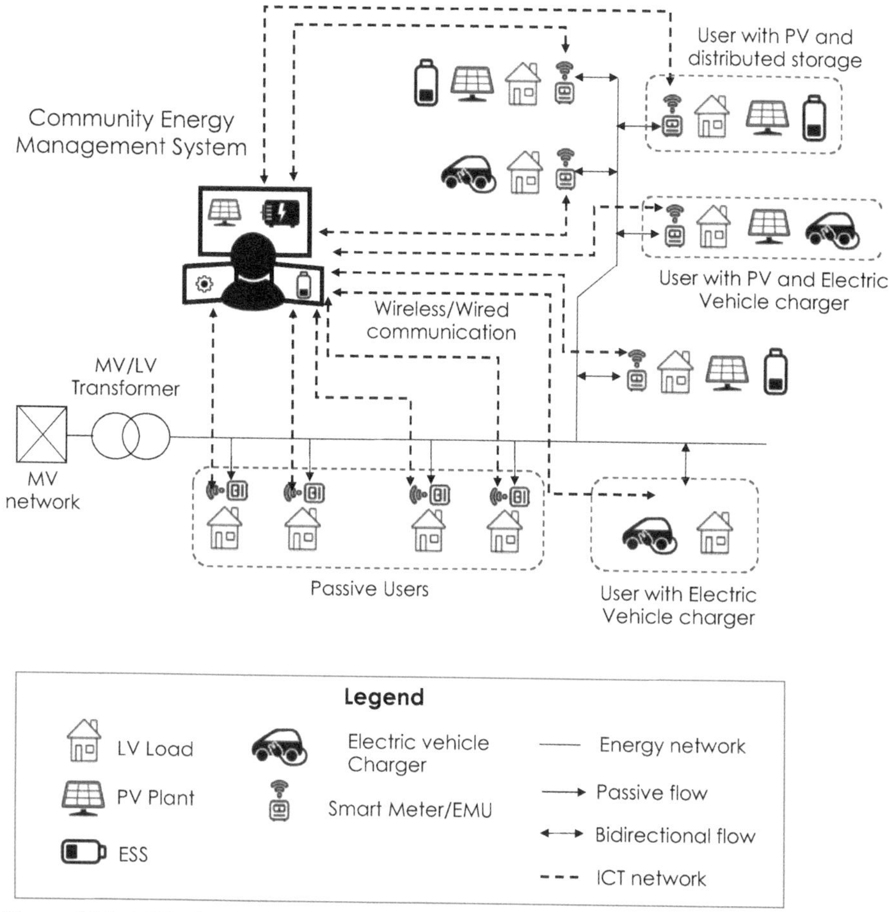

Figure 13.3 LEC scheme with low voltage users.

system operator flexibility or produce energy with specific generation units to meet the market operator's requirements. Moreover, other than satisfying the flexibility requirements of system operators, the CEMS aims at sustaining the LEC's energy self-consumption. Furthermore, the CEMS has to find a constrained optimal operation point considering network limits on voltage and power congestions, supply–demand imbalances, and reverse power flows as the DSO distribution management systems commonly do.

2.2 LEC smart metering

SMs are crucial to implementing an LEC. They increase the knowledge on energy usage and production for the purpose of energy and market management. Smart metering systems allow the implementation of effective demand response and flexibility services (Stagnaro and Benedettini, 2020). The use of nonproprietary standardized communication protocols is of utmost importance in this field to increase the level of interoperability and make the LEC constitution vendor-free.

SM/EMU[1] is crucial for coordinating the participation of LEC members. An SM/EMU is a bidirectional control and measurement device for single and three-phase systems (Olivero et al., 2021) that permit interaction among the user, third parties (e.g., Aggregators), and network and market operators (e.g., DSO and TSO). Each SM/EMU located at user premises gathers and sends data on consumption and production to the community energy management system. The near-real-time data and energy transactions are used for billing purposes. The system enables new services such as messages and notifications for efficient energy use and home/building automation (Pitì et al., 2017). Wireless technologies allowing two-way communication between users and utility suppliers, particularly with the 5G technology (Kochanski et al., 2020), will largely substitute cable.

SM/EMU can also support various interfaces such as serial peripheral interfaces, universal asynchronous receiver transmitters, high-speed data capture, and interintegrated circuit, which ensure the connection with external devices for special applications (Chakraborty et al., 2021). SMs/EMUs implement the following capabilities:

- Measurement of energy consumption and the most important electrical parameters (power, voltage, frequency);
- Real-time operation of energy;
- Monitoring of power quality and service continuity;
- Optimization of self-consumption and energy balancing; and
- Demand response function.

IoT-based sensors, intelligent devices, and communication networks are a modern solution for LEC applications, thanks to the low costs and the simple integration in brownfield development.

2.3 LEC energy management system

The LEC-EMS must exchange data with IoT systems manufactured by different brands. Thus, the use of nonproprietary standardized protocols is pivotal for interoperability. For efficient LEC operations, a user interface must supervise the users' operating parameters, forecast the daily and monthly production and consumption, and manage all economic transaction data. This user interface enables the LEC users to control and manage the operations of the SM/EMU automatically.

[1] The SM/EMU aims to save bidirectional (incoming, outcoming) data about active and reactive power and phase angle measurements in single and three-phase systems (e.g., European Measuring Instruments Directive, Directive, 2004/22/EC). The sample rate is generally high but in many applications 15 minutes for data exchange and operations is enough.

2.4 Technologies for the smart energy ecosystem

Various conceptual frameworks related to the smart grid paradigm have been proposed by both industry and academia. The most widely adopted model is the reference model proposed by the U.S. National Institute of Standards and Technology (Arnold et al., 2010). Other architectures conceptualize the smart grid as a multilayer ecosystem for devices. In particular, the home area networks, building area networks, and industrial area networks interconnect appliances with SMs and energy management devices in the bottom layer. The middle layer includes the neighborhood area networks, and at the top, the wide-area networks interconnect multiple neighborhood area networks.

This model is suitable for an energy community–based ecosystem. Within this IoT ecosystem, each device gathers and shares information with the other devices to reach the global benefit of the community. This interconnected system enables devices with Internet connectivity to exchange information with humans and machines.

Generally, devices in any communication network use a set of protocols for data transmissions. In the IoT, there will never be a single technology that meets the requirements of all applications to be deployed for low-range limits, low transmission rate, or even memory space limits (Venkatesh, 2015). Thus, a panoply of different protocols can be found beside the well-known Internet TCP/IP protocol.[2]

3. Distributed ledger technologies in P2P energy markets

Artificial Intelligence, IoT, and distributed ledger technologies are gaining a central role in digital distribution and intelligent power systems (Van Leeuwen et al., 2020; Diestelmeier, 2019). Specifically, as an essential component in energy management, energy trading is evolving from a centralized model toward a distributed scheme with the energy users that play both consumer and supplier roles. The following list highlights some of the most significant shortcomings of centralized markets:

- Energy trading highly relies on centralized third-party platforms, likely to cause a single point of failure (Aitzhan and Svetinovic, 2018; Wu and Tran, 2018), higher operating costs, low transparency, and risk of transaction data tampering are other known issues;
- Third parties may disclose the user's energy generation patterns and predict user's daily activities, leading to privacy and security issues (Guan et al., 2019);
- The lack of competition between renewable energy pricing and traditional market pricing (Ettlin, 2018);
- Vulnerability to cyber-attacks (Mylrea and Gourisetti, 2017);
- Lack of real-time models (Gregoratti and Matamoros, 2014); and
- Low system efficiency and stability due to nondistributed bidirectional energy management (Jogunola et al., 2018).

[2] Bluetooth Low Energy, ZigBee, and the IEEE 1901.2 standard support IoT environment development (Kaur, 2018).

In this scenario, the emerging distributed ledger technologies with the blockchain application are considered as one efficient way for realizing effective P2P platforms and models for P2P energy trading (Mengelkamp et al., 2018; Ahl et al., 2019; Adbella and Shuaib, 2018; Glachant and Rossetto, 2021). The P2P model makes the energy market consumer-centric and supports the prosumers' participation, which increases the energy market flexibility (Sousa et al., 2019; Murkin et al., 2016). The P2P paradigm particularly suits the energy community to obtain the maximum inclusivity of community members. In addition, the digitally connected energy system needs secure solutions for communication, automation, and documentation that cannot be easily obtained with centralized schemes. A thriving-functioning energy system is dependent on data being shared correctly, quickly, and uniquely with the relevant actors within the system. Therefore, managing large data streams from decentralized electricity feed-in, smart metering, or grid operation is crucial. In this scenario, the blockchain offers a resilient infrastructure to collect data in distributed electricity systems while allowing for a new level of transparency, tamper resistance, and security.

A P2P network can create a competitive energy market and increase the overall efficiency of the power market (Abdella and Shuaib, 2018; Di Silvestre et al., 2020). Additionally, the distributed model can also provide high-precision demand response signals (Pop et al., 2018), reduce cost, and improve speed (Sikorski et al., 2017). The P2P distributed model solves the problem of scalability since it is possible to reproduce a competitive market at a small size that benefits small-scale prosumers (Shuaib et al., 2018). The P2P structure guarantees that all nodes can work independently toward system equilibrium without central supervision (Pop et al., 2018). Finally, the application of cryptography in blockchain can solve several system security problems (Shuaib et al., 2018; Rahmadika et al., 2018) and significantly reduce security costs Hwang et al. (2017) and create open and transparent credit systems, preventing fake transactions (Gao et al., 2018).

Blockchain technology, born as a service for providing a counterparty-less banking system, ensures validated and immutable transactions on a ledger distributed among peers and allows the settlement of economic transactions reliably. Blockchain can be described as a set of shared and distributed data structures, or ledger, capable of securely and automatically storing digital transactions without the need for a central authority. Blockchain technology, first introduced in cryptocurrencies (Nakamoto and Bitcoin, 2008), has been recognized as a disruptive technology for energy communities, particularly with AI, IoT, advanced sensors, and smart metering devices. However, it is worth mentioning that some are not convinced of the need for anonymity and security of the blockchain, arguing that more straightforward and less computationally demanding approaches such as the concept of transactive energy should suffice (Barrager and Cazalet, 2021).

Blockchain enhances the role of players in the system (Hwang et al., 2017) since the customer can actively participate in the energy market and provide services to the system operator (Shayeghi et al., 2019; Mureddu et al., 2020; Luo et al., 2018). With the combination of distributed ledger and IoT, the energy community users can perform energy and power exchanges without intermediates, shifting the responsibility of energy procurement from the system/market operator toward the users themselves. According to this vision, the energy community members can autonomously operate in the system, with defined roles and responsibilities, under the system operator's supervision. The availability of an independent technology distributed among participants enables the possibility to reach a P2P consensus on the optimal use of local resources (Foti and Vavalis, 2021; Pop et al., 2018).

The technological advances in distributed ledger technologies lead to distributed ledger platforms capable of running smart contracts, immutable computer programs to operate on a distributed, counterparty-less, and trustless platform. This code changed the technological background of P2P energy trading (Buterin, 2014; Pop et al., 2018; Andoni et al., 2019; Troncia et al., 2019a). Smart contracts allow the creation of auction mechanisms (Troncia et al., 2019b; Truong et al., 2018). Zhong et al. (2018) developed an incentive auction mechanism to transfer the auction calculations to users to reduce the burden on the control center. In addition, multisignature algorithms are available with smart contracts to prove the correctness of transactions (Zhao et al., 2018).

Smart contracts applied to a local market ensure the validity of transaction data with a cryptography-based platform. Based on such validated data sources, a blockchain-based electricity marketplace can match the demand and supply side for energy purchases and immediately settle the transactions, including monitoring the electricity delivery and processing corresponding payments. Smart contracts can ensure that electricity is requested, for example, when prices fall below a price threshold or when green electricity or local power is available. The smart contract technology allows local markets to run entirely on a distributed platform with fully automated and decentralized market settlement. The rise of this technology sparked the definition of different decentralized energy community models running entirely on smart contracts (Troncia et al., 2019b; Andoni et al., 2019). Through smart contracts, it is possible to enable P2P energy sharing. The SMs and the embedded EMUs measure and optimize the consumer's energy consumptions. The smart contract acts as an optimization controller distributed among nodes. Furthermore, smart contracts allow the network to accommodate more renewable resources through the local community. New frameworks combine the double auction market model and noncooperative game theory to generate a dynamic pricing mechanism allowing geographically distributed storage units to interact and trade energy (Wang et al., 2014).

Distributed ledger technologies still have barriers that limit their practical application in energy communities due to privacy issues especially concerning the legislation for personal data protection (Eberhardt et al., 2020; Guan et al., 2021). The cost and the carbon footprint of the blockchain-based platform are crucial issues, but research on the field is progressing fast. The platform IOTA drastically reduces the blockchain costs and the carbon footprint (Popov and Labs, 2016).

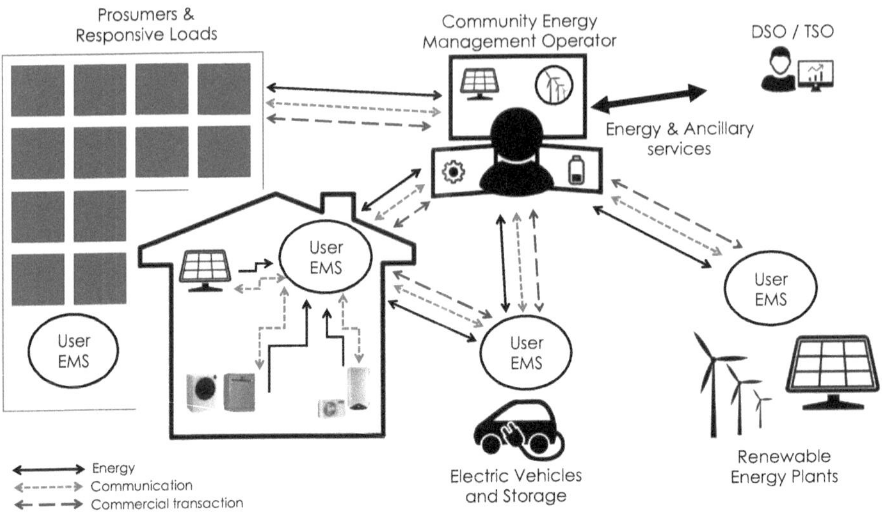

Figure 13.4 Example of aggregation of distributed energy resources with an LEC.

4. Local energy markets for energy communities

A local energy market is a market in which local power generation, energy storage, and flexible demand cooperate according to P2P models, enabling all users to participate and benefit (Teotia et al., 2016). Local energy markets are community-oriented and emphasize empowering the communities where they are used, decreasing electricity costs, and reducing electricity production by conventional sources through local renewable energy sources (Siano et al., 2019; Carreiro et al., 2017; Faria et al., 2018).

A local energy market can be achieved by aggregating DERs physically connected through a distribution network or virtually connected, as illustrated in Fig. 13.4. The virtual connection allows nongeographically contiguous small and medium-size consumers to constitute an energy community and participate in the market since, in many countries, the existing regulation does not allow LEC to own its distribution network.

5. LEC case study

This section describes an LEC functioning with a P2P market based on the application of ledger technologies. The LEC and the operational platform are depicted in Fig. 13.5. The LEC members (10) are connected to a public low voltage distribution network. 4 LEC members are prosumers with PV generators, and 6 are flexible energy consumers; 6 customers in the network are not LEC members. The laboratory-scale pilot is a digital twin of a portion real LEC (Ghiani et al., 2019) specifically designed to assess the realizability of P2P markets and the practical application of blockchain. The small number of LEC community members does not affect the general validity of the results

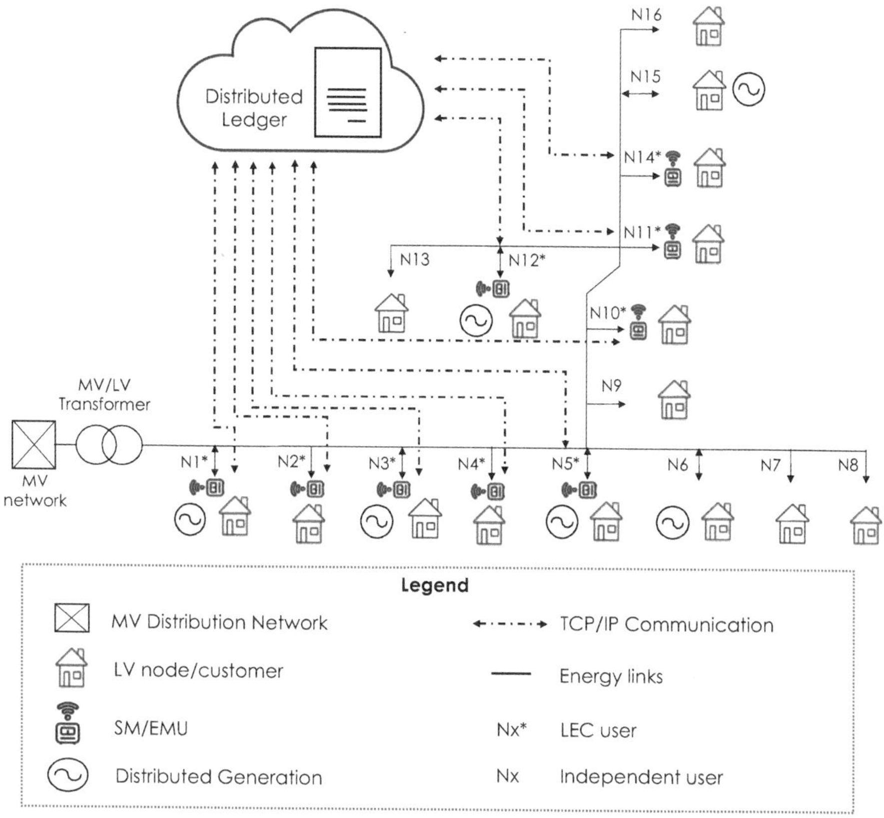

Figure 13.5 The energy community simulating the control system and the blockchain P2P market.

since the experiment aims to demonstrate that smart contracts can be implemented in cheap SMs and used for the LEC operation.

The IoT devices can autonomously place energy bids on the local market by interfacing with peers. In addition, they adapt consumption and production to the system to comply with voltage and power flows technical constraints. In this way, the system can autonomously perform energy management without central third parties.

A blockchain ledger is used to store users' transactions; the market-clearing bids and prices are also immutably stored with the designed blockchain database.

Each market participant is equipped with smart devices able to perform three main tasks:

- Place bids in the market for the relative user with an intelligent reinforced machine learning algorithm. According to the machine learning algorithm, each user actively learns to place the offer, maximizing user profit given the known external conditions. Each blockchain-capable device saves its digitally signed bid on the blockchain, making it immutable and available to the whole system.

- Once users have placed bids, each IoT device performs a collaborative computation with all the other intelligent devices in the system, aiming to clear the market by finding a solution that minimizes the total system costs and respects the technical constraints. Once the peers agree on the market solution, it is written on the blockchain and made immutable and available to everyone.
- After the market-clearing, each intelligent device reads the market-clearing data and adapts the consumption and generation accordingly.

The blockchain-capable devices execute the network optimization to find at the same time an agreement on the market-clearing that maximizes the profits of the involved payers. The current setup proves that it is possible to run a process every minute, enabling a real-time market that is much better for system security and adequacy since shorter time cycles reduce uncertainties on renewable production and demand. Fig. 13.5 shows a snapshot of the market at 5:30 p.m. on a typical weekday.

The LEC core is represented by the EMUs connected to the SMs. The EMUs have been realized with low-cost, simple components (Rabsperry devices) since their cost is crucial for LEC and P2P markets. Despite the simple architecture and the low-cost components, the hardware-in-the-loop[3] tests proved that it is possible to run a real-time P2P market and operate the network within its technical limits without complex devices. The utilization of an external platform to enable smart contracts and blockchain did not pose any constraint.

The laboratory tests performed on a digital twin of a real LEC allowed testing the feasibility of blockchain P2P markets with cheap components. In particular, the laboratory experiment successfully addressed several significant issues, including the following:

- *Modularity:* The blockchain-based platform allows the connection of an unlimited number of users—if each user possesses an internet connection. A plug-in element that exploits internet protocol communications to transmit data from the SM to distributed ledger ensures that every end-user can access this system. Indeed, exploiting the consumer's home internet connection ensures the interoperability and applicability of the system.
- *Scalability:* One common problem of distributed ledgers is scalability. Distributed ledger's scalability problem refers to the limited capability of the blockchain network to deal with large volumes of transactions in a short period. However, local private distributed ledgers can overcome this issue. In this way, only the effective participants of the corresponding physical power grid can access the energy community data and perform transactions. Consequently, as the number of elements increases, the system would still be manageable by users in terms of the complexity of the communication system and user-friendly interface.
- *Economics:* From the economic point of view, the proposed blockchain-based energy community model assumes that all energy providers exchange energy directly with consumers. One consequence would be that the intermediaries operating in the market might no longer be needed, or in any case, they would be reduced to a considerably minor role. On the other hand, there are the operating costs of blockchain platforms, including fees for blockchain transactions. The required computing power and related energy use might also have to be

[3] The Hardware in the Loop tests of the Energy Management Units was done with RTDS.

factored in. However, using private blockchains instead of public blockchains is possible to have lower transaction costs due to the simplified verification processes.

- *Barriers:* From the regulatory point of view, there are still limitations to adopting blockchain technology. Indeed, in Italy, only in 2019, the government approved an amendment to incorporate "technologies based on distributed registers such as blockchain" into the legal system and provide legal value to smart contracts. On the other hand, applying such technology to energy communities is not as straightforward as it may seem. Customers almost constantly switch suppliers in blockchain-based transaction systems, as they can find new transaction partners and contract them in short timescales—down to a few minutes. This behavior requires that energy community members are flexible in choosing their suppliers.
- *Privacy:* Another blockchain-related issue is digital consumer protection. All consumer protection efforts aim to protect consumers from the economic, digital, and health perspectives. All data processing systems should be designed to ensure that no or as little personal data as possible is used and that data should be anonymized or pseudonymized to the extent possible. In this view, the blockchain-enabling devices should verify that each piece of information must be encrypted before storing it into the ledger, a requirement not always applicable due to the computational power of the blockchain-enabling device. Consequently, the application of blockchain into energy communities could lead to considerable consumer resistance, primarily due to privacy concerns. Despite the potential for monetary savings, it is unclear whether a consumer would want its consumption from the electric shower, washing machine, or electric vehicle to be publicly recorded in a public ledger. Hence, the way the information is recorded in the ledger, such as to preserve the privacy and anonymity of individual domestic and industrial consumers, could prove to be a vital issue in blockchain implementations of IoT systems.

6. Conclusions

This chapter analyzed the technology for LECs with a special focus on digital sensors and network automation. For each technology, the chapter describes short- to long-term benefits arising from LEC as well as the minima features required for each of the examined categories. The survey showed that, despite the readiness of many technologies, an off-the-shelf solution for energy communities could not be found yet, and tailored solutions must be designed on a case-by-case basis. The standardization for interoperability and cost reduction is still a critical need. However, the maturity level of the technology for energy communities is much higher than regulation and legislation readiness, which is still the most challenging barrier in many countries.

This chapter also shows that LEC could be operated with P2P decentralized markets without high system costs. The authors propose an original low-cost EMU that interacts with a P2P market with blockchain-based smart contracts. The blockchain technology tested at a laboratory scale proved that transaction costs across the system could be reduced. The scalability is the main issue that emerged from laboratory tests. It can be solved by using proprietary ledger platforms that also reduce energy consumption and carbon footprint. Further research is necessary to assess and compare other decentralized methodologies for operating energy communities with particular attention to the robustness against cyber-attacks.

Acknowledgments

The authors acknowledge M. Mokrane, Trendy, ProSymbols, Becris, M. Riza, Vectorstall, Ibrandify, A. Arshad, Humantech, jokokerto, J. Blake, J. Fernandes, B. Kaito, M. Munir, Icon Market, W. Joseph, A. Roaayala, G. Furtado, B. Merriman, F. Couto, Linda, L. Mette, O. Panasovskyi, and A. Ahmed that produced the artworks used in visuals (thenounproject.com).

References

Abdella, J., Shuaib, K., 2018. Peer to peer distributed energy trading in smart grids: a survey. Energies 11 (6), 1560.

Ahl, A., Yarime, M., Tanaka, K., Sagawa, D., 2019. Review of blockchain-based distributed energy: implications for institutional development. Renew. Sustain. Energy Rev. 107, 200−211.

Aitzhan, N.Z., Svetinovic, D., 2018. Security and privacy in decentralised energy trading through multi-signatures, blockchain and anonymous messaging streams. IEEE Trans. Dependable Secure Comput. 15 (5), 840−852.

Andoni, M., Robu, V., Flynn, D., Abram, S., Geach, D., Jenkins, D., Peacock, A., 2019. Blockchain technology in the energy sector: a systematic review of challenges and opportunities. Renew. Sustain. Energy Rev. 100, 143−174.

Arnold, G.W., Wollman, D.A., FitzPatrick, G.J., Prochaska, D., Holmberg, D.G., Su, D.H., Boynton, P.A., 2010. NIST Framework and Roadmap for Smart Grid Interoperability Standards, Release 1.0. NIST.

Barrager, S., Cazalet, E., 2021. Transactive Energy. Public Utilities Reports, Inc, Reston.

Buterin, V., 2014. A next-generation smart contract and decentralised application platform. White Paper 3 (37).

Carreiro, A.M., Jorge, H.M., Antunes, C.H., 2017. Energy management systems aggregators: a literature survey. Renew. Sustain. Energy Rev. 73, 1160−1172.

CEDEC, E.DSO, Eurelectric, and GEODE, June 2021. Roadmap on the Evolution of the Regulatory Framework for Distributed Flexibility. A Joint Report by ENTSO-E and the European Associations Representing DSOs (CEDEC, E.DSO, Eurelectric, GEODE). https://www.edsoforsmartgrids.eu.

Chakraborty, S., Das, S., Sidhu, T., Siva, A.K., 2021. Smart meters for enhancing protection and monitoring functions in emerging distribution systems. Int. J. Electr. Power Energy Syst. 127, 106626.

De Zotti, G., Pourmousavi, S.A., Madsen, H., Kjolstad Poulsen, N., 2018. Ancillary services 4.0: a top-to-bottom control-based approach for solving ancillary services problems in smart grids. IEEE Access 6, 11694−11706.

Decision 300/2017/R/eel, May 2017. Prima apertura del mercato per il servizio di dispacciamento (MSD) alla domanda elettrica e alle unità di produzione anche da fonti rinnovabili non già abilitate nonché ai sistemi di accumulo. Istituzione di progetti pilota in vista della costituzione del testo integrato dispacciamento elettrico (TIDE) coerente con il balancing code europeo. ARERA.

Di Silvestre, M.L., Gallo, P., Guerrero, J.M., Musca, R., Sanseverino, E.R., Sciumè, G., Zizzo, G., 2020. Blockchain for power systems: current trends and future applications. Renew. Sustain. Energy Rev. 119.

Diestelmeier, L., 2019. Changing power: shifting the role of electricity consumers with blockchain technology—Policy implications for EU electricity law. Energy Pol. 189—196.

Directive (EU). 2018/2001 of the European Parliament and of the Council - of 11 December 2018 - on the Promotion of the Use of Energy from Renewable Sources. (RED II Directive).

Directive (EU). 2019/944 of the European Parliament and of the Council of 5 June 2019 - on Common Rules for the Internal Market for Electricity and Amending Directive 2012/27/EU. (IEM Directive).

Eberhardt, J., Peise, M., Kim, D.H., Tai, S., May 2020. Privacy-Preserving netting in local energy grids. In: IEEE International Conference on Blockchain and Cryptocurrency (ICBC). IEEE, pp. 1—9.

Energy European Commission, October 2017. Clean Energy for All Europeans. Available online: https://ec.europa.eu/energy/en/topics/energy-strategy-and-energy-union/clean-energy-all-europeans.

Ettlin, A., June 2018. Dynamic modeling of peer-to-peer power market making. In: 2018 15th International Conference on the European Energy Market (EEM). IEEE, pp. 1—5.

European Union, 2011. Standardisation Mandate to European Standardisation Organisations (ESOS) to Support European Smart Grid Deployment. European Commission, Brussels, Belgium.

Faria, P., Spínola, J., Vale, Z., 2018. Reschedule of distributed energy resources by an aggregator for market participation. Energies 11, 713.

Foti, M., Vavalis, M., 2021. What blockchain can do for power grids? Blockchain: Research and Applications 2, 1. https://doi.org/10.1016/j.bcra.2021.100008.

Galici, M., Ghiani, E., Troncia, M., Pisano, G., Pilo, F., 2019. A cyber-physical platform for simulating energy transactions in local energy markets. In: 2019 IEEE Milan PowerTech, pp. 1—6.

Gao, C., Ji, Y., Wang, J., Sai, X., May 2018. Application of blockchain technology in peer-to-peer transaction of photovoltaic power generation. In: 2018 2nd IEEE Advanced Information Management, Communicates, Electronic and Automation Control Conference (IMCEC). IEEE, pp. 2289—2293.

Ghiani, E., Garau, M., Celli, G., Pilo, F., Marongiu, G., June 2017. Smart integration and aggregation of nanogrids: benefits for users and DSO. In: 2017 IEEE Manchester PowerTech. IEEE, pp. 1—6.

Ghiani, E., Giordano, A., Nieddu, A., Rosetti, L., Pilo, F., 2019. Planning of a smart local energy community: the case of Berchidda municipality (Italy). Energies 12 (24), 4629.

Glachant, J.M., Rossetto, N., 2021. "A New World for Electricity Transactions: Peer-To-Peer and Peer-to-X.", Robert Schuman Centre for Advanced Studies Research Paper No. RSC, p. 56.

Gregoratti, D., Matamoros, J., 2014. Distributed energy trading: the multiple-microgrid case. IEEE Trans. Ind. Electron. 62 (4), 2551—2559.

Guan, Z., Zhang, Y., Wu, L., Wu, J., Li, J., Ma, Y., Hu, J., 2019. APPA: an anonymous and privacy preserving data aggregation scheme for fog-enhanced IoT. J. Netw. Comput. Appl. 125, 82—92.

Guan, Z., Lu, X., Yang, W., Wu, L., Wang, N., Zhang, Z., 2021. Achieving efficient and Privacy-preserving energy trading based on blockchain and ABE in smart grid. J. Parallel Distr. Comput. 147, 34—45.

Hwang, J., Choi, M.I., Lee, T., Jeon, S., Kim, S., Park, S., Park, S., 2017. Energy prosumer business model using blockchain system to ensure transparency and safety. Energy Proc. 141, 194—198.

Jogunola, O., Ikpehai, A., Anoh, K., Adebisi, B., Hammoudeh, M., Gacanin, H., Harris, G., 2018. Comparative analysis of P2P architectures for energy trading and sharing. Energies 11 (1), 62.
Kaur, G., 2018. An In-Depth Study on Power Line Communications.
Kochanski, M., Korczak, K., Skoczkowski, T., 2020. Technology innovation system Analysis of electricity smart metering in the European union. Energies 13, 916.
Luo, F., Dong, Z.Y., Liang, G., Murata, J., Xu, Z., 2018. A distributed electricity trading system in active distribution networks based on multi-agent coalition and blockchain. IEEE Trans. Power Syst. 34 (5), 4097–4108.
Mengelkamp, E., Gärttner, J., Rock, K., Kessler, S., Orsini, L., Weinhardt, C., 2018. Designing microgrid energy markets: a case study: the Brooklyn Microgrid. Appl. Energy 210, 870–880.
Mureddu, M., Ghiani, E., Pilo, F., August 2020. Smart grid optimisation with blockchain based decentralised genetic algorithm. In: IEEE Power & Energy Society General Meeting (PESGM). IEEE, pp. 1–5.
Murkin, J., Chitchyan, R., Byrne, A., August 2016. Enabling peer-to-peer electricity trading. In: 4th International Conference on ICT for Sustainability, pp. 234–235.
Mylrea, M., Gourisetti, S.N.G., September 2017. Blockchain for smart grid resilience: exchanging distributed energy at speed, scale and security. In: 2017 Resilience Week (RWS). IEEE, pp. 18–23.
Nakamoto, S., Bitcoin, A., 2008. A Peer-To-Peer Electronic Cash System. Bitcoin. URL. https://bitcoin.org/bitcoin.
Olivero, S., Ghiani, E., Rosetti, G.L., 2021. The first Italian renewable energy community of Magliano Alpi. In: IEEE CPE-POWERENG 2021.
Pitì, A., Verticale, G., Rottondi, C., Capone, A., Lo Schiavo, L., 2017. The role of smart meters in enabling real-time energy services for households: the Italian case. Energies 10, 199.
Pop, C., Cioara, T., Antal, M., Anghel, I., Salomie, I., Bertoncini, M., 2018. Blockchain-based decentralised management of demand response programs in smart energy grids. Sensors 18 (1), 162.
Popov, S., Labs, J., 2016. The tangle. IOTA. https://iotatoken.com/IOTA_Whitepaper.pdf.
Rahmadika, S., Ramdania, D.R., Harika, M., 2018. Security analysis on the decentralised energy trading system using blockchain technology. Jurnal Online Informatika 3 (1), 44–47.
Sharifi, R., Fathi, S.H., Vahidinasab, V., 2017. A review on demand-side tools in electricity market. Renew. Sustain. Energy Rev. 72, 565–572.
Shayeghi, H., Shahryari, E., Moradzadeh, M., Siano, P., 2019. A survey on microgrid energy management considering flexible energy sources. Energies 12 (11).
Shuaib, K., Abdella, J.A., Sallabi, F., Abdel-Hafez, M., April 2018. Using blockchains to secure distributed energy exchange. In: 2018 5th International Conference on Control, Decision and Information Technologies (CoDIT). IEEE, pp. 622–627.
Siano, P., De Marco, G., Rolán, A., Loia, V., 2019. A survey and evaluation of the potentials of distributed ledger technology for peer-to-peer transactive energy exchanges in local energy markets. IEEE Syst. J 13 (3), 3454–3466.
Sikorski, J.J., Haughton, J., Kraft, M., 2017. Blockchain technology in the chemical industry: machine-to-machine electricity market. Appl. Energy 195, 234–246.
Sioshansi, F.P., 2017. Innovation and disruption at the grid's edge. In: Innovation and Disruption at the Grid's Edge. Academic Press, pp. 1–22.
Sousa, T., Soares, T., Pinson, P., Moret, F., Baroche, T., Sorin, E., 2019. Peer-to-peer and community-based markets: a comprehensive review. Renew. Sustain. Energy Rev. 104, 367–378.

Stagnaro, C., Benedettini, S., 2020. Smart meters: the gate to behind-the-meter?. In: Behind and beyond the Meter. Academic Press, pp. 251–265.

Teotia, F., Bhakar, R., 2016. Local energy markets: concept, design and operation. In: 2016 National Power Systems Conference (NPSC), pp. 1–6.

Troncia, M., Galici, M., Ghiani, E., Pilo, F., Ruggeri, S., 2019a. Blockchain Local Markets for the Distributed Control of Microgrids. AIM. CIRED 2019.

Troncia, M., Galici, M., Mureddu, M., Ghiani, E., Pilo, F., 2019b. Distributed ledger technologies for peer-to-peer local markets in distribution networks. Energies 12 (17), 3249.

Truong, C.N., Schimpe, M., Bürger, U., Hesse, H.C., Jossen, A., 2018. Multi-use of stationary battery storage systems with blockchain-based markets. Energy Proc. 155, 3–16.

Van Leeuwen, G., AlSkaif, T., Gibescu, M., Van Sark, W., 2020. An integrated blockchain-based energy management platform with bilateral trading for microgrid communities. Appl. Energy 263.

Venkatesh, N., 2015. Ensuring coexistence of IoT wireless protocols using a convergence module to avoid contention. Embedded Innovator. 32–35. Intel IoT Solutions Alliance.

Wang, Y., Saad, W., Han, Z., Poor, H.V., Başar, T., 2014. A game-theoretic approach to energy trading in the smart grid. IEEE Trans. Smart Grid 5 (3), 1439–1450.

Wu, J., Tran, N.K., 2018. Application of blockchain technology in sustainable energy systems: an overview. Sustainability 10 (9), 3067.

Zhao, S., Wang, B., Li, Y., Li, Y., 2018. Integrated energy transaction mechanisms based on blockchain technology. Energies 11 (9), 2412.

Zhong, W., Yang, C., Xie, K., Xie, S., Zhang, Y., 2018. ADMM-based distributed auction mechanism for energy hub scheduling in smart buildings. IEEE Access 6, 45635–45645.

Further reading

ARERA, March 2016. Functional Specifications Enabling Low Voltage Smart Meters and Performance of the Related Second Generation (2G) Smart Metering Systems in the Electricity Sector, Pursuant to Legislative Decree No. 102, 4 July 2014. Resolution n.87/2016/R/eel.

Enabling business models and grid stability: case studies from Germany

14

Simon Köppl[1], *Elisabeth Springmann*[1], *Vincenz Regener*[1] and *Andreas Weigand*[2]
[1]FfE Munich, Munich, Germany; [2]Stadtwerke Munich (SWM), Munich, Germany

1. Introduction

The project C/sells[1] created solutions for a reliable and intelligent energy system facing a high penetration with decentralized and renewable resources based on several living lab ecosystems in southern Germany. Two of these living labs, whose operations have since been completed, are the "Altdorfer Flexmarkt" (ALF) and "Intelligent Heating Munich" (IHM) with 20 and 70 participating households, respectively. Their objective went beyond a mere proof-of-concept but was to prove the functionality of the solutions under real-world conditions in a defined field-test environment.

Until recently, the energy supply has been centrally oriented and controlled by few actors. But growing public interest in clean energy supply and rising numbers of decentralized electricity producers call for higher civic/public participation in the organization of energy systems. Hence, besides the development of new technological solutions for improved grid efficiency, the vision of C/sells is to create a broad movement among the population and boost the energy transition bottom-up. Dividing the energy system into smaller "cells" of limited geographical scope is key to meeting the challenges of an increasingly complex supply structure plus promoting community involvement. The individual cells featuring unique organizational structures were introduced to meet the specific requirements of different C/sells living labs. First and foremost, however, these smaller organizational entities are intended to bring together citizens with complementary energy interests and motivate them to take responsibility. As the cells foster the visibility of energy infrastructure and synergies between participants emerge, the prospect of communal benefits will become a driving force for a renewable, decentralized energy future.

C/sells' fundamental principles of cellularity, participation, and diversity build upon a collective effort within the so-called "participation cells" to generate solutions

[1] The activities described within this paper are part of C/sells, a joint project carried out as part of the Federal Ministry for Economic Affairs and Energy (BMWi) funding program "Smart Energy Showcase - Digital Agenda for the Energy Transition" (Funding code: 03SIN121 (FfE), 03SIN143 (SWM)) and https://www.swm.de/magazin/energie/intelligente-waerme-muenchen information see www.ffe.de/csells and https://www.swm.de/magazin/energie/intelligente-waerme-muenchen.

Energy Communities. https://doi.org/10.1016/B978-0-323-91135-1.00010-9

that, in return, benefit the participants as well as the entire community in the operating area. With ALF, in which citizens supported system operations by providing flexibility services on an online platform, it was possible to improve utilization of the local electricity grid (Zeiselmair et al., 2019; Köppl et al., 2019; Heilmann et al., 2020). Intelligent coordination of electricity feed-in and consumption to prevent grid congestions alleviates the need for expensive traditional network expansions (Colle et al., 2019). The concomitant savings can be passed on to the platform's participants via compensations for their flexibility provision or via reduced network charges for all community members (Heilmann et al., 2020; Köppl et al., 2019; Zeiselmair et al., 2019).

Besides economic returns, C/sells specifically seeks environmental benefits for the participants of the different living labs. The project IHM has shown that surplus electricity from renewable sources can be utilized in electric storage heaters, heat pumps, and refrigeration systems. By the intelligent aggregation of these flexible assets and by shifting the demand for electric heating, this concept helps system operators to integrate more renewables safely and efficiently (Greif and Weigand, 2019). The improved utilization of renewable producers decreases their plants' amortization time and overall carbon pollution. Since reduced electricity production costs and clean air are not exclusive to the group of participants, the benefits of including distributed renewable producers are also shared with those unable to contribute. Hence, the intentions of both projects—ALF and IHM—are in line with fundamental energy community motives (REScoop.eu, 2019). Their primary objective is not to generate financial profits but to provide environmental, economic, or social benefits for shareholders, members, or the areas of operation in general. Fig. 14.1 visualizes the location of the C/sells platform approach within the energy community context.

This chapter focuses on different levels of community designs and proposes a concept for the intelligent interaction of the congestion management mechanism implemented in ALF and the aggregation of flexible assets via a virtual power plant (VPP) implemented in IHM. Following up on the individual C/sells projects, this chapter outlines a structure combining the living lab-tested approaches into a common federated concept to create the organizational basis for citizen empowerment in the energy system. As part of the conceptual work, questions regarding communal entry barriers, the geographical scope of communities, and fair reimbursement

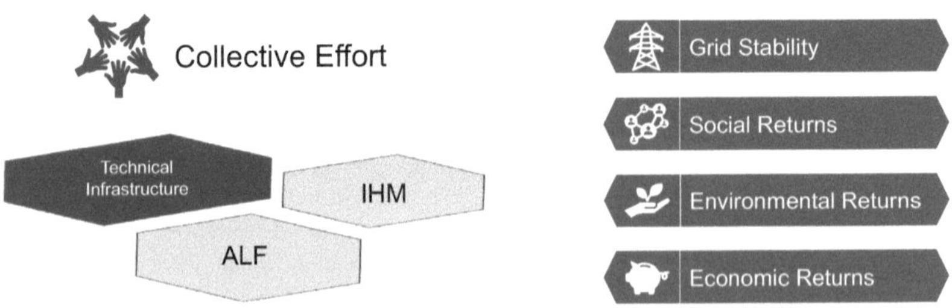

Figure 14.1 C/sells platform concepts as providers of technical infrastructure to utilize collective effort for various communal benefits.

systems are discussed. Therefore, this chapter's main body is divided into the following sections:

- Section 2 provides an overview of the main actors and technical components of the two exemplary C/sells living labs and structures them into four levels of interaction;
- Section 3 builds upon this structure and outlines a potential dovetailing of the concepts developed for ALF and IHM;
- Section 4 discusses the proposed system design and compares it against alternative approaches; and
- Section 5 summarizes the chapter and its conclusions.

2. Levels of coordination within distributed energy systems: basis for energy communities?

With the increasing decentralization of the energy system, organization among the actors is also becoming increasingly difficult. Today, simple top-down approaches are no longer sufficient to ensure efficient operations management and grid stability. As a response to this, new, locally constrained coordination mechanisms have been implemented and tested in C/sells. The idea is to utilize a tight geographical scope, aggregation, and distributed grid operation to resolve instabilities before they escalate to a higher grid level. While these mechanisms are de facto implemented top-down by energy suppliers or grid operators, they aim at leveraging the potential of communities and establishing a sustainable bottom-up movement.

2.1 End customers and neighborhoods

Calling end customers "the lowest organizational unit" would fall short of their importance within a distributed energy system. Due to the increasing numbers of privately owned flexible generation and consumption facilities, the relevance of end customers for reliable grid operation is higher than ever. With their ability to not only consume, but also feed in, store on demand, or share/trade electrical energy, their role is evolving from pure consumers to so-called "flexumers" (Westphal et al., 2019). To harness the end consumers' full grid-stabilizing potential, it is vital to implement new communication channels, recognizing that flexumers will be at the center of future energy systems. Originally, the term flexumer relates to owners of single-family homes. For an application in an urban context, it seems sensible to extend the term to owners of residential blocks or other units bigger than single-family homes (Westphal et al., 2019).

One approach to incorporate flexumers into distributed grid operation measures is a Digital Network Connection Point (DiNCP). It allows for a maximum power limit depending on the current grid load to be transferred as a setpoint to a building. This transfer happens via an enhanced Smart Meter Gateway (SMGW) as a secure communication device (Weigand et al., 2021). The corresponding communication channel is depicted in Fig. 14.2. Besides the contractually agreed maximum connection value (comparable to a service fuse), a dynamic setpoint value P_{lim} is introduced by the connection network operator (CNO) to ensure grid stability. The CNO represents

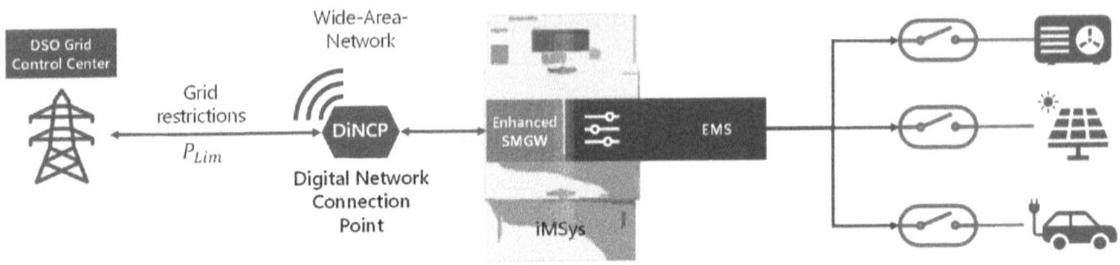

Figure 14.2 Communication channel between distribution system operator (DSO) and domestic flexible assets via intelligent Measurement System (iMSys).

the distribution system operator (DSO) of the lowest order in the area. In reaction to imminent grid congestions, the DSO can adjust this dynamic setpoint within specified limits. For the building, the P_{lim} setpoint constitutes a maximum power limit, under which all flexible assets can operate freely.

To optimize operation under these grid-related restrictions, an energy management system (EMS) orchestrates the buildings' available flexible assets, considering customer preferences regarding their use. Among others, orchestrating the assets allows for an increase in the domestic use of solar power or adjusting the operating mode to dynamic price signals to name just two use cases.

In the previous paragraph, the DiNCP was related to a single building, but the concept can be easily transferred to bigger units, such as multiuser properties, quarters, or other smart neighborhoods. As long as the power lines within the units do not represent a bottleneck themselves, the crucial advantage of control at the lowest level is retained: Fluctuations in power demand or supply can be balanced locally without burdening the surrounding grid (Kalz et al., 2018). The merging of smaller neighboring units entails another advantage: While electric storage solutions are often not profitable for individual customers, they amortize more quickly in a neighborhood with several PV systems or electric heat pumps (Knoefel, 2019). With several and especially different types of flexible assets connected to an EMS, the optimization scope can be expanded significantly. Comparable to a shared bandwidth, given power limits can be flexibly distributed and adjusted among the end customers to meet individual demands. Thus, the DiNCP is a technical concept for housing communities and neighborhoods for connecting and optimizing flexible assets collectively to meet network requirements (Kalz et al., 2018).

2.2 Expanding the cellular boundaries: communal EMS, virtual power plants, and transregional aggregators

However, forming communities that optimize their energy flows is not limited to the immediate neighborhood. Even if grid restrictions play a role in larger dimensions, the interconnection of several smaller entities (cells) can bring benefits for the electricity grid and all parties involved. At this scale, EMS are often enhanced with the possibility to trade power on the electricity market. Since they aggregate the capacities of heterogeneous distributed producers and flexible consumers to perform the tasks of traditional power plants, they are often referred to as VPPs (Braun, 2007). It is only through the cooperation of individuals in the spirit of the community idea that VPPs can provide specific services and thus open up new business models for their participants (Braun, 2007).

For their IHM project, the Stadtwerke München (SWM), Munich's municipal utilities company, expanded its existing VPP by storage heaters, electric heat pumps, and refrigerator systems located in their area. The respective units of the participating customers in the test area received the control commands via iMSys. Power-to-Heat-/Power-to-Cool-Technologies (P2H/P2C) were included in the pool of decentralized assets since heat is much cheaper to store than electricity. This helps to utilize temporary power surpluses from renewable energies, which would otherwise, with no alternative consumers, endanger grid stability. Therefore, the VPP balances supply and demand at a municipal level. In this way, communal EMS deployment entails increased energy autonomy in the affected grid area. Additionally, VPPs can also offer additional capacities on the operating reserve or Intraday market across regional boundaries.

For the owners of small-scale facilities, the participation in VPPs allows for additional revenue streams, as the intelligent aggregation leads to economic returns and therefore complies with common definitions of an energy community (see REScoop.eu, 2019). While in the case of IHM the SWM on one side provides the VPP-service for the community as a communal company, they are at the same time collectively owned by the community itself. The project is thus representative of a current trend toward increasing remunicipalization of the energy system (Kinkel, 2014; REScoop.eu, 2019).

The compensation options for participants range from a flat-rate service price to unit rates for individual call-offs, depending on the degree of flexibility and the forecasting capability. The plant operator chooses the option that best suits the asset's requirements based on the individual risk aversion or profit orientation. However, with powerful VPPs or other aggregators representing the only market entry ticket for operators of small plants, detrimental one-sided dependencies can arise quickly. To mitigate the risk of regional monopolies, the aggregator market should be open to free competition. Thereby, plant operators retain their freedom of choice to market their assets either via a regional aggregator like SWM or in transregional associations.

Companies like Sonnen or Viessmann, both providers of virtual energy sharing communities, do not base the community idea on regional vicinity but rather on the ownership of equipment (PV systems, battery storage, or heat pumps) from the same manufacturer. While the key benefits of an energy community, like efficient energy management, also apply to these concepts, they are run by profit-oriented companies whose nationwide copperplate approach undercuts real network constraints (Park and Yong, 2017). Matching demand and supply at two distant nodes without proper handling of physical limitations can increase the grid load and cause congestions, even if the aggregator commands a well-balanced portfolio. Also, transregionally organized collectives face comparably high transmission losses and do not comply with the C/sells cellularity principle. This states that cells are usually geographically self-contained and operate according to the principle of subsidiarity:

the cell supplies itself autonomously as far as possible; if necessary, it cooperates with other cells (Park and Yong, 2017).

2.3 Wholesale markets and market for system stability

As mentioned above, the increasing number of small electricity generators and consumers and their intelligent networking at home or community level open the possibility of trading electricity on energy markets. This can be done via a broker that aggregates and trades many small assets and/or systems together in one product. For the sake of brevity, only the standardized products of the Day-ahead and Intraday markets are considered in this chapter, as they play a crucial role in the concept proposed in Section 3. On the German Day-ahead market, quarter-hourly to hourly products can be traded until 3:00 p.m. on the day before delivery. On the Intraday market, electricity is traded on the day itself up to 5 minutes before delivery (EPEX SPOT SE, 2021). Both markets thus enable the spontaneous intraday trading of electricity.

Another market in Germany is the balancing power market. Here, offers are allocated weekly (primary and secondary balancing power) or daily (tertiary control reserve) (50Hertz Transmission GmbH et al., 2020). Thus, high predictability alongside a mandatory prequalification is required to generate offers with the necessary reliability. Such high reliability is only realizable by combining specific assets. It was thus not implemented in ALF and IHM but should be considered in future developments to tap further marketing opportunities for energy communities.

Products with different time frames allow brokers to sell available amounts of energy and flexibility both spontaneously as well as with longer-term planning. Thus, a broad revenue potential can be exploited. Through the aggregation of different assets, e.g., on a communal level, entry barriers for specific markets, for example, the minimum power of an offer, can be reached, enabling a wide range of actors to generate revenue.

2.4 New instruments for grid operators: flexibility platforms for congestion management

With rising volatility and the decentralization of energy supply, grid operators are facing challenges during the energy transition. Electrification in the heating and transport sectors is causing increasing loads and several simultaneities. For grid operators, this means that the power to be transported and distributed will increase further and that there may also be more peak loads. This results in a need for measurability and controllability of components in the grid. Prospectively, consumers and producers will be interacting and cooperating with grid operators to ensure grid and system stability. For this purpose of congestion management, different kinds of approaches for local flexibility are conceivable (Radeke et al., 2019).

The flexibility platform ALF adds a market-based instrument to the established tools of congestion management. On ALF, flexibility is traded to resolve grid congestions. Different from the markets described in Section 2.3, not only the price but also the effectivity of the unit's flexibility on specific grid congestions in lines or grid nodes are considered in the matching process.

By using their flexibility for grid stabilization, ALF offers market entry to owners of small-scale units which are currently without market participation, as heat pumps or small PV systems. Owners can choose between two different models of contracting options (see Heilmann et al., 2020) for further details on the contracting and product design): Either, the owners choose the short-term contracting model and create a flexibility bid on the platform when flexibility becomes available or the long-term contracting model is chosen. For the long-term contracting, the platform assumes the forecasting and determination of a plant's flexibility capacity, considering additional information like the plant type and weather forecast data. So, a flexibility offer is generated. Hence, in the long-term contracting option, no additional efforts close to the flexibility call-off from the owners of flexibility options are needed.

ALF offers an additional tool for grid congestion management for DSOs: flexibility demands can be submitted to the platform for specific network components. This enables DSOs to avoid bottlenecks and overloads without controlling a high number of flexibility options. The basic mechanism is described in Fig. 14.3.

Flexibility offers and demands have been submitted to the flexibility platform; ALF then undertakes the matching of flexibility offers and demands. As described in Zeiselmair and Köppl (2021), both the plants' effectivity on the affected network components and the costs are considered in the matching process.

In the context of the C/sells project, ALF was put into practice and tested with participants at distribution grid level (Estermann et al., 2019). During the field test, participants were able to register their flexible assets online and offer flexibility at a certain price. Depending on their location in the power grid and their associated costs, ALF makes quarter-hourly allocations of flexibility supply and flexibility demand to resolve occurring congestions. As described above, the resulting remuneration of the

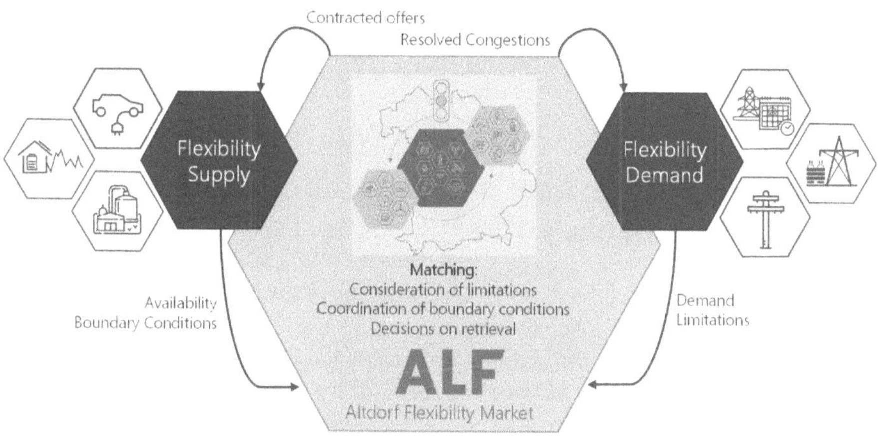

Figure 14.3 Mechanism of the flexibility platform ALF with flexibility supplier and demander and their matching on the platform.

flexibility call-off depends on the selected contracting model. When schedules are transmitted in the short-term contracting model, a price for the call-off in €/kWh is included. In the long-term contracting model, a fixed amount is remunerated that is independent of the calls, and a kilowatt-hour price is paid additionally for individual call-offs. The field test further included far-reaching technical analyses of the smart meter infrastructure to evaluate reliability, performance, and associated data volume of the deployed iMSys devices.

Besides technical issues, civic participation was one of the key factors in the successful realization of the living lab. Therefore, potential participants were approached well in advance to convey a feeling of appreciation and establish trust. Additionally, an app was also developed for direct dialogue, and local decision-makers were involved to strengthen the participants' understanding and generate coverage.

As for today, however, bottlenecks rarely occur within low voltage networks, so that the concept is to be extended to higher voltage levels where grid congestions can be observed. Striving for interaction between network levels, the concept is also to be expanded and scaled in this regard. Moreover, ALF's concept of utilizing small-scale units' flexibility potentials should be combined with the DiNCP concept that prevents grid overloads caused by flexible assets (Estermann et al., 2019).

However, it must be considered that the scalability of the platform concept depends not only on the occurrence of bottlenecks but also on the availability of flexibility. The availability varies widely between regions, such that suitable areas for flexibility platforms must be defined depending on the regional conditions. To utilize as much of the existing potential as possible, i.e., to motivate plant owners to offer their flexibility, concepts with low entry barriers are required.

The concepts of DiNCP, VPPs, and ALF were developed in the C/sells context with the ulterior motive of creating new organizational forms for a diverse, cellular, and community-driven energy future. While the concepts as such are only architectural structures, they allow for the formation of energy communities from household to transregional level with sufficient citizen participation. In the following section, a concept is described, which emphasizes communal interests and benefits on different organizational levels. It connects the concepts of DiNCP, IHM, and ALF to realize an interactive, networked, and participatory smart energy system.

3. The big picture: how different coordination entities interact with each other

The previous Sections 2.1–2.4 describe four levels of coordination of the usage of flexibility, which played a substantial role within the C/sells project. While they proved expedient to foster grid stability and enable new business models, a coupling of the different concepts was beyond the scope of the living labs. In the subsequent sections, a holistic coordination approach focusing on the interlinking of the existing concepts is put forward. It features the layout of a common system landscape, including end customers and the different coordination mechanisms discussed in

Section 2. Starting from a static system landscape, a multilevel sequence diagram is derived, which arranges all relevant interactions into a chronological coordination process. The resulting concept is intended to be an architectural–technical basis for energy communities, enabling them simultaneously to design business models within the energy community and ensure grid stability even with high penetration rates of renewable energies.

3.1 Energy communities as part of a common system landscape

Fig. 14.4 depicts the system landscape of a cellular energy system, which combines the afore-described concepts. To gain a better understanding of the relationships and the technical interfaces between the main actors and components, they are explained in more detail in the following.

The right-hand side of Fig. 14.4 shows the end customers. In the upper right corner, an operator of an electric heat pump is depicted. The electric heat pump is marketed as a flexibility option on ALF via the SMGW infrastructure. Since no marketer or aggregator operates in the owner's area, or the customer prefers to market their system independently, they use a long-term contracting model to offer flexibility directly on ALF without any intermediaries. As explained in Section 2.2, flexibility platforms show their strengths in a supraregional scope where they can access a larger pool of flexibility options and cover a higher number of possible grid congestions. In the common system landscape, ALF operates on a nationwide basis and coordinates flexibility from and for the entire German grid area to resolve congestions in the distribution grid. Besides DSOs, transmission system operators can also demand flexibility via a second

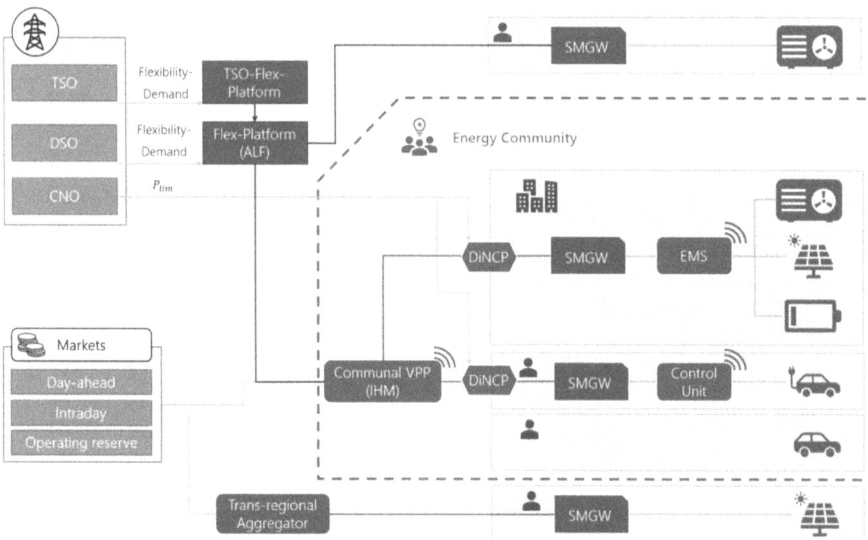

Figure 14.4 Interlinking community levels of different C/sells-concepts in a comprehensive system landscape.

platform on ALF. This two-stage concept has already been demonstrated by the successful coupling of ALF and TenneT's COMAX platform (Bayernwerk Netz GmbH, 2021). The CNO complements the grid entities and is responsible for compliance with P_{lim} specifications at the house connection.

Below the previous end customer, a communal VPP is shown. Its portfolio comprises a neighborhood with several flexible assets and the electric vehicle of a separate end customer. Communal boundaries were introduced to the operation area of the VPP to avoid invalid copperplate simplifications transregional aggregation would entail. In the neighborhood, a cascaded optimization of energy flows is present since the optimization by the communal VPP is followed by another optimization by the local EMS. The VPP can trade electricity on behalf of its customers on the Day-ahead- or Intraday-market or participate in the operating reserve. Due to its communal limitation, the VPP can also market the aggregated flexibilities on ALF without provoking grid congestions. However, this only applies if sufficient grid capacities can be assumed within its operating area of the energy community. Since transregional aggregators, depicted at the very bottom of Fig. 14.4, cannot guarantee not to cause grid congestions, they can only operate on the three markets (Day-ahead, Intraday, operating reserve), but not on the flexibility platform. For end customers without flexible assets, who thus neither participate in the long-term contracting of ALF nor a communal VPP (depicted in Fig. 14.4 below the communal VPP and above the end consumer involved with a transregional aggregator), this marketing concept can still benefit from the side effects of an energy community (often attractive mobility or housing concepts).

In the Directive (EU) 2018/2001 on the promotion of the use of energy from renewable sources of November 12, 2018, the EU includes a limited geographical scope in its definition of energy communities (see also the chapter by Spasova and Braungardt). Others like Pahkala et al. (2018), however, do not bind communities to the immediate vicinity of generation and consumption. In our case, a geographic boundary was introduced by the communal limitation in the cellular context of IHM and ALF. As a result, in Fig. 14.4, the scope energy community, framed by a dashed line, is congruent with the operational area of the VPP.

3.2 The chronological process—a multilevel sequence diagram

In the following, the chronology of the concept is presented. For the sequencing of the process, the four levels of Section 2 and the grid operator level, representing the different actors in the energy system, are considered. Their roles, tasks, and relations have been described in Section 3.1. The process sequence is cascaded, cascading from the lowest level of the property units, via the municipality to the markets or flexibility platforms, and finally to the grid operators. Fig. 14.5 shows the timeline of 1 day in the introduced system landscape. The process starts the day ahead of delivery day.

First, the local EMS (see Section 2.1) creates a forecast for the overall schedule of the connected assets (1). It includes weather forecasts, heat demand, mobility requirements, and further relevant empirical data. The schedules of the local EMSs are transmitted to a communal EMS (see Section 2.2), which then creates an overall schedule for the following day for the whole community (2). Depending on the revenue

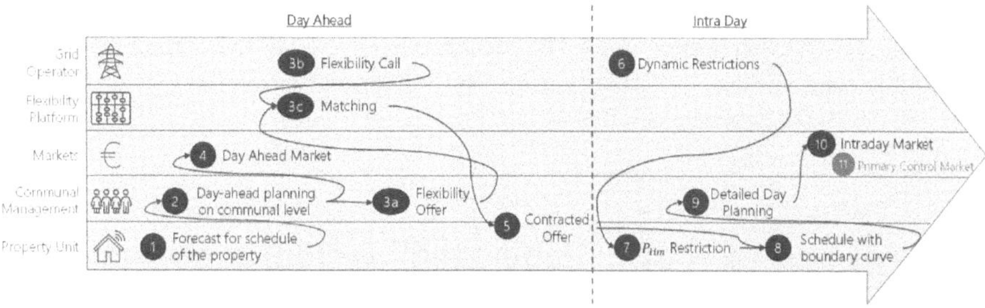

Figure 14.5 Chronological sequence of the concept.

prediction, the communal EMS then offers capacities on the Day-ahead market (4) (see Section 2.3) or the flexibility platform (3a) (see Section 2.4).

The network operator, who is simultaneously informed about the load curves by the EMS operators, daily prepares forecasts for the upcoming day and identifies potential congestions and overloads. Based on these forecasts, flexibility demands are defined for individual network nodes or line sections (3b). At 3:00 p.m., the matching of flexibility offers and demand takes place on the flexibility platform (3c). Depending on the matching result, the community's flexibility offer is contracted via ALF or traded on the Intraday market.

On the day of operation, network operators adjust their forecasts for the day. If necessary, they impose grid restrictions by distributing limiting power values (P_{lim}) for individual grid connection points (6). This process takes place before 9:00 a.m. and includes hours from 11:00 a.m. of the same until 10:45 a.m. of the following day. These power limits indeed constitute a top-down engagement of the network operators, which seems contrary to the community idea. However, they are necessary to allow for all bottom-up activities at the different markets in the prior stages as otherwise, these activities could cause grid overload.

The power limits P_{lim} are submitted to the lowest level of the property unit (7). Based on the new information of the limiting power value P_{lim}, updated information on the acceptance or rejection of the assets' flexibility offer, and more detailed information of the load and generation profile of the connected assets, the EMS generates a new, adjusted schedule for the day, including the P_{lim} restrictions (8). Now, the final overall power scheduling for the community within the boundary curves of the individual EMSs can be carried out by the communal management (9). In case of potential flexibility being unused, it can be traded on the Intraday market (10). Here, a relatively low minimum trading volume and gate closure 5 minutes before delivery (Bundesnetzagentur and Bundeskartellamt, 2020) offer low entry barriers for flexibility or energy available at short notice within the energy community.

As an alternative to participating at the spot markets, the community could offer balancing power. As described in Section 2.3, in Germany this requires very high generation availability and a minimum of 1 MW of disposable power. Since the auctioning of secondary and tertiary power takes place the day before delivery, this marketing option would already have to be taken into account at the beginning of the process

(50Hertz Transmission GmbH et al., 2021a). Primary balancing power, on the other hand, is marketed in 4-hour intervals, which allows energy communities to trade remaining capacities, if sufficiently large, on the primary balancing power market (11) (50Hertz Transmission GmbH et al., 2021b).

4. Pros and cons of a "federated" system architecture

In addition to the concept presented in Section 3, it was debated to combine the functionalities of ALF and IHM's VPP in one entity or to coordinate them in parallel, to construct a comprehensive system landscape. While both approaches would on the one hand streamline coordination structures, they would on the other hand entail complications: both for the formal combination as well as the parallel coordination on one level, various grid and system services would be in direct competition. To prevent conflicting controls or duplicate utilization of assets, both approaches require the implementation of a prioritization logic, which, in turn, would be close to cascading the process as in the previous sections.

For the same reason, the concept presented in Section 3 excludes the parallel marketing of assets via a communal VPP and a transregional aggregator. In principle, different plants connected to one local EMS could do so (market assets both via a communal VPP and a transregional aggregator), but it would hamper the optimization of energy flows within the EMS/at EMS level.

To accentuate the communal benefits of the various concepts from DiNCP to IHM to ALF, a layered coordination structure was developed instead. This way, all three approaches can optimally utilize their strengths and dovetail to form an efficient, comprehensive system architecture. At the lowest cellular level, the DiNCP regulates the setpoint specifications for end customers' and neighborhoods' local energy optimization. Above that, the regional VPP optimizes the balancing of generation and consumption at communal or municipal level. In case of acute critical network conditions, the CNO can still prevent equipment overload via P_{lim} defaults for individual grid connection points. Finally, at the national level, ALF coordinates the deployment of the distributed flex options and the regional VPPs enlisted. In summary, the proposed concept combines three technical architectures to promote energy communities at the immediate neighborhood-, communal- and transregional level. At each level, the driving force behind the community formation will be the prospect of various communal benefits. As enablers of business models and grid stability, the C/sells concepts provide the architectural foundation for this development.

From the three-level organizational structure, parallels to the principle of federalism can be drawn. Comparable to a central government unit with several provincial and other subordinate units, the proposed coordination structure also increasingly imposes restrictions for the sphere of action at the lower levels but does not affect the concrete behavior. This design approach facilitates keeping complexity manageable even in highly distributed energy systems. Nevertheless, it raises the question of why the boundary of the Energy Community in Fig. 14.4 was drawn as shown. Following the federalism idea, equally smaller or larger entities within the landscape can also be defined as energy communities equally as well. Neither does the respective literature exclude extreme forms, such as single buildings on the one or transnational concepts at the other end of the scale, in its definitions of energy communities (Fina et al., 2019). Considering that, the positioning of the dashed community border in Fig. 14.4 is therefore debatable. Variants of the figure with one displaced or several nested lines are also conceivable. For the C/sells community, however, the perimeter as drawn hits a sweet spot and most strongly fulfills the three guiding principles cellularity, diversity, and participation. It is both narrow enough to satisfy the requirements for cellularity and diversity and large enough to allow participation in a movement bigger than the immediate neighborhood.

5. Conclusion

The C/sells project has shown the necessity to include small-scale assets for congestion management and other system services in the organization of a cellular energy system. To realize this reorganization, citizens must be empowered both technically and organizationally to provide these services. The living labs ALF and IHM could already demonstrate the successful commercialization of small-scale assets in different use cases and at different scales. Thereby it became evident that targeted and appropriate compensation mechanisms strengthen community cohesion and drive the energy transition.

In contrast to virtual communities working with copperplate simplifications, the proposed federated concept considers grid constraints and facilitates the balancing of fluctuations. These processes take place at different community levels before instabilities propagate to the surrounding power grid. This way, the strengths of energy communities are harnessed not only to increase renewable penetration but also to ensure grid stability.

However, the mere top-down implementation of the outlined organizational structures does not create an energy community. In addition to new, functioning coordination mechanisms and appropriate compensation options, it is often participatory aspects that determine the success of projects with strong civic involvement. Within C/sells, it was therefore essential to choose appropriate communication methods within the "participation cells" to create broad approval for new intelligent solutions.

The living labs are completed, but the C/sells community will use their key takeaways to keep pushing new ideas forward. For future projects, living labs should be set up with even more participants to better understand complex market dynamics and economies of scale. Therefore, the dialogue with local politicians and the deployment of an interactive app to track community benefits can be beneficial to boost participation and accelerate the local energy transition, especially in larger field test environments. In addition, it must be ensured that legal innovation zones are set up for such projects that allow new solutions to be tested without regulatory obstacles.

In terms of content, future projects in the field should also focus stronger on collective platform models to leverage community benefits even more for participants and less for profit-oriented platform operators. Only this way, a self-sustaining bottom-up movement can be created that will strengthen the community motive on several levels and become the driving force for a clean energy future.

References

50 Hertz Transmission GmbH/Amprion GmbH/TenneT TSO GmbH/TransnetBW GmbH (Eds.), (2021a). Übertragungsnetzbetreiber Deutschland, Ausschreibung Minutenreserve. Europäisches Parlament; Europäischer Rat. Available online at: https://www.regelleistung.net/ext/static/mrl.

50 Hertz Transmission GmbH/Amprion GmbH/TenneT TSO GmbH/TransnetBW GmbH (Eds.), (2020). Übertragungsnetzbetreiber Deutschland: Präqualifikation für die Vorhaltung und Erbringung von Regelleistung. Available online at: https://www.regelleistung.net/ext/static/prequalification (accessed 5/28/2021).

50 Hertz Transmission GmbH/Amprion GmbH/TenneT TSO GmbH/TransnetBW GmbH (Eds.), (2021b). Übertragungsnetzbetreiber Deutschland, Gemeinsame Ausschreibung Primärregelleistung. Available online at: https://www.regelleistung.net/ext/static/prl.

Bayernwerk Netz GmbH, 2021. Kleinanlagen im Verteilnetz ergänzen Großkraftwerke. Regensburg, Pressemitteilung vom 22.01.2021. Available online at: https://www.mynewsdesk.com/de/bayernwerk/pressreleases/kleinanlagen-im-verteilnetz-ergaenzen-grosskraftwerke-3066728.

Braun, M., 2007. FENIX –Virtuelle Kraftwerke und Möglichkeiten der Bereitstellung von Netzdienstleistungen Hanau, 5. In: Hanauer Dialog "Strom aus Biomasse: Der Systemdienstleister der Erneuerbaren".

Bundesnetzagentur für Elektrizität, Gas, Telekommunikation, Post und Eisenbahnen/Bundeskartellamt (Eds.) (2020). Monitoringbericht 2020. Monitoringbericht gemäß § 63 Abs. 3 i. V. m. § 35 EnWG und § 48 Abs. 3 i. V. m. § 53 Abs. 3 GWB. Bundesnetzagentur für Elektrizität, Gas, Telekommunikation, Post und Eisenbahnen; Bundeskartellamt. Bonn. Available online at: https://www.bundesnetzagentur.de/SharedDocs/Mediathek/Berichte/2020/Monitoringbericht_Energie2020.pdf?__blob=publicationFile&v=5(accessed 3/4/2021).

Colle, S., Micallef, P., Legg, A., Horstead, A., Breakell, J., 2019. Where does change start if the future is already decided? DSOs must keep pace with the new ways to generate, distribute and consume energy... and become catalysts of these changes. Available online at: https://www.eurelectric.org/media/3637/ey-report-future-of-dsos.pdf.

Directive (EU) 2018/2001 on the Promotion of the Use of Energy from Renewable Sources of 12/11/2018, 2018.

EPEX SPOT SE (Eds.), 2021. Basics of the power market. Available online at: https://www.epexspot.com/en/basicspowermarket (accessed 5/28/2021).

Estermann, T., Bruckmeier, A., Köppl, S., Wohlschlager, D., 2019. Field test to demonstrate a smart market platform via smart meter infrastructure. In: Thomas, A., Uta, B. (Eds.), Digital Proceedings of the 9th Solar & Storage Integration Workshop. 15–16 October 2019 | Dublin, Ireland. Energynautics GmbH, Darmstadt.

Fina, B., Auer, H., Friedl, W., 2019. Profitability of PV sharing in energy communities: use cases for different settlement patterns. Energy 189, 116148. https://doi.org/10.1016/j.energy.2019.116148.

Greif, S., Weigand, A., 2019. Flexibilität elektrischer Heizsysteme im Feldtest. Energiewirtschaftliche Tagesfragen (06/2019). Available online at: https://www.ffe.de/themen-und-methoden/waermeversorgung/903-flexibilitaet-elektrischer-heizsysteme-im-feldtest.

Heilmann, E., Klempp, N., Wetzel, H., 2020. Market design of regional flexibility markets: a classification metric for flexibility products and its application to German prototypical flexibility markets. In: Philipps-University Marburg, School of Business and Economics. MAGKS Joint Discussion Paper Series in Economics. Marburg.

Kalz, D., Klein, K., Palzer, A., Schlösser, T., Schumacher, P., Sterchele, P., Stinner, S., Yu, Y.J., Kallert, A.M., 2018. Netzdienliche Gebäude und Quartiere. Gebäude entlasten Stromnetze. Bonn. BINE-Themeninfo I/2018.

Kinkel, K., 2014. Rekommunalisierung der elektrischen Energieversorgung: Herausforderungen am Beispiel dreier nordhessischer Energieversorgungsunternehmen. Wuppertaler Studienarbeiten zur nachhaltigen Entwicklung, No. 8. Wuppertal.

Knoefel, J., 2019. Energiespeicherdienste für smarte Quartiere. Ergebnisse aus dem Projekt ESQUIRE Berlin, Strommarkttreffen, 10.05.2019.

Köppl, S., Lang, C., Bogensperger, A., Estermann, T., Zeiselmair, A., 2019. Altdorfer Flexmarkt - decentral flexibility for distribution networks. In: Rehtanz, C. (Ed.), Internationaler ETG-Kongress 2019. Das Gesamtsystem im Fokus der Energiewende: 8.-9. Mai 2019, Neckar-Forum, Esslingen am Neckar. VDE Verlag GmbH, Berlin, pp. 477−482.

Pahkala, T., Uimonen, H., Väre, V., 2018. Flexible and Customer-Centred Electricity System: Final Report of the Smart Grid Working Group. Publications of the Ministry of Economic Affairs and Employment 39/2018, Helsinki.

Park, C., Yong, T., 2017. Comparative review and discussion on P2P electricity trading. Energy Proc. 128, 3−9. https://doi.org/10.1016/j.egypro.2017.09.003.

Radeke, J., Hefele, J., Hirth, L., 2019. Markets for Local Flexibility in Distribution Networks. Working Paper. Kiel, Hamburg. Available online at: http://hdl.handle.net/10419/204559.

REScoop.eu, 2019. In: Q&A: What Are 'citizen' and 'renewable' Energy Communities?. Available online at: https://www.rescoop.eu/toolbox/q-a-what-are-citizen-and-renewable-energy-communities. (Accessed 28 May 2021).

Weigand, A., Köppl, S., Rogg, K., Springmann, E., 2021. Digitaler Netzanschluss: Schnittstelle zwischen Gebäude und Stromnetz neugedacht. BWK−Das Energie-Fachmagazin (5/6).

Westphal, E.L., Köppl, S., Kießling, A. & Mauch, W., 2019. Flexumer als Gestalter der digitalen Energiezukunft − Eine Begriffseinordnung. Energiewirtschaftliche Tagesfragen (7−8), 68−71. Available online at: https://www.ffe.de/veroeffentlichungen/flexumer-als-gestalter-einer-digitalen-energiezukunft-eine-begriffseinordnung/.

Zeiselmair, A., Köppl, S., 2021. Constrained optimization as the allocation method in local flexibility markets. Energies 14 (13), 3932. https://doi.org/10.3390/en14133932.

Zeiselmair, A., Köppl, S., Estermann, T., Lehmann, N., Kraft, E., Klempp, N., 2019. Netzdienlicher Handel als Element des zellulären Energiesystems am Beispiel des Altdorfer Flexmarkts (ALF). In: Institut für Energiesysteme und Elektrische Antriebe/Austrian Association for Energy Economics. IEWT 2019. 11. Internationale Energiewirtschaftstagung "Freiheit, Gleichheit, Demokratie: Segen oder Chaos für Energiemärkte?", vol. 13. Wien.

The path to energy communities via local energy management and digital customer care

15

Tina Hadler
Coneva GmbH, Munich, Germany

1. Introduction

Energy communities (ECs) can contribute to fighting climate change and encouraging the energy transition based on the three Ds of digitization, decentralization, and decarbonization. However, the perspective of where to start with ECs and who will take the lead can vary.

A regulator might be guided by governmental goals following a national commitment made under the Paris Agreement (United Nations Framework Convention on Climate Change, 2016). An energy supplier might care about changing requirements for distributed energy portfolio management and the new needs to manage both commodity and noncommodity offers for their clients. A distribution system operator (DSO) responsible for the local distribution grids might care more about the challenges of grid extension, organization, and maintenance with a continuously increasing level of renewable energy capacity at a low voltage level and the growing need for connectivity and communication infrastructure for controlling purposes. For an energy service provider like Coneva, the motivation to deal with ECs is the opportunity to develop new business models by creating platform-based services to handle complex data transactions.

This chapter focuses on the perspective of energy service providers, in particular how ECs can be implemented as a business model in today's energy markets. The chapter is structured as follows:

- Section 2 introduces Coneva, providing an overview of the roles and archetypes in ECs;
- Section 3 lays out Coneva's approach and concept of implementing ECs as a business model through local energy management systems (EMS) and digital customer care; and
- Section 4 provides two examples from implementation projects in which Coneva demonstrates how their approach translates into practical projects followed by the chapter's conclusions.

Energy Communities. https://doi.org/10.1016/B978-0-323-91135-1.00011-0

2. Roles and archetypes in energy communities

Coneva is a digital energy service provider, founded in 2018 as a corporate startup of the well-known producer of PV inverters, SMA Solar Technologies. Coneva's main capabilities include the handling of local energy management data along the energy value chain and the connection of consumer and prosumer installations behind, at and beyond the meter that acts as the grid access point.

Targeting on ECs, Coneva focuses on the following:

- Optimizing within a decentralized value chain by managing a system of multiple distributed energy supply installations based on a platform;
- Creating a space to connect people and enhance participation with the help of a social networking space for prosumers and consumers; and
- Integrating value-added services by developing a marketplace for related services and products for ECs.

Coneva is enhancing and supporting different types of ECs that are described in Sections 2.2–2.4. The following section starts by explaining the different roles involved in ECs.

2.1 Roles in energy communities

Three new parties are engaged in setting up ECs:

Community members can be prosumers or consumers with varying installed capacities and different distributed renewable plants (mainly focused on PV). Under the Coneva approach, producers only disposing of distributed (renewable) energy resources to feed their energy into the community without actively participating in the social network structure of an EC are classified as assets managed directly or indirectly by the community provider, e.g., in a portfolio.

The *community provider* establishes and manages the relationship with community members, offering direct support to make the EC come to life. Their business model might comprise different forms of energy supply and energy services. Community providers come from a variety of backgrounds: Coneva has identified classic energy suppliers but also real estate companies, communal governments, original equipment manufacturers (OEMs), as well as other new players in the field of digital service models. The variety is assumed to result from the changing market circumstances driven by digitization, competition in new business models, and increasing cooperation possibilities along the value chain.

The *community platform provider* develops, owns, and operates a platform to serve both the community provider in a business-to-business relationship and the community member(s) in a business-to-business-to-customer relationship. Coneva takes over this role and furthermore defines itself as a community-as-a-service provider.

The EC is a platform-based energy sharing concept that enables the generation from local energy resources plus sharing and trading among the participants on a local level. This requires localization of the entire energy value chain—generation, feed-in to the system, distribution, transmission, balancing, metering, and invoicing. Thus, for every future EC, a distributed value chain must be implemented and integrated into the existing energy supply system. To enable describing different community archetypes, the community platform needs to ensure the following:

- Energy produced and consumed within a community can be differentiated from energy being supplied from outside the community;
- With connected local EMS, optimization with the community supply can be assured and will be aligned with individual optimization of prosumers, taking into account the amount of self-consumption and the extent to which they are self-sufficient (as opposed to relying on the system);
- All relevant energy flows within the community can be invoiced, apart from the energy supplied via the grid; and
- All energy flows can be visualized for and controlled by both the community provider and the community member.

One of the core contributions of Coneva's platform provider services is to keep track of all transactions among and between EC members plus the interface with the grid for settlement, billing, etc.—as schematically laid out in Fig. 15.1.

Fig. 15.1 shows the relationship between the three roles connected via the community platform established and run by the community platform provider. This arrangement offers the ability to handle energy produced and supplied within the community between the members, separately from the energy delivered from the grid. The community provider can ensure the connection to the public energy grid in terms of both grid and retail services, including grid stability at a certain voltage level, portfolio management, contract management, metering, or invoicing. The community member concludes a contract with the community provider via the platform to benefit from conditions created within the community, e.g., a lower price, subsidies, regional or local

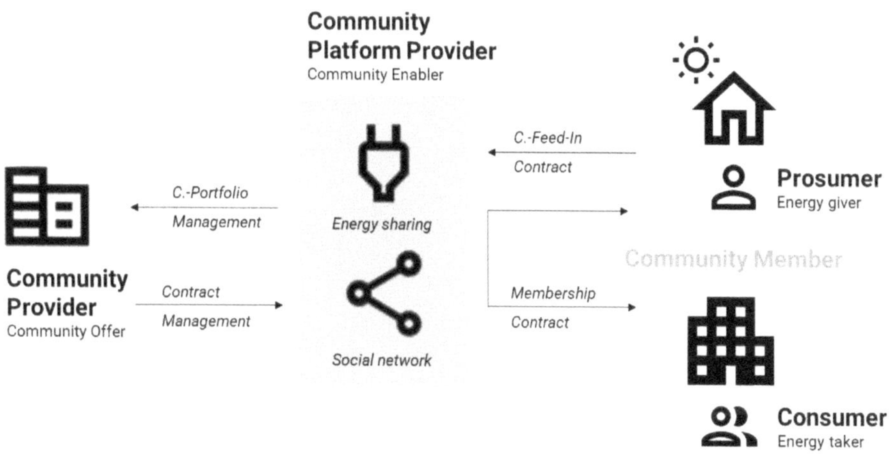

Figure 15.1 Roles in an energy community.
Source: Coneva (2021).

energy supply, and carbon dioxide reduction. Depending on their role as a prosumer or consumer, members need to comply with specific prerequisites in terms of metering, hardware investments, or transparency and control applications or devices.

2.2 Archetypes in energy communities

To implement ECs in a platform economy, data handling is a primary issue, whereas the legislative framework forms a basis for the basic taxonomy and conceptual options in the respective markets, as described in the chapter by Rossetto et al. Coneva's platform concept is mainly predicated on the taxonomy of the legislation of the European Union (EU) based on the "Clean Energy for all Europeans Package" and two EU directives as outlined in detail in the chapter by Spasova and Braungardt in this volume. First, the recast Renewable Energy Directive (RED II) of December 2018 (Directive EU 2018/2001, 2018) applies in particular to Renewable Energy Communities and to jointly acting renewables self-consumers (e.g., collective self-consumption). Second, the recast Electricity Directive on common rules for the internal market for electricity, referred to as the Energy Market Directive, EMD (Directive EU 2019/944, 2019) defines the regulation for Citizen Energy Communities.

The energy produced and supplied in an EC can be organized directly via the grid or indirectly as part of a portfolio. This leads to Coneva's definition of physical versus virtual communities (Fig. 15.2):

- Physical communities can be structured either as collective self-consumption, based on a private grid infrastructure without feeding into the public grid, or as community self-consumption, which allows the use of both private and public distribution networks; and
- Virtual communities are portfolio based, with either a geographical focus as a regional community or no geographical relation. The latter can be considered to be a cloud community, pointing to common interests among the community members.

Figure 15.2 Summary of energy community archetypes.
Source: Coneva (2021).

2.3 Physical energy communities

Corresponding to a "cradle-to-grave principle," physical ECs are based on a physical grid connection between the energy produced and supplied (Fig. 15.3). By defining a grid balancing point, it builds the relation to the prosumers and consumers and optimizes the energy flow over local and physical grid connection. Usually, these grid balancing points are at a low voltage level to ensure a discharge of less transformation activity with high grid levels.

Collective Self-Consumption: In the case of ECs using a private grid infrastructure, the production capacity and the participants are all located behind the same access point to the public grid. This may be the case in multitenant buildings, quarters, or areal grids. Typically, they have the following characteristics:

- Energy is produced and supplied over one private grid infrastructure;
- Mainly consumers participate, consuming from local (renewable) production units;
- The production capacity jointly supplies the community, i.e., there is usually no individual self-consumption;
- An oversupply or the provision of additional demand from the community will be collectively organized via the public grid and supply chain;
- The community's target is to optimize the self-consumption by minimizing the grid supply and maximizing the self-sufficiency over all community members with the help of certain metering concepts and optimization algorithms, i.e., local EMS;
- Therefore, additionally incorporated flexibilities like battery storage, electric vehicles (charge points or wallboxes), or heat pumps can increase the self-sufficiency; and
- An additional monetary incentive scheme making energy supplied in communities more attractive compared to conventional supply can enforce the development by means of tariffs and subsidies offered by the community provider, or the energy supplier via specific legislation triggered by RED II and the subsequent national regulations in the EU.[1]

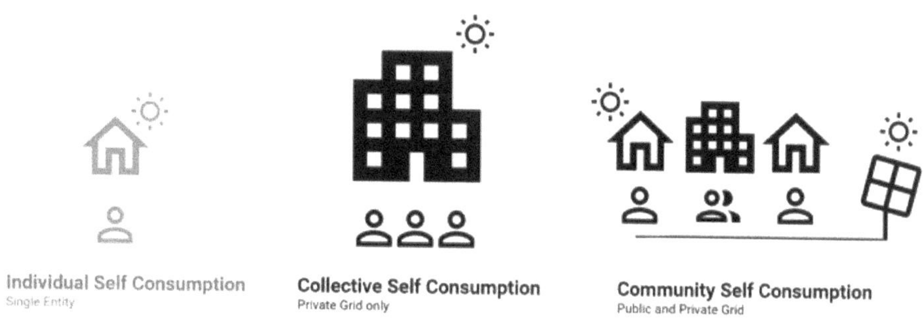

Figure 15.3 Physical energy communities.
Source: Coneva (2021).

[1] Compare status in Decide4energy.eu (2021).

Community Self-Consumption: When ECs use the public grid, grid balancing becomes a prerequisite. A joint grid access point needs to be defined, usually a substation on low to mid voltage level. Further characteristics are typically as follows:

- Participants and their assets are connected to the grid at the predefined voltage level;
- The optimization of the system's production and supply capacity is realized at that voltage level;
- Participants can be consumers and prosumers. In this context, the prosumer might pursue maximizing individual self-consumption before contributing to the EC; and
- Further subsidies for installed capacity or incentives incorporating avoided grid service costs or tax reliefs can support the model; for example, legislation schemes like the "Decreto Milleproroghe" in Italy (article 42-bis[2]) or the approaches of energy market integration of §14a EnWG referring to controllable load management in Germany.

In the next section, virtual ECs are explained.

2.4 Virtual energy communities

To enable regional or national ECs without a physical and grid-based connection, Coneva also supports virtual ECs. In this case, distributed prosumer and consumer installations are not balanced within a specific area behind a substation or a grid access point. The energy balancing is portfolio based. Offers and tariff schemes focus on marketing energy produced in community production or storage installations. If regulation allows, feed-in subsidies or other incentives are incorporated in these tariffs. Coneva offers two types of virtual communities—called regional virtual communities and cloud communities—as illustrated in Fig. 15.4.

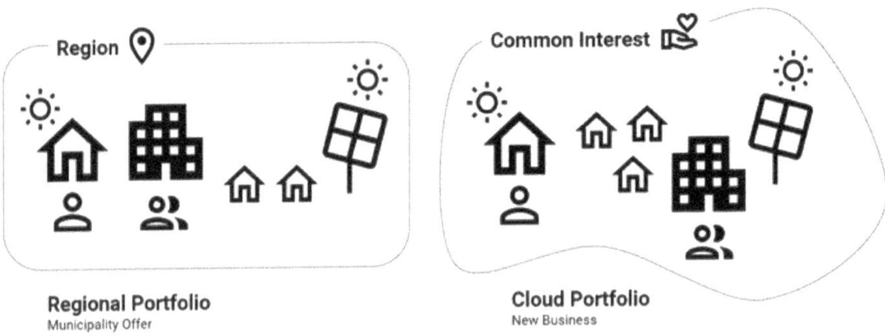

Figure 15.4 Virtual energy communities.
Source: Coneva (2021).

[2] www.comunitaenergetiche.net, 2021.

Regional virtual communities connect prosumer and consumer installations in a specific region. Providers or mediators of such models are mainly regional utilities or municipalities, focused on the differentiation of their offering or solution portfolio.

Specific tariffs guarantee that the community members are supplied with a specific characteristic of energy reflecting on the local origin. The portfolio comprises production resources situated in the same predefined region. This might also include prosumer installations if market models enable certain participation. The balancing between demand and supply is realized with the help of virtual balancing groups. Providers tend to prefer the development of a regional virtual community concept over a regional physical community. German local energy suppliers (municipalities or "Stadtwerke") often follow this approach based on their well-established brand within a certain region or community. For example, in 2019, Allgäuer Überlandwerke launched a special regional tariff offering electricity from renewable energy plants located in their distribution grid and marketed over a direct marketing scheme without a subsidy.[3]

Cloud communities are also portfolio based. The main difference is the value proposition for community members: instead of a physical grid connection or regional vicinity, cloud-based communities follow a content-driven common interest or goal.

In the German market, there is a wide range of community providers for cloud communities, ranging from energy service providers to nationally operating energy suppliers or OEMs such as sonnen, SENEC, E.ON, or EWS Schoenau.

3. Implementation of energy communities

For the implementation of ECs, the concept of data handling based on the digital community platform is key. In this context, from the community members' viewpoint, ECs need to ensure services capable of the following:

- Visibility:
  - Community members gain visibility on the community platform by having an ID or account and
  - The account needs to be connected to the assets that a prosumer or consumer integrates in the EC, i.e., the grid access point, the smart meter, and local EMS.
- Measurability:
  - For this purpose, a physical connection for prosumer and consumer installations needs to be installed and operated to enable energy services and
  - In order to develop specific energy services, data must be measurable to ensure individual and community energy services, as well as unidirectional or bidirectional energy services (see Section 3.2 for details).
- Invoicing
  - To ensure the invoicing and remuneration of the energy supply within an EC, the community provider needs a customer relationship management (CRM) tool and eventually interfaces with third party systems.

[3] AllgäuStrom 100%; www.auew.de/privatkunden/strom/allgaeustrom-100/, 2021.

- Controllability
  - To enable ECs to contribute to the stability of the energy grid control, installations need to be addressed via nonlocal signals and
  - The community assets must be technically connected to a central control to receive and send event-based signals, e.g., from a DSO.
- Trading
  - The aggregated installed capacities and captured flexibilities in a community can be used for further marketing purposes in the energy market, e.g., in virtual power plants.

To address these issues, Coneva follows three basic questions to first build and then meet the necessary capabilities within ECs.

- How to obtain the data?—This addresses the question of interconnectivity of community members, their field devices, and production units behind the grid connection point.
- How are the data processed?—This part deals with the programming and implementation of data services to the EC.
- How are the data provided?—This focuses on the interaction with and between the community members and providers.

The idealized architecture model behind this is shown in Fig. 15.5 and further explained in the following sections.

3.1 How to obtain the data? Connectivity

Gathering the necessary data for ECs means interconnecting assets and systems in different constellations. First, this includes the technical connectivity of the assets to the platform. Second, the systems—for example, the EMS—of all production and storage assets behind the meter need to be compatible. Third, the type of connection needs to fit it either unidirectional only, or bidirectionally for additional controlling and communication purposes.

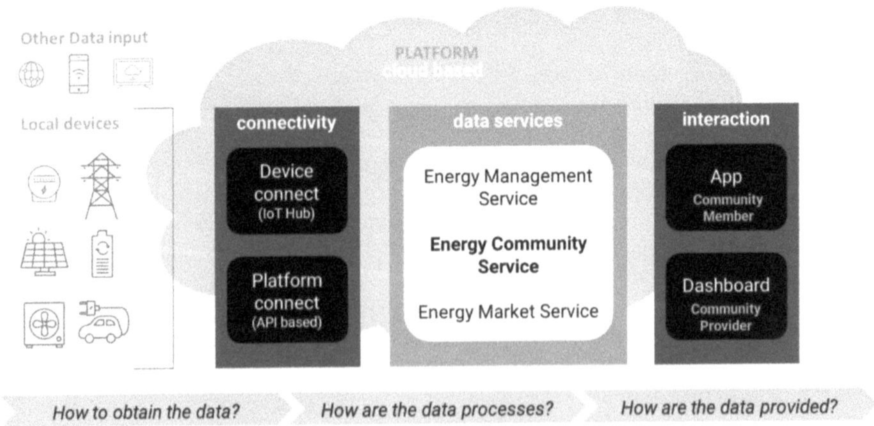

Figure 15.5 System architecture for energy communities.
Source: Coneva (2021).

Coneva offers the following three technical solutions to achieve a connection with the local device:

- *Energy management gateway*: This solution is a proprietary gateway mainly used for enabling complex prosumer installations, comprising a production capacity—usually photovoltaic—battery storage, electric vehicles, or heat pumps. The gateway contains an energy management software stack, which serves to optimize self-consumption and self-sufficiency for the prosumer. The software communicates with a certain technical protocol with the devices of the installation to be connected to the local energy management system (EMS). The communication of the EMS with the local devices can be unidirectional or bidirectional.
- *Virtual energy management gateway:* The idea of virtual energy management gateway is a special form of the prosumer gateway. To substitute hardware, in the context of the smart meter rollout, the energy management software stack is virtualized and integrated on the smart meter infrastructure. Such a solution is implemented on the controllable local system channel in the context of the smart meter rollout in Germany (Stoetzel et al., 2019).
- *Basic digital energy meter*: To connect all types of community members, especially consumers, to the community platform, the use of a complex EMS as described above would be technically possible but too expensive. Connecting simple prosumer installations or especially consumers' metering points, a "low-cost energy meter" is used to collect—unidirectionally only—a basic set of data points necessary for managing the community supply. Coneva uses sensor-based probes on digital meters, which vary depending on the metering technology on site. Two examples in the German market are from Powerfox with the product "Poweropti" or IK Elektronik with the "MCA Cloud Module." Another solution for the Italian market—provided by the company MAC Italia—works in a similar way using a powerline specific protocol "Chain2Gate" to connect with the Italian smart meter (Mac Italia, 2019).

All three solutions for connectivity can be integrated to the Coneva community platform in two ways: a direct connection to an Internet of Things (IoT) hub solution or connection based on an Application Programming Interface (API). Moreover, a combination of the two ways is possible to extend the functional options.

3.2 How are the data processed? Data services

Once the data are gathered from the individual entities in the community, they need to be processed. As mentioned above in the list of capabilities, the *measurability* regards both individual energy services to the community members as well as energy services for the EC as a whole.

3.2.1 Energy services to the community members

Within the EC, the model needs to deliver individual data-based services to prosumers and consumers that are essentially structured as follows:

- *Visualization services* for both consumers and prosumers containing the following:
  - Visualization of actual, historic, and forecasted production and supply data for information purposes;
  - Provision of data to external parties, e.g., over APIs;

  - Basic digital energy manager as necessary hardware; but
  - Not ensuring a controlling ability to the installations and connected devices.
- *Services for a local self-consumption optimization* for prosumers with an installation for self-consumption, e.g., PV plant with battery storage or other flexibilities containing the following:
  - Management and controlling of production plants and connected devices over local EMS algorithms to ensure self-consumption optimization;
  - Installation of additional hardware devices, like the energy management gateway or the smart meter with a virtual energy gateway, to grant access to local data for involved parties;
  - Integration of additional access to direct data connection for external third parties, apart from the local EMS provider, to give access to portal functions; and
  - Visualization services as described above.

The services for local self-consumption optimization require a bidirectional platform connection over the local EMS for controlling purposes, whereas the first category (visualization services) is unidirectional data communication from the connected installation to the data hub.

In the context of ECs, these individual services to prosumers or consumers are amended or extended following their role as community members and their necessary contribution to the EC defined by the community provider. In terms of the visualization services, the data collected from the installations are set into the context of the energy supply model in the EC (see archetypes) and are further used for contractual purposes, like invoicing by connecting to the appropriate CRM system of the community provider.

For an extended self-consumption optimization in the context of ECs, the reaction to central controlling parameters from outside the EC must be enabled. These central control signals may be triggered by short-term shortages in the distribution grid, balancing purposes, or price-triggered signals coming from the spot markets.

3.2.2 Energy services to the energy community

Certain additional data services for the EC as a collective are necessary, based on connectivity. For Coneva, the key capability behind the EC data service is the optimization of local energy supply for the connected entities. Community providers usually require these additional services for their business models.

To define the community, individually collected energy data from community members' installations need to be handled to enable the following services:

- The aggregation of load and production profiles of the community members for a community balance;
- A basic (nonmonetary) distribution logic for the community supply to community members; and
- A benchmarking function to visualize community members' own contribution to both the EC and other members.

In Section 4, an example of the aggregation service within a virtual EC is explained. Referencing the archetypes, it is worth noting that the data model itself for the basic EC services is valid for all EC archetypes described in Section 2.

To optimize and differentiate the services and character of ECs, the data service needs to be continuously refined. Coneva differentiates between three optimization approaches incorporating controlling and prediction measures:

- Optimization via pricing signals:
  - Develop tariff models or other monetary incentives for a certain production or supply behavior or during fixed time schedules;
  - Offer visualization services and the connection of subsequent invoicing to enhance reactions to pricing signals; and
  - Introduce automatic control of local devices based on price signals to be integrated.
- Optimization regarding grid bottlenecks:
  - Offer controlling access for DSOs to the connected entities in the EC by means of bottom-up balancing at a low voltage distribution level; and
  - Grid-based optimization differs between individual and aggregated stabilizing services.
- Optimization via energy and flexibility trading:
  - Incentivize tradeable goods to several spot markets, based on a controlling access to the connected entities.

The structure of the related data services in ECs is summarized in Fig. 15.6.

3.3 How are the data provided? Interaction

Once all the data points from the entities in an EC are connected, captured, controlled, and processed in data services as described in Sections 3.1 and 3.2, the final question of

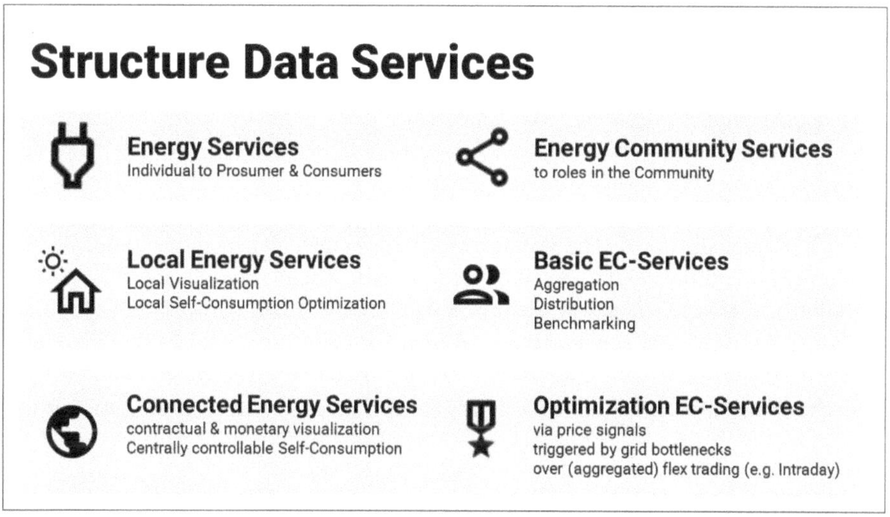

Figure 15.6 Overview on the data service structure in ECs.
Source: Coneva (2021).

the Coneva EC implementation model deals with the data provision to both the community provider and the community members over interactional interfaces.

Interaction over an app service to the community members: Community members are usually end users who are used to using digital applications for almost every part of their life. Therefore, the Coneva app service named "Energy Buddy App" provides them the central interface to the EC. It mainly focuses on visualizing the status of both the individual and the collective energy situation. An appropriate branding suite provides a change to a corporate design of the community provider.

Interaction over web-based dashboard services for the community provider and other professional stakeholders: Having the Community Provider as the business partner for the successful marketing, scaling, and operating of the EC, an appropriate monitoring and controlling service is necessary. With the dashboard service, all of the above-described options to set up an EC can be administered by the community provider, along with professional stakeholders, like investors and regulatory bodies.

4. Case study of energy communities

Based on these classifications and descriptions of the necessary capabilities for realizing ECs, two exemplary projects in which Coneva is involved are described. The first is an example of a virtual community following a portfolio-based approach of a green energy supplier in Germany, while the second is an example of a community self-consumption project based on recent regulation in Italy. These examples also illustrate the range of approaches to planning and operation of ECs.

4.1 Virtual community concept for a green energy supplier in Germany

The challenges of global climate change and the energy transition in Germany present ECs opportunities for new business approaches. This holds true although there is not yet an appropriate supportive regulatory as seen in other European countries (see Spasova and Braungardt in this volume).

The energy supplier with which Coneva is working is a green energy pioneer founded in the late-1980s. The supplier has recently developed a new vision focusing on sustainable infrastructure development for local energy markets and new product offerings for their customers. It has developed a comprehensive offering for local energy management to prosumers and consumers. Prosumers should be able to supply each other with their own electricity in several regionally clustered ECs, and consumers should be able to purchase that electricity. The necessary electricity must be delivered exclusively from plants that do not receive state-aided subsidies, like the feed-in tariffs following the Renewable Energy Act (EEG).

In order to create new business approaches and react to dynamics in the German market like the smart meter rollout, focus was placed on the development of Coneva's virtual community concept (see Section 4.2). All members of a community were to be connected and a visualization of all energy flows in real time was to be realized both individually and on an aggregated basis. Apart from the interactional app- and web-based front ends, the leverage to form the EC is the technical connection solution for the heterogeneous customer base of the energy supplier. The underlying data model ensures transparency of the direct allocation of the energy flow, combining all feed-in and exit points within a balancing group. The aim of the balancing group is a portfolio-based supply to meet residual electricity demand not met within the community and marketing the EC's energy surpluses.

To start the implementation of a first EC, Coneva built a technical proof-of-concept with approximately 40 prosumers and consumers, including several PV systems, various battery systems and several small, combined heat and power plants.

Coneva supports the implementation, connection, and integration of distributed prosumer and consumer installations into their cloud-based platform model. Thus, the energy supplier is extending its role as provider to the EC to also be the interface to the energy market via a virtual power plant approach. New business models are in development as minimal viable products, like aggregated controlling of charging infrastructure for electric vehicles or heat pumps heatings based on optimized schedules or new feed-in tariffs for unsubsidized small-scale PV.

The regulated smart meters are to be incorporated as the key infrastructure to replace proprietary EMS, like the Sunny Home Manager of SMA or comparable device controllers (see Section 3.1 and Stoetzel et al., 2019).

4.2 Collective self-consumption in Italy

Based on the EU Clean Energy Package, at the beginning of 2020, the Italian Government established a new energy regulation, including subsidies for local energy supply structures based on renewable energy production (see chapters by Lo Schiavo et al. and Del Pizzo et al.). This regulation supports a new market approach, establishing a connection of local energy production from renewable resources to local energy consumption at a certain grid level by paying a 20-year incentive of 110 €/MWh for self-consumed electricity. By 2025, experts estimate that this will lead to 30,000 renewable ECs in Italy.

To receive the incentive, a future *community provider* must provide proof—on an hourly basis—of how much of the energy being produced within a community was consumed within the same community. This central key performance indicator (KPI) is called community self-consumption.

Creating a business model based on this KPI and the legal framework requires the following:

- The definition of a legal entity with a *community statute*, describing the responsibilities, duties, and benefits of the roles involved;
- A *smart metering concept* for capturing the generation and consumption units involved; and
- A *data analytics framework* for analyzing, optimizing, and providing the community self-consumption.

Coneva provided technical and conceptual consulting to a collective-self-consumption project with an installed capacity of 55 kWp PV capacity and 40 distributed private households in the vicinity of the PV plant, connected by the public distribution grid.

To enable this EC concept, Coneva created an end-to-end solution for community providers and subsequent investors in the Italian market, focusing on the capturing and subsequent optimization of the community self-consumption. This solution was a first step toward realizing ECs as legal entities based on the Italian regulation. It follows the Coneva system architecture as presented in Section 3. It first comprises a digital metering concept to collect the necessary production and consumption data of community members and resources. The main challenge is compatibility with the Italian smart meter—called open meter—and in particular complying with a specific protocol (Chain2Gate) that can be read over specific powerline communication device (PLC-C) (see Section 3.1 and Mac Italia, 2019). Moreover, the platform connection to set up the community model needs to be defined and implemented and combined with the captured data from the metering. To fulfill the requirements of the Italian community concept, the integration of the data model regarding the metering must follow an aggregating and analytical logic to provide the community self-consumption on an hourly basis (Zulianello et al., 2020). Finally, the provisioning to certain user front ends to support the visualization and analysis of the data needs to be realized. Front ends in the Coneva solution are web and app based.[4]

Although the regulatory framework in Italy is promising for a scalable business model, the main challenges on which Coneva focuses are discrepancies with an investment subsidy called "Superecobonus 110%" (Governo Italia, 2020). The bonus gives residential and small commercial properties a one-off monetary investment subsidy for energy-efficiency investments. This remuneration cannot be combined with the community tariff scheme (Barroco et al., 2020) and leads to delays in the development of ECs because the tariff scheme is valid until the end of 2021.

5. Conclusion

Realizing the vision of ECs as the real-time optimization of distributed energy production to distributed energy supply in a predefined user group marks a promising step in the energy transition and the related digitization of the energy sector. The room for new business opportunities with platform-based approaches has been growing in recent years and can be judged as already being technically feasible. Nonetheless, there are regulatory and societal barriers to realizing the full potential of ECs to create value in the energy market and to realize that potential requires a clarification of standard roles in ECs and moving from pilots and small-scale projects to broader and full-scale businesses. For an energy service provider like Coneva, the trend toward local energy supply ecosystems is considered to be a major opportunity. However, the uncertainty in the market currently hinders the larger scale required to make a positive business case.

[4] Example of a web widget to community members: https://community.cev.energy.

References

Barroco, F., Borghetti, A., Cappellaro, F., Carani, C., Chiarini, R., D'Agosta, G., De Sabbata, P., Napolitano, F., Nigliaccio, G., Nucci, C.A., Corredor, C.O., Palumbo, C., Pizzuti, S., Pulazza, G., 2020. Le comunità energetiche in Italia. Green Energy Community. https://doi.org/10.12910/DOC2020-012.

Decide4energy.eu, July 2021. DECIDE Energy Community Monitor DECIDE_D3.1_final.pdf (decide4energy.eu).

Directive (EU) 2018/2001, 2018. Directive (EU) 2018/2001 of the European Parliament and of the Council of 11 December 2018 on the promotion of the use of energy from renewable sources (Text with EEA relevance.). http://data.europa.eu/eli/dir/2018/2001/oj.

Directive (EU) 2019/944, 2019. Directive (EU) 2019/944 of the European Parliament and of the Council of 5 June 2019 on common rules for the internal market for electricity and amending Directive 2012/27/EU (Text with EEA relevance.). http://data.europa.eu/eli/dir/2019/944/oj.

Governo Italia, 2020. Superbonus 110%. https://www.governo.it/it/superbonus. (Accessed 10 July 2021).

MAC Italia, 2019. It's Now Possible to Communicate with the 2nd Generation Open Meter Through C2G. https://www.mac-italia.com/now-possible-communicate-second-generation-open-meter-e-distribuzione-macs-innovative-device-chain2gate/. (Accessed 10 May 2021).

Stoetzel, M., Hadler, T., Britzl, F., Berndlmaier, T., Kittlaus, B., Roessler, K., et al., 2019. Energiemanagement und Steuerung einer PV-Anlage über die Smart-Meter-Gateway Infrastruktur: Ein Proof of Concept der Coneva GmbH und Stadtwerke München GmbH. coneva GmbH. https://Coneva.com/fileadmin/PDFs/Coneva-SWM-White-Paper-SMGW-Integration.pdf.

United Nations Framework Convention on Climate Change, 2016. Key Aspects of the Paris Agreement. https://unfccc.int/process-and-meetings/the-paris-agreement/the-paris-agreement/key-aspects-of-the-paris-agreement (Accessed on 8 June 2021).

Zulianello, M., Angelucci, V., Moneta, D., et al., 2020. Energy Community and Collective Self Consumption in Italy. IEEE.

Governing energy communities: the role of actors and expertise in business model innovation

16

Jake Barnes and Paula Hansen
Environmental Change Institute, University of Oxford, Oxford, United Kingdom

1. Introduction

Since its founding in 2013, the European Federation of renewable energy cooperatives[1] has grown into a network of 1900 cooperatives across Europe, jointly representing over 1.25 million citizens. This impressive growth can largely be reduced to two complementary explanations. The first concerns the consolidation of a particular business model (BM), in which citizens collectively invest in, and then jointly own and manage, a renewable energy generation technology such as wind or solar. Most renewable energy cooperatives in Europe have adopted this simple, scalable, and low-risk BM. The reason for it being low risk concerns the second factor: the provision of favorable national policies toward the generation of electricity from renewable sources. The removal of these policies in several European countries in recent years has subsequently undermined this previously successful BM. To survive in the next stage of the energy transition—in which policy attention, and therefore energy market design, is shifting from the deployment of renewable generation technologies toward managing higher levels of renewables in systems (Burger et al., 2020) —energy communities (ECs) are having to innovate.

Of some comfort to aspiring ECs is the range of activities open to them, at least on paper. Recent work by (Brown et al., 2019; Burger and Luke, 2017; Reis et al., 2021; see also the Chapter 1 by Rossetto et al.) indicate a wide variety of BM archetypes as being possible. Such work provides insights into generic EC BM patterns. Yet, they say little about the favorable policies or wider energy market contexts on which they are built. They also obscure understanding of the actors involved in the design and operation of each archetype. Questions about who is involved in the delivery of specific archetypes, or of the activities they undertake, are suppressed. Too often the search for clarity over higher level archetypes has resulted in the unreflective allocation of BMs to single (community) actors. Yet, single actors are rarely sole proprietors. Many new BMs are delivered through interdependencies between multiple actors (Brown, 2018; Hellström et al., 2015). The frequent presentation of such archetypes

[1] www.REScoop.eu.

Energy Communities. https://doi.org/10.1016/B978-0-323-91135-1.00014-6

thus gives rise to a sense of ECs having a free hand in how they develop. It is this simple idea, seemingly valid on paper, that is contested and dismissed in this chapter.

This chapter builds on a systems perspective on BMs (Bolton and Hannon, 2016) that recognizes BMs as involving a range of activities which may span the boundaries of multiple actors (Zott and Amit, 2010), and of BMs as evolving in response to the markets they are embedded within (Nolden et al., 2020). This perspective emphasizes the governance of EC BMs both in terms of the actors involved in the design and management of a BM, and in its realization through sets of interdependent activities.

To examine the impact of EC BM governance arrangements on what ECs can achieve, this chapter employs three ideal-typical representations of BM governance in ECs. The first representation, encapsulated by the idea of *doing it yourself,* places emphasis on collective decision-making, and coordination and management of business activities within a single collectively controlled organization. The second, embodied within the idea of *forming an alliance,* emphasizes the rise in the strategic and instrumental formation of alliances to deliver new energy activities. The third, encapsulated within the idea of *someone else doing it for you*, is premised on the development and implementation of replicable BMs by others to solve systemic issues.

This chapter thus examines the emergence and prevalence of differing BM governance arrangements, the BM archetypes achieved under each, and their characteristics. The chapter demonstrates that communities do not have a free hand when it comes to BM innovation. The realization of specific archetypes is instead influenced by the energy landscape they are embedded within and the BM governance arrangements adopted, whether they do it themselves, form an alliance, or let someone else do it for them.

This chapter has five sections, in addition to this introduction:

- Section 2 introduces BMs in the context of ECs, and sets out a focus on BM governance arrangements; and
- Sections 3–5 explore each ideal-typical governance arrangement in turn followed by the chapter's conclusions.

2. Business models and energy communities

Scholarly attention to BMs is widespread and rapidly growing. Multiple definitions exist, while a wide variety of approaches, linked to different literatures, seek to identify and explain specific aspects (Ritter and Lettl, 2018). Three approaches to studying BMs are worth highlighting here.

A common understanding of BMs focuses on the elements deemed essential to explaining what they do. Boons and Lüdeke-Freund (2013) identify four core elements of a generic BM: the value proposition, supply chain, customer interface, and financial model. Osterwalder and Pigneur (2010) include nine elements in their BM canvas. Such approaches emphasize how BMs create value. They place less emphasis on the actors involved, and do not describe the logic behind each model: this logic is implicit in how the elements are assembled.

A second approach seeks to identify generic logics about how firms create value through the identification of BM archetypes. For example, Bocken et al. (2014) identifies eight sustainable BM archetypes applicable across industry boundaries. Within the field of ECs, there have been multiple attempts to identify emerging yet generic BM archetypes (e.g., Brown et al., 2019 or Reis et al., 2021).

A third approach, employed here, views BMs as a description of the activities necessary to achieve a firm's goals (Chesbrough, 2007). Under this view, BMs describe "a system of interdependent activities that transcends the focal firm and spans its boundaries" (Zott and Amit, 2010, 216).

This systems-based approach has two strengths. It emphasizes the interdependencies and interactions between BM activities, and the extent to which they might be undertaken by a variety of actors. It also opens the door to analysis of wider system components, like markets, policies, and regulations, that are increasingly recognized as important in explaining the decline of incumbent, and growth of new BMs in the energy system (Bolton and Hannon, 2016; Hall et al., 2020,b). The perspective thus emphasizes the governance of EC BMs both in terms of the actors involved in their design and management, and in their realization through sets of interdependent activities.

BM governance, in the sense used here, thus concerns the design, management, and delivery of business activities (Brown, 2018; Zott and Amit, 2010). It includes design of the product or service, identification of activities and actors involved, and of how they are linked. As a result, BMs may be contained within a single, typically hierarchical actor, or they may include a constellation of actors that interact to provide a product or service (Boons and Lüdeke-Freund, 2013). This may lead to interdependencies between actors, maintained through trust or contractual relations. Governance arrangements can thus occupy a continuum, from hierarchical firms at one end, to networked relationship at the other, with hybrid, mixed approaches sitting in between.

Applying a BM governance lens in the context of ECs requires first defining the focal actors involved. This is of central importance for understanding who makes up the "community" and who participates in decision-making (Hicks and Ison, 2018). Focal actors may or may not be constituted as business entities. They may be local to the site of the EC and may include individuals, municipalities, SMEs, or businesses; or they may be non-local, in the sense that they have no connections or attachment to the site of the proposed EC. Non-local actors may include licensed suppliers, technology start-ups, or intermediary organizations. Whether they are local or not is likely to influence the extent of knowledge actors have on local cultural, social, and economic conditions. Their position in relation to the EC is also likely to affect their motivations. Local actors are likely to be more interested in developing solutions to local needs, while non-local actors are likely to be more interested in identifying locally beneficial solutions that are replicable elsewhere (Devine-Wright, 2019).

These characteristics form the basis of three ideal-typical representations of BM governance in ECs (top half of Table 16.1), which are employed in the remainder of this chapter to examine how governance arrangements influence what BM activities (and therefore archetypes) can be achieved. In the following sections, each ideal-typical representation is introduced and contextualized with a number of examples before the range of BM activities realized under each is discussed.

Table 16.1 Ideal-typical BM governance arrangements and associated activities of ECs.

		Ideal-typical governance arrangements of ECs		
		Do it yourself	**Form an alliance**	**Someone else does it for you**
Business model governance	*Actors*	Single local organization, comprised of individuals acting collectively	Networks of local and nonlocal actors including individuals acting collectively, SMEs, municipalities, and commercial enterprises	Single or networks of non-local institutions working in partnership across sectors
	Motivation	Multiple and diverse, addressing community needs and tackling wider environmental challenges	Overlapping and competing	Financial gain through identifying locally beneficial solutions that are replicable elsewhere
	Attitude to governance	Emphasis placed on participation and consensual decision-making	Largely based on trust, underpinned by contracts	Fundamentally market driven
Business model activities	*Associated BM archetypes*	Collective generation; collective self-consumption;	Local supply arrangements; microgrids; private wire; energy service company (ESCo)	Peer-to-peer; virtual power plant; local energy market; aggregator
	Position in relation to infrastructure	Typically, behind the meter	Across multiple meter points but at low-voltage levels	Across meters points (no scale limit)
	Interaction with value chains	Limited and discrete; adjunct	Potentially disruptive but reliant upon continued functioning of existing value chain	Potentially disruptive across full value chain
	Relation to energy markets	Largely separate	Protected engagement: Resulting from market rules (e.g., Dutch "postcoderoos" scheme) or relation to infrastructure (behind the meter)	Fully integrated

3. "Do it yourself" energy communities

The first ideal-typical form of governance arrangement, encapsulated within the idea of *doing it yourself,* places emphasis on collective decision-making, and coordination and management of the value proposition and BM activities within a single, collectively controlled organization.

Historically, community-based activity on energy or "community energy" has been the preserve of activist communities pursuing alternative, often more socially progressive futures (Berka, 2018; Ornetzeder and Rohracher, 2013). With limited institutional or mainstream support, activists developed activities that were managed within their own networks. In practice, this translated into governance arrangements that concentrated power in a single organization. Horizontal, open, and collective decision-making structures, typically based on one-person-one-vote, were privileged, structuring how decisions are made and by whom. This strong focus on participation is the reason why ECs are often extolled as holding the potential to democratize energy systems (Devine-Wright, 2019; Radtke et al., 2018). To this day, ownership and control of renewable energy installations by a single collective organization is thought to be by far the most common form of ECs in Europe (Caramizaru and Uihlein, 2020; Hewitt et al., 2019).

An example of a community project with roots in the "first wave" (Hewitt et al., 2019) of ecological activism of the 1960s and 1970s is Dalby solar village. The project was initiated in the late 1970s by activists from Lund, Sweden. Over the following decade, activists located a site, secured financial support, set themselves up as a cooperative housing association, and constructed 50 apartments as well as a communal garage, laundry, kindergarten, and storehouse. The first members moved in 1987/88. Passive solar gain techniques were employed in the design of the housing. More than a quarter of a century later, solar thermal collectors were installed (in 2014) followed by solar PV (in 2019) on shared buildings, having previously been researched but deemed economically unviable (Barnes et al., 2022c).

In contrast, Zuiderlicht energy cooperative emerged in what Hewitt et al. (2019) have identified as the third and most recent wave of community-based activism on energy, which emerged as a reaction to the global financial crisis of 2008 and distrust in neoliberal capitalism. Zuiderlicht energy cooperative was founded by a group of local activists in Amsterdam in 2013, many part of a wider houseboat community, who were motivated to contribute to, and accelerate, the energy transition (Hansen et al., 2022a). A renewable energy cooperative was subsequently set up through which community members could invest in, own, and manage roof-mounted PV systems on local buildings. Their first project went live in 2013, was supported by the Dutch premium feed-in scheme (SDE), and was "open to all Amsterdamers." Since then, a further 17 projects have been developed, all but two supported by the SDE.

Zuiderlicht's two latest projects have been developed under the Dutch reduced-rate regulations, commonly referred to as the "postcoderoos" (PCR) scheme. The scheme offers a partial tax exemption for owners of distributed renewable generation technologies who live in the postcode area of a generation asset (the central postcode), or in postcode areas adjacent to it (the "rose petals"). PCR regulations were introduced in 2014 and have resulted in a reasonably standardized EC BM (de Bakker et al.,

2020; Hall et al., 2020a): community members invest in cooperatively owned distributed generation assets in their own or a neighboring postcode; a portion of members' electricity imports are subsequently matched to their share of power generated by the cooperative installation, which is then eligible for a tax deduction of approximately 12 c/kWh (at 2020 tax rates) up to 10,000 kWh/year for 15 years. Enacting any PCR scheme requires engaging with a licensed supplier, who covers billing and provides all residual power required outside of collective self-consumption.

In both Zuiderlicht and Dalby, energy projects were developed by small volunteer groups, who set about researching possible technologies, their requirements, and potential utility, before financing and commissioning installations through cooperative governance arrangements. In BM archetype terms, both examples resulted in a form of collective generation project in the first instance, Dalby Solby then collectively self-consuming power onsite within communal buildings, while Zuiderlicht energy cooperative self-consumes over the public grid under specific national regulations.

This represents a narrow realization of BM archetypes when compared against the menu of possibilities set out in Brown et al. (2019) or Reis et al. (2021). Yet, given the prevalence of renewable energy cooperatives across Europe, which specialize in collective generation projects, the result is unsurprising. On the contrary, Zuiderlicht's PCR project stands out as being the exception. To explain this result requires examining the activities within these archetypes, the actors involved, and wider energy landscape they have emerged in.

Taking a first look at the activities involved in the development of these projects provides an entry point. What becomes clear is how BM activities are concentrated in the design and construction phases. Their operation is largely passive: once a PV system is commissioned, subsequent activities are largely administrative. Where collective consumption occurs, it is typically achieved behind a single meter point, as in the case of Dalby Solby, and often in the form of collective *communal* consumption through schools, churches, or community buildings. To achieve collective *private* consumption via individual households, Zuiderlicht energy cooperative has to contract with a licensed supplier for its PCR-backed projects. Such projects may be attractive to communities reliant upon volunteer labor (Herbes et al., 2017) because they require concerted effort in their design and implementation but few ongoing activities.

Looking at the wider energy landscape, it is clear that *do it yourself* community renewables projects have only emerged in significant numbers when public subsidies were available (Hewitt et al., 2019). That is not to belittle existing achievements. Getting ECs to work has never been easy. Pragmatism and determination combined with enthusiasm and entrepreneurial spirit have been required (Nolden et al., 2020). Yet, given the risk-averse nature of most community organizations (Herbes et al., 2021), it becomes clear how projects governed by single cooperative organizations have largely been achieved by tapping into existing, easily identifiable, and relatively stable revenue streams. The supportive role played by national feed-in tariffs is widely recognized. The Dutch PCR regulations have achieved the same result, derisking particular EC BMs to the point where community-led actors are confident in financing and managing such projects.

Collectively, this suggests EC BMs achieved under *do-it-yourself* governance arrangements are typically discrete energy projects that bolt on to existing energy systems. The BM archetypes achieved are limited and principally concern collective generation projects, at times coupled with onsite use. Such business activities typically occur behind the meter, making projects largely separate from national or local energy markets (even if they are shaped by them).

4. "Form an alliance" energy communities

Forming an alliance encapsulates the second ideal-typical governance arrangement explored here, in which a range of actors, including community members, are involved in the design and delivery of business activities (top half of Table 16.1).

The formation of alliances is perhaps more widespread than first appears and has a precedent in partnership working. In 2013, one third of community renewable energy capacity in the United Kingdom was found to be jointly owned (Harnmeijer et al., 2013). In 2016, de Bakker et al. (2020) found over 70% of renewable energy cooperatives in the Netherlands had entered a form of contractual partnership with an incumbent energy utility. One reason for the rise of partnerships concerns the extent to which communities are endowed with different levels of skills, knowledge, and expertise for undertaking projects (Seyfang et al., 2013). Indeed, not all communities can deliver projects themselves. At the same time, and equally important, has been the promotion of joint ownership projects by national governments in order to foster community acceptance of generation assets. Denmark was the first country to implement such a requirement in 2009 (Olsen, 2018). The United Kingdom followed suit in 2014, with the Netherlands adopting a similar policy in recent years. While these two drivers remain important today, contemporary alliances are thought of as being formed for instrumental and strategic reasons (de Bakker et al., 2020). Increasing revenue from existing collective generation projects is one motivation. Finding a way to supply directly to members and thereby achieve collective self-consumption is another. The result is new BM governance arrangements in which multiple actors are involved in the design and delivery of EC BMs.

The range of contemporary alliances to draw on is large. For example, in the United Kingdom, the social enterprise Energy Local has received much attention after winning a prestigious Ashton Award in 2018. Heralded as a new approach to managing generation and use at a community scale, Energy Local connects local generators and consumers through the creation of local "clubs" (Barnes et al., 2021). Within each club, locally generated renewable electricity is shared equally among the members and paid for on a "match tariff" agreed by the club annually. Smart meters record half-hourly generation and demand. A "fair share" algorithm allocates the distribution of locally generated electricity to members under the "match tariff". Additional electricity required by members is provided by a licensed supplier through a time-of-use tariff. Members can see forecast local generation and track their consumption and expenditure via online displays, building awareness, and know-how.

To achieve this unique relationship between actors, clubs partner with a licensed supplier and operate under a "complex site" arrangement: a little-known regulatory exception designed to facilitate the take up and use of renewable energy on small-scale industrial sites where generation facilities cannot be colocated behind a single meter point. As such, electricity is shared over the public grid and balanced through a formal agreement with the licensed supplier. Energy Local coordinates the creation of local clubs and their relationship with a licensed supplier. In doing so, it coordinates the EC BM but does not undertake activities that would normally be considered as vital to the operational delivery of the BM.

Use of this regulatory exception for the creation of an EC is unique in the United Kingdom. However, the basic arrangement, regularly understood as a form of local supply arrangement, is common in the Netherlands. As mentioned above, the Dutch reduced tax regime, known as the PCR scheme, facilitates the development of ECs in which generation is collectively owned and consumed. Just under 400 solar projects had been supported by the scheme by early 2020 (Hier Opgewekt, 2020). Zuiderlicht energy cooperative is one of these, and one that has managed to develop generation assets and negotiate contractual relations with a supplier while retaining overall control of the EC. For this reason, Zuiderlicht's PCR-backed projects are best described under *do it yourself* governance arrangements. For many other ECs (see for example (Hansen et al., 2022b)), realizing a local supply arrangement under seemingly favorable policy contexts remains challenging. The regulation's complexity and associated administrative burdens have presented significant barriers to communities (de Boer et al., 2018). For many, this has meant the more active involvement of licensed suppliers in the design and delivery of the BM, with suppliers, such as GreenChoice, adapting their own BMs in order to specialize in the delivery of PCR-backed projects (de Bakker et al., 2020).

In southern Europe, a wholly different BM archetype is being developed in Messina, Sicily, as part of a large-scale multipronged urban regeneration project. Here, attention is being directed toward developing an equitable way of generating and sharing renewable electricity between socially disadvantaged tenants. SO_EN Social Housing is a pilot project being developed by the Community Foundation of Messina in collaboration with SO_EN S.p.A. (an energy service company), and a range of research and other partners. The pilot's main objective is to test novel energy generation and storage technology with an innovative accounting system which together enable the equitable sharing of solar energy. The pilot, delayed in part by the global pandemic, but due to be launched in 2021, is expected to operate following an energy service company (ESCo) archetype in which onsite generation and storage assets will be managed to provide energy services to residents. The project will join six domestic units behind a single meter point through a microgrid. Residents will be billed via energy service contracts according to the accounting system (Hansen et al., 2022c).

These ECs involve networks of local and non-local actors coming together in their design and management and/or their realization through sets of interdependent activities. They achieve a wider range of BM archetypes than under *do-it-yourself* governance arrangements, including local supply arrangements, the development of microgrids supplying multiple users, and ESCo models. The associated high-level activities involved in the design, implementation, and delivery of these BM archetypes are also wider.

All examples involve some form of local generation and a range of activities, required to design, install, and commission generation assets. Yet, all go beyond this to achieve collective self-consumption, and do so in different ways. By adopting a local supply archetype, Energy Local and PCR-backed projects operate in front of the meter on the low-voltage public network. This, by law, requires them to operate through a licensed supplier. In practice, this implies a power purchase agreement between the generator and supplier on one side, and modified supply contracts between households and supplier on the other. This contractual arrangement allows generation to be pooled before being shared out equally in the case of Energy Local, or according to ownership rights under PCR-backed projects. By contrast, SO_EN Social Housing avoids partnering with a licensed supplier by operating behind a single meter point. The relation of ECs to energy infrastructure therefore has important implications for the actors involved.

A focus on activities highlights how EC BM archetypes achieved by *forming an alliance* are more complicated than the do-it-yourself projects previously outlined. What unites them is a focus on achieving novel arrangements for collective *private* consumption: each example discussed here goes beyond onsite self-consumption to supply power to multiple end users. This is a significant step change in activity. It requires metering equipment and billing platforms, alongside personnel and skills to provide customer service. Energy Local and PCR-backed projects navigate this issue by partnering with a licensed supplier. They access existing metering and billing arrangements of the supplier. In addition, ECs that operate in front of the meter also have to adhere to a wider set of energy market rules and regulations, for instance, around customer safeguarding. Partnering with licensed suppliers again offers one means to achieve this, as they already have knowledge of, and expertise to handle, all regulatory compliance issues. SO_EN Social Housing evades the need for metering and billing through use of energy service contracting where customers are billed monthly for the service they receive (heat, light) rather than the kilowatt hours delivered. Yet, they are only able to do this by operating behind a single meter point, which in turn provides protection from incumbent market rules and regulations. The relation of ECs to existing markets, witnessed here as a form of partial, protected engagement, thus provides a further reason for the formation of alliances.

A third and final reason for the increased formation of alliances concerns the increasing complexity of EC BM archetypes and the operational activities required to realize them. For SO_EN Social Housing, for example, collective generation and storage systems need to be managed to meet demand instantaneously, while for Energy Local supply needs to be allocated to demand in half hour periods through a so-called fair-share algorithm. This in turn requires specialist knowledge, skills, and equipment that are likely to be beyond the capacity of largely volunteer-based organizations (the chapter by Hadler provides a good example of how commercial companies are specializing in being platform or service providers for ECs).

5. "Someone else does it for you" energy communities

The third ideal-typical BM governance arrangement is encapsulated in the idea of *someone else doing it for you,* and is premised on individual non-local or networks of non-local institutions working in partnership across sectors (top half of Table 16.1). Typically motivated to develop locally beneficial solutions that are replicable elsewhere, the governance of such ECs is driven by rent-seeking behavior, is rooted in commercially orientated market relationships, and seeks to engage members as rational, self-interested consumers (Devine-Wright, 2019). Such ECs are a relatively new but rapidly growing phenomenon, driven by the decentralization and digitalization of energy systems.

Two examples shed light on the kinds of ECs emerging under this ideal-typical BM governance arrangement. The first is well established and widely known. In 2016, Germany's leading residential energy storage provider, sonnenBatterie, launched a new community energy trading platform. The sonnenCommunity sought to provide participants with the opportunity to cover 100% of their electricity needs through a combination of self-consumption and community supply (Barnes et al., 2022a). The concept was simple: the new trading platform would link sonnen customers across Germany and allow them to trade surplus energy from their domestic PV and battery systems. The process would be automated by sonnen software, itself enabled by advances in digital connectivity. Participants would need only to sign up to receive benefits. The rebranded company, sonnen, had high ambitions. It aimed to help participants "determine their own future," by "declaring independence from traditional energy suppliers."[2]

A second example, comprised of 30 households, also comes from Germany. Project Z, a pilot project of a multinational electric utility company, aims to trial the use of local energy trading to retain privately generated renewable electricity within a predefined geographical area (Barnes et al., 2022b).[3] The trial was initiated in response to repeated customer inquiries about accessing regional renewable electricity and regarding what happened to surplus domestic solar production. For the team behind Project Z, the trial also responds to the declining role of incumbent utilities and the growing need for "smart," innovative solutions for local excess generation that does not include domestic batteries. Project Z was subsequently designed as a digital service that supports collective self-consumption of locally generated renewable electricity in suburban neighborhoods. The pilot uses privately owned residential solar PV systems, advanced metering, and a range of platform technologies to facilitate the exchange of electricity between households. This set up creates a virtual private network through which excess prosumer generation is exchanged between participants. Nonetheless, no direct energy trading occurs between households. Instead, each household has a bilateral contract with Project Z, which details import and export tariffs that are slightly more favorable than those currently available.

These ECs have been designed and managed by focal firms—sonnen and a multinational electricity utility company respectively—yet have been realized through interdependent activities of multiple actors. For instance, the team behind Project Z has also worked strategically with others to implement distributed ledger technologies. What

[2] See https://sonnengroup.com/sonnenbatterie.

[3] Due to reasons of commercial sensitivity and at risk of revealing the business strategy of those involved both the name of the project and of the central actors involved, have been removed. The project will simply be refereed as Project Z.

unites these examples under this representation is the extent to which their associated community members have no direct control over, or operational role in, the BM. All associated actors are non-local. Members are passive, playing no role in the design, management, or operation of the BM. Instead, they are treated as customers and for sonnen, encouraged to act as unpaid ambassadors.

Identifying the BM archetypes achieved within each case is not as straight forward as first seems. Cases merge or layer multiple archetypes. At its core, sonnen appears to operate a virtual power plant through its new trading platform. The platform links and controls privately owned generation and storage units through cloud-based software to maximize collective self-consumption. Yet, it also provides a range of ancillary services to the public grid and thereby acts as an aggregator. In doing so its BM targets multiple value streams. Project Z was designed as a peer-to-peer pilot, has been described as a local energy market, but is perhaps best articulated as a local supply arrangement or simply, a regional electricity tariff. In each case, the focus is less on generation than in other governance representations. Attention is directed toward creating arrangements that facilitate collective consumption.

The activities involved in each case are again more complex than the previous two governance arrangements. Most activities focus on managing electricity flows across meter points. Underlying this shift from previous governance arrangements, neither EC BMs own the generation and storage assets. Project Z's BM is described as "asset free", its activities residing solely in front of the meter, linking members, and sonnenCommunity's activities involve the control and management of domestic PV and battery equipment across meter points. Crucially, both involve significant ongoing operational activities: for the EC to function, a variety of activities need to be performed on a regular basis. This includes metering and billing but also the use of advanced digital technologies to track and allocate flows of electricity in real time. Sonnen relies on cloud-based computer software, while Project Z employs distributed ledger technologies. The extent and range of ongoing operational activities, often linked to the use of advanced digital technologies, suggests a wider repertoire of skills and knowledge needed to coordinate and manage the EC in practice (Hansen et al., 2021). As the chapter by Chudoba and Borges demonstrates, many existing utility companies do not have such skills and expertise in house and therefore turn to specialist start-ups.

This ideal-typical governance arrangement has the potential to include the widest range of EC BM archetypes, which may operate behind or in front of meter points. As the sonnenCommunity demonstrates, such ECs are not constrained when working across meter points to operating at low-voltage levels, as appears to be the case under *forming an alliance* governance arrangements. On the contrary, the sonnenCommunity suggests there are no scale limits. Moreover, such ECs appear to be fully integrated with existing energy markets, and the range and layering of BM archetypes suggests EC BMs are targeting multiple value streams (bottom of Table 16.1). While contemporary cooperatives view the idea of "value stacking" with hesitancy, differing perceptions of risk, combined with more agile decision-making structures, may make this less of an issue for commercial firms.

6. Conclusions

The central argument in this chapter is that the BMs of ECs are shaped by the BM governance arrangements they adopt and the energy markets they operate in. Too often, the presentation of EC BM archetypes results in the assumption that single actors create, deliver, and capture value. In this chapter, this simple idea has been examined and dismissed. Instead, this chapter has demonstrated how the BM archetypes open to ECs are largely conditioned by the wider energy landscape in which they operate and the extent to which they are willing to form alliances with others and/or let someone else do it for them.

Examining three ideal-typical governance arrangements of ECs, and the contexts in which they have emerged, results in a simple conclusion: The range of BM archetypes open to ECs expands significantly when other non-local actors are involved.

The range of archetypes open to ECs in which members retain sole control over the design, management, and operation of their BMs is small (bottom half of Table 16.1). Historically, *do it yourself* governance arrangements have resulted in collective generation projects at times combined with use, onsite, and behind a single meter point. Such activities have largely occurred separate from existing energy markets, an adjunct to existing value chains. On rare occasions, and where supported by favorable national policies like those in the Netherlands, it has also been possible for individual households to consume from a local collective supply and as a form of collective *private* consumption.

Forming an alliance opens up a wider array of BM archetypes, including local supply arrangements, the development of microgrids supplying multiple users, and ESCo models. Forming an alliance can allow ECs to achieve collective *private* consumption via individual households, either by working behind a single meter point or, more commonly, working across meter points at low-voltage levels. Such ECs engage with, but are protected from, existing energy markets by market rules (e.g., PCR regulations in the Netherlands) or by operating behind the meter. Such archetypes continue to rely on existing value chains to provide residual power required.

Finally, the BM archetypes achieved when *someone else does it for you* are the most numerous, likely to cover the full spectrum of possible archetypes. Examining two examples reveals that EC BMs led by someone else are unlikely to be limited to working within low-voltage networks, may be fully integrated with existing energy markets, and are potentially disruptive across the full value chain. As a result, the activities of each ideal-typical representation can be broadly mapped on to existing energy value chains (Fig. 16.1).

Thus, the value of the present analysis is threefold: it highlights the range of stylized governance arrangements adopted in EC BMs; it challenges the idea that single actors are typically responsible for creating, delivering, and capturing value, and it contributes to existing debates about the range of BMs open to ECs.

Of final note is the extent to which EC BMs are a product of the wider energy landscapes in which they develop. The contemporary shift toward the forming of alliances can be read as much as a result of policy interventions seeking to scale up and roll out

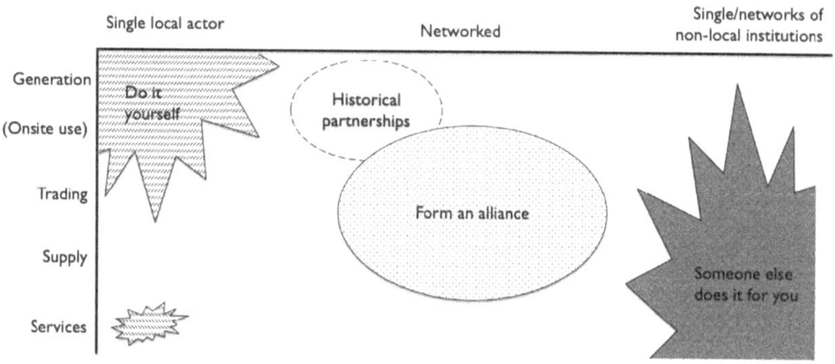

Figure 16.1 Ideal-typical BM governance arrangements of ECs and the range of activities typically achieved.

community orientated energy activities (as outlined in Section 4, above), as a shift from one phase of the energy transition to another. Where previously policy and practitioner attention was directed toward the generation of power from renewable sources, in many European countries the rapid deployment of renewable generation technologies is shifting emphasis toward the effective integration of variable renewable power. For many ECs, this means a shift from BMs focused on the realization of "simple" renewable energy generation projects toward increasingly smart and dynamic BMs that better integrate renewable energy generation with demand in real time. In practice, this entails a wider range of skills and expertise, and it means engaging more widely in energy markets and associated value chains. This, in turn, points to the shifting BM governance arrangements of ECs, with alliances being formed for the simple reason that engaging in energy markets is hard: it requires the expertise typically found in incumbent energy utilities.

Acknowledgments

The authors thank Jenny Palm and Sarah Darby for their comments on previous drafts. This research was funded by the European Union's Horizon 2020 research and innovation program under grant agreement No 837752.

References

Barnes, J., Hansen, P., Darby, S., 2021. Smart and fair: introducing energy local. Available at: https://www.newcomersh2020.eu/materials-and-deliverables.

Barnes, J., Hansen, P., Darby, S., Sommer, S., Andor, M., 2022a. NEWCOMERS summary case study report: the sonnenCommunity. NEWCOMERS Project grant agreement No 837752, 9 February 2022a. Available at. https://www.newcomersh2020.eu/materials-and-deliverables.

Barnes, J., Hansen, P., Darby, S., Sommer, S., Andor, M., 2022b. NEWCOMERS summary case study report: Project Z. NEWCOMERS Project grant agreement No 837752, 3 February 2022b. Available at. https://www.newcomersh2020.eu/materials-and-deliverables.

Barnes, J., Hansen, P., Darby, S., Palm, J., Reindl, K., 2022c. NEWCOMERS summary case study report: Dalby Solby. NEWCOMERS Project grant agreement No 837752, 3 February 2022c. Available at. https://www.newcomersh2020.eu/materials-and-deliverables.

Berka, A.L., 2018. Community renewable energy in the UK: a short history. In: Holstenkamp, L., Radtke, J. (Eds.), Handbuch Energiewende und Partizipation. Springer, pp. 1013−1036.

Bocken, N.M.P., Short, S.W., Rana, P., Evans, S., 2014. A literature and practice review to develop sustainable business model archetypes. J. Clean. Prod. 65, 42−56. https://doi.org/10.1016/j.jclepro.2013.11.039.

Bolton, R., Hannon, M., 2016. Governing sustainability transitions through business model innovation: towards a systems understanding. Res. Pol. 45 (9), 1731−1742. https://doi.org/10.1016/j.respol.2016.05.003.

Boons, F., Lüdeke-Freund, F., 2013. Business models for sustainable innovation: state-of-the-art and steps towards a research agenda. J. Clean. Prod. 45, 9−19. https://doi.org/10.1016/j.jclepro.2012.07.007.

Brown, D., 2018. Business models for residential retrofit in the United Kingdom; a critical assessment of 5 key archetypes. Energy Efficiency 2, 1−26. https://doi.org/10.1007/s12053-018-9629-5 (Accepted).

Brown, D., Hall, S., Davis, M.E., 2019. Prosumers in the post subsidy era: an exploration of new prosumer business models in the UK. Energy Pol. 135. https://doi.org/10.1016/j.enpol.2019.110984.

Burger, C., Frogatt, A., Mitchell, C., Weinmann, J., 2020. Decentralised Energy a Global Game Changer. Ubiquity Press Ltd.

Burger, S.P., Luke, M., 2017. Business models for distributed energy resources: a review and empirical analysis. Energy Pol. 109, 230−248. https://doi.org/10.1016/j.enpol.2017.07.007.

Caramizaru, A., Uihlein, A., 2020. Energy Communities : An Overview of Energy and Social Innovation. Joint Research Centre. https://doi.org/10.2760/180576.

Chesbrough, H.W., 2007. Why companies should have open business models. MIT Sloan Manag. Rev. 1−22.

de Bakker, M., Lagendijk, A., Wiering, M., 2020. Cooperatives, incumbency, or market hybridity: new alliances in the Dutch energy provision. Energy Res. Social Sci. 61, 101345. https://doi.org/10.1016/j.erss.2019.101345. December 2019.

de Boer, J., Zuidema, C., van Hoorn, A., de Roo, G., 2018. The adaptation of Dutch energy policy to emerging area-based energy practices. Energy Pol. 117, 142−150. https://doi.org/10.1016/j.enpol.2018.02.008. February.

Devine-Wright, P., 2019. Community versus local energy in a context of climate emergency. Nat. Energy 4 (11), 894−896. https://doi.org/10.1038/s41560-019-0459-2.

Hall, S., Brown, D., Davis, M., Ehrtmann, M., Holstenkamp, L., 2020a. Business Models for Prosumers in Europe. PROSEU−Prosumers for the Energy Union: Mainstreaming active participation of citizens in the energy transition (Deliverable N°D4.1).

Hall, S., Mazur, C., Hardy, J., Workman, M., Powell, M., 2020b. Prioritising business model innovation: what needs to change in the United Kingdom energy system to grow low carbon entrepreneurship? Energy Res. Social Sci. 60. https://doi.org/10.1016/j.erss.2019.101317.

Hansen, P., Barnes, J., Darby, S., 2021. First things first: laying the foundations for smart energy communities by exploring concepts and realities in European case studies. ECEEE Summer Study Proc. 5−073-21.

Hansen, P., Barnes, J., Darby, S., van der Grijp, N., Petrovics, D., 2022a. NEWCOMERS summary case study report: Energiecoöperatie Zuiderlicht. NEWCOMERS Project grant agreement No 837752, 8 February 2022a. Available at. https://www.newcomersh2020.eu/materials-and-deliverables.

Hansen, P., Barnes, J., Darby, S., Petrovics, D., 2022b. NEWCOMERS summary case study report: Buurtmolen Herbaijum. NEWCOMERS Project grant agreement No 837752, 8 February 2022b. Available at. https://www.newcomersh2020.eu/materials-and-deliverables.

Hansen, P., Barnes, J., Darby, S., Nicita, A., Musolino, M., 2022c. NEWCOMERS summary case study report: SO_EN. NEWCOMERS Project grant agreement No 837752, 9 February 2022c. Available at. https://www.newcomersh2020.eu/materials-and-deliverables.

Harnmeijer, J., Parsons, M., Julian, C., 2013. The Community Renewables Economy. ResPublica. September.

Hellström, M., Tsvetkova, A., Gustafsson, M., Wikström, K., 2015. Collaboration mechanisms for business models in distributed energy ecosystems. J. Clean. Prod. 102, 226−236. https://doi.org/10.1016/j.jclepro.2015.04.128.

Herbes, C., Brummer, V., Rognli, J., Blazejewski, S., Gericke, N., 2017. Responding to policy change: new business models for renewable energy cooperatives − barriers perceived by cooperatives' members. Energy Pol. 109, 82−95. https://doi.org/10.1016/J.ENPOL.2017.06.051.

Herbes, C., Rilling, B., Holstenkamp, L., 2021. Ready for new business models? Human and social capital in the management of renewable energy cooperatives in Germany. Energy Pol. 156 (16), 112417. https://doi.org/10.1016/j.enpol.2021.112417.

Hewitt, R.J., Bradley, N., Compagnucci, A.B., Barlagne, C., Ceglarz, A., Cremades, R., McKeen, M., Otto, I.M., Slee, B., 2019. Social innovation in community energy in Europe: a review of the evidence. Front. Energy Res. 7 (April). https://doi.org/10.3389/fenrg.2019.00031.

Hicks, J., Ison, N., 2018. An exploration of the boundaries of 'community' in community renewable energy projects: navigating between motivations and context. Energy Pol. 113, 523−534. https://doi.org/10.1016/j.enpol.2017.10.031.

Hier Opgewekt, 2020. Lokale Energuie Monitor 2019. Available at: https://www.hieropgewekt.nl/lokale-energie-monitor.

Nolden, C., Barnes, J., Nicholls, J., 2020. Community energy business model evolution: a review of solar photovoltaic developments in England. Renew. Sustain. Energy Rev. 122, 109722. https://doi.org/10.1016/j.rser.2020.109722. May 2020.

Olsen, B.E., 2018. Community wind in Denmark. In: Lars, H., Radtke, J. (Eds.), Handbuch Energiewende und Partizipation. Springer, pp. 1037−1046.

Ornetzeder, M., Rohracher, H., 2013. Of solar collectors, wind power, and car sharing: comparing and understanding successful cases of grassroots innovations. Global Environ. Change 23 (5), 856−867. https://doi.org/10.1016/j.gloenvcha.2012.12.007.

Osterwalder, A., Pigneur, Y., 2010. Business Model Generation. John Wiley & Sons.

Radtke, J., Holstenkamp, L., Barnes, J., Renn, O., 2018. Concepts, formats, and methods of participation: theory and practice. In: Holstenkamp, L., Radtke, J. (Eds.), Hanbuch Energiewende und Partizipation. Springer, pp. 21−42.

Reis, I., Goncalves, I., Lopes, M., Antunes, C., 2021. Business models for energy communities: a review of key issues and trends. Renew. Sustain. Energy Rev. 144 (March). https://doi.org/10.1016/j.rser.2021.111013.

Ritter, T., Lettl, C., 2018. The wider implications of business-model research. Long. Range Plan. 51 (1), 1–8. https://doi.org/10.1016/j.lrp.2017.07.005.

Seyfang, G., Park, J.J., Smith, A., 2013. A thousand flowers blooming? An examination of community energy in the UK. Energy Pol. 61, 977–989. https://doi.org/10.1016/j.enpol.2013.06.030.

Zott, C., Amit, R., 2010. Business model design: an activity system perspective. Long. Range Plan. 43 (2–3), 216–226. https://doi.org/10.1016/j.lrp.2009.07.004.

Grid-friendly clean energy communities and induced intracommunity cash flows through peer-to-peer trading[1]

17

Reinhard Madlener and Robert Crump
RWTH Aachen University, Aachen, Germany

1. Introduction

Globally, climate change mitigation is considered the main challenge of the 21st century. With 866 million metric tons (t) of CO_2 equivalents, Germany emits the most greenhouse gases (GHGs) in Europe. The energy sector accounts for 34.5% of these (UBA, 2019). While the German energy sector has progressed toward decarbonization, many conventional fossil fuel power plants are still active. Thus, reducing GHG emissions in the energy sector remains a strong lever to meet emission reduction targets. Especially against the background of electrification of energy services (e-mobility, electric heating via heat pumps) and sector coupling (i.e., coupling the energy, heating, mobility, and industry sectors), providing low-carbon electricity becomes ever more important.

However, Germany faces several challenges in advancing its sustainable energy transition ("*Energiewende*"). First, the expansion of the transmission grid, responsible for transporting electricity over large distances, happens slowly. In 2013, the German government passed a law to accelerate its expansion. By 2019, however, only 600 km (out of 5900 km planned) had received approval and only 300 km of new power lines were in operation (BNetzA, 2019a). Secondly, the interest in auctions for onshore wind energy converters (WECs) decreases: In 2019, only 50.3% of the envisaged new capacity were awarded (i.e., 1847 MW) (BNetzA, 2019b). Two main causes for this are the lack of available areas for WECs and legal obstacles combined with lacking social acceptance (e.g., manifested by active citizen resistance) (BMWi, 2019).

The European Union (EU) has recognized the importance of a citizen-supported energy transition. In May 2019, it adopted the last pieces of the "Clean energy for all Europeans" package, aimed at increasing citizen participation *in* and acceptance *of* the energy transition. To this end, the EU included provisions regarding "renewable energy communities," whose primary purpose is "to provide environmental, economic,

[1] This chapter is a shortened and updated version of Crump and Madlener (2019), where further details—especially regarding the methodology and assumptions—are available.

Energy Communities. https://doi.org/10.1016/B978-0-323-91135-1.00016-X

or social community benefits for its shareholders or members or for the local areas where it operates rather than financial profits" (RED, Art. 2(16c)). The potential of clean energy communities (CECs)—i.e., local communities running their own energy-efficient and renewable energy system (for definitions, see Chapter 1)—is substantial. By 2030, they "could own some 17% of installed wind capacity and 21% of solar" (EU, 2019: 13).

In Germany, the energy transition has reached a point where it is now necessary to look at electricity generation and consumption in a more holistic, systemic manner. Hence, energy communities are expected to play an increasingly important role in further decarbonizing the energy sector. With regard to intracommunity cash flows involved through local peer-to-peer (P2P) trading, these may not be enough to justify the purchase of P2P-enabling infrastructure. From a grid perspective, it is crucial to exploit the complementarity of CEC measures as well as the necessity to provide CECs with access to sufficient funding.

This study examines three types of CECs, thus building on qualitative work of Gui and MacGill (2018). It aims to quantify how the inception of different types of CECs and their actions affect the energy system, and examines two research questions:

- How do different CEC types influence the *grid-friendliness* of a community? This question aims at energy communities from a technological viewpoint. In the current phase of the sustainable energy transition, it is necessary to examine electricity generation and consumption in an integrated, systemic manner (BMWi, 2018a). For the German government and grid operators, it will be instructive to quantify how CECs shape grid requirements and whether CECs show an intrinsically grid-friendly behavior, or if policy-makers have to adapt their policies accordingly; and
- What sizes are the induced intracommunity *cash flows*? Here, the focus is on CEC-dependent cash flows among community members. For instance, increasing local electricity generation from renewable energy sources (RES) can create new income sources for community members (RED, 2018). In particular, intracommunity cash flows can help retain capital in the community.

The model consists of three elements:

- It simulates the electricity generation of solar photovoltaic (PV) systems and WECs;
- It generates load data for different household types using a behavior-based load profile generator (LPG); and
- The household types are combined to form a local community.

This chapter then analyzes how the inception of a CEC changes the status quo in terms of self-sufficiency ("autarky") and grid-friendliness (dynamics of aggregate load changes, see below). The model has several restrictions, related to the consideration of only private household loads, the focus on PV and wind energy, and a simplistic model of the electrical grid. Still, the model is useful and has its merits: the input data has a high resolution, the model can easily be applied to other locations, and the results are replicable using the MATLAB code (available from the authors upon request). Hence, the model meets all requirements of a system-analytic approach, in that its results are transparent, comprehensible, and verifiable (BMWi, 2018a).

The remainder of this chapter is structured as follows:

- Section 2 introduces the methodology, presents the reference networks, and explains the measures that CECs can implement;
- Section 3 provides the scenarios considered;
- Section 4 presents and discusses the results obtained; and
- Section 5 reports on a sensitivity analysis followed by the chapter's conclusions (Section 6).

2. Methodology

2.1 Overview

The model presented refers to the findings of Kerber and Witzmann (2008), who analyzed 87 low-voltage grids in Bavaria (Germany). According to them, it is possible to generate typical reference networks by combining typical transformer outputs with an average number of connection points. Hence, statements derived from such reference networks should be applicable to a broad range of real-world networks. Table 17.1 reports on the total number of households by category, setting, and building type. Despite poor data availability, the model aims to create representative living situations. For instance, it assumes that the share of retirees is higher in rural than in village or suburban areas (66.7% vs. 50.0% vs. 37%) and that employees prefer to live in the suburban setting (48.1%; village 25.0%, countryside: 16.7%), while families prefer the village setting (25.0%; countryside: 16.7%, suburban: 14.8%).

A further step in establishing the reference networks is to consider existing RES. Here, the focus lies on estimating the number of existing, small-scale, roof-mounted PV systems:

- The model estimates how many roof-mounted PV systems exist in each federal state;
- It uses the federal state population to calculate the penetration rate of roof-mounted PV generators; and
- It considers the community size in the reference networks and multiplies it with the penetration rate.

Table 17.1 Community composition of the three settings.

	Countryside	Village	Suburb
Households living in detached houses	1x Family 1x Employee 4x Retiree	2x Family 2x Employee 4x Retiree	3x Family 3x Employee 3x Retiree
Households living in apartment buildings	– – –	8x Family 8x Employee 16x Retiree	9x Family 36x Employee 27x Retiree
Total share of the household categories [%]	Family: 16.7 Employee: 16.7 Retiree: 66.7	Family: 25.0 Employee: 25.0 Retiree: 50.0	Family: 14.8 Employee: 48.1 Retiree: 37.0

Estimating the total number of roof-mounted PV systems requires knowledge of the total and average capacity of roof-mounted PV systems. By using the *Marktstammdatenregister (MaStR)*, it is possible to estimate the average capacity of roof-mounted PV systems for each state (see Crump and Madlener 2019 for further details). Combining the total and average capacity and using the state population then allows to estimate the total number of roof-mounted PV systems and the penetration rate in each of the federal states.

Grid-friendly communities exhibit low net community load (NCL) gradients (i.e., how fast a local community's aggregate demand changes in a time interval Δt). Conversely, high gradients imply that the grid has to react more quickly and to withstand higher stress (increased resilience). Hence, if a measure reduces the maximum absolute NCL gradient, this is positive from a grid perspective. Yet, it is insightful to also examine how the 99% quantile changes. This allows to verify whether high NCL gradients are an exception or occur regularly. Table 17.2 summarizes the parameters considered and explains how a grid-friendly community behavior is indicated.

A main goal of the investigation is to examine how the inception of different CEC types (*centralized*, *distributed*, *decentralized*, cf. Gui and MacGill, 2018: 100 f.) and implementation of the following four measures influences the communities' grid-friendliness:

- Energy efficiency;
- Bulk purchases;
- Community-scale RES; and
- P2P trading.

Members of *centralized CECs* are directly connected with each other and share social norms. Nonetheless, they do not necessarily form a community of locality. In a centralized CEC, all threads run together at the representative governing body which controls and manages the community rules and activities. Energy cooperatives are a prime example of centralized CECs. In contrast to centralized CECs, most members of *distributed CECs* are not directly connected with each other. The controlling entity is often referred to as the hub organization. Distributed CECs are hub-dominated structures and technology companies, due to a heavy reliance on technological infrastructure in

Table 17.2 Overview of the parameters used to indicate grid-friendly behavior.

Criterion	Unit	Grid-friendly behavior	Indicator of
Degree of autarky	[%]	Higher degree = grid-friendlier	Grid independence
Maximum absolute NCL gradient	[kW]	Lower gradient = grid-friendlier	Load volatility
99% quantile of the absolute NCL gradient	[kW]	Lower quantile = grid-friendlier	Load volatility

this type of CEC, often establish and operate the hub (e.g., they may provide a platform for community members to trade electricity with each other, while ensuring that community rules are obeyed). *Decentralized CECs*, in contrast to the other two types of CECs, have the capacity for energy autonomy (autarky).[2]

2.2 Energy efficiency

Compared to 2007, the EU aims to become 32.5% more energy efficient by 2030 (CEC, 2020), aiming at reducing the overall energy demand and thus the need for new RES and the losses in the transmission and distribution grids. Simply put, efficiency can be a cost-effective approach to reducing the EU's carbon footprint. In fact, in 2018, rising energy efficiency was the most important factor in mitigating the increase of CO_2 emissions (BMWi, 2018b). However, households often lack information and contact partners for efficiency measures in Germany (BMWi, 2019). Here, CECs can play a vital role, using their communication channels to inform members "about rational energy use [and] energy-efficient technologies" (Bauwens and Eyre, 2017: 10). In this regard, CECs profit from their trustworthiness, an important prerequisite for citizen participation in energy efficiency projects (Laskey and Syler, 2013). In the model, only centralized and decentralized CECs can implement energy efficiency measures.

The implementation is modeled as follows. In the load simulation, the LPG allows the user to set the household's energy intensity to "Random devices," "Energy saving," or "Energy intensive." By default, the "Random devices" option is selected. If a household participates in the energy efficiency measure, the "Energy saving" option is selected, translating into a 30% lower electricity consumption.

Judging from the documentation, the LPG does not model behavioral changes when switching between two energy intensities. Yet, there is extensive proof that switching to energy-efficient devices or performing environmentally friendly actions can induce a moral licensing or rebound effect (Tiefenbeck et al., 2013; Harding and Rapson, 2019).

2.3 Bulk purchases of small-scale solar PV

In the model, centralized and decentralized CECs can also organize bulk purchases of distributed generation assets. Instead of buying individually, community members pool their orders and purchase (but only in the case of the centralized CEC use them) collectively, achieving a higher bargaining power (Koirala et al., 2016). While

[2] This study follows the energy autarky definitions of McKenna et al. (2014) who differentiate between "soft" and "strict" autarky: a community or region exhibits 100% soft autarky if, over the course of a year, it self-generates as much energy as it consumes. It may use supraregional energy infrastructure to feed in excess electricity or draw electricity from the wider grid, so that the energy community does not have to meet its energy demand at all times. In contrast, a community with 100% *strict* autarky is able to cover the energy demand locally at all times. This study only considers those (i.e., the term "autarky" here always refers to *strict* autarky).

Table 17.3 Specifications of the small-scale PV systems.[3]

	Unit	Model S	Model M	Model L	Model XL
Capacity	[kWp]	3.25	5.2	7.8	9.75
Required area	[m^2]	17	27	41	51
Retail price	[€]	6900	9500	12,500	14,900
Recommended annual consumption	[kWh]	<2500	<4000	<5500	≥5500

only the bulk purchase of small-scale PV systems is considered, other applications are conceivable. Based on their annual consumption, community members choose between four PV system sizes—S, M, L, and XL (Table 17.3). The available capacities are close to the average capacity of small-scale roof-mounted PV systems. It is assumed that all building types can accommodate all PV systems.

In Germany, a trend can be observed toward also purchasing a battery when buying a PV system. In 2018, every second new PV system under 30 kWp came with a battery (AEE, 2020). Using the battery, households can store excess electricity for later use. In a recent survey, 83% of German battery owners stated that (energy) cost reduction—mainly due to the high costs of electricity purchased from the grid compared to the self-generated one—was a main purchase motivation (KfW, 2019). The model reflects this development by assigning a battery to every second new PV system, starting with the largest PV system (XL). The corresponding batteries are 5.1, 6.4, 7.7, and 9.0 kWh in size and have a C-rate of 1.[4] These storage capacities are close to the average capacity of batteries of 8.40 kWh put into operation in 2018 (BNetzA, 2019c).

Similarly to Zepter et al. (2019), the model does not consider any form of performance degradation for the batteries. This assumption is also extended to the PV systems. The model assumes that all PV systems face South and are tilted 35 degrees. The South-facing orientation is usually the optimum for PV systems in the northern hemisphere (Sarbu and Sebarchievici, 2017). In the present model, the charging strategy is to maximize self-consumption.

2.4 Community-scale RES

In the model, centralized and decentralized CECs can also install community-scale RES. Through member cooperation "more energy options become feasible at a community level due to economies of scale" (Koirala et al., 2016: 729). The output power of

[3] Data retrieved from https://www.eigensonne.de/preise/. The PV systems consist of Heckert NeMo 60M 325 W_p modules with $\eta_{\text{Module, STC}} = 19.4\%$ and $C_{\text{P, Temperature}} = -0.40\%/^\circ\text{C}$.

[4] The C-rate is a measure of the rate at which a battery is charged or discharged, defined as the current that flows through the battery divided by the theoretical current draw under which the battery would deliver its nominal rated capacity in 1 h.

such RES is typically quite large. As such, they may be connected to the medium-voltage level if they have a rated power of between 200 kWp and 20 MWp (Konstantin, 2017). Hence, community-scale RES are considered not as part of the community's low-voltage grid but as part of a different voltage level within close proximity.

In Germany, "there has been a notable shift toward wind projects" (Kahla et al., 2017: 15, own translation). Thus, the model only provides the option to install a community-scale WEC, neglecting other technologies. Although constructing new energy infrastructure often causes controversies, citizen involvement and codetermination can provide relief (IZES, 2015). For wind farms in Germany, Liebe et al. (2016) find that the possibility to own the wind park has a positive effect on the local acceptance of wind power. Also, active citizen approval is necessary for distributed RES because citizens have to be willing "to provide space for the installation of these technologies, capital investment, and behavioral changes." (Bauwens, 2013: 11) In some cases, it may be that "only by connecting the locally based citizens, certain areas are rendered available for wind energy projects." (IZES, 2015: 44, own translation) Also, literature finds wind power to be less viable in urban settings (Walker, 2008: 4405). In part, this is due to the density of constructions and thus lower wind speeds (UBA, 2013; Grieser et al., 2015). Hence, the model provides the option to install a community-scale WEC only in the countryside and village settings. To calculate the WEC output, the model uses the characteristic power curve of an ENERCON E-70 (for details, see Crump and Madlener 2019). This WEC model has a rated power of 2.3 MWp. Note that this is somewhat lower than the average output power of WECs installed in Germany in 2018, which was 3.3 MWp (AEE, 2020).

2.5 P2P trading

Finally, distributed and decentralized CECs can introduce a P2P trading scheme. In this context, Zepter et al. (2019) argue that the price for locally traded electricity (i.e., between two actors within the same low-voltage network) should not include transmission network charges. However, the EU Renewable Energy Directive states that "community members should not be exempt from relevant costs, charges, levies, and taxes that would be borne by final consumers who are not community members, producers in a similar situation, or where public grid infrastructure is used for those transfers." (RED, Remark 71). Useful recent overviews on P2P trading are Rossetto and Glachant (2021), on theoretical considerations, and Karami and Madlener (2021), on practical P2P business model development for the case of Germany.

If P2P consumers are not willing to pay more than the standard electricity retail price, this would mean that the producer receives 7.06 €-ct/kWh (Strom-Report, 2019). Yet, this is less than 9.87 €-ct/kWh, the feed-in tariff (FiT) for small PV systems (<10 kW_p) commissioned after January 2020 (lowered to 7.14 €-ct/kWh if put into operation after October 2021). It is questionable whether producers would sell their electricity for 7.07 €-ct/kWh (2021: 7.7 €-ct/kWh), particularly because older PV systems benefit from even higher FiTs. FiTs will likely continue to decrease further, and further below the aforementioned 7.07 €-ct/kWh. Between January

Table 17.4 Overview of possible measures to adopt, by CEC type.

	Centralized	Distributed	Decentralized
Energy efficiency	X		X
Bulk purchases	X		X
P2P trading		X	X
Community-scale RES	X		X

2019 and January 2020, the FiT for PV systems <10 kWp fell by 1.60 €-ct/kWh, and from January 2020 to October 2021 by another 2.73 €-ct/kWh. So local generators have more and more incentive to sell to local consumers because the price is higher than for exports to the grid.

In reality, some households might pay a price premium for locally generated electricity. In fact, some 10% of the German population may be willing to pay price premiums for electricity generated by RES, regardless of its location (Mattes and Wittenberg, 2012). Nonetheless, the model assumes a P2P price of 7.06 €-ct/kWh. Also, note that the model only covers intracommunity P2P trading between different buildings, located in the same low-voltage grid. Table 17.4 summarizes the measures that the CEC types can implement.

3. Scenarios considered

Three distinct scenarios are considered, reflecting different levels of engagement in the four measures described above: "Business as usual," "Conservative," and "Maximum." While scenarios do not constitute a forecast, they constitute a justifiable possible future, allowing to assess the potential effects of CECs on a community's grid-friendliness.

3.1 Business as usual scenario

In the *Business as Usual* (BaU) scenario, no CEC exists and thus no measures are implemented. It serves to compare how measures conducted in the other two scenarios change the status quo. In particular, it allows to determine whether implementing a measure leads to grid-friendlier energy community behavior.

3.2 Conservative scenario

In the *Conservative* scenario (CON), the model uses moderate assumptions (cf. Table 17.5, center column). Regarding energy efficiency, KfW (2019) found that 49.9% of the German population want to save more energy. Yet, cost motivations and split-incentive barriers likely reduce the number of households that actually participate in energy efficiency measures. Taking into account these limiting factors, the

Table 17.5 Summary of the CON and MAX scenarios.

	Participation/implementation CON	Participation/implementation MAX
Energy efficiency	20% of all households	49.9% of all households
Bulk purchases	3.3% of all households who do not own a PV system yet	Such that 52% of all buildings own a PV system
P2P trading	20% of prosumer buildings (min. = one) 40% of consumer buildings (min. = two)	100% of prosumer buildings 100% of consumer buildings
Community-scale RES	No	Yes (in the settings countryside and village)

model assumes that 20% of all households participate in the efficiency measure in the CON scenarios. In the model, the households with the highest annual electricity consumption are the first ones to participate in this measure.

The households can also participate in bulk purchase of solar PV systems. While in Germany 35.9% of the population want to self-generate more energy, only 2.2% of all households without a PV system plan to purchase one within the next 12 months (KfW, 2019). Since CECs can buy PV systems at lower prices, the CON scenario assumes that CECs can increase the share of households who purchase a PV system within the next 12 months by 50% (i.e., 3.3% of all households participate). Here, buildings with the highest gross load and that do not have a PV system already installed are the first ones to receive one.

Regarding the community-scale WEC, Gancheva et al. (2018) point out that some energy cooperatives may install RES only at a later stage, once they are more stable from a financial point of view. Following this rationale, the model assumes that CECs do not install a community-scale WEC in the CON scenario.

Lastly, CECs can introduce a P2P trading scheme. Note that the modeling is based on *buildings* and not individual households. Since P2P trading "is viable only when there are at least a certain number of participants" (Park and Yong, 2017: 4), the CON scenario assumes that one in five prosumer buildings locally sells its excess electricity. On the demand side, two fifths of the consumer buildings participate. The buildings with the most fed-in kWh or the highest net building load are the first one to participate. To guarantee a minimum number of participants, at a minimum one prosumer building and two consumer buildings participate.

3.3 Maximum scenario

The *Maximum* scenario (MAX) uses more extreme values than BaU and CON (Table 17.5, right column). In this scenario, CECs can be interpreted as sworn-in, environmentally friendly communities highly dedicated to increasing the community's degree of autarky. In a way, the MAX scenario investigates what would happen if communities took their efforts to the next level.

Starting with energy efficiency, the model assumes that all households who want to save more energy can do so (49.9% of the households are assumed to participate; KfW, 2019). Also, the bulk purchase adds small-scale PV systems until 52% of the buildings are equipped. This is equal to the aggregated percentages of homeowners in Germany who either already own a system, plan to purchase one, or can at least imagine buying one (KfW, 2019). Moreover, all prosumer and consumer buildings participate in the local P2P trade, and the community may install a community-scale WEC.

4. Results

The results presented below must be considered in the given context:

- First, the more households live in a community, the smaller their individual influence is on the aggregated community behavior. Hence, the potential for more extreme results is higher in smaller communities; and
- Second, when assessing grid-friendliness, the model neglects the time at which the communities feed electricity to the wider grid (ideally, they would do so at times of high demand in the wider grid).

Table 17.6 reports the main parameters that characterize the BaU scenario. The data show a positive correlation between community size, gross community load, and maximum absolute NCL gradient. Furthermore, the village and suburb communities consume more electricity locally (99.18% vs. 99.24% of the locally generated electricity) than the countryside community (83.74%). This is due to the fact that the "PV systems to citizens" ratio is higher there than in the countryside setting.[5]

Table 17.6 Results of the BaU scenario.

	Unit	Countryside	Village	Suburb
Gross community load	[kWh]	35,849	165,304	325,985
Already existing PV systems	[−]	1	3	6
Local PV generation	[kWh]	10,029	30,086	60,172
Local PV consumption	[kWh]	8398	29,840	59,712
Ratio PV consumption: PV production	[%]	83.74	99.18	99.24
Degree of autarky	[%]	23.43	18.05	18.32
Maximum absolute NCL gradient	[kW]	11.76	27.78	43.21
99% quantile of absolute NCL gradient	[kW]	6.03	12.19	19.34

[5] Using the PV penetration rate and community size, the model estimates that 0.6, 3.4, and 6.4 PV systems already exist in the countryside, village, and suburb setting, respectively. Rounding to the nearest integer results in a higher "PV systems to citizens" ratio on the countryside.

Next, the influence of implementing measures in terms of the community's grid-friendliness in the different settings is analyzed. The figures presented follow a specific color code. The BaU scenario is always presented in black, results of the CON scenario have a lighter color, and results of the MAX scenario have a darker color. In the figures, the following abbreviations are used: "Eff" (efficiency measure), "Bulk" (bulk purchase), and "Wind" (community-scale WEC).

The model assumes that communities try to maximize their degree of self-sufficiency (autarky), which is a core preference of energy communities (Koirala et al., 2016; Gancheva et al., 2018). For communities, a higher degree of autarky is desirable because it reduces the expenses for electricity drawn from the wider grid. Assuming that a higher level of grid-friendliness reduces the energy system costs (i.e., costs are not simply shifted from the CEC to non-CEC consumers), and that CEC investment costs do not outweigh the local cost savings, the overall welfare implication might well be positive (obviously, this would require some further research to be sure). Since the proposed measures can require one-off and recurring payments, they differ in their financial attractiveness. For the purpose of this analysis, however, these payments are neglected and all measures increase the community's degree of autarky.

4.1 Centralized (cooperative based) CECs

Beginning with the *countryside* community, Fig. 17.1A depicts the maximum absolute NCL gradient and the degree of autarky for each measure, while Fig. 17.1B depicts the 99% quantile of the maximum absolute NCL gradient. It is clearly visible that the BaU exhibits the lowest degree of autarky. This is intuitive since all measures either reduce the community load or increase the local electricity generation.

In the *CON scenario*, the efficiency measure is the only available measure since the combination of community size and participation rate is not big enough to motivate the purchase of a single PV unit or a WEC. Therefore, Fig. 17.1A (right plot) and Fig. 17.1B (right plot) do not include data points for bulk purchases or wind in the CON scenario.

In the *MAX scenario*, the efficiency measure and the bulk purchase of small PV systems reduce the 99% quantile (i.e., the net CEC load volatility), whereas installing a WEC significantly increases it compared to the BaU (see Fig. 17.1A, right plot). If communities can choose only one measure and maximize their degree of autarky, they choose the energy efficiency measure (CON) and the WEC (MAX). In the CON scenario, this has positive effects on the 99% quantile (−0.05 kW) and negative effects on the maximum absolute NCL gradient (+1.09 kW). Also, it increases the degree of autarky (+0.74%). In the MAX scenario, installing a community-scale WEC increases the degree of autarky markedly (+57.93%). Yet, it also makes the NCL more volatile: The maximum absolute NCL and its 99% quantile increase by 7.97 and 1.02 kW, respectively. Thus, for the chosen measures in the countryside setting, there is a grid-friendliness trade-off in both scenarios.

Turning next to the *village* setting, it is of interest to check whether a different community size and composition will have an influence on the community's choice. To

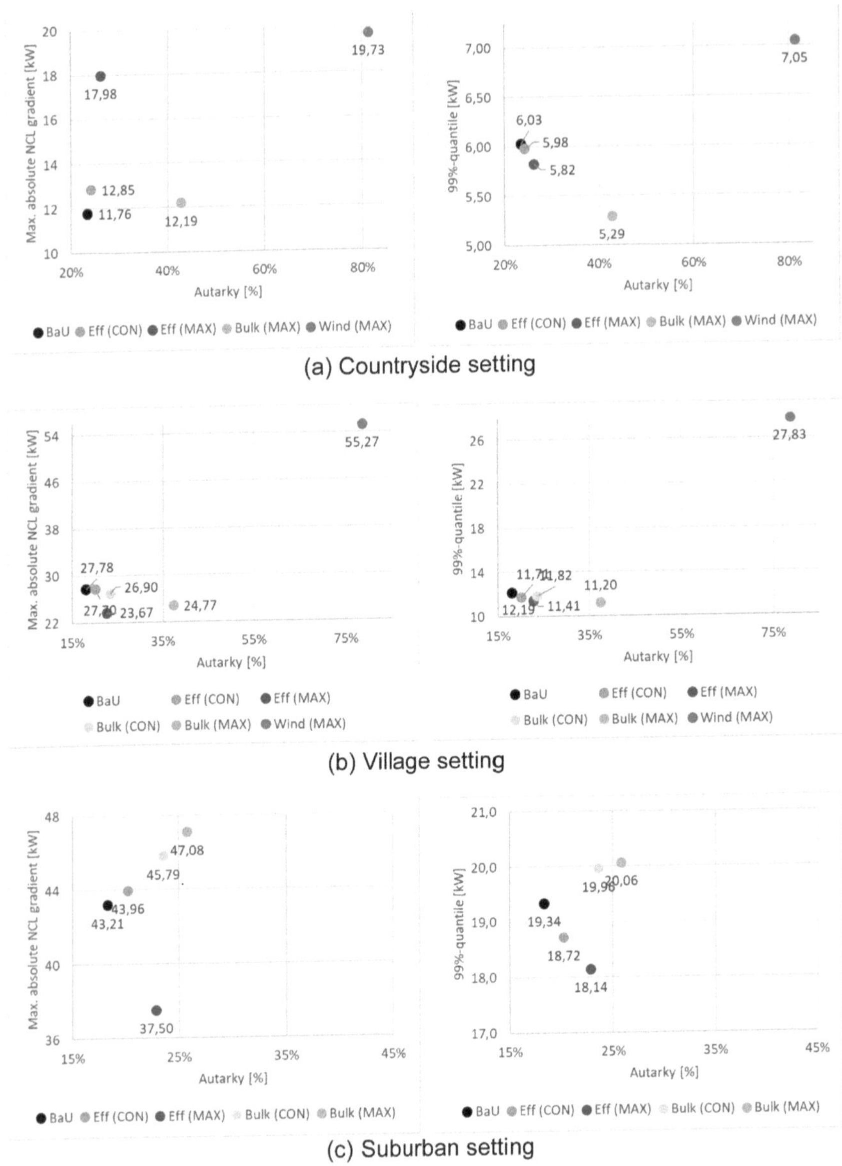

Figure 17.1 Level of autarky versus max. abs. NCL gradient (*left column*) and level of autarky versus 99% quantile of the abs. NCL gradient (*right column*) for the case of a centralized CEC.

this end, Fig. 17.1B (left plot) shows the maximum absolute NCL gradient for the village community. Here, only the community-scale WEC increases the maximum absolute NCL gradient. In contrast to the countryside setting, the efficiency measure and bulk purchase do not increase the maximum absolute NCL gradient, but reduce it. Moreover, they also decrease the 99% quantile (Fig. 17.1B, right plot). Sticking with the autarky-maximizing strategy, the community will pick the bulk purchase (CON) and the WEC (MAX). In the CON scenario, pursuing the bulk purchase is grid-friendly in all three dimensions. More specifically, the bulk purchase comprises one PV system and one PV battery in the CON scenario (five PV systems and three batteries in the MAX scenario). Moreover, installing a community-scale WEC in the MAX scenario has the same effects in the village setting as in the countryside setting: Although it significantly increases the degree of autarky, it also increases the NCL volatility. Hence, based on its decision, the centralized CEC in the village setting acts grid-friendlier in the CON scenario, whereas a trade-off exists in the MAX scenario.

Lastly, examining the *suburb* setting may show what happens when the number of buildings increases only slightly (from 16 buildings in the village setting to 18 buildings in the suburb setting), while the number of citizens increases considerably (almost doubling from 83 to 158 citizens). Note that the community composition again changes and that the share of working people increases significantly (from 27.7% in the village setting to 43.0% in the suburb setting). For the suburb setting, Fig. 17.1C depicts the maximum absolute NCL gradient and its 99% quantile, respectively. Since the model does not allow the suburb community to install a community-scale WEC, it is not able to achieve a degree of autarky comparable to the countryside or village community.

Also in the suburb setting, by assumption, centralized CECs choose to conduct bulk PV system purchases in both scenarios. In the CON scenario, this adds two PV systems and one battery, slightly increasing both the maximum absolute NCL gradient (+2.58 kW) and its 99% quantile (+0.62 kW). In the MAX scenario, the bulk purchase leads to the installation of three new PV systems and two batteries, also increasing the maximum absolute NCL gradient (+3.87 kW) and its 99% quantile (+0.72 kW). Interestingly, in the MAX scenario, the degree of autarky increases much less in the suburb setting when conducting a bulk purchase. Whereas in the village setting a bulk purchase increases the degree of autarky from 18.05% to 37.37% (MAX), the suburb community can only increase it from 18.32% to 25.85% (MAX). Here, the higher number of citizens and thus a higher load, as well as the limitation of equipping only 52% of all buildings with PV systems, lead to a smaller increase. To conclude, a grid-friendliness trade-off exists in both scenarios.

Table 17.7 summarizes the decisions of the centralized CECs. Light gray shading indicates a grid-friendlier behavior, while dark gray shading indicates a less grid-

Table 17.7 Summary of the results for centralized CECs.

		Countryside			Village			Suburban		
	Unit	BaU	CON	MAX	BaU	CON	MAX	BaU	CON	MAX
Measure adopted		-	Eff	Wind	-	Bulk	Wind	-	Bulk	Bulk
Autarky	[%]	23.43	24.17	81.36	18.05	23.46	78.65	18.32	23.60	25.85
Max. abs. NCL grad.	kW]	11.76	12.85	19.73	27.78	26.90	55.27	43.21	45.79	47.08
99%quant. abs. NCL grad.	[kW]	6.03	5.98	7.05	12.19	11.82	27.83	19.34	19.96	20.06

friendly behavior. In the CON scenario, only the village community increases its grid-friendliness in all three regards, by conducting the bulk purchase. While conducting the bulk purchase in the suburb setting also maximizes the degree of autarky, it increases the maximum absolute NCL gradient and its 99% quantile. This suggests that the effects of the bulk purchase depend on the number of buildings and the community composition.

While in the MAX scenario the chosen measures maximize the degree of autarky, they also increase the maximum absolute NCL gradient and its 99% quantile in all settings. Consequently, a clear trade-off exists between energy independence and NCL volatility.

4.2 Distributed (P2P trading) CECs

This section analyzes how much electricity is traded locally if the community uses a P2P trading scheme. Subsequently, "prosumer buildings" ("consumer buildings") refers to buildings with (without) a PV system.

Since the analysis builds on the status quo, no batteries exist in the reference networks. If the PV generation exceeds the aggregated load of a prosumer building, it can offer electricity to other buildings. These buildings either consume all the electricity (i.e., local demand exceeds supply) or the prosumer building exports some PV electricity to the wider grid (i.e., supply exceeds local demand). Since the model only compares excess generation and consumer demand, it does not change the community's degree of autarky. Currently, the challenge in low-voltage grids is to "[match] renewable energy supply to domestic energy demand, locally if possible." (Smale et al., 2017: 133) In this regard, it will be insightful to examine to which extent the communities automatically consume local excess generation.

In the P2P trading scheme, the only difference to the status quo is that occupants of consumer buildings directly pay those of prosumer buildings for the locally generated electricity they consume, instead of paying electricity retailers selling through the wider grid (by assumption, prosumers are at least as well off selling locally, see Section 2.4). With the given constellation, however, consumer buildings may (by default) consume local surplus PV electricity, regardless of whom they pay (i.e., introducing P2P trading causes only a change in the cash flow, while the electricity flow stays the same). Also, the model does not consider a change in the load pattern although consumers could try to shift their demand to times of high PV generation to increase the consumption of locally generated electricity.

For all settings, Fig. 17.2 shows the electricity volume that participating consumer buildings purchase ("P2P consumption") and the electricity volume that participating prosumer buildings feed into the grid ("P2P feed-in"). As can be seen, the P2P trading volume is by far highest in the MAX scenarios for the village and suburb setting, reaching around 21 and 45 MWh p.a., respectively. Since in the MAX scenario all buildings participate, the "P2P consumption" there is equal to the total community consumption of local PV electricity and the "P2P feed-in" is equal to the total PV feed-in. Thus, the difference between "P2P feed-in" and "P2P consumption" in the MAX scenario describes the electricity export to the wider grid.

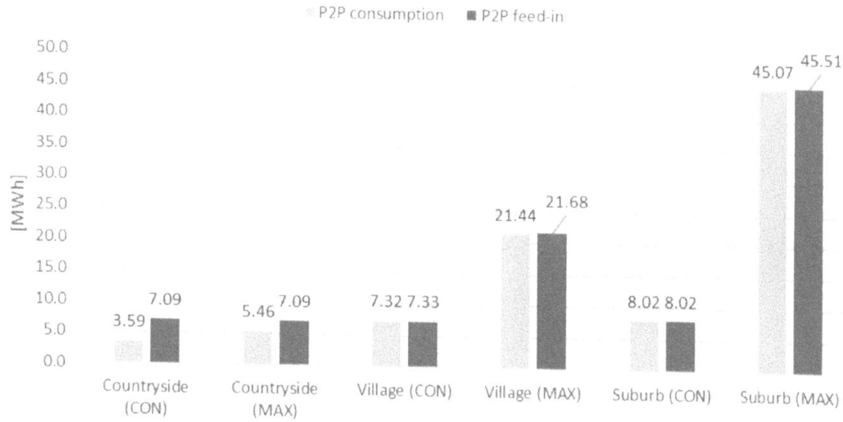

Figure 17.2 P2P consumption versus P2P feed-in (for all three distributed settings, contrasting for each the CON vs. MAX scenario).

In the countryside, the total PV feed-in is 7.09 MWh, of which the community self-consumes 5.46 MWh (CON and MAX scenario). The surplus of 1.63 MWh (or 23% of the total feed-in) is exported to the wider grid. In the CON scenario, the P2P consumption volume is 3.59 MWh, inducing a cash flow of €253.35 (at 7.07 €-ct/kWh). The results in Fig. 17.4 also imply that 5.46−3.59 = 1.87 MWh of electricity are delivered to nonparticipating consumer buildings, who would buy the electricity from their retailer otherwise.

In the MAX scenario, the P2P feed-in remains the same, since the *countryside* setting has only one PV system and thus no additional prosumer buildings can join the trading scheme. Still, the P2P consumption increases to 5.46 MWh, inducing an intracommunity annual cash flow of €385.61.

By comparison, due to the increased local consumption via P2P trading, the export of surplus PV electricity is significantly lower in the village and suburb settings. In the *village* setting, the community exports 1.10% of the surplus PV electricity. While in the CON scenario the purchase volume is 7.32 MWh (€517.01), it almost triples to 21.44 MWh (€1514.00) in the MAX scenario. Also, 14.12 MWh (i.e., 21.44−7.32 MWh) are consumed locally in the CON scenario, but the revenues flow to the electricity retailer (local residual load not met by local production). In the *suburb* setting, the community exports 0.96% of the surplus generation. Here, the P2P purchase volumes amount to 8.02 MWh (CON) and 45.07 MWh (MAX). This induces intracommunity cash flows of €566.01 (CON) and €3182.06 (MAX), respectively.

To conclude, the induced intracommunity cash flow is higher in the MAX scenario than in the CON scenario in all settings. Moreover, it correlates positively with the number of PV systems and the number of consumer buildings. Should a community decide to install more small-scale PV systems, one can expect that the prosumer buildings will have to export more electricity to the wider grid. Note that the P2P trading results likely differ when considering batteries since the feed-in of surplus electricity would occur only when the batteries are fully charged (the assumption is that prosumer households always self-supply first, then charge their batteries, then supply others, and finally feed the remainder into the wider grid). Consequently, the P2P purchase volume would depend on how much electricity consumer households participating in P2P trading consume once the prosumer batteries are full.

4.3 Decentralized (autarky oriented) CECs

Finally, it was examined how communities can completely self-supply themselves. As the previous sections illustrate, conducting only one measure cannot provide 100% autarky. In contrast to centralized CECs, decentralized CECs can subsequently pursue more than one measure. This also allows to examine the interplay between different measures.

For the *countryside* setting, Fig. 17.3A shows the maximum absolute NCL gradient and the corresponding 99% quantiles. Here, square data points indicate that the community conducts several measures. The figures indicate that no combination can provide 100% autarky. Even in the MAX scenario, combining energy efficiency, bulk purchase, and WEC is not enough. Thus, the community would still have to change its behavior and/or install more generators or batteries. The literature finds that community energy storage "facilitates energy sharing and local self-supply of locally generated renewables." (Koirala et al., 2018a: 573) So what if the community used its batteries collectively to store not only the electricity of the corresponding PV system but also from other generators? In other words, what would happen if it was possible to charge the batteries from power generated "outside" the building but within the community? Subsequently, this thought experiment is referred to as a "virtual community battery" (VCB). A VCB would allow the community to stock up on electricity in times of excess PV power generation within the community and excess wind power generation outside the community and to use that electricity in times of underproduction inside the community. As Wiest et al. (2014) point out, PV systems usually feed in a significant amount of electricity even if they are set up to maximize self-supply. Using a VCB, excess generation of PV systems without batteries could be retained in the community, at least to some extent. If available, the community-scale WEC can also charge the VCB so that, especially at nighttime, the VCB and the WEC can satisfy the community load together.

Fig. 17.3A depicts combinations including a VCB by using a triangular shape (the VCB is only available in the MAX scenario). Still, even using a VCB is not enough to reach 100% autarky. Adding a VCB can increase the degree of autarky only to 98.80% ("Eff + Bulk + Wind + VCB"), inducing an additional €238.74 in intracommunity cash flows.[6] Hence, the countryside community is physically not capable of reaching 100% autarky in any scenario. That said, the figures provide some additional insights for the MAX scenario. On the one hand, when installing a community-scale WEC, also

[6] This figure is based on how much less energy the community needs to draw from the wider grid when activating the VCB option, also taking into account the battery round-trip efficiency.

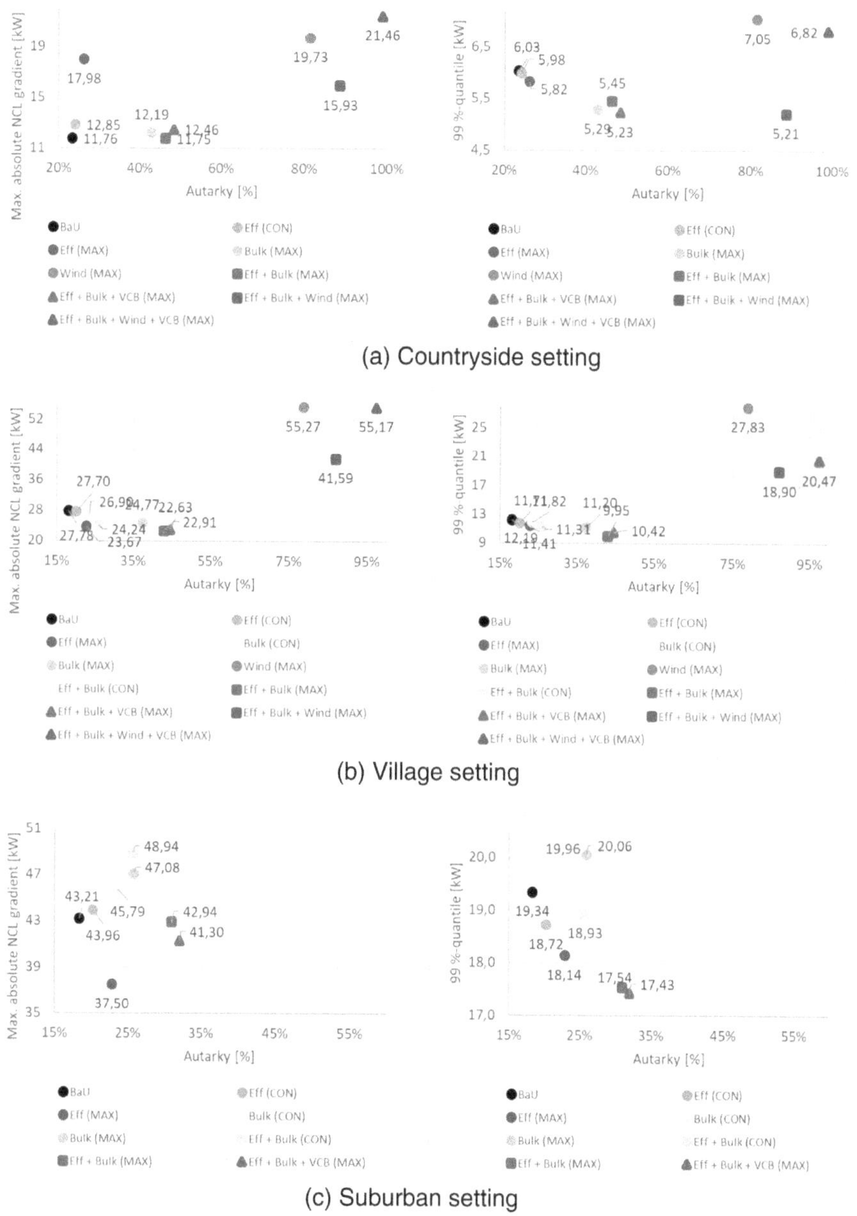

Figure 17.3 Autarky versus grid-friendliness (max. abs. NCL gradient, *left* plots; 99% quantile of the abs. NCL gradient, *right* plots), decentralized CEC types.

conducting the efficiency and bulk purchase measures results in an overall grid-friendlier behavior. On the other hand, combining the efficiency and bulk purchase measures is grid-friendlier than only implementing the efficiency measure. For the CON scenario, it is not possible to make similar inferences since in that scenario no PV systems are bought in the course of the bulk purchase measure.

In the *village* setting, the community is also unable to reach 100% self-sufficiency (i.e., autarky). Here, combining energy efficiency, bulk buying, and community-scale WEC yields an 86.97% degree of autarky (Fig. 17.3B). Thus, the degree is very close to the one in the countryside community but below 100%.

In the *CON scenario*, implementing not only the efficiency measure or bulk purchase but rather combining both yields a higher degree of autarky (25.84%), a lower maximum absolute NCL gradient (24.24 kW), and a lower 99% quantile (11.31 kW) of the absolute NCL gradient. The same is true for the MAX scenario. While using a VCB increases the degree of autarky, as already shown for the countryside setting, where the effect is even more pronounced, it may also add volatility at certain times and under specific conditions (e.g., if it is fully charged). Note that using a VCB in the "Eff + Bulk + Wind" combination increases the degree of autarky by a lot more than in the "Eff + Bulk" combination. This is partly due to the fact that the WEC generates electricity throughout the entire day and can thus charge the VCB also after sunset. This increases the average state of charge during the night and hence the VCB can better help to satisfy the community load. When conducting the efficiency measure and bulk purchase, adding the VCB induces an additional €157.35 in intracommunity cash flows. If the community also adds the community-scale WEC, this increases to €1037.21 (Fig. 17.4B).

Lastly, Fig. 17.3C shows the results for the *suburban* setting. In contrast to the previous settings, the suburb is a lot further away from reaching 100% autarky. In part, this is due to the fact that the suburb community cannot install a community-scale WEC. Also, for the bulk purchase, equipping 52% of the buildings with PV systems results in a smaller "PV systems to household" ratio since the population density is higher (MAX).

In the CON scenario, the suburb community can at most reach 25.7% autarky ("Eff + Bulk"). While this decreases the 99% quantile, it increases the maximum absolute NCL gradient. Here, a trade-off also exists for adding the efficiency measure to the bulk purchase measure, which increases both the degree of autarky and the maximum absolute NCL gradient while reducing the 99% quantile (compared to only conducting the bulk purchase).

In the MAX scenario, the community can reach a 30.9% degree of autarky ("Eff + Bulk"). Using a VCB would improve this to 31.95% while increasing the grid-friendliness and inducing an additional €207.96 in intracommunity cash flows. Moreover, conducting not only the bulk purchase but also the efficiency measure improves the community's grid-friendliness in all three regards.

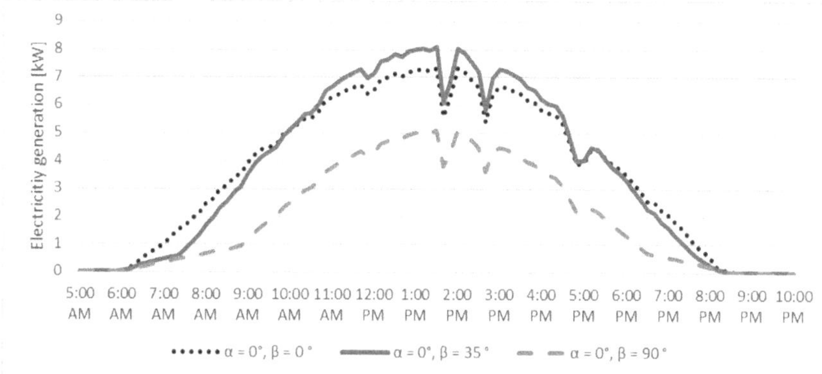

(a) PV inclination β = 0°, 35° and 90° (with unchanged orientation α = 0° = due South)

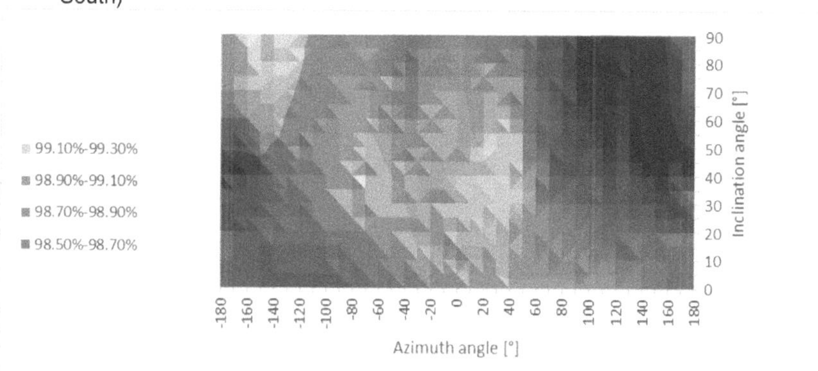

(b) PV inclination β and orientation (azimuth angle) α (legend: maximum autarky levels achievable, considering all types of CEC investigated and a location in central Bavaria, i.e. the south of Germany)

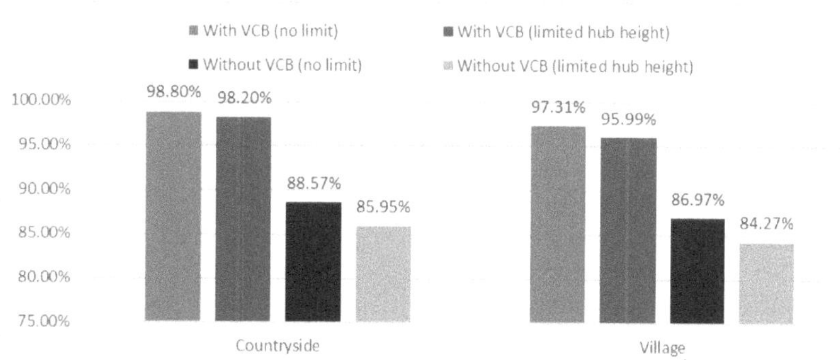

(c) Maximum hub height

Figure 17.4 Sensitivity analysis with regard to key parameters.

5. Sensitivity analysis

The sensitivity analysis conducted focuses on the MAX scenario and whether a community can achieve 100% autarky in any setting by tweaking the electricity generation parameters. For PV systems, changing the orientation and inclination angles results in a different generation curve. For three PV systems facing South, Fig. 17.4A depicts how varying the inclination angle changes the generation curve on a clear summer day. It shows that reducing β_{PV} to 0 ° increases the electricity generation in the early morning and the late afternoon. While this can result in an overall lower annual PV generation, it could help communities reach 100% autarky since it influences the interplay between PV and WEC electricity generation.

Consequently, the sensitivity analysis examines whether varying the orientation and inclination angles of the PV systems can enable 100% autarky. To this end, Fig. 17.4B displays the maximum degree of autarky when conducting all measures and using the VCB (MAX). Still, even under these circumstances no community is able to reach 100% autarky in any of the three settings. This suggests the existence of a brief period with insufficient generation. Another reason may lie in the fact that in the model all PV systems have the same angle and thus electricity generation curve. Positioning them at different angles could result in generation curves that better complement each other and enable 100% autarky. Fig. 17.4B also shows that modifying the angles of all PV systems can enable over 99% autarky.

Moving on to WECs, limiting the maximum hub height can change the degree of autarky. While the chosen hub height is very high in the original results (141 m), in Bavaria, WEC operators have to actively involve municipalities in the process of installing a WEC if the distance between the nearest residential building and the WEC is less than 10 times the height of the WEC (tip height, i.e. hub height plus the rotor radius; '10H rule'). Thus, choosing a lower hub height might be necessary to avoid conflicts in antiwind power areas. To this end, Fig. 17.4C depicts the degree of autarky for a WEC with a hub height of less than 100 m. In the *countryside*, the autarky decreases by 0.60% (with VCB) and 2.62% (without VCB). In the *village*, the decreases amount to 1.32% and 2.70%, respectively. Thus, by using the VCB, the lower WEC electricity generation can be partially absorbed. In both settings, the communities would choose a Nordex N117, but with a hub height of only 91 m (the model considered two WECs, a Nordex N117 and an Enercon E-70). This reduces the full-load hours (FLH) from 3090 to 1818 FLH.

Lastly, three additional and major cities were examined using the original setup to examine the location-dependence of the results: Hamburg (North), Berlin (East), and Cologne (West). None of them reaches 100% autarky, regardless of the setting, and the model overestimates the PV electricity generation also at these locations (see Crump and Madlener, 2019, for more details).

6. Conclusions

The model-based CEC study presented investigates the trade-offs between self-sufficiency (autarky) and grid-friendliness for different types of settings and types of CECs, rather than the trade-off between autarky and economic incentives to engage in a CEC. It is assumed that local prosumers have an incentive, or are at least not worse off, to supply themselves and then other community members (their "peers") locally in P2P trading, and trading only the rest to non-CEC members via the wider network. Likewise, local consumers have an economic incentive to buy from the local prosumers, rather than buying electricity from the grid. It is further assumed that, in the case of positive grid-friendliness effects relative to the BaU, system costs are reduced, so that energy expenditures saved by the local CEC are not simply shifted to the wider society instead. A more detailed analysis that includes how taxes and charges are shifted would be necessary. However, this is often location- and country dependent, which makes generalizations of results difficult.

As the analysis has demonstrated, quantitative modeling of CECs is a complex task. This is reflected in the challenges that arose during the simulation of the electricity generation. In particular, the model overestimates the generation of PV and WECs, calling for more accurate models. It is necessary to improve the approximation of the wind speed at hub height, which may be viable using light detection and ranging devices in the near future. With this in mind, one can now return to the two main research questions.

The *first research question* of this chapter is to determine how the emergence and actions of different types of CECs influence a community's grid-friendliness. As the results of the *centralized* CECs show, conducting only one measure almost always leads to a grid-friendliness trade-off. While the proposed measures decrease the grid dependence, they often increase NCL volatility in both scenarios. Here, the bulk purchase in the village community (CON) constitutes the only exception, improving all three parameters considered. Also, the results of the *decentralized* CECs suggest that combining measures can increase the communities' grid-friendliness. Nearly in all cases, adding the efficiency measure and/or bulk purchase to the already chosen measure improves the grid-friendliness. Only in the suburb, adding the efficiency measure to the bulk purchase increases the maximum absolute NCL gradient (CON).

To conclude, the results suggest that the different measures can complement each other quite well, increasing the community's grid-friendliness in most cases. This suggests that policy-makers should continue to promote energy efficiency measures *and* the installation of small-scale PV systems, particularly in communities where a WEC is present.

The *second research question* is to quantify the extent of the induced intracommunity cash flows. For *distributed* CECs, the CON scenario estimates annual induced cash flows of €253.35 (countryside), €517.01 (village), and €566.01 (suburb). Paired with the community size, these figures allow to estimate the potential of local P2P trading in Germany. At the same time, it is questionable whether these cash flows alone can justify the purchase of P2P-enabling infrastructure (e.g., smart meters). Also, the MAX scenario likely indicates the maximum cash flow potential. Here, local P2P trading induces intracommunity cash flows of €386 (countryside), €1514 (village), and €3182 (suburb) p.a., respectively (with VCB).

Furthermore, it is conceivable that some consumers are willing to pay more than the proposed P2P price. Future research could examine the willingness-to-pay for "local" electricity generated by energy communities. Additionally, further research could extend the scope of the model to a regional level. This would allow to assess how and when neighboring communities can help each other to act in a grid-friendlier manner.

It is important to point out that real-world communities likely differ from the reference communities investigated here. For example, they may prefer certain technologies or face particular challenges. Also, neighboring CECs may interfere with each other in a way that may (positively or negatively) affect their relative grid-friendliness. Hence, it is crucial to consider the specific local context along various dimensions having an impact on the local and wider benefits of CECs.

References

Bauwens, T., 2013. What Roles for Energy Cooperatives in the Diffusion of Distributed Generation Technologies? Center for Social Economy, University of Liège, Liège. URL: https://papers.ssrn.com/sol3/Delivery.cfm/SSRN_ID2382596_code2054653.pdf?abstractid=2382596&mirid=1&type=2. retrieved January 6, 2020.

Bauwens, T., Eyre, N., 2017. Exploring the links between community-based governance and sustainable energy use: quantitative evidence from Flanders. Ecol. Econ. 137, 163−172.

BMWi - Bundesministerium für Wirtschaft und Energie [Federal Ministry for Economic Affairs and Energy], 2018a. 7. Energieforschungsprogramm der Bundesregierung—Innovationen für die Energiewende [7th Energy Research Program of the Federal Government—Innovations for the Energy Transition]. URL: https://www.bmwi.de/Redaktion/DE/Publikationen/Energie/7-energieforschungsprogramm-der-bundesregierung.pdf?__blob=publicationFile&v=11. retrieved December 18, 2019.

BMWi, 2018b. Die Energie der Zukunft—Sechster Monitoring-Bericht zur Energiewende [The Energy of the Future—Sixth Energy Transition Monitoring Report]. URL: https://www.bmwi.de/Redaktion/DE/Publikationen/Energie/sechster-monitoring-bericht-zur-energiewende.pdf?__blob=publicationFile&v=37. retrieved December 18, 2019.

BMWi, 2019. Zweiter Fortschrittsbericht zur Energiewende—Die Energie der Zukunft [Second Progress Report—The Energy of the Future]. URL: https://www.bmwi.de/Redaktion/DE/Publikationen/Energie/fortschrittsbericht-monitoring-energiewende-kurzfassung.pdf?__blob=publicationFile&v=15. retrieved December 18, 2019.

BNetzA—Bundesnetzagentur [Federal Network Agency], 2019a. Monitoring des Stromnetzausbaus [Monitoring of the electricity grid expansion]. URL: https://www.netzausbau.de/SharedDocs/Downloads/DE/Vorhaben/Gesamtbericht.pdf?__blob=publicationFile. retrieved December 18, 2019.

BNetzA, 2019b. Statistiken zum Ausschreibungsverfahren für Windenergieanlagen an Land [Statistics Regarding the Tendering Process for Onshore Wind Energy Converters]. URL: https://www.bundesnetzagentur.de/SharedDocs/Downloads/DE/Sachgebiete/Energie/Unternehmen_Institutionen/Ausschreibungen/Hintergrundpapiere/Statistik_Onshore.xlsx. retrieved December 18, 2019.

BNetzA, 2019c. Veröffentlichung der EEG-Batteriespeicher—01/2017 bis 01/2019 [Publication of the EEG Batteries—01/2017 to 01/2019]. URL: https://www.bundesnetzagentur.de/SharedDocs/Downloads/DE/Sachgebiete/Energie/Unternehmen_Institutionen/ErneuerbareEnergien/ZahlenDatenInformationen/VOeFF_Registerdaten/2019_01_Batteriespeicher.xlsx?__blob=publicationFile&v=2. retrieved January 6, 2020.

CEC (Commission of the European Communities), 2020. Homepage > Energy > Topics > Energy Efficiency > Targets Directives and Rules > Energy Efficiency Directive. URL: https://ec.europa.eu/energy/en/topics/energy-efficiency/targets-directive-and-rules/energy-efficiency-directive. retrieved January 6, 2020.

Crump, R., Madlener, R., 2019. Modeling Grid-Friendly Clean Energy Communities and Induced Intra-community Cash Flows. FCN Working Paper No. 26/2019. Institute for Future Energy Consumer Needs and Behavior, RWTH Aachen University (December).

EU—European Union, 2019. Clean Energy for All Europeans. URL: https://op.europa.eu/en/publication-detail/-/publication/b4e46873-7528-11e9-9f05-01aa75ed71a1/. retrieved December 18, 2019.

Gancheva, M., O'Brien, S., Crook, N., Monteiro, C., 2018. Models of Local Energy Ownership and the Role of Local Energy Communities in Energy Transition in Europe. URL: https://cor.europa.eu/en/engage/studies/Documents/local-energy-ownership.pdf. retrieved January 6, 2020.

Grieser, B., Sunak, Y., Madlener, R., 2015. Economics of small wind turbines in urban settings: an empirical investigation for Germany. Renew. Energy 78 (June), 334–350.

Gui, E., MacGill, I., 2018. Typology of future clean energy communities: an exploratory structure, opportunities, and challenges. Energy Res. Social Sci. 35, 94–107.

Harding, M., Rapson, D., 2019. Does absolution promote sin? The conservationist's dilemma. Environ. Resour. Econ. 73, 923–955.

IZES—Institut für ZukunftsEnergieSysteme gGmbh, 2015. Nutzeneffekte von Bürgerenergie: Eine wissenschaftliche Qualifizierung und Quantifizierung der Nutzeneffekte der Bürgerenergie und ihrer möglichen Bedeutung für die Energiewende [Benefits of citizen energy: A scientific qualification and quantification of citizen energy benefits and their possible significance for the energy transition]. URL: https://www.greenpeace-energy.de/fileadmin/docs/pressematerial/IZES_2015_09_10_BuE-Nutzen_Endbericht.pdf. retrieved December 18, 2019.

Karami, M., Madlener, R., 2021. Business models for peer-to-peer energy trading platforms in Germany based on households' beliefs and preferences. Appl. Energy 306 (Part B), 118053. https://doi.org/10.1016/j.apenergy.2021.118053.

KfW—Kreditanstalt für Wiederaufbau [Credit Institute for Reconstruction], 2019. KfW-Energiewendebarometer 2019 [KfW Energy Transition Barometer 2019]. URL: https://www.kfw.de/PDF/Download-Center/Konzernthemen/Research/PDF-Dokumente-KfW-Energiewendebarometer/KfW-Energiewendebarometer-2019.pdf. retrieved January 6, 2020.

Koirala, B.P., Koliou, E., Friege, J., Hakvoort, R.A., Herder, P.M., 2016. Energetic communities for community energy: a review of key issues and trends shaping integrated community energy systems. Renew. Sustain. Energy Rev. 56, 722–744.

Koirala, B.P., van Oost, E., van der Windt, H., 2018. Community energy storage: a responsible innovation towards a sustainable energy system? Appl. Energy 231, 570–585.

Konstantin, P., 2017. Praxisbuch Energiewirtschaft—Energieumwandlung, -transport und -beschaffung, Übertragungsnetzausbau und Kernenergieausstieg, 4. Aufl. [Practice book energy economy—Conversion, transport and sourcing of energy, the transmission grid expansion and nuclear phase-out], fourth ed. Springer-Verlag, Berlin.

Laskey, A., Syler, B., 2013. Chapter 23—the ultimate challenge: getting consumers engaged in energy efficiency. In: Sioshansi, F.P. (Ed.), Energy efficiency—towards the End of Demand Growth. Academic Press, Waltham, pp. 591–612.

Liebe, U., Bartczak, A., Meyerhoff, J., 2016. A Turbine Is Not Only a Turbine: The Role of Social Context and Fairness Characteristics for the Local Acceptance of Wind Power. Faculty of Economic Sciences, University of Warsaw, Warsaw. URL: https://www.wne.uw.edu.pl/files/1914/8051/6018/WNE_WP223.pdf. retrieved January 6, 2020.

Mattes, A., Wittenberg, E., 2012. "Nur wenige wechseln den Stromanbieter": Sechs Fragen an Anselm Mattes ["Only Few Change Their Electricity Supplier": Six Questions for Anselm Mattes]. URL: https://www.econstor.eu/bitstream/10419/58110/1/685565742.pdf. retrieved January 6, 2020.

McKenna, R., Jäger, T., Fichtner, W., 2014. Energieautarkie—ausgewählte Ansätze und Praxiserfahrungen im deutschsprachigen Raum [Energy autarky—selected approaches and practical experiences in the German-speaking areas]. UmweltWirtschaftsForum 22 (4), 241–247.

Park, C., Yong, T., 2017. Comparative review and discussion on P2P electricity trading. Energy Proc. 128, 3–9.

RED—Renewable Energy Directive, 2018. URL: https://eur-lex.europa.eu/legal-content/EN/TXT/PDF/?uri=CELEX:32018L2001&from=EN. retrieved December 18, 2019.

Rossetto, N., Glachant, J.-M., 2021. New transactions in electricity: peer-to-peer and peer-to-X. Econ. Environ. Energy Policy 10 (2). https://doi.org/10.5547/2160-5890.10.2.jgla.

Sarbu, I., Sebarchievici, C., 2017. Solar Heating and Cooling Systems. Academic Press, Cambridge, Mass.

Strom-Report, 2019. Homepage > Strompreise > Strompreiszusammensetzung. URL: https://strom-report.de/strompreise/strompreis-zusammensetzung/. retrieved January 6, 2020.

Tiefenbeck, V., Staake, T., Roth, K., Sachs, O., 2013. For better or for worse? Empirical evidence of moral licensing in a behavioral energy conservation campaign. Energy Pol. 57, 160–171.

UBA—Umweltbundesamt [German Environmental Agency], 2013. Modellierung einer vollständig auf erneuerbaren Energien basierenden Stromerzeugung im Jahr 2050 in autarken, dezentralen Strukturen [Modeling an electricity generation in the year 2050 in autarkic, decentralized structures based completely on renewable energies]. URL: https://www.umweltbundesamt.de/sites/default/files/medien/376/publikationen/climate_change_14_2013_modellierung_einer_vollstaendig_auf_erneuerbaren_energien.pdf. retrieved December 18, 2019.

UBA, 2019. Emission of Greenhouse Gases Covered by the UN Framework Convention on Climate. URL: https://www.umweltbundesamt.de/sites/default/files/medien/384/bilder/dateien/en_indicator_klim-01_greenhouse-gas-emissions_2019-04-25_0.pdf. retrieved December 18, 2019.

Walker, 2008. What are the barriers and incentives for community-owned means of energy production and use? Energy Pol. 36, 4401–4405.

Wiest, M., Finkel, M., Engel, B., 2014. Innovatives Energiemanagement bei Haushaltskunden—ein Beitrag zur Netzstabilität? [Innovative energy management for household customers—a contribution to grid stability?]. URL: https://www.tugraz.at/fileadmin/user_upload/Events/Eninnov2014/files/lf/LF_Wiest.pdf. retrieved January 6, 2020.

Zepter, J.M., Lüth, A., Crespo del Granado, P., Egging, R., 2019. Prosumer integration in wholesale electricity markets: synergies of peer-to-peer trade and residential storage. Energy Build. 184, 163–176.

Further reading

Broering, H., Madlener, R., 2016b. Simulating the Spatial Heterogeneity of Cloud-Based Prosumer Battery Systems. FCN Working Paper No. 15/2016, Institute for Future Energy Consumer Needs and Behavior, RWTH Aachen University (November).

Madlener, R., Broering, H., 2016a. The Benefits of Perfect Foresight in Prosumer Battery Clouds. FCN Working Paper No. 16/2016, Institute for Future Energy Consumer Needs and Behavior, RWTH Aachen University (November).

Olukan, T.A., Emziane, M., 2014. A comparative analysis of PV module temperature models. Energy Proc. 62, 694–703.

Pflugradt, N., 2016. Modellierung von Wasser- und Energieverbräuchen in Haushalten [Modeling water and energy consumption in households]. Fakultät für Maschinenbau, Technische Universität Chemnitz, Chemnitz. URL: https://nbn-resolving.org/urn:nbn:de:bsz:ch1-qucosa-209036. retrieved January 6, 2020.

Tamizhmani, G., Ji, L., Tang, Y., Petacci, L., Osterwald, C., 2003. Photovoltaic Module Thermal/Wind Performance: Long-Term Monitoring and Model Development for Energy Rating. URL: https://www.nrel.gov/docs/fy03osti/35645.pdf. retrieved January 7, 2020.

Italian energy communities from a DSO's perspective

18

Anita Del Pizzo [1,*], *Giuseppe Montesano* [1,*], *Carlo Papa* [1,*], *Marina Artipoli* [2] *and Mariangela Di Napoli* [2]
[1]Enel Foundation, Rome, Italy; [2]Enel, Rome, Italy

1. Introduction

This chapter addresses the function of distribution networks and, more specifically, the role of Distribution System Operators (DSOs) in enabling the development of energy communities. In Italy, as in all EU countries, the DSO is a regulated business which provides a number of services and acts as neutral market facilitator and innovator, driving the transition of the energy system toward a more sustainable future. As long as they are endowed with enhanced digital capabilities, DSOs can make available key technologies for smooth, cost-efficient, and reliable operation of energy communities. This chapter draws in particular on the experience of E-Distribuzione[1] for illustration.

This chapter explores the nature of incentives for communities, identifying their effects and ways to allocate subsidies consistently with policy objectives. This chapter also discusses criteria to define the geographical limits of communities.

Orientation to customers, respect of market principles, and the importance of technologies and business models are topics covered throughout this chapter, in alignment with the book's overall theme. Regulatory aspects are equally important and, since this chapter focuses specifically on Italy, this creates a natural link in particular to the chapter by Lo Schiavo et al.

The chapter consists of four sections in addition to the Introduction:

- Section 2 provides a short description of the main features of power distribution networks and DSOs operating in Italy;
- Section 3 covers the role DSOs can play in enabling the development of energy communities and achieving high resilience;
- Section 4 explores how to design cost-efficient incentives; and
- Section 5 addresses the criteria for defining the territorial extension of energy communities followed by the chapter's conclusions.

[*] Enel Foundation is a not-for-profit organization promoted and financed by Enel S.p.A. with the key objective of carrying out and supporting studies and research mainly in the energy field.
[1] E-Distribuzione is the largest Italian DSO and is owned by Enel; https://www.e-distribuzione.it.

Energy Communities. https://doi.org/10.1016/B978-0-323-91135-1.00012-2

2. Power distribution networks and DSOs in Italy

In Italy, the distribution of electricity is mainly carried out at medium voltage standardized level of 20 kV (even if there are still lines at lower levels such as 9–15 kV) and at low voltage at 230 and 400 V. 69% of the national distribution network is operated at low voltage.

The 2020 Annual Report by ARERA, the Italian Regulator,[2] reveals that as of December 31st, 2019, in Italy there were 127 operating DSOs, managing a network of 1,273,000 km and almost 37 million connection points (see also Fig. 18.1).

The three largest DSOs are spinoffs of former vertically integrated corporate groups, that also generate and supply electricity: E-Distribuzione (ex Enel Distribuzione), Unareti (ex A2A Reti Elettriche), and Areti (ex Acea Distribuzione). This complies with the rules on accounting, functional, and legal unbundling,[3] according to

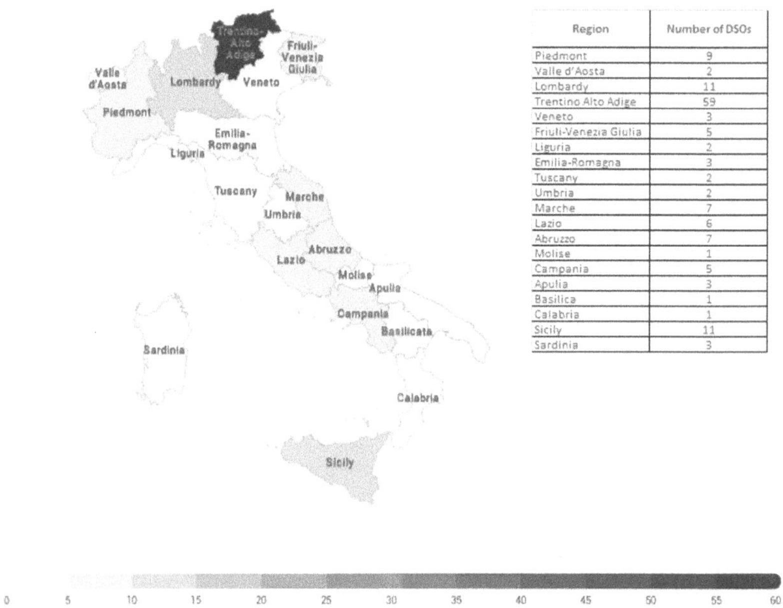

Region	Number of DSOs
Piedmont	9
Valle d'Aosta	2
Lombardy	11
Trentino Alto Adige	59
Veneto	3
Friuli-Venezia Giulia	5
Liguria	2
Emilia-Romagna	3
Tuscany	2
Umbria	2
Marche	7
Lazio	6
Abruzzo	7
Molise	1
Campania	5
Apulia	3
Basilica	1
Calabria	1
Sicily	11
Sardinia	3

Figure 18.1 DSOs in Italy.
Source: Elaboration by authors based on ARERA, 2020. Annual Report.

[2] ARERA: Italian Regulatory Authority for Energy, Networks and Environment.

[3] Delibera ARERA of 22 June 2015 n. 296/2015/R/com "Testo integrato delle disposizioni … in merito agli obblighi di separazione (unbundling) funzionale per le imprese operanti nei settori dell'energia elettrica e del gas (TIUF)," and Delibera ARERA of 24 March 2016 n. 137/2016/R/com "Testo integrato delle disposizioni … in merito agli obblighi di separazione (unbundling) contabile per le imprese operanti nei settori dell'energia elettrica e del gas e per i gestori del servizio idrico integrato e relativi obblighi di comunicazione (TIUC)."

which, in 2016 they also changed their names to be distinguished from the other companies in the group in terms of identity, branding, and communication policies. Altogether they distribute about 93% of total electricity and serve more than 34 million connection points.

In the Region of Trentino-Alto Adige only, there are as many as 59 DSOs, although together they manage about 2.2% of the national distribution network for a total of approximately 28,000 km of grid. Other regions with a considerable number of DSOs are Lombardy and Sicily, with 11 companies each, followed by Piedmont with 9 DSOs, and Marche and Abruzzo with 7 each.

In 2019, out of 36.7 million customers, almost 30 million were households and just over 7 million were businesses. Despite that, households only account for 22% of total electricity consumption, i.e., 58.5 TWh compared to 210.2 TWh for business customers. The areas with the highest energy consumption are in Lombardy, Veneto, Emilia-Romagna, Piedmont, and Lazio. Tuscany, Campania, and Sicily are the next largest consuming areas, while the remaining 11 regions together account for only a fifth of the energy distributed nationwide.

As also illustrated in the chapter by Koltunov et al., the presence of many small distribution companies is linked to a long tradition of energy cooperatives in particular in South Tyrol (Trentino Alto Adige) and in Northern Italy.

Other forms of energy cooperatives have emerged in recent years. Sometimes cooperatives offer energy services, such as energy efficiency consulting and district heating, in addition to electricity supply. This happens in a defined regulatory framework, which allows commercial and industrial cooperatives and consortia to own and manage their distribution network. However, each consumption and production unit is responsible for its imbalances (i.e., differences between actual and planned hourly withdrawals from or injections into the grid), with netting of imbalances between production and consumption not allowed.

Consortia and cooperatives benefit from partial exemptions from system and network charges since the variable components of such charges are applied only to electricity withdrawn from public networks with third-party connection obligations. As a result, the cost burden is transferred to the rest of consumers. This feature is critical from the point of view of both fairness and ability of the system to collect the necessary revenues and must be duly considered when designing the incentives for the new energy communities, as described in Section 4 of this chapter. It is therefore more than appropriate that the Electricity Directive introduces the notion of compensation for the DSO when realizing energy transfers within an energy community.

3. Role of DSOs

The energy communities are an opportunity for the empowerment of citizens in the context of the energy transition and for increasing awareness about the efficient use of energy.

They also represent an important potential added-value for the system, as they can enhance both demand-side flexibility and storage in times of abundant renewable generation through the development of renewable self-generators, electric vehicles, and stationary storage systems.

When they are active consumers or self-generators, energy communities can contribute to reduce grid losses and improve local flexibility, thus efficiently supporting the decarbonization of the energy system through electrification.

Efficiency requires avoiding unnecessary investments. In this respect, the provisional "experimental" virtual model adopted in Italy for Renewable Energy Communities in anticipation of the expected final implementation of the European Directive on renewables is inspired by the principles of cost efficiency and simplicity of implementation. In fact, this model, described in detail in the chapter by Lo Schiavo et al., prohibits energy communities from building new networks, or, in general, managing distribution activities. This prohibition prevents duplication of the grid infrastructure and the possibility of lower service quality and weaker climate proof features.

Under such a model, DSOs should of course have the obligation to provide data concerning the infrastructure they operate for the activities related to registration and facilitation of energy exchange. This task should, in principle, be properly remunerated under the regulatory framework, even though the existing model does not provide for that.

A model involving new physical grids would make sense only in exceptional cases, for instance, communities in areas not yet electrified. In any case, a cost–benefit analysis, comprising climate proof consideration, should assess this possibility against an extension of the existing distribution network.

A shift toward a more distributed energy model based on the generation of electricity from renewables is possible today not only thanks to the continuous reduction of the costs of power plants and to the improvements in their efficiency and reliability, but also—and even more so—thanks to ongoing innovation in the digitalization and control of networks and generation plants, which are increasingly connected to distribution networks.[4]

The development of energy communities amplifies the need for a profoundly innovative role for DSOs, and a change of paradigm, where DSOs must have advanced and sophisticated technological capabilities enabling cost-effectiveness, high service quality, and resilience of the power system.

Power distribution must evolve from a linear system to a utility-as-a-platform model making best use of digital capabilities. In this context, E-Distribuzione plays an important role in Italy for multiple reasons. Firstly, it is the largest DSO in Italy, present in around 7500 out of 8000 Italian municipalities, managing a network of over 1 million km with over 2000 HV/MV substations, and almost 450,000 MV/LV substations. Secondly, E-Distribuzione has been a pioneer in the digitalization of electricity grids. Back in the 1980s, the technological development of the network through remote control systems had already begun, followed in 2000 by the launch of the first

[4] https://www.eurelectric.org/connecting-the-dots.

generation of smart meters. In 2009, when the first European Directive (2009/72/EC) encouraging the diffusion of innovative measurement systems was issued, E-Distribuzione had already replaced most of the old electro-mechanical meters with smart meters.

Since then, the strong need to evolve toward an increasingly smart electricity grid has required the development of new functionalities able to introduce additional improvements for all players in the electricity market: customers, energy distributors, and other market operators (Resolution ARERA 87/2016/R/eel, Resolution ARERA 594/2017/R/eel).[5] Open Meter, the second-generation smart meter by E-Distribuzione (CE2G), has been enriched with important new features.[6] It provides customers with full awareness of their own energy consumption (Near Real Time) and real opportunities to play an active role in the energy transition. Open Meter has a dedicated Power Line communication channel to the customer's home (Chain 2), facilitating the development of energy management and home automation services to improve rational management of domestic consumption and renewable generation plants. Open Meter allows the end customers to understand their energy footprint and interact increasingly with other market players, for example, enabling competing retailers to provide "tailor-made" commercial offers with customized tariffs based on the customer's consumption behavior. In this respect, the customer is the owner of his/her consumption data and, by signing the electricity supply contract, indirectly authorizes the supplier to use these data. The new functionalities also allow the development of Active Demand services that benefit the customer. In addition, with the new system, energy service providers or aggregators will be able to offer Demand–Response services, rewarding a customer's flexibility according to system-balancing needs and market prices for ancillary services, such as load shifting to flatten the consumption curve and adapt it to the daily production of renewable energy sources (Fig. 18.2). The functionalities of the Open Meter are very useful for energy communities, offering members the possibility to receive near real-time data with a high granularity and facilitating the energy settlement of each member and of the whole Energy Community. Notably, this can be done through the already existing system, avoiding the extra cost of any additional dedicated device.

The scalability of innovative technological solutions is essential to improve the performance of power systems as a whole. The massive installation plan for the Open Meter is under way. As of the first half of 2021, more than 20 million CE1Gs have been already replaced by new CE2Gs, with a target of 32 million new units planned by 2024 (E-Distribuzione, 2017).

In order to be smart, the grid must be able to provide real-time information, facilitate the connection of new generation from renewable sources, and minimize outages on the grid. This implies the installation of devices equipped with digital communications systems based on specific protocols, smart metering, and control and monitoring of the connected infrastructures. However, this must be done while also preserving

[5] https://www.arera.it/it/docs/16/087-16.htm; https://www.arera.it/it/docs/17/594-17.htm.

[6] https://www.e-distribuzione.it/open-meter.html.

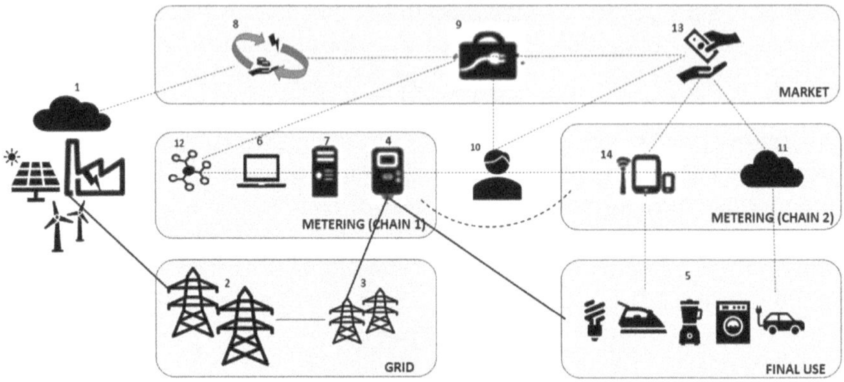

Figure 18.2 Architecture of the second-generation smart metering system. Notes: 1. Generation Plants; 2. Transmission Network; 3. Distribution Network; 4. Smart meter; 5. User Appliances; 6. Head-End System; 7. Data Concentrator; 8. Electricity Spot Market; 9. Retailers/aggregators; 10. Customer; 11. Internet; 12. Data Management Hub; 13. Service Provider; and 14. In-Home Devices.
Source: Elaboration by authors based on Ceneri, G., et al., 2019. Smart Metering 2G—Evolution of a Smart Metering Experience.

safe, reliable, and cost-effective service for all customers. Smart functions can be fully and effectively integrated if accompanied by traditional activities such as replacement of transformers with larger or new ones, and of overhead and underground wires with bigger cables, or construction of new line segments (E-Distribuzione, 2020).

Testing and prototyping projects covering the entire spectrum of smart features is a necessary step to ensure that these can actually be implemented at large scale. Examples are several projects carried out by E-Distribuzione such as "Puglia Active Network—NER300" "PON Projects" "REPLICATE," and "POR Project" which have the common goal of identifying the best technologies to increase the hosting capacity of the distribution network while also ensuring a continuous improvement in the quality of service (E-Distribuzione, 2019).

While the creation of new hosting capacity for distributed generation of electricity hinges of course on network upgrades such as new substations and the expansion of MV networks, advanced digital features can dramatically enhance performance and possibly limit the need for investing in wires. The installation of optical fiber in the substations, for example, enables new communication networks, Smart Fault Selection on the MV network, and remote controls at MV substations.

Such an always-on communication system enables real-time monitoring of the energy produced by renewable energy plants connected to the MV-LV grid and prepares the distribution network for new flexibility services, including for energy communities, the opportunity to better match consumption with the generation from their own plants. These are features included in the "DSO 4.0—Digital Network' project by E-Distribuzione, which, among others, will increase the hosting capacity of distributed renewable sources (i.e., the volume of distributed generation that can be connected to the grid with no infrastructural intervention).

These are all examples of how DSOs are advancing their technological and system management capabilities. The operation of energy communities does not imply a change in the DSO's role in the ownership and management of the grid. On the contrary, a suitably open and technologically ready DSO is the best enabler to accelerate the development of energy communities. Energy communities and DSOs working in a cooperative framework, each within the limits of its own responsibilities, would significantly contribute to an efficient and cost-effective accelerated energy transition toward decarbonization goals.

Italian and European projects, such as the ComESto "Community Energy Storage" and the Eu-SysFlex projects, described below, are indeed showing the way.

The Community Energy Storage project involves 15 partners among Universities, Research Institutions, Large Companies, and SMEs, and is led by E-Distribuzione.[7] Starting from the idea that Energy Communities equipped with storage systems represent an important opportunity to make the transition from a fossil-based energy system to a fully renewable and sustainable one, the project aims at developing an integrated management of generation from renewable sources and distributed storage with the goal of facilitating the active and conscious participation of small-scale customers in the energy markets. In brief, consumers and prosumers will be "aggregated" into "communities" (Community Energy Storage), within which they will gain a higher level of knowledge and awareness of their consumption trends and needs, as well as the benefits that come from the distributed and widespread use of renewable sources. The implemented tool is a platform that develops forecasting, production and load models, and economic and environmental sustainability analyses of the examined technologies. In particular, through the application of Artificial Intelligence and Machine Learning algorithms, automated technical/economic analyses can be performed to plan activities on the electricity grid, aimed at the optimal identification of interventions, providing the grid manager with support in the decision-making process as well as greater flexibility in adapting the calculation model to future grid requirements (E-Distribuzione, 2017). Such a tool can in principle support Energy Communities to improve their capability to increase the amount of shared energy.

In the first half of 2021, the test phase to verify the tool's effective operation on real networks was launched. The network under study is a real low voltage network with mainly residential customers and the presence of PV Power plants. A pilot case will start in January 2022, which will consist of the deployment of three nanogrids from the prototypes of conventional and unconventional storage technologies. The project is expected to end by May 2022.

The Eu-SysFlex project, funded by the European Commission through the H2020 framework program, aims to ensure an efficient and adequate level of system services to facilitate the achievement of global RES integration goals while preserving system resilience. This requires the identification of specific levels of flexibility to best manage the network. The project includes six pilots in France, Finland, Germany, Ireland, Italy, and Portugal, respectively. E-Distribuzione is leading the demonstration

[7] http://www.comesto.eu.

project in Italy, which is implemented in Emilia Romagna Region (Forlì-Cesena) where a data exchange model between TSO and DSO is being tested to improve measurement, load, and generation forecasting also for balancing purposes, as well as active and reactive power modulation in HV-MV substations for TSO network regulation. The project will test the provision of flexibility services from distributed MV-LV resources (included Energy Communities) both to TSO and DSOs (E-Distribuzione, 2017).

Addressing a specific aspect of the Directive on the internal electricity market, Member States are given the possibility to grant Citizens Energy Communities the right to manage a distribution network in their area of operation. In such cases, they face the same responsibilities and obligations (such as metering, connection, billing, and licensing) as other DSOs. However, previous considerations suggest that the option for Member States to preserve DSO's ownership and management of the network is beneficial for both the system and the communities. This is because this option creates new synergies by providing flexibility services to the grid while allowing the communities to rely continuously on the very same quality of grid services (grid infrastructure and metering services), particularly important in the context of the energy transition coupled with the higher probability of extreme weather events related to climate change. In this respect, an open approach can be very fruitful. For example, E-Distribuzione has made its smart grid laboratories in Milan and Bari available to all flexibility operators and stakeholders to foster the opening up of flexibility services on distribution networks.

Working together to develop, integrate, and test flexibility services, products, devices, and resources, particularly for congestion management and voltage regulation, helps to ensure that the electricity grid continues to be reliable, resilient, and able to meet customers' needs. The laboratories are equipped with state-of-the-art instrumentation and best-in-class grid simulation technologies to simulate real-world, complex grid management scenarios. More specifically, in these labs, it is possible to test resources, devices, and systems for flexibility within a simulated or emulated interconnected power grid that can be configured with different advanced test scenarios. Several similar services exist also in countries such as Spain and the United Kingdom. In the United Kingdom, in fact, the flexibility market has already started, the scenario is more mature than in EU countries where several pilot projects are underway in light of the transposition of Directive 2019/944, in which it is made explicit that DSOs must be incentivized to purchase flexibility services.

In summary, this section makes the point that DSOs can play a key role as enablers of new functioning models able to capture the best opportunities for citizens' empowerment and decarbonization without generating unwanted costs and unnecessary risks resulting from the duplication of infrastructures. In that respect, the resilience of distribution grids and high service quality offered by DSOs are key factors supporting the transition from a centralized model based on fossil fuels to a decentralized model based on renewables.

The introduction of open standards will also facilitate communication and trade among participants of energy communities, as well as of energy communities with power exchanges and with TSO-DSO service platforms, for example, for congestion management and/or balancing purposes. Open standards could therefore lead to increased local optimization of supply and demand and system-wide efficiencies. Possible benefits include reductions in network expansion and operation costs, increased integration of renewable energy volumes, and additional revenue streams for participants of Energy Communities.

4. What incentives?

This section and the following one draw on a study about innovation and resilience-related aspects of energy communities deployment in Italy that Enel Foundation has carried out in partnership with Legambiente, an Italian environmental NGO (Legambiente, 2021).

Communities can leverage the efficiencies and capabilities offered by DSOs. Such an opportunity would be facilitated by appropriate incentives. In this section, three questions are addressed.

- First, how (and how much) should regulation remunerate the communities and modulate incentives in a cost-effective way in the context of a virtual model[8] such as the one provisionally being implemented in Italy?
- Second, how should the framework support the development of renewable energies with a view to meeting the 2030 target?
- Third, how should regulation be designed so that the business opportunities for ESCOs are also supported, thus exploiting the potential represented by the latter?

Let us start by restating a couple of reasons why incentives should be given to energy communities and clarifying some principles that should be respected in designing the incentives.

The first reason is that energy communities can facilitate the development of decarbonized power generation. The 2030 target of reducing greenhouse gas emissions by 55%, as confirmed by the European Council, requires an acceleration of the transition to clean energy. Renewable energies should increase their share to 38%–40% of final gross energy consumption, well above the binding target (32%) set by the latest Directive. Renewable Energy Communities, or, more generally, energy communities including renewable generation power plants, can play a significant role to fulfill this increased ambition. Therefore, incentives for energy communities in line with the support provided to other renewable operators appear logical, especially because the remuneration offered to communities for avoided costs (e.g., grid costs) is generally insufficient, as well as variable and uncertain over time. The guiding principle is therefore offering explicit incentives, instead of exemption from bill components (implicit incentive). This is a fundamental difference compared to what typically happens now in the case of individual self-consumption, where energy that does not flow on the

[8] The model previously mentioned in this chapter and described in detail in the chapter by Lo Schiavo et al. uses existing distribution infrastructure.

public network is not charged the variable grid and general system charges components of the tariff.

Secondly, energy communities are also an opportunity for the empowerment of citizens and for increasing awareness about an efficient use of energy and of the energy system. In order to encourage efficient behavior and technologies, the incentive should therefore be carefully directed to self-consumed energy and to the energy shared by the participants in the community, thus helping decarbonized heating and mobility through electrification.

Delving a little more into network charges, focus should be kept on the efficient use of the network. As for any other type of network tariffs, customers' charges for energy communities should reflect grid cost while providing incentives for efficient behavior. In Italy's virtual model, energy flows on existing distribution networks justify the compensation for the value of reduced network losses. However, an exemption from general system charges appears inappropriate since this would not be related to avoided production costs or to positive externalities for the system. Moreover, an exemption from paying these costs would place an unfair burden on other consumers who would pay more.

The type of support—even more than its level—is a decisive element for orienting market choices. In the case of fiscal incentives for investments in generation capacity, they represent an incentive for community members to own the generation plants, rather than buying from a service provider. In the case of incentives on energy, contracts for difference and feed-in tariffs remove any risk connected to energy prices on the market. On the contrary, a price risk remains in the case of an incentive based on a premium on top of the energy market price. In any case, in order to be as consistent as possible with the actual costs, incentives should be modulated according to the dimension of the plant as the economics of bigger plants are generally more favorable. This deserves particular attention as the capacity limits for energy communities are raised in order to avoid risks of over- or underremuneration.

Another important aspect is that regulation should avoid predetermined methods to distribute benefits among community participants. For example, deducting the energy shared from the energy component of the bill, in addition to complicating the tracking of energy flows on both wholesale and retail markets, would automatically direct a significant share of the revenues toward the highest consuming members, de-facto imposing not necessarily desirable criteria to share the benefits inside the community, be it residential, commercial, or composed of small and medium enterprises. The choice of the criteria should be left to the members of the community, consistent with their business model and the characteristics of members' investment. Imposing sharing criteria could, instead, make unattractive or even impossible a number of business models, including ESCO ones, that would be beneficial for the community.

Let us finally mention that, if they are able to align consumption and renewable generation profiles, energy communities could contribute, although in a limited way, to balance the system and therefore reduce balancing costs for the Transmission System Operator. However, this aspect is still being studied in order to assess to what point such a benefit can really be captured and how it can be quantified.

4.1 Case study: a quantification of incentives in Italy

At the moment of writing this chapter, the process of transposing the Renewable Energy Directive into the Italian law is still ongoing. At this stage, the incentive framework for Renewable Energy Communities in Italy includes the following remuneration for a period of 20 years:

- A premium tariff of 110 €/MWh on top of the market price on the energy generated and shared inside the community[9];
- An additional amount, estimated at approximately 10 €/MWh on the energy generated and shared inside the community, rebated on the electricity bill and corresponding to the benefits brought to the system in accordance with an assessment carried out by the Regulator; and
- An energy market price on generated electricity not shared inside the community.

Assuming energy prices in the order of 50 €/MWh, remuneration would sum to approximately 170 €/MWh for the quota of shared energy and 50 €/MWh for the total energy fed into the grid.

Fiscal incentives can be added, ranging from a rebate of 50% of the investment cost of photovoltaic plants up to 200 kW (with a limit of 96,000 €) to an additional rebate up to 110% of the investment cost up to 20 kW if combined with other energy efficiency investments on the building. The latter however excludes the premium tariff of 110 €/MWh on the energy generated by the 20-kW capacity (Legambiente, 2021).

5. The territorial extension of energy communities

Thanks to new smart and clean technologies and new services allowing their integration into energy markets, energy communities have the potential to bring benefits both to citizens and the electricity system itself, by boosting the development of distributed energy renewables solutions and bringing generation closer to consumption.

In order to materialize this potential, rules should allow for an effective development by avoiding excessive constraints which could undermine potentially promising business models. Allowing for a more active role for energy service companies (ESCOs) could significantly improve the economics of the projects, accelerate their development and contribution to the system, or even make them possible at all. Power Purchase Agreements or the installation of renewable assets at a commercial or industrial site in the neighborhood are examples of opportunities which could be developed.

Energy communities can indeed be a business opportunity for ESCOs. The latter can in fact offer various services, as well as play the role of producer.

A crucial aspect may also be the scale of the energy communities and how the geographic limits are designed. A specific aspect discussed in this section is the territorial extension and how to overcome the present limits linked to the technical

[9] The premium tariff is defined by the Ministry as an "incentive" (i.e., premium on top of market price). Therefore, due to how the "virtual model" works, the amount of energy that is paid as "shared" is fed into the grid and is thus entitled to receive also the hourly zonal price.

constraint of being connected to a single secondary substation. To a certain extent, this aspect is also related to the issue of making it possible to exceed the present limit of 200 kW renewable generating capacity.

In the Italian virtual model mentioned in Sections 3 and 4, communities are presently constrained by an "electrotechnical" criterion. Members of a community must all be served by the same MV/LV substation. Virtual energy flows are thus consistent with physical flows on the distribution network. Shared energy has therefore an impact on the system which is similar to the one generated by physical self-consumption. One reason to apply such a criterion is that it makes it possible to clearly identify the positive externalities to be recognized. However, an equivalent application can take place at the higher level of HV/MV substations: if members of the community are localized in a conventional area referred to the same substation, they do not use the transmission network for the shared energy and therefore should be entitled to receive back the relevant tariff component.

As for the explicit incentives for the electricity generated from renewable plants, these can be relatively easily modulated according to the schemes described in Section 4 by referring to the electricity consumed (or stored) at the same time by community members.

The issue with the electrotechnical criterion applied to MV/LV substations is that, in some cases, that criterion makes it impossible for neighbors to be in the same community because of their connection to a different substation. Moreover, especially in highly dense population areas, the hosting capacity of a substation could be saturated, requiring the connection of generation plants to other sections of the network. The latter is especially critical in the case of plants with a generating capacity exceeding 150 kW. There is a risk of substantially jeopardizing the contribution energy communities can make to decarbonization and the declared objective to increase generation plant size in the implementation of the EU Directive.

Geographical criteria could overcome these limits to the development and the dimensions of RECs, by moving away from the correspondence between physical and virtual energy flows, heading instead for a system which is mainly a mechanism for commercial settlements. The most immediate solution could be to adopt administrative criteria, namely how the territory is subdivided by public administrations like municipalities or provinces or zip codes.

In the case of Italy, a criterion based on municipalities would create an enormous heterogeneity. Compared to an average slightly above 4000 members, Renewable Energy Communities could range between less than 10 to approximately 850,000 members. On the contrary, referring to zip codes appears more balanced, as the same code gathers multiple small municipalities, while larger municipalities are divided into multiple codes. Considering that there are currently about 5000 zip codes in Italy, the variance in the number of potential community members associated to the same zip code is depicted in Fig. 18.3.

It is important to point out that, of the 3% with potentially more than 30,000 members, less than one third would potentially have more than 50,000 members.

The adoption of a criterion based on zip codes would typically open the door to urban communities with approximately 1000 members, sharing the energy generated by

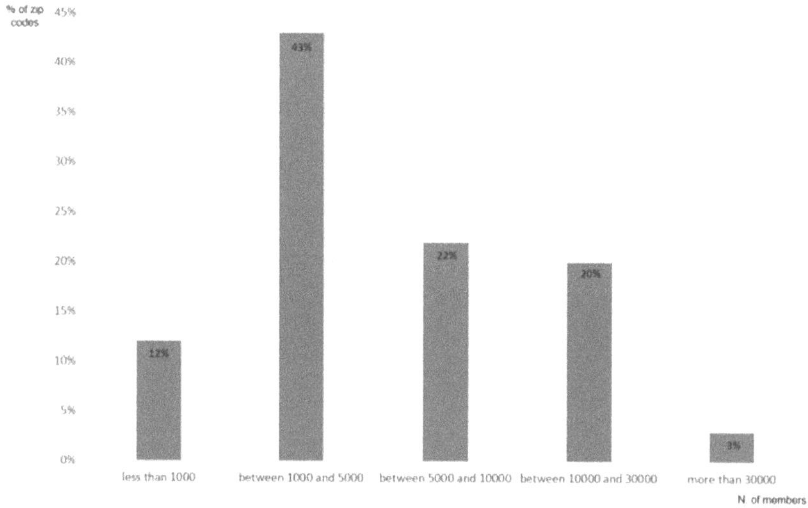

Figure 18.3 Distribution of the number of potential community members associated to the same zip code (elaboration by authors).

multiple rooftop plants. In the case of an average individual annual consumption of 3 MWh, a total PV capacity of approximately 2.5 MW would be necessary to reach 50% of consumption from shared energy. At the other extreme, in rural communities located in relatively large territories covered by the same zip code, 50% of consumption from shared energy could be met by installing a total capacity of 20–30 MW, assuming that the community would include 10,000 members and that there would be availability of land for ground-mounted PV plants or even wind power plants. Also depending on the ownership of the generation plant, the latter case may shift toward utility-scale configurations and complex commercial configurations, which would deserve careful legal consideration, including reduced incentives.

It is clear that the adoption of geographical criteria (like zip codes) would lead to an increase of the potential dimension of the communities, compared to the electrotechnical criterion of all members being served by the same MV/LV substation. A benchmark can be established analyzing the networks of the six largest Italian DSOs, responsible for approximately 96% of the energy distributed on national networks. These DSOs operate approximately 480,000 MV/LV substations serving approximately 35 million clients, averaging slightly more than 70 clients per substation. The variance is of course significant, and the territorial extension is diversified, typically depending on the environment, whether urban or rural. However, the order of magnitude is clearly smaller than in any of the configurations depicted above as representative of the adoption of geographical criteria. An additional comparison can be made with a possible electrotechnical criterion applied to HV/MV substations, which, in Italy, would lead to a potential number of community members up to about 10,000. Any criterion, including a geographical one, can of course be complemented by appropriate caps on the number of members and on installed generation capacity.

A geographical criterion could also be based on the distance from the community "center," which could be represented by the generation plant or geometrically determined by weighing generation capacities of multiple plants or consumption points. In the case of Italy, such a criterion would probably be more suitable for urban areas, where the number of members would reach the order of 1000 in the case of 1 km maximum distance from the "center." It would probably be less practicable for less densely populated agricultural areas.

An equivalent approach can be applied at HV/MV substations, providing that community members are located in a conventional area related to the same substation. In this way, the advantages of simplicity in the adoption of the geographical criterion and of coherence from the electrical point of view would be combined.

6. Conclusions

The main propositions brought forward in this chapter are the following:

- DSOs are key enablers of a system capable of developing energy communities in the simplest and most cost-efficient and reliable way, by facilitating energy transfer among their members and peer-to-peer transactions thanks to regulated data services supported by advanced digital technologies.
- The role of neutral market facilitator of DSOs brings economic value to the energy system as a whole, as well as higher resiliency coupled with a lower infrastructure footprint.
- Support schemes to communities should be carefully designed to avoid improper subsidies and/or cross-subsidies and, at the same time, to represent an adequate incentive to install power plants, helping to reach the 2030 target on energy generated by renewables.
- Criteria defining the perimeter of communities can have a significant impact on the opportunities to operationalize them and can be rethought to increase their potential.

Bibliography

ARERA, 2020. Annual Report.
Ceneri, G., et al., 2019. Smart Metering 2G—Evolution of a Smart Metering Experience.
E-Distribuzione, 2017. Open-meter, piano di messa in servizio, dicembre 2016.
E-Distribuzione, 2019. Piano di Sviluppo e-Distribuzione 2019–2021.
E-Distribuzione, 2020. Piano di Sviluppo e-Distribuzione 2020–2022.
Legambiente – Enel Foundation, 2021. Le comunità energetiche come motore di innovazione e resilienza del sistema energetico. https://www.enelfoundation.org/news/a/2021/07/discussing-energy-communities-as-driving-force-of-innovation-and.

Further reading

E-Distribuzione, 2021. Piano di Sviluppo e-Distribuzione 2021–2023.
Servizi A Rete. https://www.serviziarete.it/levoluzione-rete-enel-distribuzione.

Community energy design models in Brazil: from niches to mainstream

19

Richard L. Hochstetler[1] *and Paulo H.S. Born*[2]
[1]Instituto Acende Brasil, São Paulo, São Paulo, Brazil; [2]Independent Consultant

1. Introduction

Recent technological innovations have opened a whole new range of distributed energy resources that can be tapped to meet the customer's power needs. These distributed energy resources also enable consumers to engage in the supply and management of their power demand. However, autonomous provision of electricity is still much costlier than collective provision. Integrated networks, connecting generators and loads, provide substantial economies of scale and scope. Effective power interaction through the electric grid requires coordination that is difficult to be achieved by individual prosumers.

This is where energy communities come into play. As described in this chapter by Robinson and del Guyao, the pooling together of consumers' distributed energy resources in energy communities can facilitate coordination of electric power network transactions to reduce generation and network costs. Energy communities can also enhance the discovery and deployment of distributed energy resources, and empower consumers to collaboratively engage in the provision of their needs.

While energy communities have the potential of providing net benefits for society, it is difficult to motivate citizens to form new energy communities. Substantial leadership and organizational effort is required to identify the potential of local energy resources and to mobilize a group of electricity consumers to invest in a common venture. Thus, public policies are needed to stimulate the establishment of energy communities.

Incentives can be used to jump-start the creation of energy communities. The rise of successful energy communities will spur others to adopt similar approaches. Once energy communities are firmly established and their benefits are better understood, policymakers can calibrate the incentives offered to new energy communities to reflect the positive externalities they provide.

These incentives must be carefully designed to ensure the promotion of welfare-enhancing energy communities. Otherwise, the incentives can simply, at best, lead to zero-sum gain that benefit some consumers at the expense of others, and, at worst, stimulate rent-seeking activities that result in inefficient solutions that lower social welfare.

Energy Communities. https://doi.org/10.1016/B978-0-323-91135-1.00001-8

The challenge for policymakers at this point in time is to evaluate what are the public policies that are most likely to foster effective energy communities. That is the focus of this chapter.

Based on early initiatives in the development of energy communities in Brazil, this chapter seeks to provide insights regarding the pros and cons of different energy community models and the public policies that are most likely to foster welfare-enhancing energy communities.

Energy communities in Brazil, as elsewhere in Latin America, have emerged in three different contexts (Vásquez-León et al., 2017; Lucas and Gómez, 2017; Mejdalani et al., 2018; Estévez, 2014):

- *Rural communities* not serviced by public utilities, in which consumers come together, usually in the form of rural cooperatives, to build their own distribution network, generation plant, or connection to nearest local public utility;
- *Isolated communities* that are too far to connect to a public utility and thus must establish their own microgrids and generation supply; and
- *Prosumers* serviced by public utilities that engage in the exchange of electricity with the public utility under a *net metering program.*

The first two forms of energy communities have existed for many years and have been a means to provide electricity to niches of the population that have not been reached by traditional power suppliers.

The third form of energy community is a recent phenomenon that became viable with recent generation technologies—particularly photovoltaic panels. The attractiveness of this technology is due the fact it is simple to install, can be deployed at any scale, require little maintenance, and its costs have been dropping rapidly. This technological innovation coupled with net metering programs—which provide a simple way for prosumers to transact energy with the local grid—is enabling the emergence of energy communities among mainstream consumers and has the potential of inducing major transformations to the electric power industry.

The chapter is composed of three sections in addition to this introduction:

- Section 2 evaluates various early energy community initiatives that have been pursued in Brazil to meet the needs of specific niches of society;
- Section 3 considers the potential of widespread development of energy communities fostered by Brazil's net metering program, examining how the various forms of net metering foment efficient and effective community engagement in the procurement and deployment of distributed energy resources; and
- Section 4 provides reflections on these experiences and discusses public policy followed by the chapter's conclusions.

2. Energy community initiatives in Brazil

Given that most of Brazil's electric power is derived from large hydropower plants, the system evolved early on to a single integrated network with the most notable exception of isolated communities in the Amazon region. Electric power provision is regulated by Aneel, the federal regulatory agency, which ensures uniform regulation across the whole country.

2.1 Electrification cooperatives

The earliest energy communities in Brazil emerged in the 1940s in the form of electrification cooperatives. Citizens of small towns and rural areas formed cooperatives to provide their own power supply. Most of these communities were located in areas far from the existing power grids, so they had to provide for both the distribution lines and power generation.

These cooperatives followed the same principles established by cotton mill artisans who established the first modern cooperative organization in Rochdale, England, in 1844 (International Co-operative Alliance, 2015):

- Voluntary and open membership;
- Democratic member control (one member, one vote);
- Member economic participation (surplus distributed in proportion to their transactions);
- Autonomy and independence;
- Education, training, and information;
- Cooperation among cooperatives; and
- Concern for the community.

These principals provide good fundamentals for community building and empowering citizens to meet their local needs.

While the first electrification cooperatives in Brazil were grassroots initiatives, most electrification cooperatives emerged later, prompted by state and the federal governmental programs introduced in the late 1950s (Manuaretto, 2015).

Inspired by the United States' Rural Electrification Act, several governmental initiatives were established to promote rural electrification, such as the *Serviço Especial de Eletrificação Rural* program, launched in 1959 by the state of São Paulo, and the *Eletrificação Rural de Minas Gerais* program initiated in 1962 by the Minas Gerais state-owned utility, Cemig (Pelegrini, 2003).

The largest boost to the electrification cooperatives came in 1960s when the Federal Government passed the Land Statute Law (*Lei 4.504*), which defined rural electrification as a key priority to be promoted by means of a program of subsidized loans (art. 90).

These initiatives led to a proliferation of electrification cooperatives in the 1970s, but governmental support for rural electrification cooperatives remained ambivalent after the initial expansion due to

- Conflicts with public utilities;
- Their higher costs; and
- Their legal status.

The lack of clear boundaries for the electrification cooperatives led to conflicts with public utilities as networks grew and began to overlap.

Consumers generally favored the public utilities over rural electrification cooperatives, because the former generally offered lower tariffs rates. Although the rural electrification cooperatives benefited from subsidized loans and numerous exemptions, their costs were still usually higher than those of the nearest public utilities (Oliveira, 2001; Pelegrini, 2003), due to the fact that:

- The latter serviced lower cost areas (i.e., areas with higher density loads—i.e., urban areas and large industrial customers);

- Benefited from the economies of scale of operating larger networks; and
- From lower generation costs derived from the larger power plants connected to the integrated grid.

Part of the governmental ambivalence was also due to the legal status of electrification cooperatives. Unlike the public utilities that had a mandate to offer nondiscriminatory services to all potential customers in their concession area under regulated tariffs, the rural electrification cooperatives' operations were deemed for "exclusive use" (*Decreto 1.033*/1962) or "private use" (*Decreto 62.655*/1968). This meant electrification cooperatives had full autonomy on pricing and no obligation to service nonmembers of the cooperative (Souto and Loureiro, 1999).

In 1999, Aneel determined that rural electrification cooperatives operate as public utilities, with set boundaries for operation, and regulated tariffs, providing a transition period for them to conform to the new electric power legislation (*Lei 9.074/1995* and *Lei 9.427/1996*) (*Resolução Normativa 333/1999* and *12/2002*).

As a result of these trends, rural electrification cooperatives have dwindled in the last few decades: from 260 in 1980 (Munaretto, 2015) to 51 today. Thus, the Brazilian experience suggested rural electrification cooperatives were effective in accelerating electric power coverage on the frontier, but not necessarily in the promotion of the most efficient long-term solutions.

2.2 Electricity inclusion programs

The shift to consolidation of power supply became even more pronounced in recent years as the federal government increasingly turned to the public utilities to execute their electricity inclusion programs, like the "Light in the Countryside Program" (*Programa Luz no Campo—Decreto 2/1999*) and even more so in the "Light for Everyone Program" (*Programa Luz para Todos—Decreto 4.873/2003* and *7.520/2011*). Both programs sought to promote electrification primarily by expanding existing distribution networks to reach the unserved communities.[1]

While these programs have community-building features that have contributed to the development of the local economy, and improved access to information and education (Andrade et al., 2011; Ribeiro et al., 2017), they do not qualify as fostering the development of energy communities in the sense that they do not aim to promote consumer engagement in energy provision. These are top-down programs executed by public utilities to meet customers' energy needs.

[1] The *Light in the Countryside Program* focused on expanding power supply in rural communities. One of the main features of the program was the adoption of a simpler technology (single-phase distribution) to lower the connection costs. The program also employed subsidized, long-duration loans to make it more attainable (Oliveira, 2001). The *Light for Everyone Program* sought to promote social inclusion, education, and economic development by subsidizing electric power distribution expansion in lower-income, undeveloped communities (Freitas and Silveira, 2015).

2.3 Isolated systems

While the expansion of distribution networks was effective to provide electrification in rural areas in most of the country, this was not viable approach in the immense and sparsely populated northwestern part of the country, where much of the unserved population lived in small-isolated communities. Many of these communities are only reachable by boat. Connecting these small communities to the grid would be prohibitively expensive.

To meet the electrification needs of the communities in this region, the federal government opted for local provision by mean of autonomous systems which consisted of microgrids powered by small generating plants. These "Isolated Systems" are primarily located in the Amazon Region, as shown in Fig. 19.1.[2]

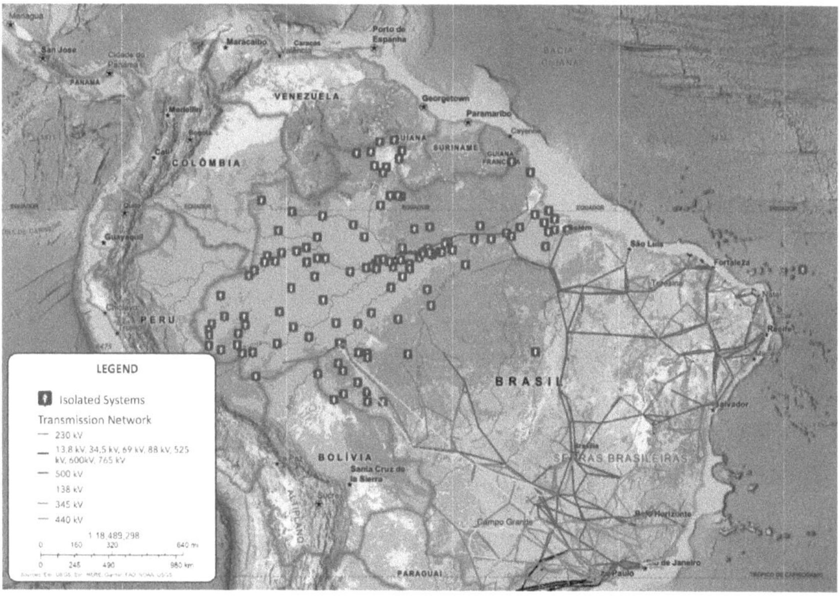

Figure 19.1 Location of the isolated systems in Brazil.
Source: Aneel (2022).

[2] The operating costs of generators in isolated communities are much higher than the large power plants connected to the interconnected transmission grid. To make the electricity more affordable to these communities, the federal government subsidizes the fuel costs of these power plants by means of cross-subsidies funded by electricity consumers connected to national grid (Lei 8.631 and Lei 12.111). Even with these subsidies, the benefited communities often cannot afford continued electricity supply and must make do with provision only a few hours each day (Andrade et al., 2011; Matiello et al., 2018; Neto et al., 2011; Nascimento et al., 2019).

Today there are 235 isolated systems in Brazil, which together meet an average load of 483 MW. The size of each system varies considerably: from 6 kW to 271 MW. Almost all these isolated systems are serviced by diesel generators (97.5%),[3] but there are a few that employ natural gas (1.5%) and biomass (1.0%) (ONS, 2019).

Looking ahead, new technologies such as solar, wind, and biomass generation provide attractive alternatives to meet the energy needs of remote communities.

In 2012, Aneel introduced two new forms to serve isolated communities (*Resolução Normativa 493/2012*), which were denominated:

- *Isolated Generation and Distribution Microsystems* (*microssistema isolado de geração e distribuição de energia elétrica—MIGDI*) in which the public utility constructs an autonomous microgrid and generation to service the community and
- *Individual Intermittent Generation Systems* (*sistema individual de geração de energia elétrica com fonte intermitente—SIGFI*) in which the public utility equips individual households with renewable energy generation to meet their individual needs.

These alternatives have allowed public utilities to meet needs of isolated communities with differentiated quality of service requirements, to enable the use of local renewable intermittent generation, reduce the reliance on imported fuels, and reduce the cost of service. Such alternatives not only favor the use of local energy sources, but also promote closer interaction between the public utility and customers. The regulation requires the public utility to hold public hearings in the community to explain the system and decide key features, such as the time period of supply.

Nevertheless, engagement with customers remains marginal. The interaction with consumers is mostly aimed at explaining technical details of power systems, while supply decisions remain fully under the control of the public utility.

3. Brazil's net metering program

In 2012, Aneel introduced a net metering program (*Resolução Normativa 482/2012*). The program's aim is to incent consumers to install generation derived from renewable energy sources or cogeneration on their property.

The program allows for the installation of *microgenerators* (up to 75 kW capacity) or *minigenerators* (between 75 kW and 5 MW capacity)—the difference being that latter must comply to an Operational Agreement, pay for the new meter, the new grid connection, and, if necessary, grid reinforcements to accommodate the power injection at that point of the grid, while microgenerators are exempt from these obligations.

These customers are billed based on their net consumption during each monthly billing cycle. Consumers are not allowed to sell their surplus production. If the consumer's total production in a particular billing period exceeds his or her consumption,

[3] Historically, one of the draws of diesel generation has been the fact it is a well-known technology in riverside communities where diesel motorboats are the main mode of transportation, which means there are established fuel supplies and technical know-how for maintenance and repairs of the generators.

one receives a credit that can be utilized to compensate net consumption in the following 60 months.

The restriction on the sale of the prosumers surplus production has two justifications:

- The legal characterization of the net metering program as a form of barter trade (*empréstimo de mútuo*), where transactions are registered as loans to avoid taxation on net metering transactions; and
- The regulator's intent of incentivizing small-scale, dispersed generation to reduce load on the distribution networks.

Unlike the programs discussed in the previous sections, this program is completely customer centered. It is up to the customer to take the initiative to engage in the program and to seek out his or her supply alternatives.

Initially, the program only admitted *local production* (*autoconsumo*),[4] case in which the generator is located at the consumer's property but, in 2015, the program was expanded to admit three other possibilities:

- *Dispersed production* (*autoconsumo remoto*),[5] in which the micro- or minigenerator is located in another property of the consumer (as long as it is located in the same distribution-concession area);
- *Local collective self-consumption* (*empreendimento com múltiplas unidades consumidoras*), in which a group of consumers located on the same or contiguous properties share micro- or minigeneration at the same location; and
- *Community supplier* (*geração compartilhada*), in which a group of consumers share generation located at a different property in the same distribution-concession area.

Fig. 19.2 shows the four categories of net metering arrangements that are currently allowed in Brazil.

The net metering categories that are of most interest in this book are the two categories on the right: *local collective self-consumption* and *community supplier*, since these are the categories that involve shared resource energy deployment and coordination among groups of consumers—situation that leads to the formation of energy communities. But first, it is convenient to reflect on a few issues.

3.1 Considerations regarding the net metering program

3.1.1 Cost shift

The net metering program is currently a major source of contention, because of the implicit cross-subsidies involved (also referred to as "cost shift").

Currently, the net metering compensation is made on the full tariff rate, which includes not only the generation costs, but also the grid costs (transmission and

[4] The Brazilian regulators refer to this form of net metering as "auto-consumption" but is hereby referred to as "local production" to harmonize with the taxonomy utilized in this book.

[5] This is form of net metering is called "remote autoconsumption" in the Brazilian regulatory context but is referred to as "dispersed production" as in other chapters of this book.

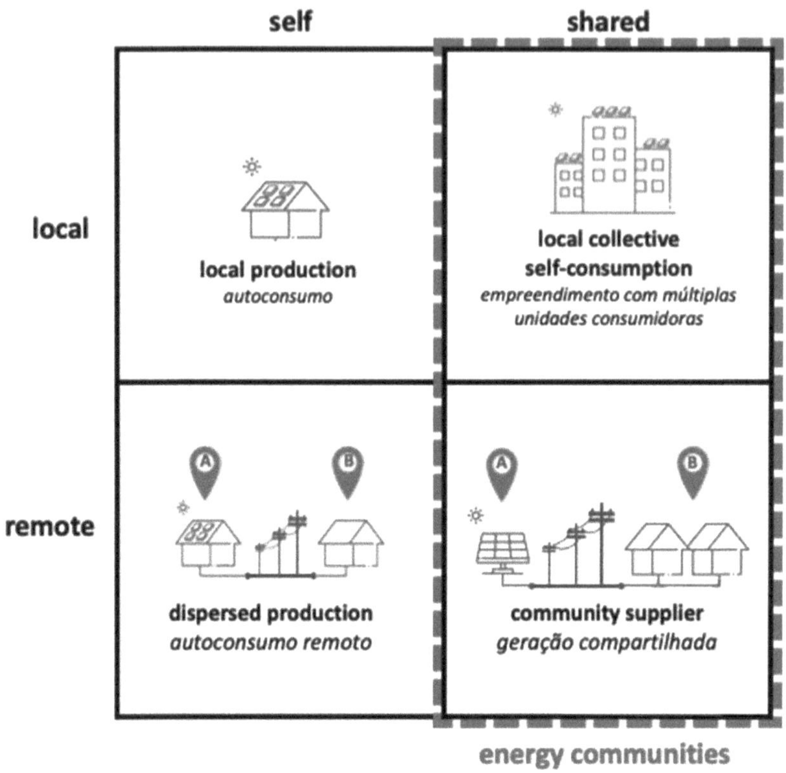

Figure 19.2 Forms of net metering admitted in Brazil.

distribution), losses, and the surcharges that are levied by the government to fund various sectorial costs and subsidies to particular customer classes.

Fig. 19.3 shows the current breakdown of the average residential customers' electric tariff before taxes. Generation costs, which is the expense that the consumer is substituting with mini- or microgeneration, account for just 39% of the total cost; the remaining 61% are associated with the cost of transmission (7%), distribution (29%), losses (9%), and surcharges (14%).

When a prosumer opts to participate in the net metering program, she not only benefits from exchanging her surplus energy for energy provided by other power plants when her demand surpasses her own generation (39%), but also is exempted from contributing to cover her share of the grid costs, losses, and surcharges (the other 61%).

At the first instance, this results in a revenue shortfall for the distribution company. Then, at distribution's next rate review process, when the diminished net demand from the prosumers that participate of the net metering program is factored in, rates are increased to compensate the reduced customer base. Thus, the revenue short fall ends up being covered by the customers that do not participate of the net metering system.

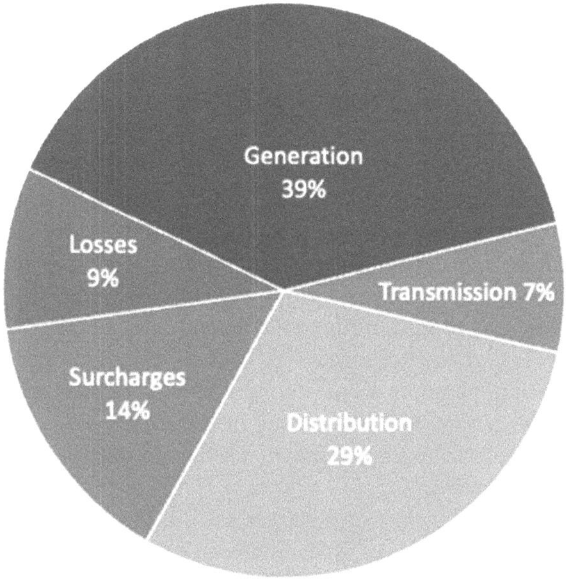

Figure 19.3 Electric tariff breakdown.
Source: Aneel (2020).

This is the most acute problem associated to energy communities built around net metering regimes in Brazil and elsewhere as discussed in Section 4.

This cost shift is particularly detrimental in Brazil, because of its great income inequality. Participation in the net metering program requires investment in micro- or mini-generation, which is prohibitive for most of the lower-income households since access to credit in Brazil is limited and expensive. Industrial, commercial, and wealthy residential customers, on the other hand, are likely to take advantage of the net metering program, thus resulting in an income transfer from the poor to the wealthy.

Aneel is aware of these cross-subsidies. They were intentionally put in place to incent the consumers to invest in micro- and minigeneration in the early stages, but were to be phased out over time once they were deemed unnecessary.

The Regulatory Agency has faced great backlash since it proposed a gradual phase out of these incentives in 2019. Lobbies have mobilized Congress to legislate on the matter. As in the writing of this chapter, Congress is debating a bill on the issue (*Projeto de Lei 5.829/2019*). The bill is very contentious, but most likely the final bill will call for a gradual reduction (if not elimination) of the cross-subsidies over the coming 8 years. The current bill requires Aneel to estimate the benefits attained by the deployment of distributed energy resources and to internalize those benefits in the tariff structure employed in net metering program.

3.1.2 Local versus remote and dispersed versus community

One of the main benefits associated with distributed generation is the unloading of the transmission and distribution networks, which reduces losses and investments in grid expansion. Thus, one would think that, among the net metering alternatives presented in Fig. 19.2, *local production* and *local collective self-consumption* are the most effective forms to minimize network costs.

Likewise, one would expect *dispersed production* to be preferable to *community supplier* at reducing power flows on the grid.

However, it is important to consider not only the geographic dimension but also the time dimension. The prosumers' hourly production and consumption profiles often differ, thus demanding grid services to distribute the prosumer's surplus energy to other consumers during parts of the day, and to supply the prosumer's net demand with energy derived from other generators connected to the grid at other times.

When one examines the hourly load pattern of high-voltage and low-voltage customers,[6] one often observes that the high-voltage customers' load profile aligns better with the production profile of solar generation. Furthermore, low-voltage lines typically are characterized by radial distribution, which is more prone to bottlenecks, while high-voltage lines tend to be better interconnected to major substations, which facilitates the distribution of energy across the grid.

Fig. 19.4, for example, shows the load pattern of high- and low-voltage customers do CPFL Piratininga, a distribution company in the state of São Paulo, that exemplifies this situation.

The low-voltage consumption profile on Fig. 19.4 is driven primarily by residential customer behavior. What explains large part of peak demand is the use of electric showers at the end of the day. Most residences utilize electric shower heads that provide continuous-flow water heating. This technology is cheap and economical but, given typical usage patterns, results in an accentuated coincident peak demand for electricity at the end of the day.

Historically, hourly load variations have not been a major concern in Brazil, because most of the electricity is provided by hydroelectric plants, which can modulate production with relative ease and negligible cost variations. However, the higher peak demand requires more grid investment.

Given the better match of solar production and high-voltage consumption profiles, larger *community suppliers*, connected to the distributors' high-voltage network, may actually demand less of the grid than *dispersed generation* or *local collective self-consumption* when energy storage is not a viable option.

Furthermore, because of the high correlation of production by solar generators, power sources that present different production profiles or that may be dispatched when needed will become increasingly valuable. Often these other power sources are restricted to particular sites or require a minimum viable scale (such as small hydropower, biomass thermal generation, and wind power)—situations that most likely can only be met as a *community supplier*. Again, a situation that would favor the remote and communal alternatives.

[6] Low-voltage customers are defined as those that are connected to the distribution grid at voltages of up to 69 kV.

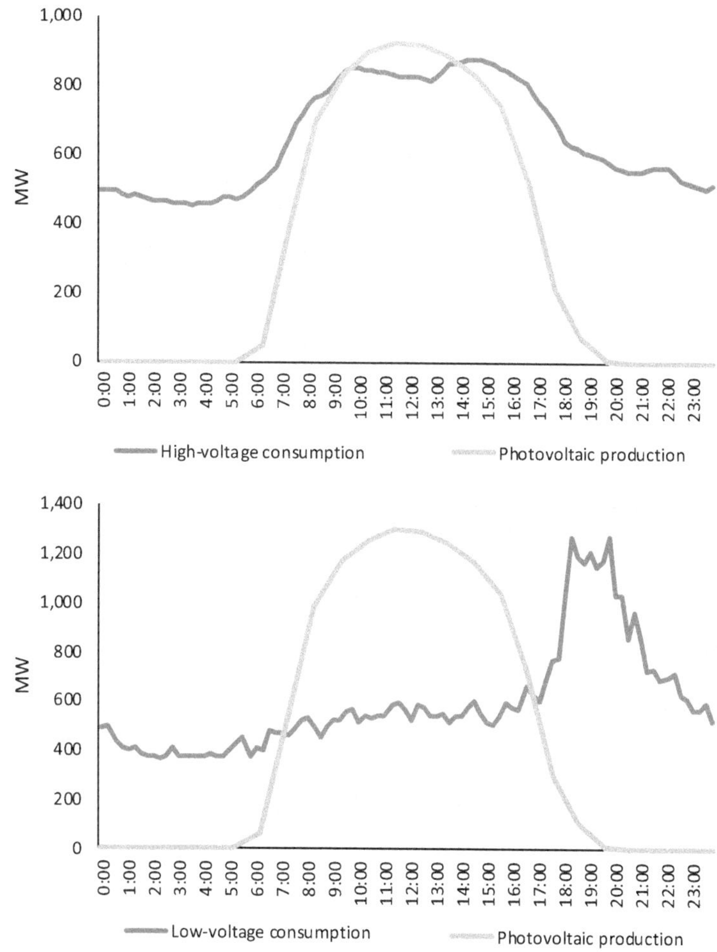

Figure 19.4 Comparison of solar production profile with high- and low-voltage customer load profiles.
Source: Aneel (2019)—CPFL Piratininga.

Alternatively, energy communities could provide storage or demand side flexibility to enable further deployment of solar generation. This would involve communal solutions utilizing both local and dispersed resources to meet the system's power needs (as discussed later when addressing the potential of aggregation services).

3.1.3 Consumers' revealed preferences

One of the strong points of the net metering program is its simplicity, which greatly facilitates consumer engagement. This empowers consumers to procure their own energy supply, providing consumers an alternative to the centralized rate-making process. This structurally alters the balance of power of the regulated electricity market.

Over the past few decades, taxes and surcharges have been the components that have most contributed to the elevation of electricity tariff rates in Brazil.[7] Affordability of electricity is a major concern for most consumers. The net metering program, in its current form, has provided an opportunity for consumers to avoid a significant portion of these surcharges and taxes. Thus, the program provides consumer with a backstop to resist the steady encroachment of taxes and surcharges.

Given the threat of a dwindling base, due to the consumers opting to have an ever-increasing share of the energy needs met by mini- and microgeneration in the net metering program, policymakers will have to contain their impetus to raise taxation and surcharges on electricity in the coming years.

3.1.4 Qualifying requirements for community supply

A close examination of the net metering regulation (*Resolução Normativa 482/2012*) reveals the regulator's concern in promoting consumer engagement, rather than just fostering other power service providers. The requirements to become a *community supplier* are stringent and include the following:

- First, a community supplier must be a prosumer (i.e., the micro- or minigeneration must be in the premises of a consumer; stand-alone generation is not allowed);
- Second, the prosumers' surplus energy cannot be sold to other consumers; it can only be shared with other coowning consumers united in a consortium or cooperative; and
- Third, the grouping of consumers in coownership of mini- or microgeneration must be formalized by the establishment of a consortium or cooperative.

These restrictions demonstrate the regulator's intent was not merely to promote distributed generation, but rather to have consumers actively procure their own power supply by directly investing in micro- and mini-generation.[8]

Although the regulation allows members of the consortium or cooperative to rent the generation assets, it specifies that rental payments and the division of the power produced must be made in fixed proportions (rather than by kWh consumed). The clear implication being that the regulator's intent is that the net metering program induces

[7] Electricity provides a wide and stable revenue base, thus becoming one of the main sources of revenues for the federal and state governments. Likewise, subsidies funded by surcharges on the electricity tariff are one of the easiest ways for congressional representatives to benefit constituents bypassing the Congressional Budgetary Process. Dozens of programs enacted by Congressional bills—such as the rural electrification, electricity inclusion, and electricity provision in the Isolated Systems, mentioned in the previous sections—are funded with surcharges on electricity tariffs.

[8] The legal requirements for the establishment and operation of consortiums and cooperatives in Brazil are not trivial. The formation of a consortium must be approved by a Trade Board (*Lei 6.404/1976*) or, in the case of financial consortiums, by the Brazilian Central Bank (*Lei 11.795/2008* and *Circular BACEN nº 3.433/2009*). The consortium's objectives must be specified as well as the rules concerning the rights and obligations of its members. Financial consortiums must also meet minimum capital requirements and present their business plan for regulatory approval. Cooperatives also must meet various requirements, such as have at least 20 members, each having equal voting power; constitute and maintain contingency reserves; and abide by complex rules for the exclusion or substitution of members (*Lei 5.764/1971*).

coownership of generation assets among consumers rather than just the purchase of electricity from a third-party provider.

Another interesting aspect of the regulation is that it allows condominiums to invest in micro- and minigeneration on site—in the form of *local collective self-consumption*—but not as *community supplier*. Again, this restriction demonstrates the regulator's concern of not having consumers just purchase electricity from another entity but to become effective investors in generation.

3.2 Consumer response to the net metering program

The installation of mini- and microgeneration distributed generation has grown exponentially in Brazil since the introduction of the net metering program. The cumulative average growth rate between 2012 and 2020 has been of 205% per annum.

As can be observed on Fig. 19.5, most of the generation has been in the *local production* category, followed by *local collective self-consumption*. However, many initiatives taking form at this moment suggest that in the coming years a large part of mini- and microgeneration investment will be provided by *community suppliers*, as discussed further ahead.

The overall majority of the mini- and microgeneration is derived from photovoltaic panels (97%). The other sources are thermal (2%), small hydro (1%), and wind (less than 1%). Investment in storage by prosumers in Brazil has so far been negligible.

Regarding the consumer class, most of the participants of the net metering program are low-voltage residential (39%) or commercial (37%) customers, followed by rural (14%) and industrial (9%) customers, as shown on Fig. 19.6.

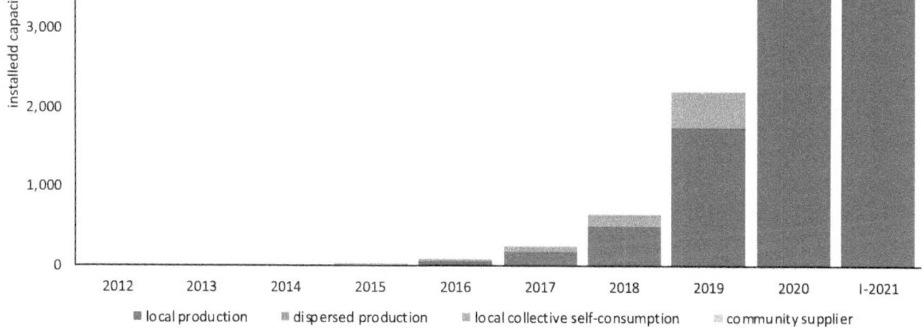

Figure 19.5 Installed capacity of mini- and microgeneration in Brazil.
Source: Aneel (2021a).

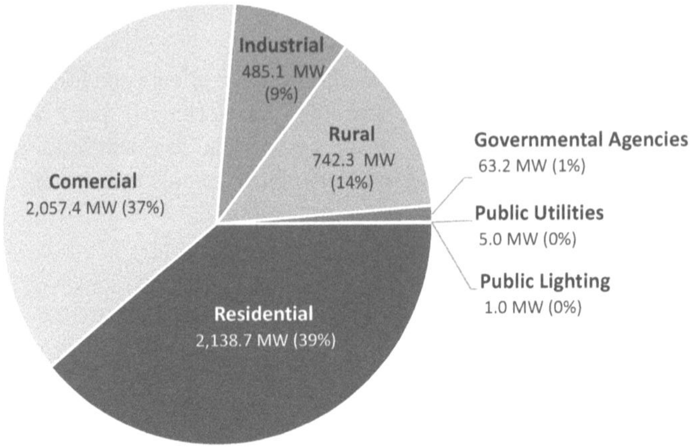

Figure 19.6 Participation in the net metering program by customer type.
Source: Aneel (2021a).

While most participants of the net metering program so far are from high-income households or businesses, emerging business models suggest that in the coming years *community suppliers* will play a much larger role and will make net metering accessible to a wider group of consumers.

One of the main barriers for the adoption of net metering is the lack of knowhow and capital to make the initial investment in generation. This void is gradually being overcome by small firms that enter the market providing the knowhow and capital.

Many service providers have entered the market in the last few years to help consumers engage in the opportunities provided by the net metering program.

Although the Brazilian regulatory framework only allows prosumers to engage in self-production, by grouping consumers in cooperatives or consortiums with distinct cost and risk sharing arrangements, various business models have emerged.

Utilizing the archetypes described in the chapter by Kolesar, one can identify at least four business models:

- Energy Cooperatives;
- Community Collective Generation;
- Third-Party Sponsored Communities; and
- Community Energy Service Companies.

Energy Cooperatives have existed for many decades in Brazil since the first rural electrification cooperatives established in the mid-20th century, but most recently constituted energy communities cannot be characterized as Energy Cooperatives. Even though many are legally characterized as cooperatives in the net metering program, they are better characterized by the other three business models because they are initiated and managed by third-party service providers.

Mecatron Solar (https://mecatronsolar.com.br/about-us/) is an example of a service provider for *Community Collective Generation*. The firm provides turnkey solutions for the planning, development, implementation, operation, and maintenance of photovoltaic generation.

Cosol (https://cosol.com.br) is best characterized as *Community Energy Service* provider. These are collaborative ventures in which the service provider assists consumers interested in participating in the net metering program but that do not have power generation to find prosumers with excess generating capacity to purchase or rent generating capacity. They are then organized as cooperatives or consortiums in the net metering program.

Alsol (https://www.alsolenergia.com.br/sobre/) is an example of a *Third-Party Sponsored Community* since it takes the initiative of constructing, operating, and maintaining the generating facilities on the premises of the prosumer to be rented to consumers Alsol enlists to participate in the program. In 2019 Alsol was acquired by Energisa, one of Brazil's major public utilities, which demonstrates that distributed generation and the net metering program have become mainstream.

The differences between the business models are subtle but significant. At one extreme are the Energy Communities in which most of the risks and benefits of implementing distributed energy resources are borne by the prosumers; at the other extreme are the Third-Party Sponsored Community, where most of the gain and the risks are absorbed by the service provider; and the Community Collective Generation and Community Energy Service are situated in between.

These firms have fomented the formation of energy communities, but they tend to target higher-income consumers that present a lower credit risk. Thus, participation in energy communities has remained unattainable for large portions of the Brazilian population.

However, there are initiatives by nongovernmental organizations and social entrepreneurs that provide the same type of services but targeted at lower-income consumers. Examples of these types of initiatives may be found in IDEAAS (*Instituto para o Desenvolvimento de Energias Alternativas e da Auto Sustentabilidade*) and PopLuz.

IDEAAS is a nongovernmental organization that seeks to promote social inclusion by implementation of renewable energy technologies. IDEAAS has promoted numerous programs aimed at meeting the specific needs of low-income communities.[9] IDEAAS also provides technical assistance (elaboration of business plans, development of prototypes, technical appraisals ...) to meet the particular needs of the community (www.ideaas.org.br).

[9] For example, the "Iluminate Your Life" program ("*Ilumine sua Vida*") which provides high-efficiency LED lighting of public spaces in low-income urban communities or rural communities without electricity, in which case solar generation is also included. The program seeks to foster community interaction by lighting plazas, school gyms, and community centers. Another example is the "Sun Lighthouse" program ("*Farol do Sol*") in a fishing village on the northeastern coast of Brazil, which consists of the installation of a system that collects, stores, and distributes energy from solar photovoltaic panels to special lamps that are used to attract shrimp to the fishing area. The system substitutes the gas lanterns that were previously used by the fisherman.

PopLuz can also be characterized as a Community Energy Service provider focusing in providing minigeneration that can be rented to lower-income households by means of energy consortiums. PopLuz estimates that there are approximately 28 million vulnerable households in Brazil that could be serviced in the net metering program, at a cost 20% lower than their current electricity bill. A key to success of the program is the composition of consumers in each consortium so as to mitigate overall credit risk (https://www.popluz.com.br). PopLuz is still in a preoperational phase and a pilot project is about to be implemented in the city of Belo Horizonte.

The emergence of organizations like these suggests that the benefits of energy community building can also reach poorer portions of the population.

4. The future of energy communities in Brazil and their impact on social welfare

As pointed by Verde and Rossetto (2020), energy communities can take on many forms. They can be delimited by location ("communities of place") or by economic, environmental, or relational interests ("communities of interest"). A common thread in most European energy communities are the environmental concerns.

In Brazil, the main motivation is economic, although it is difficult to isolate the environmental and economic motivations, since the most competitive small-scale generation alternatives are based on renewable sources.

While energy communities can play an important role in promoting the use of the most suitable distributed energy resources, the arena in which energy communities are likely to provide the largest contribution is in coordinating demand response. As consumers come to possess mini- and microgeneration, batteries, electric vehicles, and programmable electric appliances, they will be increasingly able to respond to the electric power system needs. However, this requires integration and coordination, a role specialist usually refers to *aggregation services*.

The three forms of aggregation services are as follows:

- *Local aggregation* in which the customers' distributed energy resources located at the same property are collectively managed to operate as "virtual plants" to meet their needs, render ancillary services, or sell energy in the wholesale market (these distributed energy resources include not only generators but also batteries and major electric appliances that can be remotely controlled, such as water heaters, heat pumps, and electric vehicles connected to their home recharging station);
- *Microgrids* in which a group of customers in particular area build and operate their own microgrid that is managed independently from the local utility grid; and
- *Dispersed aggregation* in which customers dispersed throughout the network share generation resources or collectively manage distributed energy resources.

These are roles energy communities could come to execute.

Fig. 19.7 shows the various ways consumers can engage in the provision of energy services, adding the various forms of aggregation services to those previously presented in Fig. 19.2.

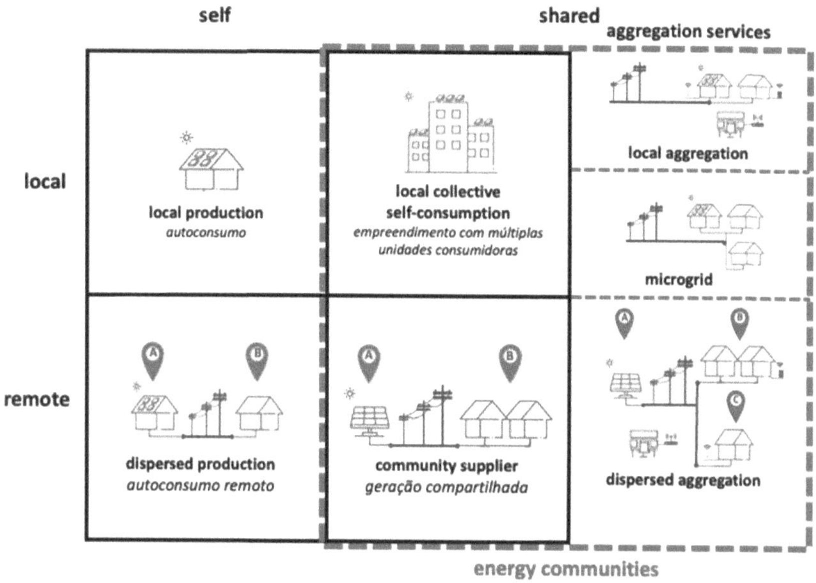

Figure 19.7 Forms of consumer engagement in the electric power sector.

The five categories encompassed in the middle and right columns (in the dotted-line box) are the forms of collective engagement that are of interest in this book, for these are the cases in which energy communities can play a major role. These are situations in which consumers share resources or coordinate the operation of their distributed energy resources for a common cause.

While Brazil has yet to implement the aggregation services on commercial scale, it is likely these types of services will be provided in the coming years as electricity pricing evolves to better reflect system conditions and regulations are passed to allow for such services. The possibility of rendering aggregation services expands the set of benefits energy communities can provide.

As one examines the initiatives arising from the net metering program, several questions come to mind based on the previous experiences that are discussed in the following paragraphs.

> *Will the net metering program result in effective consumer engagement in power-supply decision-making process or will energy continue to be managed by the public utilities and other third-party providers as in the case of Electricity Inclusion and Isolated System programs previously implemented in Brazil?*

The net metering program differs from those two previous programs in a fundamental way: it is consumer centric. The consumer is the one that must take the initiative to participate in the program. While third-party providers may offer services and know-how, it is the consumer that makes the decisions—and this is a key to the formation of energy communities.

Seeking to foster long-lasting energy communities, Aneel requires consumer groups to take the form of consortiums or cooperatives. While the aim of this requirement is to ensure governance, it also has become a significant hurdle to the establishment of energy communities.[10]

In spite of the regulatory requirements, however, most energy communities in Brazil have been initiated and administrated by third-party service providers, with little interaction between consumers. Over time it could be that these requirements will induce more interaction among consumers and lead them to actively engage in their own energy management, but so far it is still an open question if consumers will take on a central role in the deployment of distributed energy resources or if it will remain a space dominated by specialized service providers who often reap most of the program's benefits.

Will the energy communities supported by the net metering program be short-lived? Will they eventually be incorporated by the public utilities or other corporations further ahead as in the case of the rural electrification cooperatives?

While the possibility of future consolidation exists, this should not be seen as a failure of energy communities. The consumer awareness and learning achieved in the process of creating and operating energy communities will provide lasting benefits to consumers. The benefits of developing distributed energy resources will have a transformative impact that will forever change the electric power industry. Furthermore, just the mere possibility of forming energy communities has a powerful impact of providing an alternative to consumers not satisfied with their service providers.

How should policies be adjusted over time to avoid inefficient or distorting incentives?

The economic incentives provided by the net metering program in Brazil (as in many other countries) are currently much higher than the net welfare gain obtained from the deployment of distributed energy resources. As demonstrated in the chapter by Biggar and Hesamzadeh, the main motivation for consumers to invest in generation in the net metering programs is most often the reduced payments to the retailer and distributor in amounts that surpass the economies obtained from the deployment of distributed energy resources.

This mismatch of costs and benefits is due to the combination of net metering programs with retail rate structure that do not properly reflect marginal costs. Retail rates are typically based on time-averaged and geographically averaged pricing, which give rise to many opportunities for arbitrage by prosumers, which benefit those that join the program, but do not always improve social welfare.

[10] Recently, Aneel's attorney-general issued a legal opinion (*Parecer 00081/2021/PFANEEL/PGF/AGU*) stating that in addition to consortiums and cooperatives, the formation of "associations" would be an admissible form to consumer grouping to participate as community supplier in the net metering program. Since the regulations for the formation of associations are much simpler and flexible, this would significantly lower the barriers to the formation of energy communities.

Such distortions cannot be tolerated beyond the very early stages of development of energy communities. It is important to fine-tune the incentives to a level that results in sustainable growth of the industry and, in the long run do not surpass the positive externalities associated with the deployment of distributed energy resources.

The uncoordinated growth of distributed generation, together with already-contracted centralized generation expansion, may lead to supply outstripping demand, which would result in excessive capacity—a situation that would be detrimental to consumers.

The approach described in the chapter by Lo Schiavo et al. is the most prudent way to promote energy communities. Starting with a pilot program, policymakers can observe its effects prior to widespread adoption, which can avoid costly mistakes and frustration of investments due to policy reversals. Furthermore, promoting the program by means of explicit incentives based on estimated (or expected) systemic benefits, that can later be calibrated based on the results obtained, is much superior to the implicit cross-subsidies provided by the net metering programs.

What policy adjustments are needed to ensure effective and efficient distributed energy resource deployment?

The largest risk associated with the widespread use of distributed energy resources is possibly poor coordination between system expansion and operation. The antidote to this undesirable outcome is better network pricing. Tariffs need to evolve to better reflect system costs—both in temporal and spatial terms. Net metering program must consider the value of electricity in the various seasons of the year and times of day. It is also crucial that grid usage tariffs properly reflect grid conditions, providing price signals that induce investment in locations where distributed energy resources are most beneficial and discourage investment where it is least suitable. Only with proper price signals will consumers and energy communities be able to internalize system costs and benefits.

5. Conclusions

The development of energy communities is not easy, but the gains from combining and coordinating efforts can be substantial.

The deployment of distributed energy resources by consumers has the potential of reducing generation costs, reducing transmission and distribution losses, and reducing grid investment. However, if not properly coordinated, net metering programs could have the opposite effect.

Power systems present many economies of scale and scope, which make integration through grids very beneficial. Consumer loads are smoothed out as they are aggregated; requiring less capacity then would be required if the demand of each consumer were to be supplied autonomously. Integration also improves reliability, providing more alternatives in the case of contingencies.

Furthermore, the unbridled deployment of distributed energy resources can give rise to:

- Incompatible generation and consumption profiles;
- Distribution bottlenecks; and
- Excess investment in generation.

To avoid these pitfalls, it is necessary to calibrate incentives and adopt a rate structure that better reflects system marginal costs.

Currently, Brazil's net metering program does not meet these criteria, but regulatory changes are expected in the coming years to rectify its main distortions.

In addition to promoting the implementation of distributed generation, energy communities can also help consumers coordinate their actions to attain the full benefits of distributed energy resources. While these services can be managed by third-party service providers, such as public utilities or other specialized firms, energy communities involve another dimension which is also desirable: consumer engagement.

Getting citizens directly involved in the quest for solutions on how best to meet their needs can be enlightening and powerful. Successful energy communities not only help identify local energy resources and coordinate their deployment but also strengthen the community by forging a common set of vested interests among community members.

It is not a simple undertaking, as the African proverb states: "Alone a youth runs fast, with an elder slow, but together they go far" (Luo proverb).

References

Andrade, C., Rosa, L., Silva, N., 2011. Generation of electric energy in isolated rural communities in the Amazon region a proposal for autonomy and sustainability of the local populations. Renew. Sustain. Energy Rev. 15, 493–503.

Aneel, 2019. RTP CPFL Piratininga. https://www2.aneel.gov.br/aplicacoes/tarifa/arquivo/TR_v2.8-CPFL%20Piratininga_v2019%20pos%20AP.xlsx. (Accessed March 2021).

Aneel, 2020. Luz na Tarifa. https://app.powerbi.com/view?r=eyJrIjoiOTY0NWQzOGItMmQ3ZS00MWUzLTllNmMtNTA5NTYxODdhYTkzIiwidCI6IjQwZDZmOWI4LWVjYTctNDZhMi05MmQ0LWVhNGU5YzAxNzBlMSIsImMiOjR9. (Accessed March 2021).

Aneel, 2021a. Micro e Minigeração Distribuída. https://app.powerbi.com/view?r=eyJrIjoiZjM4NjM0OWYtN2IwZS00YjViLTllMjItN2E5MzBkN2ZlMzVkIiwidCI6IjQwZDZmOWI4LWVjYTctNDZhMi05MmQ0LWVhNGU5YzAxNzBlMSIsImMiOjR9. (Accessed April 2021).

Aneel, 2022. Sigel. http://sigel.aneel.gov.br/portal/home/index.html. (Accessed January 2022).

Estévez, G., 2014. Distributed energy resources as a solution for universal energy access in Latin America. IEEE Smart Grid Newslett. (August, 2014).

Freitas, G., Silveira, S., 2015. Programa Luz para Todos: uma Representação da Teoria do Programa por Meio do Modelo Lógico. Planej. Polit. Publicas 45, 177–198.

International Co-operative Alliance, 2015. Guidance Notes to the Co-operative Principles. Coop, London.

Lucas, H., Gómez, J., 2017. Net-metering Mechanism for the Promotion of the Decentralized Electricity Generation from Renewable Sources in Latin America and the Caribbean (Funded by the Global Environment Facility). Factor, Bilbao.

Matiello, S., Cerri, F., Pagani, C., Moret, A., Cemin, A., 2018. Energia e desenvolvimento: alternativas energéticas para comunidades isoladas da Amazônia. Revista Presença Geográfica VII (1), 11–21.

Mejdalani, A., Chueca, J., Soto, D., Ji, Y., Hallack, M., 2018. Implementing Net Metering Policies in Latin America and the Caribbean: Design, Incentives and Best Practices. Inter-American Development Bank, Washington.

Munaretto, L., 2015. Um estudo relativo ao processo histórico das Cooperativas de Eletrificação: O caso do Brasil e Estados Unidos. Revista de Gestão e Organização Cooperativas 2 (3), 83–96.

Nascimento, B., Catelan, T., Chaves, G., Celeste, W., 2019. Evaluation of the viability of implementation of renewable hybrid systems for energy access in the Amazon region. Res. Soc. Dev. 8 (10), e448101415.

Neto, J., Costa Junior, C., Bitar, S., Junior, W., 2011. Forecasting of energy and diesel consumption and the cost fo energy production in isolated electrical systems in the Amazon suing a fuzzification process in time series models. Energy Pol. 39 (9), 4947–4955.

Oliveira, L., 2001. Perspectivas para a Eletrificação Rural no Novo Cenário Econômico-Institucional do Setor Elétrico Brasileiro (Doctoral Thesis). COPPE - Universidade Federal do Rio de Janeiro, Rio de Janeiro.

ONS, 2019. Plano Anual da Operação Energética dos Sistemas Isolados para 2020 – PEN SISOL 2020. DPL-REL-0248/2019 Operador Nacional do Sistema Elétrico, Brasília.

Pelegrini, M., 2003. A Regulação das Cooperativas de Eletrificação Rural (Doctoral Thesis). Escola Politécnica - Universidade de São Paulo, São Paulo.

Ribeiro, F., Teixeira, G., Soares, S., 2017. Investimentos em Infraestrutura e Desenvolvimento Socioeconômico: o Caso do Programa Luz para Todos. Planej. Polit. Publicas 49, 321–356.

Souto, C., Loureiro, G., 1999. O novo modelo do setor elétrico brasileiro e as cooperativas de eletrificação rural. First Edition. Porto Alegre: Livraria do Advogado Editora, 164p.

Vásquez-Leon, M., Burke, B., Finan, T., 2017. Cooperatives, Grassroots Development, and Social Change: Experiences from Rural Latin America. University of Arizona Press.

Verde, S., Rossetto, N., 2020. The Future of Renewable Energy Communities in the EU: An Investigation at the Time of the Clean Energy Package (Research Report). European University Institute, Florence.

Further reading

Aneel, 1999. Resolução Normativa 333/1999 (*REN 333/1999)*. (Rural Cooperatives, with Revisions Introduced by *REN 12/2002*). Agência Nacional de Energia Elétrica, Brasília.

Aneel, 2012a. *Resolução Normativa 482/2012*. (Net Metering Program, with Revisions Introduced by *REN 517/2012, 687/2015,* and *786/2017*). Agência Nacional de Energia Elétrica, Brasília.

Aneel, 2012b. *Resolução Normativa 493/2012*. (Alternative Service Alternatives for Isolated Systems). Agência Nacional de Energia Elétrica, Brasília.

Aneel, 2021b. *Parecer 00081/2021/PFANEEL/PGF/AGU* (Legal Opinion on Admissible Forms of Consumer Grouping to Engage in Community Supply). Advocacia-Geral da União - Procuradoria Federal junto à Aneel, Brasília.

Brazil, 1962. Decreto 1.033/1962. (Rural Electricification Cooperatives). Brasília: Casa Civil - Presidência de República.
Brazil, 1964. *Lei 4.504.* (Land Statute). Casa Civil - Presidência de República, Brasília.
Brazil, 1968. Decreto 62.655/1968. (Rural Electrification services). Brasília: Casa Civil - Presidência de República.
Brazil, 1971. *Lei 5.764* (National Policy for Cooperatives). Casa Civil - Presidência de República, Brasília.
Brazil, 1976. *Lei 6.404* (Joint-Stock Corportation). Casa Civil - Presidência de República, Brasília.
Brazil, 1993. Lei 8.631 (Cross-Subsidy Program to Promote Power Supply in Isolated Systems). Brasília: Casa Civil - Presidência de República.
Brazil, 1999. Decreto 2/1999. ('Luz no Campo' Program). Brasília: Casa Civil - Presidência de República.
Brazil, 2008. *Lei 11.795* (Consortiums). Casa Civil - Presidência de República, Brasília.
Brazil, 2009. *Lei 12.111* (Power Provision in Isolated Systems). Casa Civil - Presidência de República, Brasília.
Brazil, 2011. Decreto *7.520/2011* (with revisions introduced by *Decretos 7.560/2011, 8.387/2014, 9.537/2018*). (*'Luz para Todos'* Program). Casa Civil - Presidência de República, Brasília.
Central Bank of Brazil, 2009. Circular BACEN Nº 3.433/2009 (Regulation of Financial Consortiums). Central Bank of Brazil, Brasília.

Part Four

Case studies and implementation

Institutional and policy context of energy communities in France and Italy: how to increase the welfare-enhancing capacity of the sector

20

Maksym Koltunov[1], *Valentina-Miriam Cittati*[2] *and Adriano Bisello*[2]
[1]University of Trieste, Trieste, Italy; [2]Eurac Research, Institute for Renewable Energy, Bolzano/Bozen, Italy

1. Introduction

Climate change and economic inequalities induce policymakers to take action and look for new models of interaction that could be more efficient for the utilization of available technological advancements in the energy sector. In this context, intercountry analysis enables a deeper understanding of the models behind energy communities (ECs). Understanding the background of existing ECs, in turn, supplies fertile ground for innovation.

Institutional relationships are usually anchored in history, and they cannot be ignored in the analysis of the energy transition (Poupeau, 2020). The institutional context of the markets where ECs operate influences either ECs' deployment or their absence, as well as the choice of the business model. Today, ECs have many organizational forms (Moroni et al., 2019a) whose main purpose is providing people with energy-related services (Koltunov and Bisello, 2021). Beyond institutional forces, policy support determined by political will is another factor explaining the recent development of ECs. A survey from 2014 (Klagge et al., 2016) showed that nearly 80% of all regional ECs in Germany and more than 80% of supraregional ECs relied on the Feed-in-Tariffs (FiTs) policy for their revenue stream. For France and Italy, such incentive schemes have also been a decisive factor. Nevertheless, in many cases, the institutions interested in renewables and more precisely distributed energy lobbied for support schemes. Such institutions were already involved with or saw the potential of ECs. An interplay of market forces and state support has determined the past of, and will continue to determine the future of ECs.

ECs represent a unique opportunity for enhancing community welfare because in most cases returns stay in the locality. A French study conducted by Énergie Partagée (2019) showed a ratio of 1€: 2.57€ for equity invested in a project and the value returning to the territory within 20 years. In addition to local economic benefits, community ownership and decision-making in ECs significantly enhance local democracy and

Energy Communities. https://doi.org/10.1016/B978-0-323-91135-1.00007-9

acceptance of new green technologies (Brummer, 2018; Devine-Wright, 2005; Rosanvallon, 2011; Süsser et al., 2017; Yildiz et al., 2015). Moreover, profits from energy savings in buildings can be redirected to additional actions (e.g., shared e-mobility schemes). Community members are more likely than the regional/national authorities to support such a local circuit of investment–improvement–investment because it goes against the traditional energy producers' lobby. Also, social aspects, such as having higher citizen support for projects promoted by community members themselves rather than by external institutions, mobilize citizen endeavors and local finance for the energy transition (Koltunov and Bisello, 2021). This chapter concentrates primarily on the welfare gains of shareholders by answering the following questions:

- Which factors enhance the welfare-enhancing capacity of ECs?
- What are the main benefits ECs bring?
- How should the institutional and policy context evolve to foster ECs' further deployment?

Other chapters of this volume, including the chapters by Robinson and Del Guayo as well as Biggar and Hesamzadeh, discuss in more detail how ECs could also enhance the welfare of the broader community, and not only of shareholders.

The chapter is organized as follows:

- Section 2 presents Italian and French ECs' movements;
- Section 3 describes benefits that three case studies in France and Italy bring to the respective communities; and
- Section 4 discusses the main findings followed by the chapter's conclusions.

2. The energy communities' movements in France and Italy

This section analyzes French and Italian experiences starting with statistics about the ECs' deployment and typology, followed by a description of the specific factors (state and market institutions, policies, civil society, and research centers) that influence the welfare-enhancing capacity of ECs.

2.1 France

France has a highly centralized energy sector, with nuclear energy playing a major role. However, since the 2000s, decentralization of the energy system has started to emerge, mostly in response to environmental concerns and the antinuclear movement that arose after the Fukushima disaster. Consequently, France plans to reduce the share of nuclear energy in its electricity mix from 70% to 50% by 2035 and to close its last coal plants by 2022 (IEA, 2021). Today in the country, there are 215 ECs: 118 already operating, 73 under development, and 24 emerging.[1] As Fig. 20.1 shows, most of them

[1] According to Energee Partagee SOURCE: https://energie-partagee.org/.

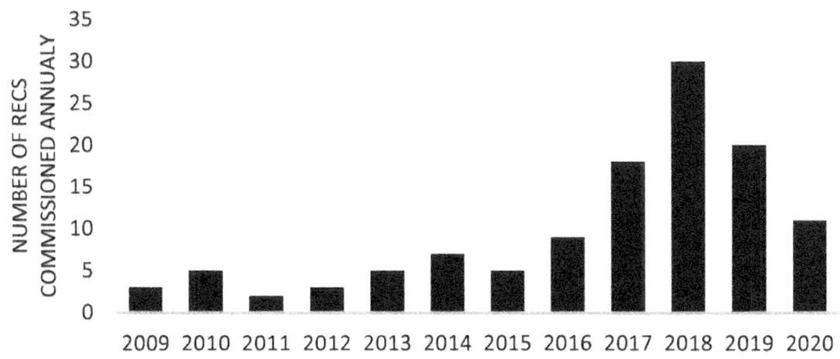

Figure 20.1 Annual deployment of ECs in France.
Source: Own illustration, based on data from EPA.

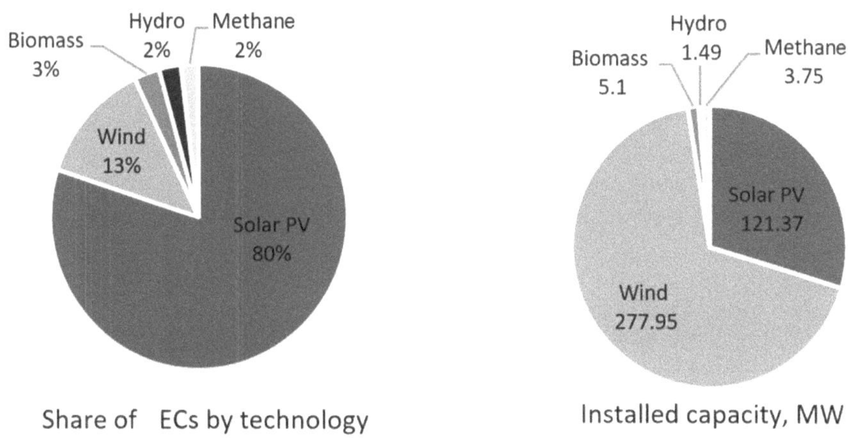

Figure 20.2 ECs by technology and installed capacity in France.
Source: Own illustration, based on data from EPA.

have appeared after 2016, while Fig. 20.2 compares the share of ECs by utilized technology or installed capacity. Sebi and Vernay (2020) classified French ECs into four types:

- The first type, "citizen PV clusters," are small individual solar rooftop projects that have been grouped into clusters;
- The second includes wind farms, microscale wood plants, and big PV projects with installed capacities from 2 to 18 MW;
- The third consists of mini-hydro projects and small biogas plants; and
- The fourth consists of big projects, such as large wind and solar PV farms or wood-chip district heating with large generation capacities.

Public and tertiary sectors have a pivotal role for ECs in France. The main public players are EDF,[2] RTE,[3] ENEDIS,[4] DGEC,[5] CRE,[6] and ADEME.[7] EDF manages all nuclear power plants and most fossil fuel and hydropower plants, together with some large renewable facilities. ECs are obliged to sell their energy to EDF or, more recently and following highly constrained criteria, Enercoop (Sebi and Vernay, 2020). Although the distribution systems belong to local authorities, 95% are subject to concessions managed by ENEDIS, with the remainder managed by local distribution companies. DGEC is responsible for implementing the government policy on energy according to the framework defined by EU directives, in particular, for the liberalization of the energy market, energy efficiency, development of renewable energy, and climate change issues. For instance, DGEC is responsible for transposing EU legislation on "energy communities" into French law and defining the incentive schemes for renewables. Among the public actors, ADEME, which covers a broad spectrum of environmental policies, is the one that supports ECs the most and has done so from the very beginning. For instance, it helped create EPA, which is a national association of ECs. ADEME provides funding for regional ECs' associations and awareness-raising campaigns; indeed, it supported the establishment of the majority of regional citizen energy associations. In 2018, together with an investment company, a cooperative bank, and a pension fund sensitive to environmental issues and solidarity economy, ADEME was also a founding partner of EnRcit, which provides loans under very preferential conditions exclusively to ECs (EnRciT, 2021). Pivotal tertiary actors in the citizen energy sector are Enercoop, EPA[8] with its subsidiaries, and Centrales Villageoises.

Enercoop was founded in 2005 by environmental and solidarity economy NGOs (Enercoop, 2021a); it is a supplier of renewable energy and a cooperative of cooperatives that includes 11 regional subcooperatives. It has 100,000 customers, 50,000 of which are members, namely, individuals, communities, and businesses, of the cooperative. Overall, Enercoop purchases electricity from 300 renewable energy producers (Enercoop, 2021b), but half of Enercoop's electricity comes from just 48 ECs (Énergie Partagée, 2021). Enercoop works very closely with the ECs and, besides purchasing their electricity, supports ECs at all stages of development. Since its inception, the cooperative created several energy-transition actors by itself, including EPA, which is a central actor in ECs' movement in France. EPA unites all ECs and has even spun off 3 regional ECs' associations out of 10 in total. Organizationally, it was established as the cooperative structure having as members its founding NGOs and all

[2] Électricité de France—dominant energy generator and utility and the former state monopoly.

[3] Réseau de Transport d'Électricité—Transmission System Operator.

[4] EDF's subsidiary, Distribution System Operator.

[5] Direction Générale de l'Énergie et du Climat—The Directorate General for Energy and Climate, Ministry for Ecological Transition.

[6] Commission de régulation de l'énergie—the national regulatory authority that delivers market regulation, tariffs, and promotes competition.

[7] Agence de la transition écologique—Environment and Energy Management Agency.

[8] Energie Partagée—National RECs' Association.

citizens participating in related ECs. The cooperative spun off EPI[9] and several other subsidiaries, such as EPS[10], a research company in the field of citizen energy, and EPO,[11] which delivers management services to ECs. They provide ECs with services such as legal structuring of projects, economic planning, crowdfunding tools, consultation, mobilization campaigns, territorial animation services, and partnership tools. Conceptual relationships among them are shown by Fig. 20.3.

Simply said, it is on the one hand a French one-stop shop for activists willing to set up an EC, but on the other hand, also an investment company. It has already intervened in 74 EC projects becoming their shareholder and simultaneously providing services to them.[12] It delivers a crowdfunding campaign, promising, on average, 4% annual gross

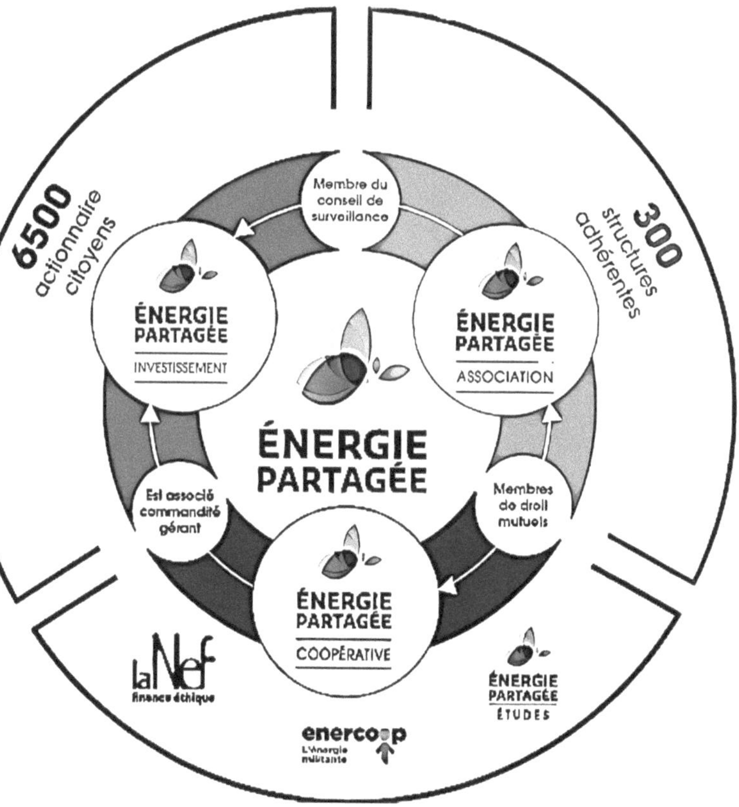

Figure 20.3 Conceptual relationships among EPA and its subsidiaries.
Source: https://energie-partagee.org.

[9] Énergie Partagée Investissement.
[10] Énergie Partagée Études.
[11] Énergie Partagée Exploitation.
[12] Data from EPA website: https://energie-partagee.org/.

profit for future shareholders (Énergie Partagée, 2019). EPI's operating income comes from new members' subscription fees, dividends from shares in ECs, and income from the organizations in which EPI is a shareholder. Subscription fees cover management expenses of EPI/EPA, whereas dividends are largely reinvested into further EC projects (Énergie Partagée, 2019). Such a scheme enables a virtuous cycle of ECs' deployment.

Other important players in France are the "Centrales Villageoises" (Village Centers in English), which are local companies with citizen governance that carry out projects to support the energy transition. The organization was established in 2010 by several regional natural parks in the Rhône-Alpes. Their ECs are distributed mostly in eastern and south-eastern France, and fall under the type "citizen PV clusters," according to Sebi and Vernay (2020). All three key tertiary actors, Enercoop, EPA, and Village Centers, as well as ADEME from the public sector, are actively lobbying state authorities for the interests of ECs.

Looking at the financial side, the main policy incentive schemes in France applicable to ECs are FiTs and "participatory bonuses." FiTs in France are fixed for 20 years, although for new installations its granting has been highly restricted since 2016.[13] For instance, only solar PV on roofs with an installed capacity of less than 100 kW or biogas and hydropower installations with less than 100 kW capacity are eligible for FiT.[14] Participatory bonus is a tool similar to Feed-in-Premium that characterizes citizen energy projects. It was offered by DGEC in 2015, awarding between 1€ and 5€ per MWh to a project participating in auctions.[15] The potential project must meet a given threshold of participatory investments for a minimum period of 3 years: 40% of equity or 40% of total financing must be raised jointly or separately from 20 investors, who could be individuals or local authorities or groups of communities. Individual investors must reside near the project or adjacent department, and the bonus was promised for 20 years. In 2020, the Directorate General for Competition of the European Commission advised that the support system of participatory bonus could no longer be maintained as it had been, so, therefore, a revised form of modified incentive is still a topic of discussion.[16]

Preferential taxes for renewable energy producers and Feed-in-Premiums for high-capacity installations are also available in France, and local incentives coming from the regions supplement the national schemes, e.g., Region Occitanie offers an economic multiplier of 1€ per 1€ of citizen-invested funds (Peullemeulle and Duval, 2017), while the Auvergne Rhône-Alps Region supports selected renewable energy projects up to 30% of the total investment fund. The summary of policy schemes benefiting ECs in France is reported in Table 20.1.

[13] The FiT for solar rooftop fluctuates, but stands at 18 c€/kw, approximately.

[14] More details about the complete description of limitations and criteria, including wind installations restrictions, can be found in Vidalic (2019).

[15] Law No 2015-992 of 17 August 2015 on "energy transition for green growth" amended the Energy Code with the new Article L.314-18.

[16] Data from EPA website: https://energie-partagee.org/.

Table 20.1 Summary of policy schemes benefiting ECs in France.

Incentive scheme	Period of use	Type of EC benefiting (incl. according to Vernay and Sebi)	Level of actual benefiting
FiT (Tarif d'achat)	2005–16–Ongoing	Type 1, Type 3 (small installations)	High
Participatory bonus (bonus participatif)	2015–21	All ECs (differently according to the installation type and share of participatory capital)	High
Simplification of legal obstacles for ECs' setting up (code de l'Énergie)	2015–Ongoing	All ECs	High
Feed-in-premium (complément de rémunération par guichet ouvert)	2017–Ongoing	Type 2, Type 4 (plants that generate power beyond the threshold eligibility of FIT capacities)	Low
Reduced VAT tax	2000–Ongoing	All ECs	Low
Income tax deduction (CITE) since 2019, it was highly modified	2005–Ongoing	All ECs	Low
Enabling of energy sales to utilities different from EDF (code de l'Énergie)	2015–Ongoing	Supplier EC (only Enercoop got the accreditation so far)	Low
Loan (Fonds chaleur)	2009–Ongoing	Type 4 (only biomass installations)	Low

Overall, 215 French ECs had 24,627[17] citizen members in 2020 and produced electricity and thermal energy equivalent to the electricity consumption of almost 700,000 people and heat for almost 13,000 people.[18] In Fig. 20.1, the rapid growth of ECs since 2015 and the decline of newly commissioned ECs starting from 2019 are depicted, with further details in Sebi et al. (2019, 2021).

2.2 *Italy*

Energy cooperatives in Italy started to appear as early as the beginning of the 20th century in the Alpine regions of Trentino–South Tyrol and Lombardy. These historical

[17] Of which 5000 citizens are members of the Centrales Villageoises network.

[18] Approximate number based respectively on 1172 kWh of electricity consumption and 4.5 MWh of heat consumption per person per year. Data from the EPA Centrales Villageoises annual reports.

ECs usually appeared in remote valley towns and villages, typically including all local inhabitants as members. Due to historical reasons and geographical constraints, they are still not obliged to be unbundled as the current national law generally requires, and they usually own generation facilities and distribution systems simultaneously, supplying locally generated energy, mainly from renewables, to their members.

New ECs in the non-Alpine part of Italy have emerged mostly since 2010, as a result of raising the environmental awareness of citizens and the development of business models as opposed to commercial energy providers, sustained by incentive schemes like FiT. There are currently 76 ECs in Italy with 47,309 members,[19] 65 of the ECs considered as historical ECs. Historical ECs are mostly organized as cooperatives with members living in the same municipality or nearby valley. They mostly utilize hydropower and biomass for their installations. On the contrary, members of new ECs do not always live in the same municipality, meaning that they are united on the premise of common interest rather than geographical proximity. New ECs utilize mostly PV solar energy and are organized as innovative business models (Moroni et al., 2019b). There are some new ECs that, in addition to generation, also supply energy-efficiency consultancy or provide consumption-related services (Koltunov, 2019), mostly operating on a national scale (Candelise and Ruggieri, 2020).

Key state authorities influencing ECs' institutional context in Italy are shown in Fig. 20.4 and further described hereafter. ARERA is an independent regulatory authority that is responsible for the energy sector and also for water supply and waste management.

Another important institutional player is the Ministry of Economy and Finance. It established GSE[20] that, in turn, functions following guidelines given by the Ministry of

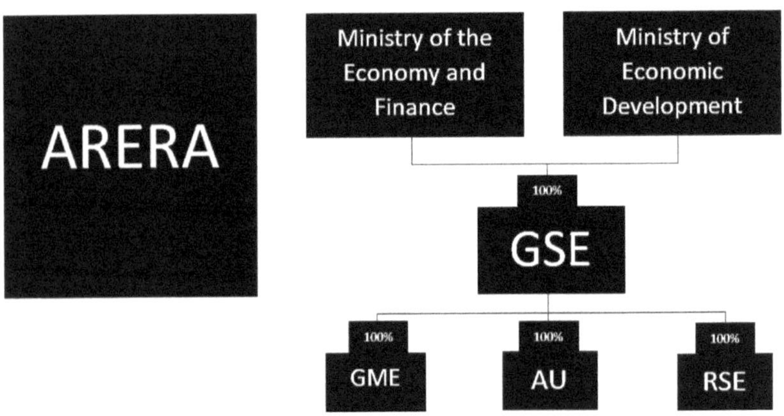

Figure 20.4 Key state authorities responsible for the energy sector in Italy.
Source: Own illustration.

[19] Key figures about Italian RECs from Koltunov (2019) adapted for 2021.
[20] Gestore dei Servizi Energetici.

Economic Development. GSE is a publicly owned company that plays a central role in the promotion and development of renewable energy sources in Italy, including through the provision of economic incentives. GSE implements policy-incentive mechanisms, qualification of plants powered by renewable energy, plant inspections, and forecasting and monitoring of the power fed into the grid. Moreover, it established three other public companies that supplement its activities (GME,[21] AU,[22] RSE[23]), all having responsibilities for the public interest in the energy sector.

Finally, there is TERNA, a joint-stock company owning the Italian national transmission grid for high- and extrahigh-voltage power. TERNA, with a few exceptions, has the monopoly granted by governmental concession as the transmission system operator.

The most visible private players are retail suppliers that usually also own distribution companies (DSOs). Among 410 active retailers in Italy, the market is dominated by ENEL, Edison, HERA, and ENI.[24] The top three cover about 50% of overall sales (Koltunov, 2019), and the biggest, namely ENEL, has been a public body since 1992.[25] Several new ECs signed agreements with retailers for energy delivery to the ECs' members on beneficial terms allow rebates for energy bills and a certified supply of 100% renewable energy (Koltunov, 2019). For instance, EC Energia Positiva (see Section 3.3) signed an agreement with Dolomiti Energia. The chapter by Del Pizzo et al. addresses in detail the Italian perspectives of DSOs' integration with ECs. Another remarkable private player for many Italian new ECs is the cooperative bank "Banca Popolare Etica" that, in accordance with its ethical–finance principles, has funded the installation of several generation plants.

A unique regional player with whom South Tyrolean alpine ECs tightly cooperates is SEV,[26] a "cooperative of cooperatives" having more than 300 members encompassing energy cooperatives, grid operators, municipalities, producers, etc. SEV acts as a service provider for all members belonging to a specific geopolitical context, including historical ECs but also some new ones (Koltunov, 2019), by taking care of activities such as lobbying, billing, provision of technical systems for grid operation, and trading (SEV, 2019).

For new ECs, private companies providing technical and managerial solutions for shared energy, smart grids, and energy efficiency are important market factors influencing their developmental opportunities. An example is the Italian company Regalgrid Srl, which created a platform for energy monitoring and sharing that enables the inclusion of consumers and prosumers into a smart grid of interconnected buildings (Moroni et al., 2019a), turning them into prosumagers (Sioshansi, 2019). Similar projects are described in detail in the chapter by Del Pizzo et al. Such companies are expected to flourish with the transposition of EU Directives (IEM, 2019; RED-II, 2018)

[21] Gestore dei Mercati Energetici.
[22] Acquirente Unico.
[23] Ricerca sul sistema energetico—Public research institution for energy systems in Italy.
[24] Data source: Annual report of ARERA 2018.
[25] Today, 23% of the company's shares are owned by the government.
[26] South Tyrolean Energy Association.

to Italian law, which partially started in 2020 with the "Milleproroghe" and "Relaunch" decrees,[27] as they technically enable what the new legislation allows. In fact, the Ministry of Economic Development currently promotes ECs organized around a block of flats or buildings connected to the same substation, pursuing a collective self-consumption scheme; the chapter by Lo Schiavo et al. contains detailed information on the most recent Italian regulation and provides an estimate of sector growth. New laws also increase the role of local authorities in joining citizen-led initiatives or acting as initiators, thus opening the energy market to new actors. An example of such an innovative approach is provided by the small municipality of Magliano Alpi, in the alpine area belonging to the Province of Cuneo in the Piedmont Region, which is an EC founder and simultaneously a member. After the successful launch of this local pioneering initiative, the mayor announced the intention to expand the model also to neighboring municipalities (Matalucci, 2021). Several other municipalities in the Piedmont region declared their interest in establishing ECs, and actors from the municipality of Pinerolo in the Province of Turin and Valle Maira in the Province of Cuneo are currently carrying their two EC projects forward. Such fertile ground was prepared by the local regulatory framework: Piedmont was the first Italian region to transpose the EU Directive RED-II into the regional law 12/2018 (Magnani and Cittati, 2022).

The main civil society actors that urged the Italian government to implement the EU Directive RED-II were NGOs ITALIASOLARE and Legambiente. Moreover, environmental NGO "Legambiente" even established the "Renewable Municipalities Association." Looking at the academic or national research centers engaged in such ambitions, the most prominent example is the RSE endeavors, active since 2018 (Zulianello et al., 2020). Also, the Energy Center Lab of the Polytechnic University of Turin recently started to realize projects in this field, not only from a theoretical perspective but also by becoming involved as the technical coordinator of the EC in Magliano Alpi (Energy Center, 2021).

As shown in Table 20.2 and explained in the following paragraphs, several policy schemes are benefiting ECs; a few of them are explicitly designed to sustain new ECs (e.g., the latest FiT dedicated to ECs), while others are anchored to specific renewable energy systems or investments, thus applicable to old or new ECs according to specific and detailed criteria.

Recent Italian regulation[28] allows energy sharing under two configurations of "jointly acting self-consumers" and "renewable energy communities," simplifying the joint management of plants up to 200 kWp. The first configuration requires all the participants (at least two) to be in the same building, while the second requires connection to the same low-voltage transformer substation. This paves the way for the emergence of what Moroni et al. (2019a) call "place-based" ECs, which are communities composed by members belonging to a specific place, similar to the French type-1 "citizen PV clusters."

[27] Milleproroghe decree (Law 8/2020) and Relaunch decree (16 September 2020).

[28] The already mentioned "Milleproroghe" and "Rilancio" decrees.

Table 20.2 Summary of policy schemes benefiting ECs in Italy.

Incentive scheme	Year of use	Type of EC benefiting	Level of actual benefiting
FiT (ECs' incentive tariff)	2020–Ongoing	New ECs	Expected high
IRPEF (income tax) rebate for installed rooftop PV during renovation (Eco-bonus—superbonus)	2007 to 20–Ongoing	New ECs	Expected high
System costs exemption	2008–Ongoing	Historical ECs	High
Exemption from unbundling	2016–Ongoing	Historical ECs	High
Simplified energy sale and purchase regime (ritiro dedicato)	2007–Ongoing	Historical and new ECs	High
IRPEF (income tax) rebate for innovative start-ups	2017–Ongoing	New ECs	High
FiT (conto Energia)	2005–13	Historical and new ECs	High
FiT (Tariffa onnicomprensiva[a])	2007–16	Historical and new ECs	High
Green certificates	2002–16	Historical and new ECs	High
Environmental regulations on manure	2001–Ongoing	New ECs in South Tyrol	Low
FiP (GRIN)	2016–Ongoing	Historical and new ECs	Low
Exemption of nonhousehold users from the state tax on electricity	No data–Ongoing	Historical and new ECs	Low
Exemption from the carbon tax on heat energy	2000–Ongoing	Historical and new ECs	Low

[a]In contrast with Conto Energia, this incentive applies only to small plants with a capacity less than 1 MW (200 kW for wind farms) and for a period of 15 years, rather than 20.
Source: Own calculations, based on data from Koltunov, M., 2019. The Impact of Renewable Energy Cooperatives on the Welfare of Local Communities. Corvinus University of Budapest, updated to the year 2021.

In addition, this regulation awards the ECs a 20-year incentive of 11 c€/kWh for self-produced and immediately used or stored energy and grants 5 cent€/kWh to the self-produced energy sent to the grid.[29] With the average cost for domestic

[29] The "Rilancio" decree increases the previous "ecobonus" income tax rebate from 50% to a "super bonus" of 110%, allowing homeowners to compensate for expenses related to PV and storage systems installation through a tax credit of 110% within 5 years (up to 48,000€ of investment). This credit may be also transferred to a financial institution in exchange for cash or to the construction company doing the installation, by receiving a discount of 100% of the full cost (Bisello, 2020).

end-users in Italy, which in 2020 stands around 21.5 c€/kWh, new incentive tariffs for ECs allow a considerable rebate, «cash-back», in the electricity bills of its members. A similar narrative on the Italian context may be found in the chapters by Lo Schiavo et al. and Del Pizzo et al.

The "green certificates mechanism" is based on the mandatory rule introduced in 2002 by the Italian Government under which all producers and importers of electricity must annually ensure an increasing minimum share of electricity in the grid produced by RES. Although this mechanism was terminated in 2016, some energy producers still benefit from their sale on the market or to the GSE because green certificates can be traded and used for 20 years from the date of issue. A similar situation applies to power plants built before 2013 (termination of the FiT called "Conto Energia") but still receiving the FiT incentive. For example, Energia Positiva, WeForGreen, and ènostra ECs benefit from such incentives because most of their plants were acquired on the secondary market (Candelise and Ruggieri, 2020; Koltunov, 2019).

Regarding the historical ECs, the state government granted them an exemption from the system costs component in the members' final bills in 2008, as well as an exemption from the state tax on electricity component, but only for nonhousehold members. Also in 2016–2017, a policy granted an exemption from the unbundling requirement for the companies supplying less than 25,000 points of delivery. Historical ECs meet that condition. Both the exemption from the system costs component and the exemption from the unbundling make historical ECs activities feasible and advantageous for their members (Koltunov, 2019). On the contrary, members of those new ECs recognized by law as innovative start-ups can benefit from an income tax rebate equal to 50% of their investment in shares.

Assuming the same calculation metrics used in Section 2.1, the result was that 76 Italian ECs, having more than 47,000 citizens in 2020 as members, produced electricity and thermal energy equivalent to the electricity consumption of more than 420,000[30] people and heat for 64,000 people. This means that members number roughly twice as many in Italy as in France, while the electricity generation is just over half and the thermal energy production is almost fivefold, which is therefore a quite different development with a more fragmented ownership structure and a preference for thermal energy.

Termination of previous FiTs has suspended EC deployment in Italy in recent years (Candelise and Ruggieri, 2020). With the transitional provision of EU Directive RED-II in June 2021, Italy hopes to revitalize the sector and engage new players, especially in electricity generation and sharing.

3. The case studies: ECs from France and Italy

Three representative ECs are highlighted in this section, focusing on the business model of each followed by an examination of the factors that influence the benefits.

[30] Three historical ECs are excluded due to the unavailability of data.

These case studies, one French and two Italian, are based on heterogeneous business models and have been chosen to show the benefits that each offers.

3.1 VercorSoleiL (France)

VercorSoleiL is a simplified joint-stock company founded in 2015 by 17 inhabitants from the "Communauté de commune Royans-Vercors"[31] in France, belonging to the Centrales Villageoises network. The original motivation of its members was to manage private funds, contracts, incentives, and bank loans for a viable energy plan. From 2015 to 2020, the company installed 25 PV plants on public and private housing, reaching an energy production of around 300 MWh/year and investing more than 800,000€. In 2020, it also implemented an electric car-sharing service.[32]

Today, the company has 124 shareholders and follows the "one person—one vote" rule without any limitation on the number of shares that each member decides to acquire. In addition, the "Communauté de commune Royans-Vercors" became the company's principal shareholder, whose role has been fundamental in boosting the activity in an early stage.

VercorSoleiL benefits from the FiT (Tarif d'achat) by selling its energy to EDF and Enercoop. It also benefits from regional government support. Since 2017, the Auvergne Rhône-Alpes Region has awarded VercorSoleiL 52,700€, which corresponds to 6.6% of total investment in PV plants.

The main motivation for shareholders to join VercorSoleiL is to participate in the community energy transition rather than the financial profit. The business model does not grant dividends or rebates in final bills (see also Archyde, 2019). However, for the individuals, joining VercorSoleiL means benefiting from being morally engaged in the energy transition. During the meetings, members frequently discuss new ideas on mobility, wind, and microhydro energy projects for their communities. In addition, the company boasts a good reputation among the local community and is promoted by local and national newspapers, which are often interested in presenting VercorSoleiL's projects. In turn, a sense of belonging to the community and doing the right thing are the main rewards for members.

Currently, VercorSoleiL intends to secure the energy autonomy of its remote territory by strengthening its renewable energy production. For this reason, its principal goals for the next 20 years are to preserve the existing installations, while investing in further renewable plants, and to encourage a sustainable lifestyle among citizens.

3.2 E-Werk Prad (Italy)

E-Werk Prad (EWP) is a historical cooperative in the Italian Alps founded in 1926, a member in turn of the previously mentioned SEV. EWP owns four hydropower plants

[31] Community of Municipalities of Vercors—Regional public institution that is responsible for the regional planning and cross-municipal economic development.

[32] The section summarizes the interview conducted by the authors with Francis Tasset (cofounder and shareholder of VercorSoleiL) in May 2021, to whom we express our gratitude for the collaboration.

with a total capacity of 4 MW; two biomass district-heating plants with 9 MWT of thermal and electric power; a 100 kW PV plant; electricity, heat, and broadband Internet distribution systems; two e-mobility charging stations; and one shared e-car in the small municipality of Prato allo Stelvio (with around 3500 inhabitants). EWP's 1442 members constitute the majority of local families, and the cooperative fully covers the energy needs of its members from April to November, also exporting any excess outside of the municipality.[33] From December to March, EWP purchases energy on the market to cover the community's consumption because hydropower stations produce less energy during winter months (Koltunov and Bisello, 2021). However, a recent research agreement between EWP and RSE has enabled experimenting to determine whether an EC could be self-sufficient, supplying its members with all their needs throughout the year with its own energy, and, for this goal, digital control systems and efficient energy storage systems are being tested.

Financially speaking, EWP benefits from various policy schemes, including an exemption from unbundling for historical cooperatives in Alpine regions, system costs exemption, simplified energy sales, and a purchase regime. Because of these advantages, EWP can offer its members electricity and heat at lower prices than external suppliers offer. Tables 20.3–20.5 illustrate the economic benefits that EC brings to the community.

Moreover, EWP assisted the local biogas cooperative in the construction of its plant, adding more social and environmental long-term benefits to the community (Koltunov, 2019). Other important benefits not yet translated into a monetary value for the community are the e-car sharing services and the bicycle route network that

Table 20.3 Electricity cost reduction to the members comparing to nonmembers.

Variable	Members	Non-members	Savings for members
Households			
Price, c€/kWh	17.39	21.35	3.96
Annual expenditure per one HH, €	422	519	97.3
Annual expenditure per community, €	608,329	750,742	142,413
Non-Households (industrial and commercial consumers, farmers)			
Price, c€/kWh	13.86	20.75	6.89
Annual expenditure per community, €	1,688,897	2,528,471	899,574
Outdoor lighting			
Price, c€/kWh	12.52	19.01	6.49
Annual expenditure per community, €	26,058	39,566	13,507
Total, €	2,322,284	3,318,729	995,494

Source: Own calculations, based on data from EWP (2018).

[33] Some community citizens are not members of the EWP because they are big consumers or engage in industrial activities. They cannot join the EC because their inclusion would increase the energy bills for the rest of the members and due to the resulting lower self-sufficiency.

Table 20.4 Heating-cost reduction to the members compared to the equivalent for oil-based generation.

Variable	Members	Equivalent for oil-based generation	Savings for members
Households			
Price, c€/kWh	9.53	16.1	6.57
Non-Households (industrial and commercial consumers, farmers)			
Price, c€/kWh	8.66	11.7	3.04
Total, €	-	-	589,145

Source: Own calculations, based on data from EWP (2018).

Table 20.5 The cumulative economic benefit to the community in 2018.

Variable	Economic benefit, €
Electricity bill reduction	995,494
Heat bill reduction	589,145
Broadband Internet bill reduction	27,853
Employees' salaries & social security contribution (10 employees)	501,672
Payments to local contractors	517,425
Own profit reinvested in future energy projects	355,038
Total	2,986,627

Source: Own calculations, based on data from (EWP, 2018).

EWP developed alongside the pipelines, both contributing to sustainable local mobility. In the future, besides experimenting to become self-sufficient by innovative solutions, EWP plans also to build a new 3-MW power plant in cooperation with a neighboring community. In 2020, EWP green energy provided an environmental benefit by avoiding 13,409 tons of CO_2 and emissions in the electricity and heat sectors, in comparison to oil-based equivalents.

3.3 *Energia Positiva (Italy)*

Energia Positiva was founded in 2015 in Torino, and since then has achieved rapid growth in its business model. In 2018, 231 members owned 16 plants, whereas, in 2021, 580 members owned 25 plants. Most of the plants utilize solar energy with small hydro and wind plants added to the ECs' generation; six energy-saving projects are also managed by the EC. According to the conceptual framework of Moroni et al. (2019a), it can be classified as a "nonplace based community of interest" because members and plants are distributed all over Italy, and online crowd-funding campaigns to raise capital or acquire existing plants are open to anyone looking for a financial return. Energia Positiva offers each member a rebate on the energy bill, up to 100%. This happens by discounting the economic return generated by a virtual renewable energy plant. The size and production of each virtual plant are directly related to the investment of the single member, who subscribes shares of real plants located in various places (Koltunov and Bisello, 2021). This type of bill rebate is possible due to the agreement signed with energy supplier Dolomiti Energia, which sells 100% green electricity to Energia Positiva's members.

The cooperative profits from three policy tools: IRPEF tax credit as an innovative start-up; a simplified energy sale/purchase regime; and the feed-in tariff related to plants acquired on the secondary market (therefore, still included in the "Conto Energia" although no longer available for new plants installed after 2013; Koltunov and Bisello, 2021). As illustrated in Table 20.6, the greatest economic benefit for members comes from the IPREF tax credit scheme.

Actually, the cooperative "EPCO," the acronym for an "energy prosumer company," a spinoff from Energia Positiva, extends the opening of this business model to small companies, shops, or freelance professionals that intend to reduce their bill, becoming a clean energy producer and profiting from a green status (Koltunov, 2019). In addition, direct supply from owned plans to members becomes possible

Table 20.6 Economic benefit for members of Energia Positiva.

Variable	Economic benefit			
The estimated annual benefit for one member				
Number of shares	1	6	13	21
Cost of shares, €	500	3000	6500	10,500
Bill reduction annual	5%	30%	60%	100%
Average electricity expenditures per HH, €	525	525	525	525
Savings due to the bill reduction, €	26.25	150	325	525
Savings due to the 30% IRPEF tax annual credit, €	150	900	1950	3150
Total benefit for one member, €	176.25	1050	2275	3675
Benefit for all members in 2019				
Number of shares in 2019	9839			
Cost of shares, €	4,919,500			
Number of members	413			
The average number of shares acquired	23.8			
Production of electricity by EC, kWh	1,554,500			
EC revenue, €	322,453			
Members' consumption of electricity, kWh	614,000			
Members' consumption of heat, m^3	113,000			
Members' electricity expenditures, €	188,488			
Savings due to the bill reduction, €	149,003.5			
Actual bill reduction	79.1%			
Savings from IRPEF tax annual credit (30% in 2019), €	1,475,850			
Total benefit, €	1,624,853			

Source: Own calculations, based on data from Energia Positiva (Gastaldo, 2019).

due to the recent Italian policy updates, making the business model even more sustainable and profitable (Gastaldo, 2020).

The three cases just reviewed are evidence of ECs' welfare enhancement for its members and communities. Place-based ECs, like VercorSoleiL and E-Werk Prad, enhance considerably the social and environmental welfare of local communities, and are built on a sense of common belonging to a specific area, whereas ECs with the geographically distributed membership and ubiquity generation facilities, like Energia Positiva, aim to contribute to economic welfare while leveraging digitalization and innovation.

4. How to increase the welfare-enhancing capacity of ECs

This section presents the main findings from the French and Italian movements. It points out how, according to Carayannis and Campbell (2009, 2010), the interplay among policy context, market context, civil society, and academia leads to the ECs' deployment process. Through the analysis of these elements, conclusions are drawn about what influences more the welfare-enhancing capacity of ECs. Then, it continues with a discussion of the similarities and differences in benefits of ECs that operate in different policy and institutional contexts. Finally, some insights are provided, addressing possible paths for the institutional and policy contexts, to contribute to the further development of ECs.

4.1 *Insights from French and Italian ECs*

Vernay and Sebi (2020) argue that a good ecosystem for the energy transition develops where no single key actor is favored by the government. Nonetheless, in France, where the public company EDF plays a pivotal role, the EC sector showed rapid growth due to the decisive role of regional associations acting as intermediaries, ranging from EPA to the Centrales Villageoises network. As for Italy, the policy does not prioritize a specific keystone actor, but rather the energy sector is more heterogeneous, having plenty of market players, and most Italian ECs are historical cooperatives in the Alps. Moreover, there is still a lack of intermediate organizations, providing both practical knowledge and technical or managerial services to small size ECs. There are only a few regional capacity builders today in Italy—SEV is unique—while in France, they exist in each region, acting as one-stop shops. This supports the conclusion of Boyle et al. (2021) that regime-based public organizations acting as intermediaries achieve higher levels of trust from citizens. In the absence of such specialized entities, market players such as DSOs could also become regional capacity builders of Italian ECs in the future, as suggested in the chapter by Del Pizzo et al. Also, the current experience shows that even historical cooperatives could spin off the new ECs, according to the evolution of the legal framework and available technologies. Even though the ECs' movement in Italy is less developed than in France, the number of single citizen members is remarkably high. This may create better general prospects for ECs' deployment, letting people choose not only among business-oriented players but including in the choice options various value propositions (Casalicchio et al., 2021). The role of sustainable investing is also expected to grow, given the increased interest in environmental, social, and governance (ESG) issues (Agrawal and Hockerts, 2019).

The state plays a crucial role in unlocking and sustaining the initial investments for the energy transition: designing supportive policies, especially financial, is of the utmost importance until ECs reach maturity. After this happens, the sector will follow the usual product life cycle and should be able to sustain itself without subsidies from the government, relying on a clear and favorable legislative framework. The growth of the number of ECs in France happened due to the "participatory bonus" scheme, which in turn allowed diversification of business models of prominent civil society actors, such as EPA and the Central Villages, and the recent decline can be similarly related to the uncertainty of its continuation, which is dreaded by the Directorate-General for Competition of the European Commission. It is hoped that this could be outweighed by the removal of some legal obstacles for setting up ECs done by the energy transition law. It is also expected that the transposition of EU Directives will probably benefit diverse business models because it is strongly lobbied by the French ECs' associations.

Similarly, due to the termination of many supportive policies in Italy, only 11 new ECs exist today. However, the initial transposition of EU Directives into Italian law is creating new opportunities thanks to allowing energy sharing and collective self-consumption, but it is still facing the uncertainties of the transitional phase and bottlenecks due to local administrative procedures.

Coordination, collaboration, and open communication are of paramount importance to solve uncertainties, remove preconceived notions, and reduce the risk of failure. In this regard, international innovation projects—as those cofinanced by the EU—are powerful tools to investigate ad test new solutions. According to the quadruple innovation helix model developed by Carayannis and Campbell (2010, 2009), when the public sector, academia, private, and tertiary sectors cooperate, they reinforce each other with new knowledge or know-how, thus increasing the successful deployment of the scheme. This process results in more educated decisions, more innovation, and increasing the welfare-enhancing capacity of the entire ecosystem. And as stressed by Heldeweg and Saintier (2020), the emergence of a new socio-legal environment is strongly connected to RECs development.

4.2 Conclusions

Despite the remarkable differences in the historical and contemporary framework conditions, leading to different developmental paths, the EC movements have significantly emerged both in France and in Italy. More than the clear predominance of the state on market forces (or vice versa), it is the balanced presence of both drivers that positively influences the welfare-enhancing capacity of ECs. Therefore, a well-designed policy system and the plurality of synergistic players are the keys to the sector's sustainable evolution.

In the three case studies, despite very different operational variables, the EC delivers considerable welfare gains for the members, although not equally distributed across the economic, social, and environmental spheres. Direct economic benefits are more evident in the Italian case studies, while, in the French example, it is the sense of belonging and the proactive contribution to local sustainable economic development that are the main levers to action. To what extent ECs bring gains to nonmembers and other local stakeholders is still a question to be investigated. Their business models will probably evolve, reflecting the interplay between market and policy, with the deployment of decentralized energy playing an important part due to political impetus from the EU climate strategy. Markets may be more important than policies in driving the sector if the profitability assessments of new projects will consider the sustainable and circular economy paradigms, thereby fostering an impact-investing approach.

Thus, the next main challenge for ECs is the termination of financial incentives because of the technological maturity of RES, storage systems, and smart-grid devices. On the other hand, relevant opportunities include the alignment of the national legal framework to the requirements of the ECs, according to the EU directives and hybridization between ECs and energy utilities to intercept a new market sector paying more attention to social and environmental impacts.

The following will enhance the formation of ECs and their successful implementation:

- Enriching knowledge circulation and collaboration across the state, markets, civil society, and academia as suggested by the quadruple innovation helix model;
- Ensuring adequate state support through legal and financial instruments until the sector reaches its maturity;
- Developing measures to create and support regional capacity builders and local one-stop shops profiting from already existing institutions or organizations active in this area;
- Allowing ECs to evolve by testing various business models that are open to multiple key actors and tailored to ensure a fair distribution of benefits for every member and to local communities; and
- Highlighting the multiple potential benefits of ECs, their positive contribution to UN Sustainable Development Goals achievement, and overall alignment to ESG evaluation criteria.

Bibliography

Agrawal, A., Hockerts, K., 2019. Impact investing: review and research agenda. J. Small Bus. Enterpren. 33 (2), 153–181. https://doi.org/10.1080/08276331.2018.1551457.

Archyde, 2019. In the Vercors, Villagers Team up to Develop Solar Energy. https://www.archyde.com/in-the-vercors-villagers-team-up-to-develop-solar-energy/. (Accessed 20 July 2021).

Bisello, A., 2020. Assessing multiple benefits of housing regeneration and smart city development: the European project SINFONIA. Sustainability 12 (19), 8038. https://doi.org/10.3390/su12198038.

Boyle, E., Watson, C., Mullally, G., Brian, O., 2021. Regime-based transition intermediaries at the grassroots for community energy initiatives. Energy Res. Social Sci. 74 (2020). https://doi.org/10.1016/j.erss.2021.101950.

Brummer, V., 2018. Of expertise, social capital, and democracy: assessing the organizational governance and decision-making in German renewable energy cooperatives. Energy Res. Social Sci. 37 (2017), 111–121. https://doi.org/10.1016/j.erss.2017.09.039.

Candelise, C., Ruggieri, G., 2020. Status and evolution of the community energy sector in Italy. Energies 13 (1888), 1–22. https://doi.org/10.3390/en13081888.

Carayannis, E., Campbell, D., 2009. "Mode 3" and "Quadruple Helix": toward a 21st century fractal innovation ecosystem. Int. J. Technol. Manag. 46. https://doi.org/10.1504/IJTM.2009.023374.

Carayannis, E.G., Campbell, D.F.J., 2010. Triple helix, quadruple helix and quintuple helix and how do knowledge, innovation and the environment relate to each other? Int. J. Soc. Ecol. Sustain. Dev. 1 (1), 41–69. https://doi.org/10.4018/jsesd.2010010105.

Casalicchio, V., Manzolini, G., Prina, M.G., Moser, D., 2021. Renewable energy communities: business models of multi-family housing buildings. In: Smart and Sustainable Planning for Cities and Regions. Results of SSPCR 2019. Springer, pp. 261–276. https://doi.org/10.1007/978-3-030-57332-4_19.

Devine-Wright, P., 2005. Beyond NIMBYism: towards an integrated framework for understanding public perceptions of wind energy. Wind Energy 8 (2), 125–139. https://doi.org/10.1002/we.124.

Enercoop, 2021a. Our History.

Enercoop, 2021b. The Production of Our Electricity.

Énergie Partagée, 2019. Les retombées économiques locales des projets citoyens.

Énergie Partagée, 2021. Key Figures for Citizen Energy.

Energy Center, 2021. The Energy Communities.

EnRciT, 2021. EnRciT Founders.

EWP, 2018. Die EWP Stromversorgung, Fernwärmeversorgung, Breitbandversorgung in der Gemeinde Prad.

Gastaldo, A., 2019. Bilancio e nota integrativa esercizio (Turin).

Gastaldo, A., 2020. Energia Positiva, Shareholder Meeting Updates.

Heldeweg, M., Saintier, S., 2020. Renewable energy communities as "socio-legal institutions": a normative frame for energy decentralization. Renew. Sustain. Energy Rev. 119. https://doi.org/10.1016/j.rser.2019.109518.

IEA, 2021. International Energy Agency Country Profile: France.

IEM, 2019. European Parliament Legislative Resolution on Common Electricity Market Rules.

Klagge, B., et al., 2016. Zukunft der deutschen Energiegenossenschaften. Raumforsch. Raumordn. 74 (3), 243–258. https://doi.org/10.1007/s13147-016-0398-3.

Koltunov, M., 2019. The Impact of Renewable Energy Cooperatives on the Welfare of Local Communities. Corvinus University of Budapest.

Koltunov, M., Bisello, A., 2021. Multiple impacts of energy communities: conceptualization taxonomy and assessment examples. In: Smart Innovation, Systems and Technologies, 178 SIST. Springer International Publishing, pp. 1081–1096. https://doi.org/10.1007/978-3-030-48279-4_101.

Magnani, N., Cittati, V.M., 2022. Combining the multilevel perspective and socio-technical imaginaries in the study of community energy. Energies 15 (5). https://doi.org/10.3390/en15051624.

Matalucci, S., 2021. Italy's first energy community for solar power sharing. PV Magazine 1. https://www.pv-magazine.com/2021/01/14/italys-first-energy-community-for-solar-power-sharing/. (Accessed 3 April 2021).

Moroni, S., Alberti, V., Antoniucci, V., Bisello, A., 2019a. Energy communities in the transition to a low-carbon future: a taxonomical approach and some policy dilemmas. J. Environ. Manag. 236, 45–53. https://doi.org/10.1016/J.JENVMAN.2019.01.095.

Moroni, S., Antoniucci, V., Bisello, A., 2019b. Local energy communities and distributed generation: contrasting perspectives, and inevitable policy trade-offs, beyond the apparent global consensus. Sustainability 11 (12), 3493. https://doi.org/10.3390/su11123493.

Peullemeulle, J., Duval, J., 2017. Guide Pratique Participatifs et Citoyens d' Énergie Renouvelable (Paris).

Poupeau, F.M., 2020. Everything must change in order to stay as it is. The impossible decentralization of the electricity sector in France. Renew. Sustain. Energy Rev. 120. https://doi.org/10.1016/j.rser.2019.109597.

RED-II, 2018. Directive (EU) 2018/2001 of the European Parliament and of the Council on the promotion of the use of energy from renewable sources. Official J. Eur. Union 2018 (L 328), 82–209.

Rosanvallon, P., 2011. The metamorphoses of democratic legitimacy: impartiality, reflexivity, proximity. Constellations 18 (2), 114–123. https://doi.org/10.1111/j.1467-8675.2011.00631.x.

Sebi, C., Doutre, J., Vernay, A.L., 2019. Mapping the energy community cooperation chains. In: ECEEE Summer Study Proceedings.

Sebi, C., Vernay, A.-L., 2020. Community renewable energy in France: the state of development and the way forward. Energy Pol. 147. https://doi.org/10.1016/j.enpol.2020.111874.

SEV, 2019. Website: https://www.sev.bz.it/en/south-tyrol-energy-association/1-0.html. (Accessed 15 July 2019).

Sioshansi, F., 2019. Consumers, Prosumers, Prosumagers: How Service Innovations Will Disrupt the Utility Business Model. Academic Press.

Süsser, D., Döring, M., Ratter, B.M.W., 2017. Harvesting energy: place and local entrepreneurship in community-based renewable energy transition. Energy Pol. 101, 332–341. https://doi.org/10.1016/j.enpol.2016.10.018.

Vernay, A.L., Sebi, C., 2020. Energy communities and their ecosystems: a comparison of France and The Netherlands. Technol. Forecast. Soc. Change 158, 120123. https://doi.org/10.1016/j.techfore.2020.120123.

Yildiz, Ö., Rommel, J., Debor, S., Holstenkamp, L., Mey, F., Müller, J.R., Rognli, J., 2015. Renewable energy cooperatives as gatekeepers or facilitators? Recent developments in Germany and a multidisciplinary research agenda. Energy Res. Social Sci. 6, 59–73. https://doi.org/10.1016/j.erss.2014.12.001.

Zulianello, M., Angelucci, V., Moneta, D., 2020. Energy community and collective self consumption in Italy. In: UPEC 2020–2020 55th International Universities Power Engineering Conference, Proceedings. https://doi.org/10.1109/UPEC49904.2020.9209893.

Energy communities in Europe: a review of the Danish and German experiences

21

Simona Benedettini[1] *and Carlo Stagnaro*[2]
[1]PwC, Milan, Italy; [2]Istituto Bruno Leoni, Milan, Italy

1. Introduction

Renewable Energy Communities (RECs) and Citizens Energy Communities (CECs) are expected to receive large attention in the legislative and regulatory framework of EU Member States over the next few years.

Lessons learned from countries with a longer tradition of citizen ownership of energy generating assets may inform the decision-making process in the rest of Europe and possibly elsewhere.

This chapter investigates the experiences of Denmark and Germany where citizen ownership of energy represents a relevant experience in the respective energy sector. In fact, 30.2% of renewable capacity for power production was held by private individuals in 2019 in Germany (German Renewable Energy Agency, 2021). Likewise, 52% of wind power capacity in Denmark was citizen owned in 2016 (Gorroño-Albizu et al., 2019).

Citizen ownership is not necessarily the same as energy communities. Both in Germany and in Denmark, citizens have had a long tradition of owning energy assets, either individually or as communities, even though they did not always share or locally consume the energy thus produced. Other types of assets, such as district heating networks, are often owned by local communities (either associated in cooperatives or through local governments) but even this may not be enough to qualify as an energy community. It should be noted that the notion of energy community is still unclear. Alternative definitions are possible, as other chapters in this book show. In fact, under the stricter framework proposed in the chapter by Rossetto et al., the early (and even more recent) citizen initiatives that will be described below may be qualified as forms of local production, local collective self-consumption, or local aggregations, rather than energy communities.

At any rate, energy communities in today's Europe also exist as a consequence of previous initiatives in several member states, particularly Germany and Denmark. The EU's own legal framework for energy communities is relatively loose and, in fact, recognizes RECs and CECs as entities that have some, but not all, of the features identified in the chapter by Rossetto et al. Only time will tell whether or not this is an intermediate step toward "real" energy communities. An EU-sponsored study on nine European countries (Germany, Denmark, the Netherlands, the United Kingdom,

Energy Communities. https://doi.org/10.1016/B978-0-323-91135-1.00015-8

Sweden, France, Belgium, Poland, and Spain) found at least 3752 cases of community energy initiatives: of these, 1750 are located in Germany and 700 in Denmark (Caramizaru and Uihlein, 2020). Most of these initiatives are not proper energy communities, even though they may fall under the EU's definition. Still, their very existence, resilience, and evolution may provide useful information regarding the future of energy communities, the challenges ahead, and the risks.

This chapter investigates the following dimensions with respect to both RECs and CECs: legal forms of energy communities; types of actors which can be involved in energy communities; rights and obligations deriving from the participation into energy communities; modalities through which energy communities' members can exercise control; activities which can be operated by energy communities; purposes of energy communities; geographical constraints affecting the operation of energy communities; support mechanisms to promote the diffusion of energy communities; the access and the ownership of both the generation and the distribution grid assets; the sources of supply of electricity to community members; and the possibility for energy communities to participate in electricity markets.

The chapter is organized as follows:

- Section 2 discusses the analytical framework employed to investigate the Danish and the German experiences;
- Sections 3 and 4, respectively, illustrate the legal framework and the most relevant experiences of energy communities in Denmark and Germany; and
- Section 5 discusses the main takeaway and recommendations emerging from the analysis of Denmark and Germany and concludes.

2. The analytical framework

According to the provisions on RECs and CECs provided respectively by the EU Directive 2018/2001 (so called REDII) and the EU Directive 2019/944 (so-called IEMD), the following relevant dimensions will be investigated with respect to citizen energy participation in Denmark (Section 3) and Germany (Section 4). These countries have a long history of community energy initiatives that dates back to the 1970s or even earlier. They still host a majority of community energy initiatives currently existing in the European Union. At the same time, not all of the German and Danish energy communities fit in the most rigorous definitions, insofar as—for example—they do not necessarily entail the sharing of energy. Furthermore, they were and to some extent still are dependent on financial transfers or other forms of regulatory incentives. Studying the reality of community energy in the countries where it is more rooted and diffused may help to identify both the opportunities for development and potential weaknesses of energy communities.

In this chapter, several dimensions of community energy are reviewed, with regard to the German and Danish experience as well as the EU common framework as derived by IEMD and RED II. The features of RECs and CECs are illustrated in the chapter by Spasova and Braungardt. Table 21.1 provides an overview of the dimensions that will

Table 21.1 Dimensions of analysis.

Dimension	Citizens energy communities	Renewable energy communities
Legal form	Any: Cooperative, partnership, nonprofit organization, SME, association	
Actors involved	Any: Individuals, local authorities (including municipalities), SMEs	
Participation	Open and voluntary. The EU directive explicitly mention that members are entitled to leave the communities	Open and voluntary
Control	Effective control limited to those actors that are not engaged in large-scale commercial activity and for which the energy sector does not constitute a primary area of economic activity	In case of private undertakings, effective control is prevented to those undertakings whose participation in RECs constitutes their primary commercial or professional activity
Activities	Generation, distribution, supply, self-consumption, consumption, sharing of the electricity self-produced, aggregation, and storage of electricity	Generation, supply, self-consumption, consumption, sharing of the electricity self-produced, aggregation and storage of electricity, promotion of energy efficiency, commercial energy services
Purposes	Social, economic, and environmental benefits for members/shareholders or the local area in which it operates, rather than financial profits	
Geographical constraints	Not defined. Electricity sharing enables members or shareholders to be supplied with electricity from generating installations within the community without being in direct physical proximity to the generating installation and without being behind a single metering point	Members or shareholders in effective control of an REC need to be located in proximity of the renewable project owned and developed by the community
Support mechanisms	The possibility for CECs to benefit of support schemes is explicitly not mentioned. However, where electricity is shared, this shall be without prejudice to applicable network charges, tariffs, and levies	Explicitly mentioned the possibility to benefit of support schemes. The latter may take different forms as support schemes for RE installations, tax incentives for RE installations, priority access to the grid for RE installation, simplified permitting procedures for small RE installations, incentives for collective self-consumption.

Continued

Table 21.1 Dimensions of analysis.—cont'd

Dimension	Citizens energy communities	Renewable energy communities
Ownership of generation assets	The generating asset can be in the ownership of both CECs and third parties	The renewable generating assets shall be owned by the members of the REC
Ownership of the electricity distribution grid	CECs may become DSOs either under the general regime or as closed distribution system operators. CECs are entitled to own, establish, purchase, or lease distribution networks	The EU directive does not explicitly mention the possibility for RECs to own or operate distribution networks. However, if self-consumption is entailed in the REC's arrangements, the latter can develop a private network to share the electricity self-produced.
Supply of electricity to community members	CECs' and RECs' members shall be granted rights as final customers as the freedom of contract, the right to switch supplier, etc.	
Participation in electricity markets	Both CECs and RECs can access all suitable energy markets both directly or through aggregation in a nondiscriminatory manner	

be investigated with respect to energy communities in Denmark and Germany and discussed in Sections 3 and 4 of this chapter. It also provides a short description of the EU law on RECs and CECs according to the same dimensions.

3. Energy communities in Denmark

3.1 The legal framework

Denmark has implemented the provisions of REDII and IEMD in its national legislation with a variety of measures. However, community energy has a long history in the country which in fact was a pioneer in citizen ownership, particularly in the areas of wind energy and district heating. Because of this legacy, though, community ownership in Denmark does not necessarily coincide with the idea of energy community as defined by the European law (see the chapter by Spasova and Braungardt) neither does it necessarily fit into the more stringent definition given by Rossetto et al. This is due to a large extent to the fact that community and individual ownership of energy assets in Denmark dates back to a preliberalization era, when the regulation of the energy sector was much different.

Two predisposing factors explain why community initiatives grew rapidly in Denmark since the 1970s. To begin with, cooperatives were already a common feature in the Danish economic environment. The first cooperatives date back to the early XIX century, when they were established in the agricultural sector and later became a widespread form of commercial activities (Bauwens et al., 2016). This made for a fertile ground for citizen participation in the energy sector, although most energy cooperatives are not registered as such—they are usually incorporated as legal partnerships. The main difference lies in the fact that cooperatives are characterized by open membership and democratic participation, whereas legal partnerships (or guilds) may have closed membership and a voting system based on the ownership of the shares (Danish Wind Turbine Owners' Association, 2009). Since the difference between legal partnerships, guilds, or cooperatives is of secondary importance within the scope of this chapter, the term "cooperative" will be used as a broad category that encompasses all these arrangements, rather than in its strict legal meaning.

Secondly, several public services such as electricity generation and district heating were required by the law to have a not-for-profit statute. This changed in 2004 for electricity generators, but not for district heating. The prices of electricity and district heating were set at a level that allowed to recover the cost of investment. This disincentivized large corporations to enter the market, at least until the liberalization in the early 2000s, and left the place open to municipal companies or small cooperatives—or even individual ownership, in the case of wind turbines.

Citizen ownership of wind power was triggered by the oil crisis in 1973. The Danish Parliament and industrial sector pushed strong to invest in nuclear power. The decision, however, raised a strong opposition, that resulted, on one hand, in the Parliament voting a ban on nuclear power in 1985, and, on the other hand, in a push toward investments in alternative energies, particularly wind power.

The tradition of cooperatives, the not-for-profit requirement, and the political commitment to switch away from imported fossil fuels created the perfect mix for a new reality: individual and community ownership of wind turbines.

3.2 Wind cooperatives

The history of individual and community wind ownership can be roughly divided in three phases: 1977–2001; 2001–09; and 2009 on. Each phase coincides with a change in the market design and underlying conditions. While individual and community ownership played a major role in the early development of wind power in Denmark, it is not clear whether it will be as important in the future.

3.2.1 Phase 1: 1977–2001

Citizen ownership of wind turbines in its early stage was rooted in the antinuclear movement. After the 1973 oil crisis and the antinuclear protests, incentives to energy savings and distributed generation were introduced (to the point that heavily energy-intensive uses were banned and energy efficiency mandates were introduced).

This resulted in a set of ground-breaking decisions, including the following:

- The right to install wind turbines on their own land was recognized for farmers and rural households;

- Ownership of wind turbines was limited to local residents;
- Local residents could associate in cooperatives and own wind turbines collectively;
- The number of shares that an individual could own in common projects was capped; and
- A maximum number of individually owned turbines were set (Mey and Diesendorf, 2018).

Support schemes included tax exemptions and, after the mid-1980s, feed-in tariffs (Fits) as well as purchase obligation by the grid, priority access, and guaranteed connection, meaning that wind turbine owners had to pay the costs for getting connected to the main grid, but they did not have to pay for grid expansions (if needed) (Gancheva et al., 2018).

The push toward local ownership went hand in hand with a considerable degree of decentralization, that entailed the creation of local energy offices that were exposed to, and often influenced by, local antinuclear, and prodistributed generation activists. This helped to create a favorable regulatory environment at the local as well as national level (Kooij et al., 2018).

The result is striking indeed. By 2002, 2897 MW of wind capacity had been installed, equal to 22.9% of the entire generation capacity in Denmark, that produced 4.88 TWh or 12.4% of the gross generation. About 150,000 households owned shares in wind power cooperatives (or, to be more precise, legal partnerships) that owned slightly less than 40% of the existing 6300 turbines (Bauwens et al., 2016).

Just a couple of years thereafter, in 2004, the number of households involved in wind cooperative had fallen to 100,000 and by 2009 it had further fallen to 50,000. What had happened?

3.2.2 Phase 2: 2001–09

In the late 1990s and early 2000s, times were changing:

- A conservative government took over from the social democrats in 2001, after the latter had been in power for virtually the whole of the 1990s.
- In 1996 the so-called First energy package had been adopted by the European Union; member states were required to transpose it in their national legislations by 1998.[1] In 2003, the directive was amended by the Second energy package, to be transposed in national legislations by 2004.[2] Both directives required member states to open up electricity markets to competition, by liberalizing the ownership of generating plants, creating wholesale markets, and granting nondiscriminatory access to power networks.
- Larger and more powerful turbines were introduced, that were more costly and increasingly difficult to manage and operate for small cooperatives or individuals.

Regulatory changes since 1999 made it less attractive for households to invest in wind turbines, either directly or by associating in legal partnerships. The existing Fit was substantially reduced and, later, substituted by a less generous feed-in premium (Fip) whereby generators were paid the Nord Pool market price plus a fixed premium (Bauwens et al., 2016). The government introduced a plan to protect the landscape,

[1] Directive 96/92/EC.
[2] Directive 2003/54/EC.

that inter alia encompassed a repowering scheme to install larger but fewer turbines. Planning zones were also introduced. Finally, pursuant to the First and Second energy packages, requirements for local ownership of turbines, limits to the number of shares per person as well as to the number of turbines per owner, and the not-for-profit rule were all removed.

The interactions between these reforms made it less attractive for citizen to start community energy initiatives and attracted larger corporations. The growing evidence of economies of scale, from both the size of turbines and utility-scale wind farms, contributed to a shift in ownership from citizens to professional firms and utilities (Mey and Diesendorf, 2018). No new cooperatives were formed and many existing ones were dissolved. On top, the expansion of wind farms triggered groups of citizens who oppose, rather than support, wind power, undermining the almost unanimous consensus that turbines had enjoyed until a few years ago (Egelund Olsen, 2014).

3.2.3 Phase 3: from 2009 on

A further reform was introduced in 2009, with the aim of stimulating investments in wind power and achieving the EU and national decarbonization targets.

Again, political changes prompted substantial developments. Under the pressure of more stringent environmental targets, the Promotion of Renewable Energy Act increased the fixed premium payment for wind energy, created a fund to support preliminary investigations by local wind turbine owners associations in order to identify the most promising sites, and provided that developers of new turbines offer 20% of the shares to local residents (meaning those living within a 4.5 km radius from the new turbine; Bauwens et al., 2016).

This spurred a new wave of investments although the outcome is mixed. On one hand, it seems that the 20% rule has been scarcely successful: the fact that shares are offered for sale prevents those with little or no financial means from participating the investments. If they do, citizens gain a role as minority shareholder but exercise no actual control over the company, neither do they have any benefit other than the financial reward from participating in the investment (Savaresi, 2019). On the other hand, participation under the 20% rule does not qualify as an energy community as defined by the RED II and IEMD directives. On top of that, a large share of the remaining citizen-owned wind turbines appears to be small projects: about 60% have five shareholders or less, and only 20% have more than 50 shareholders (Wierling et al., 2018).

More recent regulations—such as the 2018 Danish Energy Agreement—tend to favor large wind farms offshore rather than onshore; monetary incentives are increasingly awarded through tendering, which is less accessible by citizen associations. Hence, despite some increase in citizen participation, the existing community- or citizen-owned projects are aging: about 69% of citizen-owned turbines in 2016 were estimated to be at least 15 years old (Gorroño-Albizu et al., 2019). There are signs of evolution in the model of citizen ownership: the shift toward Fips and the subsequent exposure to wholesale price volatility increased the important of an efficient management of wind power. So the most dynamic cooperatives struggled to achieve a greater control over the energy produced. An example is the creation of Vindenergi

Table 21.2 Wind capacity in Denmark by type of owner (2016).

Ownership category	Ownership of installed capacity (MW)	Ownership of installed capacity (%)
Citizen ownership	2747	52
Of which individual ownership	1212	23
Of which collective ownership	507	11
Of which unidentified citizen ownership	2499	19
Large investor ownership	2499	48
Total	**5246**	**100**

Source: Gorroño-Albizu, L., Sperling, K., Djørup, S., 2019. The past, present and uncertain future of community energy in Denmark: critically reviewing and conceptualising citizen ownership. Energy Res. Social Sci. 57, 101231.

Danmark, a company that trades energy at Nord Pool—the power exchange for Northern European countries—on behalf of wind power cooperatives. Vindenergi is incorporated itself as a legal partnership, i.e., a not-for-profit company owned exclusively by its members. It is estimated that about two thirds of the Danish wind cooperatives rely on Vindenergi (Bauwens et al., 2016).

Table 21.2 shows how the ownership of wind turbines has evolved over time, depending on the changing regulatory framework and financial incentives.

3.3 District heating

District heating (DH) has a long tradition of community ownership in Denmark that is documented since the 1920s (Kooij et al., 2018). District heating provides heat to about 60% of Danish households (Chittum and Østergaard, 2014) and covers about 18% of total final energy consumption, the highest rate among the IEA members (IEA, 2017). This is the result of a long-lasting policy choice that dates back to 1979. Not surprisingly, district heating—like wind—was subject to not-for-profit or consumer-profit rules. Being a natural monopoly, this was not challenged at the time of liberalization (Kooij et al., 2018). Given its technological and economic features, DH is more practical as a community project rather than as an investor-sponsored one, even though the latter model is also possible (see also the chapter by Menke et al.).

Under the not-for-profit rule, the value DH providers create is either reinvested in the development of networks or generation plants (particularly bioenergies), or transferred to the citizens themselves (although it is not clear whether economic efficiency is taken into due account). Cost-reflective prices are charged under the supervision of local and national regulators (Eikeland and Inderberg, 2016).

Fig. 21.1 shows how steep the take-up was of district heating in Denmark since the early 1970s, although its introduction in the country is even earlier.

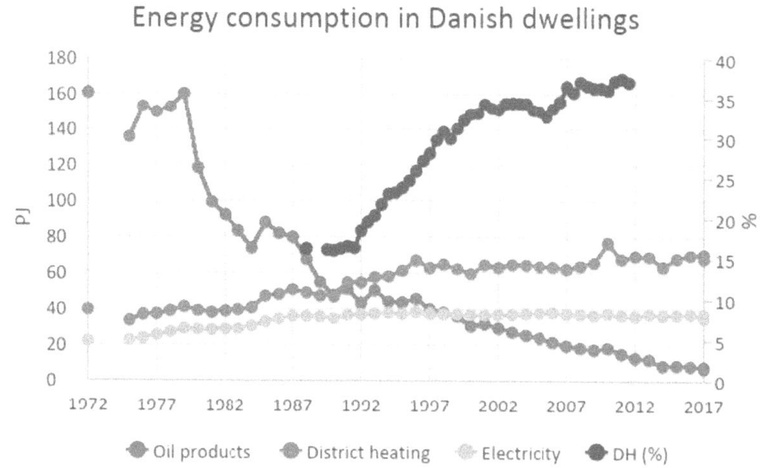

Figure 21.1 Energy consumption in Danish dwellings (1972–201).
Source: Own elaboration, based on data from the Danish Energy Authority.

In Denmark, about 400 companies provide district heating. A large majority of these (96%) is owned by citizens, either through their municipalities (especially in large cities) or by associating in cooperatives or legal partnerships (especially in rural areas) (Gorroño-Albizu et al., 2019). Both may fit into the EU's definition of RECs and CECs (depending on what the primary source of heat is) and possibly in more stringent definitions as well. Citizens have many ways to influence the decisions of district heating companies (or to exert effective control), either directly (as cooperative members) or indirectly (via their elected representatives in municipal councils). However, they do not always have a right to opt out of district heating systems, insofar as municipalities may make it compulsory to get connected to the systems themselves. Of course, where participation is compulsory, one distinctive feature of energy communities—voluntariness and citizen engagement—is missing. This notwithstanding, citizens have several ways to participate in the decision-making process, particularly when a connection obligation is being introduced:

- They need to be consulted regarding such decision;
- They may call exemptions if there is a particular reason (for example, if their building is already a low-energy home);
- They may participate in official appeal processes;
- If a new building is subject to connection obligation, but district heating is not yet available in that particular site, the district heating company must supply some temporary source of heat; and
- Citizens have a right of first refusal to buy the shares of a district heating system serving their buildings (Chittum and Østergard, 2014).

Table 21.3 summarizes the situation as of 2016 regarding the ownership of the existing district heating systems in Denmark.

Table 21.3 District heating in Denmark by type of owner (2016).

Ownership category	Number of DH systems	DH demand supply (%)
Citizen ownership	388	96
Of which municipal companies	47	60
Of which cooperatives	341	36
Commercial companies	13	4
Others	6	~0
Total	**407**	**100**

Source: Gorroño-Albizu, L., Sperling, K., Djørup, S., 2019. The past, present and uncertain future of community energy in Denmark: critically reviewing and conceptualising citizen ownership. Energy Res. Social Sci. 57, 101231.

3.4 Energy communities in practice

Table 21.4 summarizes the main features of energy communities in Denmark according to the dimensions illustrated in Section 2 of this chapter. Not all community energy initiatives that have been surveyed do necessarily comply with the requirements of CECs or RECs. In particular, citizen ownership in wind power only loosely resonates with RECs, and the limited participation in large, recent projects achieved by the 20% rule cannot be qualified as an energy community at all (for this reason it is not even considered in the Table below). On the contrary, citizen ownership of district heating may be regarded as a variation of RECs or CECs, depending on the source of heat (whether renewable or not). The aspect of self-consumption (or sharing) of energy is also lacking. In the case of wind power, one may argue that citizen- or community-owned turbines were intended as a contribution to local needs. That was more a cultural motivation for investments than a physical reality, in the sense that electricity is fed into the grid rather than used locally. Local consumption is instead an inherent feature of district heating, but in that case, the issue is that citizens hardly have a choice, insofar as they easily become captive customers in DH networks.

In evaluating the Danish experience of citizen ownership—that may be regarded as successful in the case of district heating, but ambiguous with regard to wind power—one should always remember that its rapid diffusion is also the result of a number of regulatory and legal circumstances that are not necessarily replicable, neither is it necessarily desirable to replicate. These include high monetary incentives, the not-for-profit rule that applied to wind turbines until the early 2000s and that is still in force for district heating, the requirements for local ownership and limits to the number of turbines per company and the number of shares per person (no longer in force), etc.

4. Energy communities in Germany

4.1 The legal framework

The *Erneuerbare-Energien-Gesetz* ("EEG" or "Renewable Energy Act," henceforth) has significantly contributed to the rapid development of renewable electricity over

Table 21.4 Overview of community energy in Denmark.

Dimension	Citizen ownership of wind turbines	Citizen ownership of district heating systems
Legal form	Cooperatives or legal partnerships	Cooperatives or municipal companies
Actors	Natural persons	Natural persons and local governments
Participation	Open and voluntary	Open and voluntary
Control	Effective control by members	Effective control by members, either directly (as shareholders) or indirectly (through elected officials in municipal councils)
Activities	Generation, self-consumption, supply	Heat generation and distribution
Purposes	Social, economic, and environmental benefits for members/shareholders or the local area in which it operates. No financial profits allowed until early 2000s	Social, economic, and environmental benefits for members/shareholders or the local area in which it operates. No financial profits allowed
Geographical constraints	Members need to be located in the proximity of the project owned by the individual or cooperative	Members need to be located in the proximity of the DH system
Support mechanisms	Equal to the general support to renewable generation (feed-in premium)	No explicit support. An obligation may be introduced for new buildings to become connected to the DH system.
Ownership of generating assets	The generating assets are owned by individuals or cooperatives	The heat generation assets as well as the district heating systems are owned by the DH company
Electricity distribution	Not applicable	Not applicable
Supply of electricity to community members	Yes	Yes
Participation in electricity markets	Cooperatives may sell energy to the grid or access power markets, directly or via aggregators	Not applicable

the past 2 decades in Germany. The EEG has been subject to different amendments over time. The 2017 version introduced the concept of "*Bürgerenergiegesellschaft*" or "Citizens Energy Company" (CEC) (EEG, Section 3-15).[3] According to the EEG 2017: *"citizens' energy company" shall mean every company:*

(a) *Which consists of at least 10 individuals who are members eligible to vote or shareholders eligible to vote;*
(b) *In which at least 51% of the voting rights are held by natural persons whose main residence has been registered pursuant to Section 21 or Section 22 of the Federal Registration Act for at least 1 year prior to submission of the bid in the urban or rural district in which the onshore wind energy installation is to be erected; and*
(c) *In which no member or shareholder of the undertaking holds more than 10% of the voting rights of the undertaking.*

Suggest you make this into standard bullet points with standard format.

Whereby in the case of an association of several legal persons or unincorporated firms to form an undertaking, it is sufficient if each of the members of the undertaking fulfills the preconditions pursuant to letters (a)–(c).

In addition, the EEG 2017 (Section 36-g) establishes specific support schemes for CECs. In particular:

- CECs can submit bids for up to six onshore wind energy installations with a total capacity to be installed of not more than 18 MW.
- CECs shall submit a self-declaration, to participate in the auctions, stating that (1) the undertaking is a citizens' energy undertaking at the time the bid is submitted, and the undertaking and its members or shareholders have not concluded any contracts before the submission of the bid to transfer their shares or voting rights following submission of the bid; (2) neither the undertaking nor any one of its members is holding voting rights himself or herself or is a member with voting rights of another undertaking [that?]: (i) has received a funding award for an onshore wind energy installation in the 12 months preceding the submission of the bid; (ii) has submitted other bids for the bid deadline which together with the bid exceed installed capacity of 18 MWs; and (iii) the company is the owner of the site on which the onshore wind energy installation is located.
- CECs can benefit of the possibility to correspond the financial security deposit in two tranches: (i) an initial security of 15 euro per kW of capacity to be installed, to be paid when the bid is made; (ii) a second security to be paid in the case of an award within 2 months after the granting of the approval. This second security shall be determined by the capacity to be installed in the approved installations multiplied by 15 euro per kilowatt of capacity to be installed.
- Funding to awarding CECs is established according to a pay-as-clear principle, i.e., the value of the award of the highest bid awarded funding from the same bid deadline.
- The award granted to a bid shall be tied to the rural district cited in the bid as the site.

Germany recognizes a significant role to local authorities in shaping its energy transition. For example, the EEG (Section 36-g) recognizes to the Länder "*the right to issue further rules on citizens' participation and in order to increase public*

[3] Erneuerbare-Energien-Gesetz (2017). Available at: https://www.bgbl.de/xaver/bgbl/start.xav?startbk=Bundesanzeiger_BGBl#__bgbl__%2F%2F*%5B%40attr_id%3D%27bgbl116s2258.pdf%27%5D__1622443233866.

acceptance of the construction of new installations." Local authorities play other important roles in shaping energy transition (Olhorst, 2015):

- *Overarching measures*: e.g., targets concerning the local share of renewable energy or the reduction of GHG emissions, local energy and climate action plans, cooperation with other municipalities;
- *Consumer behavior of the municipal administration:* the municipal administration can take actions to reduce its energy use and to meet its energy needs with renewable energy as well as by applying ecological criteria for its procurement;
- *Regulation and planning:* e.g., the designation of areas for renewable energy plants and facilities; urban development contracts to determine both renewable energy and energy efficiency requirements; local building codes; and
- *Support and information:* e.g., through information campaigns, educational training, consulting services, etc.

The most relevant trend with respect to public intervention concerns the so-called "remunicipalization" of the essential facilities by means of energy communities and citizen involvement more in general.

4.2 Energy communities in practice

As opposed to Denmark, Germany has substantially updated its legal framework concerning energy communities. Hence, an assessment can be made with regard to the current regulation, rather than drawing the relevant information from a range of experiences that, by and large, are rooted in the preliberalization era. However, under the German law, many citizen (or community) energy initiatives may be classified as energy communities, that would not stand to more stringent definitions. These include energy cooperatives that build and run PV or wind installations of various sizes and are incentivized to do so, while little (or no) energy sharing or self-consumption takes place. This has raised serious criticism (IZES, 2021). The analysis below shows how loose the definition of ECs is after the transposition of RED II and IEMD.

4.2.1 Legal form

The EEG 2017 does not provide limitations to the legal forms which can be adopted to operate CECs. Different legal forms have been adopted to operate CECs in Germany (Romero-Rubio and Diaz, 2015; see also the Co-Power Initiative)[4]:

- *GbR (Gesellschaft bürgerlichen Rechts)* is characterized by a low administrative burden for its constitution and has no minimum capital requirements. However, members of *GbRs* have unlimited liability for their obligations.

[4] Co-Power, 2014. *Community energy in Germany: existing models, public-private funding, and good practice example.* The CO-POWER project (Community Power: enabling legislation to increase community ownership for RES projects across Europe) is supported by the European Commission in the Intelligent Energy for Europe Program. Available at: https://www.communitypower.eu/en/publications.html.

- *GmbH & Co. KG* is usually chosen for the realization of large investment. The liability of the partners is limited to their respective share of the partnership capital.
- *eG, eingetragene Genossenschaft*, can be an alternative to GmbH & Co. KG. It is subject to the Cooperatives Act and it is at present the most common used legal entity in Germany for the establishment of energy communities. The liability of partners is limited to their respective capital contribution. The Association of Cooperatives verifies the consistency of the business plan and of the statute of the cooperative before it can be registered in the register of cooperatives. Therefore, this legal form is considered as the most insolvency-proof legal form in Germany. This type of legal entity requires a higher start-up capital and administrative costs are higher than in GbR form.

According to the German Renewable Energy Agency (2019a), energy cooperatives experienced a progressive and fast diffusion in Germany. The number of energy cooperatives skyrocketed from slightly above 60 in 2001 to 869 in 2018, mostly active in the field of renewable energy (Fig. 21.2). The registered cooperatives had more than 183,000 members. A total of 2.7 billion euro were invested in citizen-owned power plants. Most energy cooperatives generate electricity (German Renewable Energy Agency, 2019a).

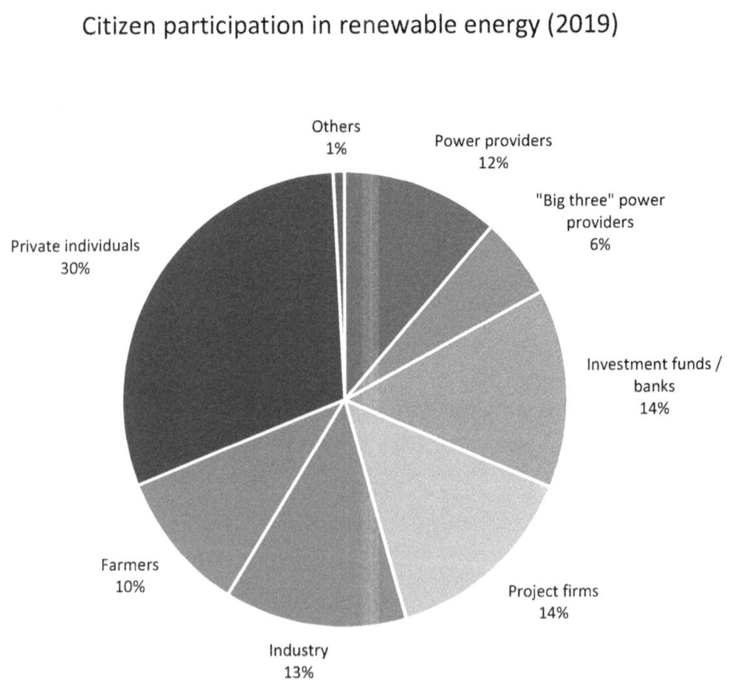

Figure 21.2 Ownership of renewable generation capacity by type of owner.
Source: Own elaboration, based on data from the German Renewable Energy Agency.

4.2.2 Actors

Under EEG 2017, only individuals can be members of energy communities.

4.2.3 Participation

The EEG does not explicitly mention the concept of open and voluntary participation into ECs. However, the national legislation does not establish any obligations with respect to either the participation in CECs or the possibility of leaving CECs.

4.2.4 Control

Control over CECs is exerted only by individuals, with a cap of 10% to each person's share of voting rights (Section 3.15 EEG). Altogether, the members of the CECs shall hold at least 51% of the voting rights to exert control shared across a minimum of 10 individuals.

4.2.5 Activities

CECs can only act as energy producers. They can participate in auctions to obtain incentives for wind generation (Section 36-g EEG). The participation of energy communities in auctions for wind onshore installation has experienced an uncertain pattern over years mostly due to legislative interventions increasing the administrative burdens for small-scale installations (German Renewable Energy Agency, 2019b).

Before 2018, energy communities could (1) bid in auctions without having already conducted an environmental impact assessment; (2) take up to 54 months to complete their wind farms, compared to 30 months for anyone else. After the third round of auctions in 2018, the conditions for energy communities were modified so they now have to (1) submit an environmental impact assessment like anyone else; (2) the 54-month construction timeframe had been reduced to 30 months. This led to a drop in participation. A further motivation for the reduced participation of energy communities in wind auctions lies in the reduction of the auctions' price cap which has been cut over years from 0.07 to 0.063 euro/kWh.

4.2.6 Purposes

EEGs do not place on CECs any obligation to pursue either social, economic, or environmental benefits to the community. However, contributing to energy transition, protecting environment, and promoting the local economy are among the most common motivations explaining the participation of citizens in energy communities (Degenhart, 2014). Similar results are found by Caramizaru and Uihlein (2020) and Verde and Rossetto (2020).

4.2.7 Geographical constraints

Individuals holding at least 51% of the shares of the CECs are required to have their main residence in the urban or rural district where the onshore wind project is located.

4.2.8 Support mechanisms

Participants in energy communities may benefit from different support mechanisms (that apply to renewable installations in general, even beyond energy community):

- Feed-in-tariffs awarded by means of auctions as stated by Section 36-5 of the EEG 2017 and confirmed by the EEG 2021. Only wind onshore installations with a capacity greater than 750 kW can participate in auctions.
- Grant aids at both the federal and regional levels, such as the Market Incentive Program of the Federal Office of Economics and Export Control for renewable heat and the National Climate Protection Initiative (NKI) of the Federal Ministry for Environment for climate protection projects.
- Loans: the government-owned development bank, KfW, provides loans to private citizens, owners, businesses, and municipalities at a low interest rate (starting from 1%) for the promotion of energy efficiency in buildings, renewable investments, and energy storage, municipal infrastructures.
- Feed-in-premiums and surcharge exemptions for self-consumption based on PV installations (see subsection below "Supply of electricity to community members").

These support schemes are not specifically intended for energy communities, but they may help to create favorable financial conditions to kick off community energy initiatives.

4.2.9 Ownership of the generating assets

Energy communities are allowed to rent renewable installations from third parties for the sake of self-consumption (EEG, 2021; Section 21). See subsection below "Supply of electricity to community members" for further details.

4.2.10 Electricity distribution

EEG does not allow CECs to operate distribution networks. Still, no indication is provided about the ownership of microgrids by the community itself. To this aim, CECs can rely on the public grid as well as on privately owned grid.

In recent years, Germany experienced a growing interest by citizens to "remunicipalize" distribution grids owned by corporate utilities as their concessions expire and the assets are tendered out. More than 900 DSOs operate the distribution networks and most of them are publicly owned and have less than 100.000 customers (Wagner and Berlo, 2017). There are several examples of energy communities owning participations in municipal utilities. Among these, EWS Schönau was the first German community to take over the grid as well as the electricity supply to local community[5]; the utility Die Stadtwerke Stuttgart in Stuttgart[6]; the Stadtwerke Union Nordhessen (SUN) in Northern Hesse.[7]

CECs can develop their own grids for the sake of collective self-consumption.

[5] https://www.ews-schoenau.de.
[6] https://www.stadtwerke-stuttgart.de.
[7] https://www.sun-stadtwerke.de.

4.2.11 Supply of electricity to community members

The EEG does not explicitly consider the possibility for CECs to operate as retailers. However, if CECs involve collective self-consumption, they have the opportunity to share the self-produced electricity with third parties not necessarily involved in the community itself.

Section 3-19 of EEG defines "self-supply" as "*the consumption of electricity which a natural or legal person consumes himself in the immediate vicinity of the electricity-generating installation if the electricity is not fed through a grid system and this person operates the electricity generating installation himself.*" The EEG 2021 establishes what follows:

- Renewable power plants with capacity lower than 100 kW can participate into collective self-consumption configurations. With respect to roof-top solar installations, EEG 2021 establishes a cap of 100 kW on the single installation, rather than the whole roof-top installation as it was in the EEG 2017.
- Self-consumption configurations are exempted from the payment of the surcharge EEG-Umlage[8] if the installed capacity is lower than 30 kW and the self-produced electricity is below 30 MWh/year.
- The self-produced electricity which is not consumed can be either injected into the grid or shared with third parties.[9]

Forms of peer-to-peer exchange of electricity are gaining momentum, too. In January 2021, the German Federal Ministry for Economic Affairs and Energy funded a 3-year project to investigate how the blockchain technology can best be used to trade electricity as part of the energy transition (IRENA, 2015; Amenta et al., 2021). At least two P2P platforms operate in Germany for P2P energy sharing[10]:

- Lumenaza allows energy sharing and communities on a local, regional, and national level. The platform includes balance group management, aggregation, billing, and visualization of energy flows. It allows energy communities to participate in electricity markets (see chapter by Chudoba and Borges).[11]
- SonnenCommunity allows the sharing of self-produced renewable power by individual consumers who are using Sonnen's own batteries. This surplus energy is not fed into the grid, but into a virtual energy pool that supplies energy to other community members during times when they cannot produce.[12]

[8] The incentives to renewable energies are funded by a surcharge known as EEG-Umlage on electricity consumers.

[9] https://www.bmwi-energiewende.de/EWD/Redaktion/EN/Newsletter/2021/01/Meldung/direkt-account.html.

[10] https://www.ledgerinsights.com/german-government-backs-build-of-blockchain-energy-trading-platform-best.

[11] https://www.lumenaza.de/en/platform.

[12] https://sonnengroup.com/sonnencommunity.

Table 21.5 Overview of energy communities' features in Germany.

Dimension	Citizens energy communities
Legal form	Any
Actors	Individuals
Participation	Open and voluntary
Control	At least 51% of the voting rights are held by natural persons in the energy community. No member of shareholder of the community can hold more than 10% of the voting rights
Activities	Production of electricity
Purposes	Mostly environmental benefits
Geographical constraints	The natural persons that hold at least 51% need to have their main residence in the urban or rural district where the wind project is located
Support mechanisms	FiT for onshore wind projects
Ownership of generating assets	The generating asset shall not be necessary in the ownership of the energy community
Electricity distribution	ECs are not authorized to operate as DSOs but can hold the ownership of the grid infrastructure
Supply of electricity to community members	Self-consumption, P2P platforms, retailers
Participation in electricity markets	Allowed through P2P platforms and auctions for wind installations

4.2.12 Participation in electricity markets

Participation in electricity markets for energy communities is allowed by means of P2P platforms and auctions for wind onshore installations.

Table 21.5 summarizes the main features of energy communities in Germany according to the dimensions introduced above.

5. Conclusions

Denmark and Germany represent relevant case studies to understand the motivations underlying citizen energy initiatives and their potential perspectives. Both countries pioneered energy communities well before the current legal and regulatory framework was devised and the energy transition became such an important driver for energy policy. This explains why many citizen initiatives do not fit with what *today* would be considered a "proper" energy community. Still, today's discourse on energy community is also, to some extent, made possible by experiences that often date back to several decades ago.

While the origins of energy communities in Denmark and Germany may have been similar, their evolution over time and—more importantly—and their recognition in the legal framework has diverged substantially. Germany gave explicit recognition to energy community and other citizen-led initiatives in its framework legislation on energy and the environment (the so-called EEG). This enabled energy communities to keep prospering and put the country in a favorable condition vis-à-vis the requirement to transpose the concepts of RECs and CECs from RED II and IEMD, respectively. However, in doing so, Germany adopted a loose definition of what an EC is, and perhaps failed to capture both the most valuable and peculiar features of, and therefore the potential benefits from, energy communities more strictly understood. On the contrary, the development of energy communities—or, rather, citizen ownership of energy assets—in Denmark was made possible by a number of nonreplicable regulatory conditions, including (but not limited to) the not-for-profit rule for wind power and district heating, the requirements that the owner of some energy assets are local residents, caps to the number of share per person and the number of wind turbines per owner, etc. For these reasons, it might be perhaps more correct to speak about community energy, rather than energy communities, in both cases.

Both in Germany and in Denmark some sort of ideological adherence to the idea of self-reliance, rather than reliance on distant corporations, for energy supply had and still has a role in motivating people's participation in energy communities and other citizen initiatives. In Denmark, in fact, this was a major determinant for the creation of the first experiments of citizen ownership of wind turbines in the late 1970s and early 1980s. In Germany, the movement toward municipal ownership of power network belongs to the same ideological wave. Within this context, seeking forms of nonfinancial remuneration explains why many people get involved in cooperatives or other forms of participation: they are not just interested in saving money but also, and possibly more, in feeling they have the control over their own energy supply, they contribute directly to reducing CO_2 emissions and other environmental impacts, etc.

The factors described above are likely to result in a continuing push toward greater citizen participation. However, other factors may not be always in place. Both in Germany and in Denmark, citizen ownership of energy assets benefitted of financial as well as regulatory incentives. The evolving regulatory framework and the technological progress have diminished these effects or the sustainability of the underlying norms. On one hand, the EU-led reforms toward market liberalization opened the door to large corporations that are best equipped to capture economies of scale (where they exist). On the other hand, the shift toward tendering and Fips, rather than Fits, rewarded the professional management of large, utility-scale plants. Even more, the increasing returns from larger plants—larger but fewer turbines instead of smaller ones, and wind fields instead of isolated turbines—promoted that shift even beyond the changing landscape of incentives.

At the same time, the decreasing cost of photovoltaic panels creates a window of opportunity for community initiatives, especially when distributed generation becomes associated with storage facilities and professional (whether or not community owned) aggregators take charge of trading the excess production of energy (Sioshansi, 2020).

The question that so far remains unanswered is whether energy communities (and/or community energy) are a potential form of individual engagement in energy transition or just a legacy from the past. This is particularly true within the context of retail competition: many cooperatives or other forms of direct involvement in energy supply and sharing (such as district heating or collective self-consumption of electricity) are rooted in a regulatory environment that left no place to individual choice. Rather, consumers were locked in the supply from the local grid operator.

Denmark seems to suggest that energy communities are by and large a legacy from the past, despite some resurgence after the 2009–12 stop. In the case of wind power, the citizen-owned capacity—albeit large—is aging and it is not clear to what extent it will be replaced by corporate- or citizen-owned turbines. Even the 20% rule—that does not fit into energy community but confirms the political importance of citizen involvement—has had mixed outcomes to say the least. In the case of district heating, obligations to connect and the not-for-profit rule make the experience at the very least not replicable.

On the contrary, Germany seems to suggest that energy communities can and do have a place in the energy transition, even though they still need a better and more precise recognition in the legal framework (at least if one takes a strict definition of what an energy community is). Energy communities and citizen involvement may become a driver for new investments, distributed generation and storage, and possibly even participation to energy markets either directly or through aggregation, both in the realm of peer-to-peer exchange of energy and in that of providing services in the balancing markets.

Differently from Germany and Denmark, in many, if not most, EU member states energy communities are a new concept being introduced as they transpose RED II and IEMD in their national legislations. Only time will tell if they will go Danish or the German way—and if ECs will remain a loose concept related to citizen participation in the energy transition, or it will become a more focused model of producing, distributing, and consuming energy.

References

Amenta, C., Riva Sanseverino, E., Stagnaro, C., 2021. Regulating blockchain for sustainability? The critical relationship between digital innovation, regulation, and electricity governance. Energy Res. Social Sci. 76, 102060.

Bauwens, T., Gotchev, B., Holstenkamp, B., 2016. What drives the development of community energy in Europe? The case of wind power cooperatives. Energy Res. Social Sci. 13, 136–147.

Caramizaru, A., Uihlein, A., 2020. Energy Communities: An Overview of Energy and Social Innovation. European Commission – JRC Science for Policy Report. JRC119433.

Chittum, A., Østergaard, P.A., 2014. How Danish communal heat planning empowers municipalities and benefits individual consumers. Energy Pol. 74, 465–474.

Danish Wind Turbine Owners' Association, 2009. Cooperatives: A Local and Democratic Ownership of Wind Turbines. https://energiayhistud.ee/wp-content/uploads/2015/01/Danish-wind-coops-2009.pdf.

Degenhart, H., 2014. Marktrealität von Bürgerenergie und mögliche Auswirkungen von regulatorischen Eingriffen. Available at: https://www.buendnis-buergerenergie.de/fileadmin/user_upload/downloads/Studien/Studie_Marktrealitaet_von_Buergerenergie_und_moegliche_Auswirkungen_von_regulatorischen_Eingriffen_BBEn_BUND_042014.pdf.

Egelund Olsen, B., 2014. Regulatory financial obligations for promoting local acceptance of renewable energy projects. In: Peeters, M., Schomerus, T. (Eds.), Renewable Energy Law in the EU. Edward Elgar Publishing, London, pp. 189–209.

Eikeland, P.O., Inderberg, T.H.J., 2016. Energy system transformation and long-term interest constellations in Denmark: can agency beat structure? Energy Res. Social Sci. 11, 164–173.

Gancheva, M., O'Brien, S., Crook, N., Monteiro, C., 2018. Models of Local Energy Ownership and the Role of Local Energy Communities in Energy Transition in Europe. European Committee of the Regions – Commission for the Environment, Climate Change and Energy.

German Renewable Energy Agency, 2019a. Energy Cooperatives on the Upswing and Downswing. Available at: https://www.unendlich-viel-energie.de/media/file/4041.AEE_RK_Energiegenossenschaften_Jul20_en.pdf.

German Renewable Energy Agency, 2019b. Community Energy in Germany. More than Just Climate Change Mitigation. Available at: https://www.unendlich-viel-energie.de/media/file/3591.89_Renews_Spezial_Community_energy_LECo.pdf.

German Renewable Energy Agency, 2021. Renewable Energy in the Hands of People. Infographic. Available at: https://www.unendlich-viel-energie.de/media-library/charts-and-data/renewable-energy-in-the-hands-of-the-people2.

Gorroño-Albizu, L., Sperling, K., Djørup, S., 2019. The past, present and uncertain future of community energy in Denmark: critically reviewing and conceptualising citizen ownership. Energy Res. Social Sci. 57, 101231.

IEA, 2017. Denmark 2017. Energy Policy Review. Available at: https://www.iea.org/reports/energy-policies-of-iea-countries-denmark-2017-review.

IRENA, 2015. Renewable Energy Prospects: Germany. Available at: https://www.irena.org/publications/2015/Nov/Renewable-Energy-Prospects-Germany.

IZES, 2021. Stand der Umsetzung der RED II-Richtlinie in Deutschland mit Blick auf die Bürgerenergie. Available at: https://www.buendnis-buergerenergie.de/fileadmin/user_upload/downloads/Studien/20210728_IZES_Kurzstudie_BBEn_RED_II_final.pdf.

Kooij, H.-J., Oteman, M., Veeman, S., Sperling, K., Magnuson, D., Palm, J., Hvelplund, F., 2018. Between grassroots and treetops: community power and institutional dependence in the renewable energy sector in Denmark, Sweden and The Netherlands. Energy Res. Social Sci. 37, 52–64.

Mey, F., Diesendorf, M., 2018. Who owns an energy transition? Strategic action fields and community wind energy in Denmark. Energy Res. Social Sci. 35, 108–117.

Olhorst, D., 2015. Germany's energy transition policy between national targets and decentralized responsibilities. J. Integr. Environ. Sci. 15, 303–322.

Romero-Rubio, C., Diaz, J.R.A., 2015. Sustainable energy communities: a study contrasting Spain and Germany. Energy Pol. 85, 397–409.

Savaresi, A., 2019. The rise of community energy from grassroots to mainstream: the role of law and policy. J. Environ. Law 31, 1–24.

Sioshansi, F. (Ed.), 2020. Variable Generation, Flexible Demand. Academic Press, London.

Verde, S., Rossetto, N., 2020. The Future of Renewable Energy Communities in the EU: An Investigation at the Time of the Clean Energy Package. Available at: https://fsr.eui.eu/publications/?handle=1814/68383.

Wagner, O., Berlo, K., 2017. Remunicipalisation and foundation of municipal utilities in the German energy sector: details about newly established enterprises. J. Sustain. Dev. Energy Water Environ. Syst. 15 (3), 396–407.

Wierling, A., Schwanitz, V.J., Zeib, J.P., Bout, C., Candelise, C., Gilcrease, W., Gregg, J.S., 2018. Statistical evidence on the role of energy cooperatives for the energy transition in European countries. Sustainability 10 (9), 3339.

Platform-based energy communities in Germany and their benefits and challenges

22

Christian Chudoba and Tereza Borges
Lumenaza GmbH, Berlin, Germany

1. Introduction

The energy sector is in a state of flux. In 2020, around 46% of German electricity was produced from renewable sources, making renewables the single largest contributor to the country's electricity supply. Germany has ambitious goals to increase this share to 65% by 2030. Before, electricity was produced in a centralized, inflexible manner. The boom of renewables has started a paradigm shift, reshaping how the energy system works. Electricity is increasingly supplied by millions of decentralized energy resources. The classic separation between producers and consumers is progressively disappearing, and more people and organizations are producing and consuming electricity—they are becoming prosumers and forming energy communities to share resources and maximize the benefits for all involved.

However, this shift has only just started. The developments of the last few years indicate the overall potential, but there is still a long way to go to transform the grid and the end-customer engagement. Platform-based energy communities can be a key enabler for the adoption of renewables and decentralized energy solutions.

The end-customer interest in electricity from a clean origin, as well as in finding ways to reduce their carbon footprint and energy bills, is often not fully capitalized due to hermetic and unclear propositions. Photovoltaic systems, batteries, electric vehicles, electric cooling and heating systems, multiple pieces of software, and services are among the components to be considered. However, average end-consumers are motivated by economic, social, and environmental benefits, to deviate from the status quo. They are looking for personalized offers that are convenient, simple, and transparent. As several chapters in this volume explain, energy communities offer a strong foundation for the provision of all the benefits end-customers desire.

These developments are putting pressure on energy providers. Renewable electricity and a digital customer experience must be brought together. The increasing number of decentralized assets needs to be connected and controlled, leading to an enormous increase in the amount of data collected, processed, and translated for the end-customer and actors in the value chain. Today, green energy service providers demand scalable solutions to grow and diversify their offering. Platform-based energy communities are a vital instrument to maximize the value of different hardware

Energy Communities. https://doi.org/10.1016/B978-0-323-91135-1.00003-1

components, providing their members with green electricity and accelerating the energy transition.

This chapter is organized as follows:

- Section 2 describes the historical development of energy communities in Germany;
- Section 3 discusses and defines the term energy service providers;
- Section 4 examines how a platform can remove barriers to the energy market;
- Section 5 analyzes the power of platforms; and
- Section 6 identifies threats and opportunities for green energy service providers followed by the chapter's conclusions.

Many of the services mentioned throughout this chapter are enabled by the Lumenaza energy-as-a-service (EaaS) platform.[1] They illustrate how platform-based energy communities are beneficial for energy service providers with examples found toward the end of each section.

2. The historical development of energy communities in Germany

For decades, there was a stable supply focused model in Germany which assumed a one-way, top-down distribution of electricity from large power plants to consumers. The first tentative opening of grid access for renewable energies through the 1991 Energy Feed Act did little to change this. Significant changes to the top-down principle in power generation did not begin until 2000 with the enactment of the Renewable Energy Sources Act (EEG) as detailed in the chapter by Spasova and Braungardt. In the years that followed, renewable energies experienced a veritable boom in Germany. For the first time in almost a 100 years, this gave decentralized generation an enormous boost and, over the years, even caused nonrenewable energy sources to be displaced due to the legally regulated priority of renewable energies (Bundesministerium für Wirtschaft und Energie, 2017). The chapter by Benedettini and Stagnaro provides further information about strong dissemination of energy communities in Germany and Denmark.

From the very beginning, the energy transition was driven by motivated citizens and organizations to move toward a more localized, citizen-centric sustainable energy system and reduce CO_2 emissions. The first 10 years were used to build significant renewable capacity. When in 2012 the regulation in Germany allowed for the first time the selling of electricity from the power plant to consumers, it was again PV installers, energy cooperatives, and local utilities with deep roots into their local communities who provided green electricity of clear origin to their customers with the help of digital EaaS platforms (Fischer & Wetzel, 2018). This success then attracted the interest of larger businesses and traditional utilities as well as politicians.

[1] https://lumenaza.de/en/.

Since then, prosumers have been on the rise. According to the German Solar Association, between 2019 and 2020, the number of prosumers increased by 100,000 (Bundesverband Solarwirtschaft, 2020). These prosumers began to form energy communities, which are not a new phenomenon in Germany. Generally, energy communities were often understood as energy cooperatives, providing electricity to their members. The cooperative movement has a long history, originating from agricultural cooperatives or credit associations. By the end of the 19th century, cooperatives were formed to produce energy and install and operate a distribution grid since larger companies did not do this on account of low profitability. During the 20th century, cheap and abundant fossil fuels became the core of the energy sector. More recently, the promotion of renewable energies has led to a revival of energy cooperatives. Energy cooperatives, also called citizen energy or community energy, mostly realize renewable energy projects and allow participants to engage in the decision-making of local energy policy. Since the inception of the EEG, the cooperative movement has greatly expanded. This was until the change to tender-based auctions to install more renewable capacity, which slowed down the establishment of energy cooperatives dramatically as shown in Fig. 22.1.

The strength of the cooperative movement is to align people around the common interest of delivering benefits to their community, both financially and environmentally. The mechanism of providing one vote to each member aims to guarantee that decisions are made in alignment with the views of most members, mitigating the risk of dominance by the most financially powerful parties. Based on our experience, the main motivations for establishing communities are the following:

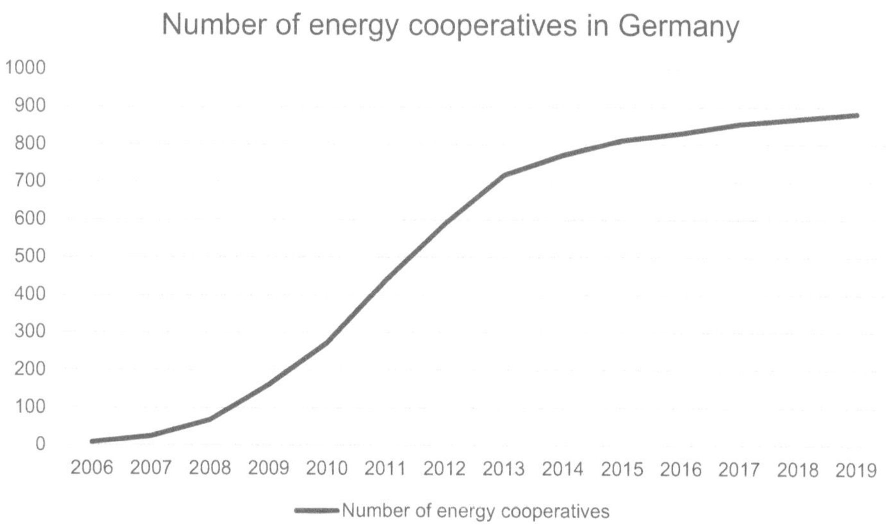

Figure 22.1 Foundations of energy cooperatives since 2006 (cumulated).
Source: DGRV, July 8, 2020. Energy cooperatives in Germany. https://www.dgrv.de/wp-content/uploads/2020/07/20200708_State-of-the-sector-2020.pdf.

- Foster local economy in supporting local production owned by the community through enabling more renewables;
- Increase local control;
- Increase the amount of real green energy usage and make this fully transparent;
- Provide attractive power tariffs to support the use of local source;
- Increase overall awareness and sense of responsibility for the grid as common good through flexible tariffs reflecting the local situation of the network; and
- Establish a communication network to raise awareness about the local energy system.

Several regulatory developments accelerated the launch of energy communities in the 2010s. One of the most significant decisions after the enactment of the EEG was the introduction of an optional market premium in 2012, designed to support demand-oriented electricity generation. The market premium spurred the realization of increased production from renewables, driving demand to consume electricity locally from a group of regional suppliers. The chapters by Spasova and Braungardt as well as Benedettini and Stagnaro provide further information on the relevant regulatory changes in Germany.

The utility of Wunsiedel, in northeast Bavaria, Germany, was one of the pioneers in this field and understood the challenges early on, establishing its regional community[2] project in 2016. It had the vision of supplying locally produced energy to attract and retain local customers, maximizing benefits for the region. Consumers and producers of renewable energy in the community are connected via a digital marketplace, providing full transparency regarding the origin of electricity. During the first months of 2021, the community produced 4.7 million kWh of green electricity, amounting to a reduction of around 2500 tons of CO_2. By creating this energy community, the local utility profited from higher customer loyalty and established itself as a truly green, credible, and resilient brand.

In 2017, the law for tenant electricity (Gesetz zur Förderung von Mieterstrom) came into force, enabling new community models for tenants. In the same year, the local utility of Karlsruhe joined forces with housing developers Hoepfner Brau to explore the opportunity created by the new law. Their joint company Badische Energie[3] enables an electricity community for neighborhoods and areas in the city. Tenants can purchase the electricity generated in their neighborhoods directly as tenant electricity, creating an emotional connection between the electricity produced and the tenants. Moreover, tenants benefit from lower tariffs and the clean origin of their electricity due to the collective self-consumption model.

Since then, Germany has been lagging behind other EU member states regarding the transposition of the EU Directive changes on Renewable Energy Communities (REC) and Citizen Energy Community (CEC) because the power of the incumbent stakeholders is too great and has not been challenged to date. There have been no changes to the regulation to facilitate or incentivize collective self-consumption, as seen in countries such as Spain and Austria, for example. These countries have

[2] fichtelgebirgsstrom.de.
[3] community.badische-energie.de.

embraced the transposition of the EU directives to their national laws as a way to maximize the potential of communities in providing not only environmental, but economic and social benefits. With the sharp growth of electric vehicles in the German market and the associated need to rethink the energy system from a flexibility-first perspective, the automotive industry can become a game changer ally in pushing for regulatory changes to accelerate the transition to green customer-centric energy offerings.

There are several motivating factors for the adoption of community-based approaches. Some market developments are driven by private profit-seeking entities, which at the same time need to shape benefits for the end customer. Others are led by groups of citizens or orchestrated by the public sector aiming to solve local problems and stimulate development. REC and CEC models provide benefits to a more diverse group of stakeholders. What they have in common is their need to have ways to bring their visions to life, shorten the time to market, and rationalize operational costs, further described below.

3. Defining energy service providers

For a long time, the energy industry was considered stable and predictable. Utilities were perceived as bureaucratic, slow institutions lacking in innovation. High investment costs and long-lasting infrastructures ensured long planning periods and a high degree of continuity. Legal frameworks supported this stability for many decades through territorial monopolies, state-owned companies, and regulated prices. However, this model has been gradually shaken for the last decade. The understanding of what a utility is has changed. It is no longer only a regulated vertical company or a distribution company, but a multitude of energy service providers, which encompasses hardware manufacturers, retailers, and electric mobility providers, supplying their customers with green energy. Throughout this chapter, the terms "utility" and "energy service providers" will be used in this context.

3.1 A state of flux

The energy system continues to transform. Decentralized, often photovoltaic generation plants now contribute significantly to electricity generation in Germany. For instance, photovoltaic systems generated around 6.9 billion kWh of electricity in April 2020 (Enkhardt, 2020). The expansion of photovoltaics in Germany continues, and prosumers have become a key contributor to the success of this energy transition.

The greater decentralization and diversity in the energy system increases its complexity, which can only be managed with the help of advanced technologies capable of remotely collecting and processing data and intervening via automated control actions, topics further explored in the chapter by Ghiani et al. Connectivity and the capacity to process enormous amounts of data have become a necessity. End-customers are changing their perspective of electricity as a commodity and guiding their decision-making toward services, convenience, value, and doing the right thing for the planet while rationalizing costs. Traditional utilities have no choice: they must reinvent themselves or face the fate of becoming legacy infrastructure holders.

3.2 Digital energy service providers

New entrants to the energy sector have been showing the way for the traditional utilities. The rise of digital energy service companies whose offerings evolve around new technologies is enabling innovative, data-based products and customer-centered services.

Some of the traditional players are progressively recognizing the inevitability of change and exploring ways to become digital, customer-oriented energy service companies. The future will belong to those players in the energy sector who do not stand still in their development and decisively lead the way in terms of innovative capability. These digital service providers invest in a strong backbone to process huge amounts of data, automate backend processes, and support seamless and attractive customer journeys. The encapsulation of the complexity behind customer-centered energy service offerings is vital for survival. Another extremely relevant aspect is the ability to continuously test new ideas and keep evolving, collecting learnings from the engagement with end-customers and adapting accordingly. In a highly interconnected world, customer centricity is naturally extendable to affinity groups or communities.

These new-age utilities or digital energy service providers have the opportunity to deliver benefits to an individual end-customer and beyond via community concepts such as sharing, collectively producing and/or consuming, collecting bonuses and coins or offering other rewards for contributing to grid stability. The various kinds energy community models have become a promising offering, enabling customers to generate, distribute, and consume energy more efficiently while strengthening customer loyalty to the service provider.

EnBW is a good example of a German utility that has adapted to the changing energy landscape. Instead of just supplying its customers with electricity, it has created prosumer communities, allowing it to target a completely new customer segment. The company's solar + product offers customers a complete PV storage solution for their household. Supply, generation, and storage of electricity are optimized by intelligent algorithms for self-consumption. Prosumers profit from tariffs that meet their needs and the possibility to share energy within the community. A family of four with an annual electricity consumption of 4000 kWh pays the utility about €1200 a year. These costs add up to about €34,000 over the next 20 years, assuming an electricity price increase of 3.5% per year. By comparison, a complete solution with EnBW solar + costs only about €20,000. Consumers secure a guaranteed feed-in tariff of €7500 for the PV surplus for 20 years, which finances the cloud contribution. Through self-sufficiency, consumers save about €21,500 with this solution. EnBW's approach increases customer loyalty and provides financial benefits to its customers because the combined solution of PV, storage, and associated electricity tariffs is comparably cheaper.

4. The power of digital platforms

Digital platforms are responding to changing market environments across different sectors. This section examines the success of digital platforms overall and more specifically for the energy market.

Over several decades, the world has seen the rise of digital platforms as organizations seek to achieve economies of scale in an ever-changing environment across a wide range of industries. Fully integrated platforms like Amazon and Google and sector-focused disruptors such as Airbnb and Uber are well known by the general public, offering customer-centric, simple solutions to highly complex problems. Airbnb's platform is an illustrative example of how the company connects travelers with hosts worldwide, ensuring trust and creating amazing experiences.

In the energy sector, digital platforms can enable new business models and handle the complexity of the decentralization of resources while ensuring compliance with the applicable rules of nature and industry regulations. Lumenaza has been the pioneer in this space since its inception in 2013.[4]

Digital platforms are a powerful response to the fast pace of change in the energy sector. The main benefit of digital EaaS platforms is that they can be used to operationalize all the tasks and processes of an energy service provider in a modular and highly automated way.

Lumenaza has been recognized as the leading residential EaaS platform provider (Guidehouse, 2019). Other companies in the space include Piclo,[5] a flexibility trading company, and Kraken Technologies,[6] Octopus' energy tech platform. Lumenaza's open and modular cloud-based platform works across the full end-customer lifecycle and is adaptable to diverse use cases. The company offers innovative ways of integrating new products within existing portfolios and also provides easy access to the energy market. The global EaaS market size was USD 17.82 billion in 2019 and is projected to reach USD 41.85 billion by 2027 (Fortune, 2021).

Digital platforms facilitate the transition of traditional utilities to decentralized operations and new business models. For companies wishing to enter the energy sector, such as hardware manufacturers, general retailers, or automotive OEMs, the use of a digital energy platform can shorten market entry time and mitigate risks. Municipalities, citizen associations, and nonprofit organizations can also bring their energy service concepts to life. EaaS platforms provide a solid framework for the setup of complex business models, integrating multiple parties and devices and ensuring compliance with industry regulations.

Energy communities are an excellent example of high complexity, as different hardware components need to be integrated, controlled, and optimized. Moreover, customer acquisition, registration, use of service, and billing processes have to support the desired customer experience. Energy services need to be billed correctly and are

[4] https://lumenaza.de/en/home/.
[5] https://picloflex.com/.
[6] https://octopusenergy.group/kraken-technologies.

often more than just for consumed electricity. Other services that need to be accurately billed include insurances, bonus schemes, the sharing of electricity, or the giving of rewards for providing flexibility services to the grid. Ultimately, energy communities are integrated into the wider energy system. Hence, they need to automate the regulatory processes and seamlessly interact with the energy market.

The diverse character of potential digital communities or marketplaces requires sophisticated algorithms, optimizing generation and consumption in each geographical region or virtual balancing group to ensure a constantly balanced electricity supply. EaaS platforms provide flexibility for the setup of innovative offerings and subsequently for carrying out continuous expansion and enhancements. Scalability and open interfaces to support easy integration to various partners and customer interfacing channels are also must-haves of digital platforms, as the overall user experience is an essential factor for customer acquisition and loyalty.

Another key feature of digital energy platforms is the ability to fit seamlessly into existing specific processes, enabling the evolution of product offerings. For this purpose, platforms need to use APIs to build best-for-purpose solutions. APIs allow the integration of already existing tools and services, again shaping the best possible solutions for end-customers.

Greencom Networks' Shine Community[7] exemplifies the potential of EaaS platforms. In Brunnthal, 20 km south of Munich, the company connects all members of its community, provides them with a live visualization of power flows and offers a bonus for every kWh of power generated and consumed within the community. This incentivizes customer behaviors that contribute to the optimization of local balancing. Greencom needed an energy platform to handle all energy management processes and billing. This platform-based community model empowers a completely new energy experience, showcasing a customer-centered proposition in practice. The benefits are being measured in terms of CO_2 savings, export/import ratio, savings on monthly bills, and self-consumption as shown in Fig. 22.2.

5. The opportunities and threats for energy service providers

Existing energy service providers can explore platform-based energy communities to drive customer acquisition and loyalty, as well as to establish a strong foundation for evolving energy services. At the same time, the introduction of such innovation may impact existing business models and value chains.

The benefits for traditional energy service providers are manifold. First and foremost, platform-based energy communities can help them to take the lead in making a serious and significant contribution to protecting the climate. As discussed earlier, platforms are a fast and powerful tool to set up customer-centered energy service models that enable engagement of a broader range of stakeholders, provide options

[7] brunnthal.shine.eco/home.

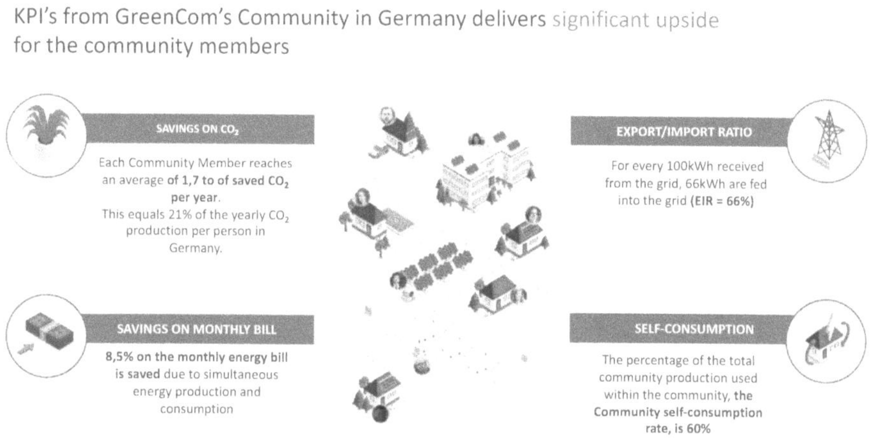

Figure 22.2 KPI's from Greencom's community.
Source: Greencom Networks GmbH.

for the roll-out of renewables, and stimulate the adoption of multivector solutions for decarbonization. Therefore, platform-based energy communities strengthen local energy transitions, increase local acceptance, and boost consumer engagement. Traditional energy service providers can also profit from an improved public perception and enhanced green brand to retain and expand their customer base.

However, the impact of platform-based energy communities goes further than enabling traditional energy service providers to retain and acquire customers. They provide the foundation for new business models and unlock opportunities for future growth. Energy service providers profit from several factors. They can start to offer new energy products at a radically lower cost. The cost savings result from the fact that new platforms are designed and built to be scalable, modular, and easy to adapt. Furthermore, existing energy service providers often operate slow and costly legacy systems. Platform-based approaches empower them to leap forward and launch the next generation of energy offerings. *Regionah Energie* has been using Lumenaza's EaaS since 2017, connecting the renewable production to local consumers in the region between Reutlingen and Ulm, as illustrated in Fig. 22.3. The most recent addition to the portfolio of offerings was the EV charging tariff, bringing further sustainable choice for the community with minimum setup effort.

Moreover, modular software platforms can support both existing and new energy service providers, no matter at what stage of the innovation cycle they are. Existing energy service providers do not have to untangle their legacy back-end systems and can instead start with a focused scope and minimal integrations. After establishing the initial setup, they are empowered to develop a roadmap to progressively transform their product offering, as EaaS platforms offer functional frameworks, APIs, and open interfaces. This lowers cost to serve and time to innovate. One recent example is the

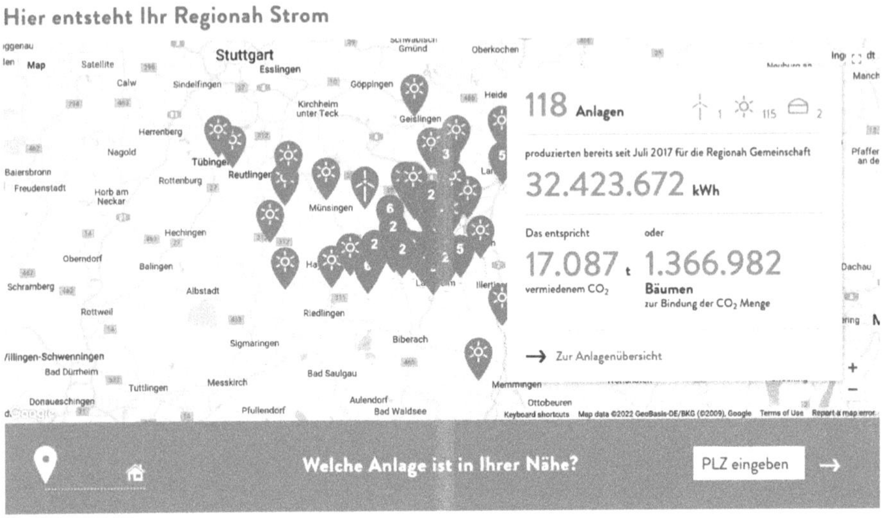

Figure 22.3 Regionah energie.
Source: regionah-energie.de.

German energy supplier eprimo,[8] a subsidiary of E.ON, which offers electricity tariffs nationwide. To broaden its offering, attract higher value customers, and increase loyalty, eprimo decided to launch a country-wide green energy community. In this scenario, eprimo opted for the use of Lumenaza's EaaS platform-based services to handle the back-end processes and payments for the producers of the electricity from clean origin.

The platform-based approach shortens time to market for new entrants to the energy market, removes barriers, and mitigates risks. Instead of building all necessary capabilities for the energy market themselves, new entrants can enter the market quickly, allowing for experimentation and adaptation to market trends as well as progressive learning.

The complexity of dealing with a high number of decentralized energy production assets is well covered with the use of an EaaS platform and services to automatically generate individual forecasts for each plant, handle billing of the electricity exported to the grid, and cover associated energy management processes. This example underscores how software-based solutions dramatically reduce cost to innovate for energy service providers while reducing time to market, allowing them to launch new propositions fast.

This is illustrated by Lumenaza's platform, which consists of several building blocks, providing a scalable, modular, and flexible end-to-end solution (see Fig. 22.4). It fits seamlessly into the processes of energy service providers and enables the evolution of their product offerings. State-of-the-art APIs allow them to build best-for-purpose solutions, while the microservices-based architecture allows them to shape market-leading customer propositions and adapt to changing demands quickly.

[8] eprimo.de/gruenstromcommunity.

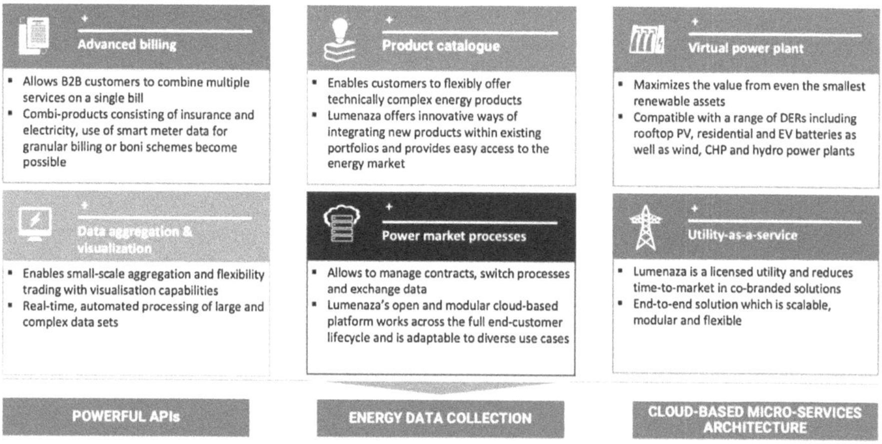

Figure 22.4 Lumenaza energy-as-a-service platform building blocks.

Energy service providers also have the opportunity to expand their current remit and move beyond traditional electricity tariffs. Instead, they can offer a wide variety of products, bundling relevant services together. Platform-based energy communities are a strong driver for promoting product bundles, as different hardware components and service concepts are combined. These digital, affordable, and attractive customer experiences attract and retain customers, offering existing energy service providers opportunities to grow and move away from low-margin pressures to more profitable segments.

Lumenaza's EaaS platform shortens the time for the setup of customer-centered energy services and reduces the operational cost by automating all the backend process to onboard, serve, bill, and terminate contracts. The virtual power plant module automates the aggregation of distributed energy resources from very small to large scale, as well as the handling of flexibility services. Currently, the platform processes 1.5 billion data points, 250.000 bills, and 5000 controllable assets per year. The platform features are flexible and scale to empower a broad range of use cases.

IBC Solar[9] is another example of a new entrant into the energy market. As a leading global system house for photovoltaics and energy storage, this company extended its offering of hardware components by launching a green electricity tariff. The tariff optimizes self-consumption and maximizes the advantages of solar systems and storage, highlighting the opportunities of energy communities for new green utilities. Making the shift from purely selling electricity or hardware to providing energy services can be very challenging. Lumenaza's energy-as-a-service platform provides IBC Solar a strong foundation and helps prevent risks, enabling the special tariff, sourcing the green electricity and running the onboarding, switching, billing, and termination processes.

[9] https://www.ibc-solar.de/efa-home-energiemanagement/.

Creating digital offerings requires a range of capabilities that are currently limited or nonexistent in traditional energy service providers. End-customer requirements are becoming more individualized and change more rapidly than they used to do in the past, but most utilities are not accustomed to thinking in a customer-centric way. This can pose a threat when creating energy communities as customers see no additional value in the new propositions. This risk can be reduced by working together with experienced partners that can guide utilities through the process.

Similarly, internal decision-making can pose a threat for energy service providers when creating platform-based energy communities. It is crucial to be fast and agile when adapting to rapidly changing market conditions. Over the last few years, the boom and breakthrough of EVs has underpinned new components entering the market, highlighting the immediate need for new offers. This risk can be mitigated, especially by using software platforms that constantly evolve and allow for new use cases.

6. Conclusion

Energy community models in various shapes and sizes have successfully engaged citizens, businesses, and the public sector and are already delivering environmental, economic, and social benefits. This trend will continue to accelerate as homes become more connected and more decentralized assets such as electric vehicles, heat pumps, and cooling systems are operated.

Electric vehicles are a leading example of how an important aspect of everyday life can be decarbonized via charging with green electricity. Moreover, the batteries found in electric vehicles can be great assets when it comes to supply peaks. The excess electricity is very cheap, while the battery helps to stabilize the grid. Connecting different assets in a community and controlling them in a granular, optimized way calls for sophisticated software platforms, especially when billing the exact amounts to each community member.

Another key trend for the next few years is the European-wide broad implementation of smart meters. Germany is lagging behind in the rollout of smart meters, while other countries such as Spain already have comprehensive coverage. Using smart meters will empower the next generation of customer-centric energy services, such as time-of-use tariffs or the optimization of collective self-consumption. Both of these examples would elevate community models and speed up their dissemination. Lastly, the number of platform-based communities will see exponential growth in Europe as regulatory, fiscal, and legal conditions mature.

Energy communities are a cornerstone of a successful energy transition. A combination of economic, environmental, and social benefits is at the heart of community models. Given their effective rollout over the last few years, platform-based communities enable consumers, prosumers, and producers of green electricity to participate actively in the energy market transformation.

Traditional energy suppliers are under pressure, while new players are entering the energy market and offering energy services combined with their core products. Each group has its own challenges, and both need capabilities to operate in a fast-changing, green, decentralized energy market. Platform-based energy communities provide a strong foundation for deployment of new customer-centric energy services at low cost with high speed. It can be a key enabler for existing utilities and new entrants alike to speed up the transition to a decentralized, green energy world.

Bibliography

Bundesministerium für Wirtschaft und Energie, 2017. Renewable Energy Sources Act (EEG 2017). https://www.bmwi.de/Redaktion/EN/Downloads/renewable-energy-sources-act-2017.pdf%3F__blob%3DpublicationFile%26v%3D3.

Bundesverband Solarwirtschaft, March 30, 2020. Photovoltaics Prosumers. Bundesverband Solarwirtschaft. https://www.solarwirtschaft.de/en/2020/03/30/nearly-100-000-new-pv-prosumers-in-germany/.

DGRV, July 8, 2020. Energy Cooperatives in Germany. https://www.dgrv.de/wp-content/uploads/2020/07/20200708_State-of-the-sector-2020.pdf.

Enkhardt, S., April 30, 2020. Fast 7 Milliarden Kilowattstunden Solarstrom im April. Pv Magazine Deutschland. https://www.pv-magazine.de/2020/04/30/fast-7-milliarden-kilowattstunden-solarstrom-im-april/.

Fischer, B., Wetzel, H., 2018. Entwicklungsdynamik und strukturelle Merkmale von Energiegenossenschaften in Deutschland in den Jahren von 2009 bis 2015 (No. 01). klimaGEN. Working Paper).

Fortune, 2021. The Global Energy as a Service Market Size. https://www.fortunebusinessinsights.com/industry-reports/energy-as-a-service-market-101204.

Guidehouse, 2019. Navigant Research Leaderboard: Residential Energy as a Service Platforms.

A community-based biomethane heat network: case study from Trier

23

Christoph Menke[1], *Achim Hill*[2], *Matthias Gebauer*[3,5], *Rudi Schöller*[4], *Daniel Gregetz*[4] *and David Lellinger*[5]
[1]Trier University of Applied Sciences, Trier, Germany; [2]Energy Agency for the Region of Trier, Trier, Germany; [3]Ecoscop GmbH, Trier, Germany; [4]Trier Municipal Utilities, Trier, Germany; [5]Climate Manager for the City of Trier, Trier, Germany

1. Introduction

In order to reduce greenhouse gases (GHGs) in local communities where local transport, power, and heat are the main sources of GHG, the main intervention areas are the supply of renewable power and the decarbonization of the heating sector, as well as changing mobility within a town.

Communities can become active, especially in the heating sector, as this is mainly a local task, as heat, unlike electricity, cannot easily be transported. This chapter provides a practical example of how communities can get active to reduce their emissions in the heating sector. In order to decarbonize this sector, one promising approach is to implement a local low-carbon district heating network that contains a high share of renewable energies. This approach was, according to a detailed study, the most promising solution for lowering heating-based GHG emissions in the inner city of Trier. The existing building stock in this area has many old buildings with high heat requirements; a large proportion of these buildings are even landmarked and are therefore not suited to other—decentralized—applications such as extensive thermal insulation of the building's exterior, decentralized heat pumps, or other energy efficiency measures. In addition, the availability of biogas generated nearby and upgraded to biomethane presented a possible energy source for a low-carbon heat network.

A part of downtown Trier with a heterogenic structure was therefore chosen, both to verify the concept that a supply with low-carbon energies is economically feasible, and to determine what is needed to cover heat and power demand in this quarter. To facilitate data collection from municipal and public sources, the area being considered (i.e., the quarter) was defined by a combination of three statistical districts: 1032 *Vereinigte Hospitien* (retirement homes), 1036 *Rathaus* (town hall), and 1037 *Mutterhaus* (hospital).

In this quarter, there are already two existing local heating networks. These are mainly powered by CHP (combined heat and power) units, which provide power and heating for a large hospital, vocational school buildings, and retirement homes.

Energy Communities. https://doi.org/10.1016/B978-0-323-91135-1.00024-9

Together with a yet-to-be-constructed new local heating network—which will supply heat to the town hall, a theater, a school and a hotel via an additional set of CHP units—the three heating networks can be combined and will supply low-carbon heat and power to the whole quarter. This covers around 800 buildings with 3500 inhabitants and a large number of public buildings. Fig. 23.1 shows the scale of the project, with an area of around 1 × 2 km. Also shown is the distribution of public buildings, commercial buildings, and residential homes in the districts.

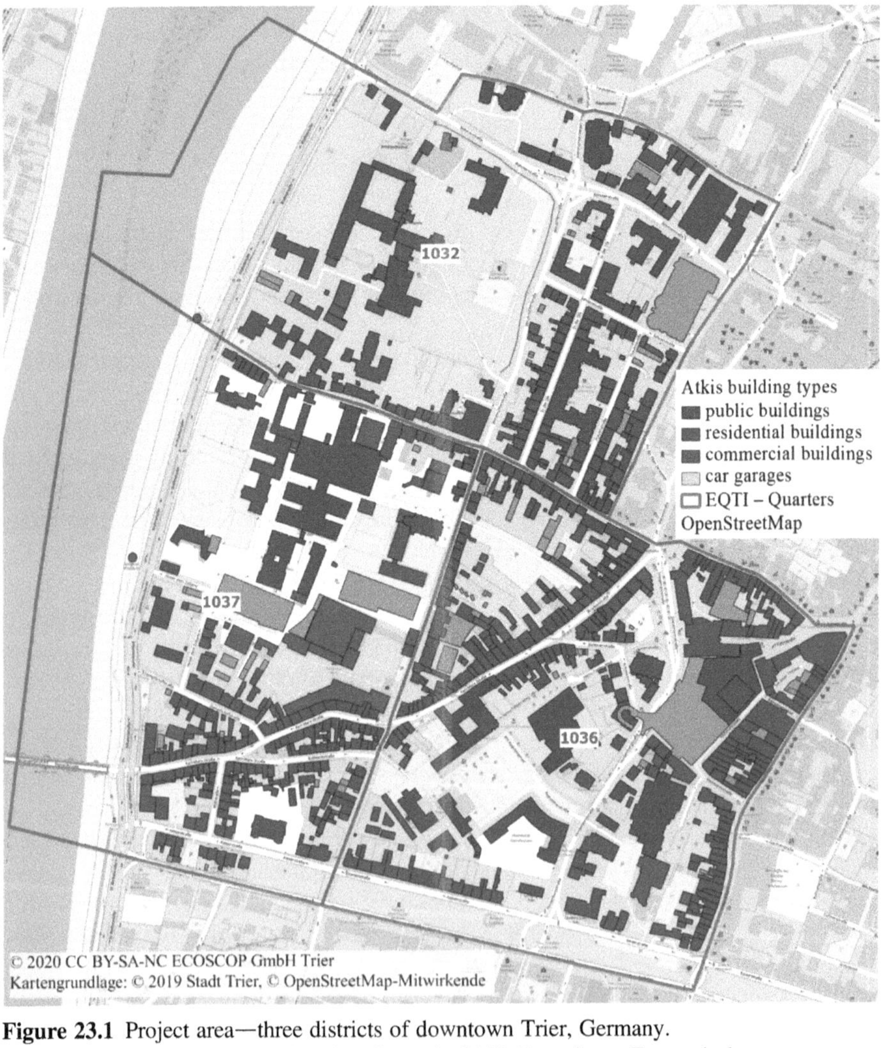

Figure 23.1 Project area—three districts of downtown Trier, Germany.
Source: Gebauer, M., Hill, A., Menke, C., et al., 2021. Integriertes Energetisches Quartierskonzept Trier-Innenstadt Südwest EQTI; Final Report. https://www.trier.de/umwelt-verkehr/klimaschutz/energiequartier/. OpenStreepMap, City of Trier.

The two existing local heating networks use biomethane, which is collected as biogas from existing biogas units around Bitburg, in the Eifel region about 30 km north of Trier, and upgraded to biomethane at a central processing plant.

Given that many cities with existing building stock in their inner-city quarters face similar challenges in decarbonizing their heating sector, the proposed approach, and strategy for building up momentum, developing a local heat network and forming an energy community may serve as an example; this is especially relevant for involving residents, which is explained in detail in Sections 5 and 6.

This chapter consists of seven sections:

- Section 2 introduces the project idea and the partners in the project;
- Section 3 looks at the biogas potential of the nearby region and upgrading it to biomethane to supply the CHP units in Trier;
- Section 4 examines the size, layout, and means for a stepwise approach to building up local heat networks and the economics of this;
- Section 5 describes the approach and methodology for involving the local community. This uses a dynamic scenario builder tool for instantly showing the possibilities and consequences of local energy planning decisions during public discussions; and
- Section 6 describes this participation process for residents and stakeholders, followed by the chapter's conclusions.

2. How the idea got started

In 2017, the newly set up KUE committee *Lenkungsgruppe für Klima, Umwelt und Energie* (Committee for climate, environment and energy) held its first meetings to discuss how to ensure an environmentally friendlier and lower-emission heat and power supply for the city. One of the first ideas that came up was a feasibility study of a renewable heat supply for a heterogeneous area in the city center.

After the meetings, an initial concept for a quarter-based pilot project was drafted. Involving several departments of the city administration, the draft was developed further and led to an application for funding from the German government, which was eventually confirmed.

The city of Trier set up a tender for the creation and execution of the concept. The Energy Agency for the Region of Trier together with the Trier University of Applied Sciences and Ecoscop GmbH formed a consortium for the tender, which was ultimately selected by the administration. The consortium was extended with the local agenda group LA21 Trier to better involve the civil society of Trier. Further to this, the municipal utility *SWT Stadtwerke Trier* and the parks and gardens department (*StadtGrün*) of the city administration also joined the team.

The partners within the consortium and their assigned roles are highlighted below:

Energy Agency for the Region of Trier (Energieagentur Region Trier): This is a regional energy agency with the task of informing and consulting with citizens, municipalities, and companies; initiating projects and activities and helping others to foster energy efficiency; making the transition to renewables; and reducing GHG. It is a key partner for involving civil society, supported by the local agenda group; developer of ideas; and the administrative and overall project management unit within the project team.

Ecoscop GmbH, Trier: This is the data gathering partner, scenario developer; also, one of the initial partners together with the energy agency project manager of the overall project; and coordinator, e.g., for the walk-through energy assessment of the quarter.

Trier University of Applied Sciences (Hochschule Trier), Germany: Master students of the energy department of the university elaborated nine master's studies to provide heat network planning, together with optimization and simulation of the local heat network. In addition, the students did a walk-through energy assessment through the whole quarter to characterize each building in terms of energy-specific data to determine their energy saving potential.

Trier municipal utilities (SWT Stadtwerke Trier): This is the partner for implementation of the concepts and ideas developed; provider of a gas grid in Trier and operator of a biogas collector grid in the region combined with a treatment site to generate biomethane from the biogas. Also, it is the operator of two existing combined heat and power plants in the chosen quarter.

Local agenda group, (Lokale AgendaLA21 Trier): The local agenda group LA21 Trier was integrated to increase the involvement of civil society in Trier. They were also in charge of moderating workshops and events.

Parks and gardens department (StadtGrün Trier): The parks and gardens department was integrated to develop and discuss measures for climate change adaptation in the quarter and (later) in the whole city.

Committee for climate, environment, and energy (Lenkungsgruppe für Klima, Umwelt und Energie KUE): This committee was set up in 2017 to improve connections between private initiatives, market players, and public organizations within the climate, energy, and environment sector and the political parties within the city council, as well as with the city administration itself. The committee meets up to six times per year and discusses topics within its scope. Moreover, experts from different fields support the committee. Together with the political parties included, the committee helps the city council to inform decisions with a deeper understanding of the items discussed.

The project was highly supported by the Trier staff unit for climate protection, environment, and energy (Stabsstelle Klimaschutz, Umwelt und Energie) and additional input came from the Trier city planning office (*Stadtplanungsamt*) and the Trier office for statistics and urban development (*Amt für Statistik und Stadtentwicklung*).

3. Biomethane from biogas units for the CHP units

The starting point is raw biogas from agricultural residues (animal and agricultural) and from energy crops like corn. The raw biogas is cleaned and cooled and then fed into the biogas pipeline and transported to the central processing plant in Bitburg. The participating biogas units are currently dropping out of their 20-year feed-in tariffs, which guarantee a fixed price for the electricity they produce and therefore force owners to look for alternative ways to continue operating profitably. The KNE (*Kommunale Netze Eifel*), a community-owned company near the city of Bitburg, and communities in the Bitburg area planned and built a 45 km-long east-west biogas pipeline system as part of a bigger 128 km pipeline system, the Westeifel integrated network

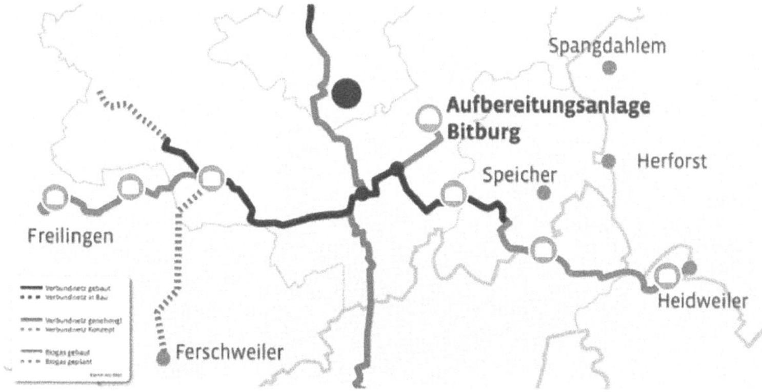

Figure 23.2 Biogas collection pipeline network in the Eifel area.[1]
Source: https://www.swt.de/p/Bio_Erdgas_aus_der_Region-15-7666.html.

(*Verbundnetz Westeifel*), as shown in Fig. 23.2. It is built to transport raw biogas from biogas plants to the pressure swing adsorption (PSA) plant in Bitburg to upgrade the biogas to biomethane.

In the upgrading process, the gas components are separated using PSA, a physical process for separating gas components under pressure by means of adsorption.[2] Finally, the biomethane is injected into the existing natural gas pipeline.

Once the biogas collection network is completed with around 48 biogas units, it will be able to collect around 10,000 m^3/h of raw biogas. Currently, seven biogas units produce 1800 m^3/h of biogas. The municipal utility (*Stadtwerke Trier*) plans to expand this biogas collection system to other locations as well. This is to ensure that even more upgraded biogas will be available for the CHP units in Trier and, if possible, for more applications like public city buses in Trier or for industrial use. Moreover, a more sustainable raw material feed for the biogas units is required in the longer run to lower the overall GHG emissions of the biogas generated.

4. Local heat network in downtown Trier

There are currently two heating networks in operation: one that heats the hospital buildings and one that heats retirement homes as well as a vocational school. In the project, each district was evaluated individually in terms of its current heat supply and demand on a building scale, any changes to the heat demand within the next 30

[1] Aufbereitungsanlage Bitburg: Bitburg processing plant; Verbundnetz gebaut: Integrated network constructed; Verbundnetz in Bau: Integrated network under construction; Verbundnetz genehmigt: Integrated network approved; Verbundnetz Konzept: Integrated network concept; Biogas gebaut: Biogas constructed; Biogas geplant: Biogas planned; Stand: Valid from.

[2] As CO_2 is more strongly adsorbed than methane (CH_4), the less adsorbable component methane is enriched in the gas phase.

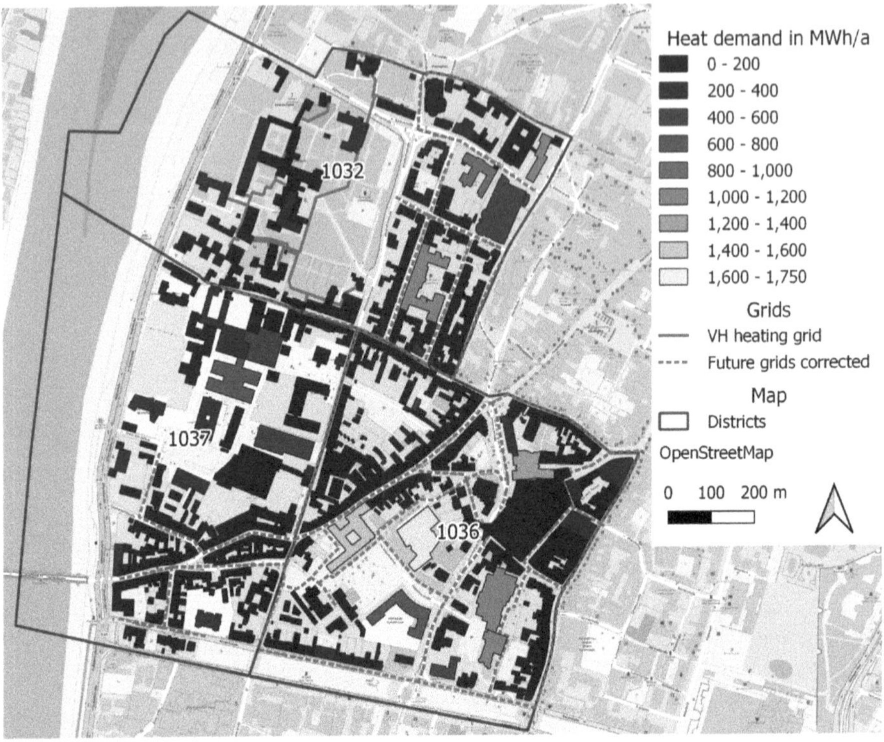

Figure 23.3 Heat demand structure within the districts.
Source: Gebauer, M., Hill, A., Menke, C., et al., 2021. Integriertes Energetisches Quartierskonzept Trier-Innenstadt Südwest EQTI; Final Report. https://www.trier.de/umwelt-verkehr/klimaschutz/energiequartier/. OpenStreepMap, City of Trier.

years, and possible heat grid layouts to accommodate these changes. The current heat demand was calculated based on gas consumption data provided by the local utility and data for each individual building. The latter was gathered from existing statistics and by visual inspection of all buildings in the quarter.

Fig. 23.3 shows the heat demand and heat network in the quarter, with an unbroken red (light gray in print version) line showing the current grid layout in district 1036 *Rathaus* (town hall) together with future grid layouts (dotted line). The plotted heat demands were aggregated for each street and assigned to each building according to area covered.

A detailed analysis of the existing and future energy demand for heating and hot water was performed for each district. To calculate present and future key parameters, the following assumptions were used for changes in heat demand, as shown in Table 23.1.[3]

[3] The assumptions from Table 23.1 were derived from a previous detailed study—"Wärmestudie (heat demand study) Region Trier/Eifel, 2017"—allowing for a renovation rate of 1.3%/a and typical reductions in heat demand for refurbishments (7).

Table 23.1 Assumptions for changes in heat demand from 2020 to 2050.

Parameter (heat demand in kWh/a)		Value (%)
Change in heat demand for residential housing	2020 → 2030	−13
	2020 → 2050	−48
Change in heat demand for business, commerce, services	2020 → 2030	−14
	2020 → 2050	−35

Source: Gebauer, M., Hill, A., Menke, C., et al., 2021. Integriertes Energetisches Quartierskonzept Trier-Innenstadt Südwest EQTI; Final Report. https://www.trier.de/umwelt-verkehr/klimaschutz/energiequartier/.

The future scenarios, for which the heat networks are calculated, are:

- *Business-as-usual (BAU) 2030:* In this scenario, around 50% of district heat demand is met by district heating by the year 2030;
- *Business-as-usual (BAU) 2050:* In this scenario, around 50% of district heat demand is met by district heating by the year 2050; and
- *Max. 2050:* In this scenario, all buildings are connected to district heating by 2050 (Gebauer et al., 2021).

4.1 *Results for district 1032* Vereinigte Hospitien *(retirement homes)*

The existing heating network is used for heating retirement homes and a nearby vocational school. It was found that the existing CHP unit can meet the additional heat demand of buildings nearby in that district, as shown in Table 23.2 (Binczyk, 2020).

Table 23.2 System characteristics for different scenarios for district 1032.

Performance indicator	Unit	Scenario			
		2020	BAU 2030	BAU 2050	Max. 2050
Thermal power of CHP	kW	1992	1992	1992	1992
Thermal energy output of CHP	MWh/a	7449	11,067	11,390	12,323
No. of CHP starts	/a		881	871	861
Full load hours for CHP unit	h/a	3700	5536	5718	6187
Thermal power of back-up unit	kW		3740	3740	3740
Thermal energy output of back-up unit	MWh/a	1085	841	1170	2569
No. of starts of back-up units	/a		220	205	168
Storage volume	m^3	200	200	200	200

Source: Binczyk, J, 2020. Extension and Optimization of the Block Heat and Power Plant of the Vereinigten Hospitien Trier as Part of the Energetic Quarter Concept for the City Center of Trier; Master's Thesis. Trier University of Applied Sciences.

Table 23.2 shows the key system results for providing heat to the network. By utilizing the current size of the CHP unit and its back-up units, all nearby buildings could be supplied by increasing their operation time. This results in an increase in thermal energy of 65% when allowing for the current thermal energy production of the CHP unit and an increase of 137% for the back-up units, without increasing the heat storage volume. This increase can be made while keeping the yearly starts of each unit to an acceptable level (Binczyk, 2020).

4.2 *Result for district 1037* Mutterhaus *(hospital)*

In district 1037, one CHP unit and three natural gas back-up heaters are used to heat the hospital, delivering around 10,500 MWh of thermal energy as shown in Table 23.3 (Dau & Jesinghaus, 2020).

Table 23.3 shows that extending the network to all buildings in this district requires upgrading of the existing cogeneration unit from 2 to 2.7 MW. This is possible, given the existing CHP unit comes to the end of its life within the next 3 years in any case. It shows that even when extending the heat network through energy conservation measures in buildings, the CHP unit is able to provide even more thermal energy in scenario Max. 2050 while maintaining its required power demand. In the final stage, the CHP units will provide around 85% of the total heat demand.

Table 23.3 System characteristics for different scenarios for district 1037.

		Scenario			
Performance indicator	**Unit**	**2020**	**BAU 2030**	**BAU 2050**	**Max. 2050**
Thermal power output of CHP	kW	2060	2800	2500	2700
Thermal energy output of CHP	MWh/a	8883	14,948	13,287	15,084
No. of CHP starts	/a		1571	517	585
Full load hours for CHP unit	h/a	5347	5339	5315	5587
Thermal power output of back-up units	kW	6900	7450	7450	7450
Thermal energy output of back-up unit	MWh/a	1977	2496	1731	2613
No. of starts of back-up units	/a		359	297	358
Storage volume	m^3	210	210	210	210

Source: Dau, S., Jesinghaus, M., 2020. Technical and Economic Design of a Local Heating Grid to Supply the Hospital Mutterhaus Trier and the Nearby Urban District by Means of Biomethane CHP in the Context of the EQTI Project; Project Study. Trier University of Applied Sciences.

4.3 *Results of district 1036* Rathaus *(town hall)—new heat network*

As the local municipality wants to decarbonize public buildings within district 1036 as well—mainly the town hall, a school, and an events venue—a heating grid is planned to connect all these buildings as shown in Fig. 23.4. This is considered the base scenario for this district. To drive carbon emissions down across the whole building stock within district 1036, an additional more comprehensive—extended—scenario is evaluated[4] as well. The two scenarios are shown in Fig. 23.4, with the base scenario showing a grid design only incorporating public buildings within a close vicinity and the extended scenario showing what the district heating grid could look like if almost every building were connected (Lellinger & Lölsberg, 2020).

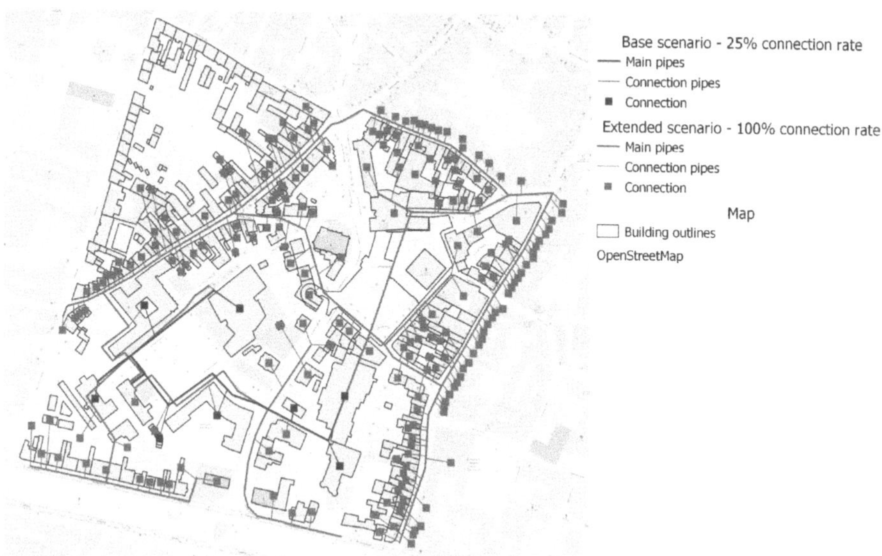

Figure 23.4 Base and extended grid layouts for district 1036.
Source: Gebauer, M., Hill, A., Menke, C., et al., 2021. Integriertes Energetisches Quartierskonzept Trier-Innenstadt Südwest EQTI; Final Report. https://www.trier.de/umwelt-verkehr/klimaschutz/energiequartier/. Lellinger, D., Lölsberg, T., 2020. Technical and Economic Evaluation of District Heating Designs in the Context of the Trier District Energy Project EQTI; Project Study. Trier University of Applied Sciences. OpenStreepMap, City of Trier.

[4] To rapidly evaluate different scenarios, a GIS-based python script was written that randomly connects buildings that are within a range of 30 m of a pipeline until the predefined percentage rate of customer connections is achieved.

Table 23.4 System characteristics for different scenarios within district 1036.

Performance indicator	Unit	Scenario		
		BAU 2030	**BAU 2050**	**Max. 2050**
Thermal power output of CHP	kW	1563	1563	1930
Thermal energy output of CHP	MWh/a	8255	8543	12,593
No. of CHP starts	/a	287	277	284
Thermal power output of back-up unit	kW	1600	1000	1600
Full load hours for CHP unit	h/a	5282	5466	6525
Thermal energy output of back-up unit	MWh/a	660	894	2527
No. of starts of back-up units	/a	224	85	182
Storage volume	m^3	150	150	150

Source: Lellinger, D., Lölsberg, T., 2020. Technical and Economic Evaluation of District Heating Designs in the Context of the Trier District Energy Project EQTI; Project Study. Trier University of Applied Sciences.

With the resulting heat demands, the BAU and Max. 2050 scenarios are simulated using a simulation tool for heat networks,[5] as displayed in Table 23.4.

Table 23.4 shows that through the improving and accelerated energy efficiency renovation standard for buildings, the additional heat demand of a grid extension between 2030 and 2050 is almost balanced out, thus enabling continuous operation of the initial CHP unit size but more operation hours per year. The CHP unit size has to be increased only when all buildings within district 1036 are connected to the network (2050 Max.). A hot water storage volume of 150 m^3 is considered sufficient for all scenarios in this district (Lellinger & Lölsberg, 2020).

4.4 Summary energy parameters for all three districts in the quarter

Table 23.5 shows that in total, a maximum heat demand of around 48,000 MWh/a is expected if all buildings are supplied with district heating, of which around 85% is supplied by the CHP units. Considering a thermal efficiency of 43%, the cogeneration units will require around 93,000 MWh/a of biomethane gas, while the back-up units, expected to achieve 90% thermal efficiency, will require around 8600 MWh/a of gas. With a calorific

[5] Simulation tool Sophena: "The open source software 'Sophena' (software for planning heating plants and local heating networks) offers the possibility to carry out the technical and economic planning of a heating supply project quickly and in a well-founded manner. At the heart of Sophena is a boiler and buffer storage simulation, from which annual duration curves and energetic key figures are determined. CHP plants and heat pumps can also be taken into account. Based on these calculations, an economic efficiency analysis is carried out according to VDI 2067. Further results include a greenhouse gas balance and the heat occupancy density of the network"; https://www.carmen-ev.de/service/sophena.

Table 23.5 Summary of energy parameters for all three districts in the quarter.

Performance indicator	Unit	Scenario		
		BAU 2030	BAU 2050	Max. 2050
Thermal power output of CHP	kW	6355	6055	6622
Thermal energy output of CHP	MWh/a	34,270	33,220	40,000
Thermal power output of back-up unit	kW	12,790	12,190	12,790
Thermal energy output of back-up unit	MWh/a	3997	3795	7709
Total biogas demand	MWh/a	84,140	81,472	101,590
	m^3/a	8,413,880	8,147,250	10,158,881

value of 10 kWh/m^3, this translates into a total biomethane requirement of around 10,160,000 m^3/a or 102,000 MWh/a of biomethane. Compared with the supply from currently seven biogas units, it becomes clear that at least 3–4 more biogas units need to be connected to supply all biomethane demand to the three CHP units in the long run.

The biomethane-driven CHP network will replace individual—fossil—natural gas heating systems in the buildings of the quarter and so reduce GHG emissions. As a consequence, the existing gas grid will be dismantled in the corresponding quarter in the long run and only the CHP units will supply low-carbon heat and power.

4.5 Economics of the local heat network

The economic evaluation[6] of the local heat network was done from a consumer perspective with the LCOH (levelized cost of heating) of heat from a district heating network compared with heat from a conventional gas boiler in each building. The operational costs for district heating were calculated using current prices and without considering any grants or investment subsidies; these are currently available for up to 40% or more of investment cost in heat networks using a renewable energy and CHP approach.

The results are displayed in Fig. 23.5. They show that in 2030, low-carbon heat is costlier than gas when the CO_2 price is around EUR 80/t and no investment subsidy for the low-carbon heat network is assumed for all consumer groups. With a theoretical CO_2 price of EUR 195/t, CO_2 in district heating in 2050 will have lower costs for all consumer classes regardless of grid size assuming no investment subsidy. Given that substantial investment subsidies for renewable energy heat networks are available in Germany, the CO_2 price actually required will be much lower. A summary of the economic modeling of heat demand in the quarter and the design of the CHP system shows that locally produced heat from CHP units will be cost competitive in the long

[6] Both gas and district heating costs are assumed to increase by 1%/a.

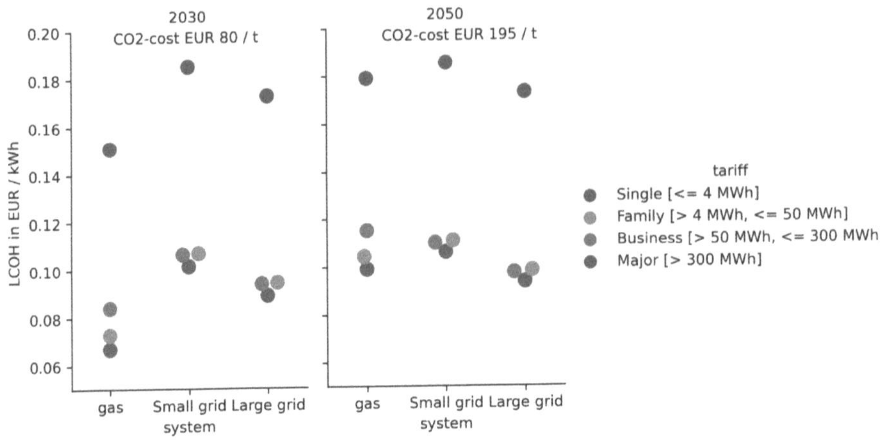

Figure 23.5 Economic costs of life cycle cost of heat (LCOH) for individual gas heating systems and for two different district heating scenarios in 2030 and 2050, assuming different CO_2 costs but no investment support.
Source: Own calculations.

run, as compared with conventional natural gas heating systems in individual houses in the quarter. More detail information is available in the final report of the project (Gebauer et al., 2021; Lellinger & Lölsberg, 2020).

However, to achieve this low-carbon, biomethane-based heat network, residents in the corresponding quarter need to connect step-by-step to the heat network. This requires participatory workshops and intensive discussion around why this option is the best solution and, for example, why there is not enough biomethane available for individual heating systems; the latter is in response to requests from many citizens in the quarter to get a direct supply of biomethane for their own boilers so as to avoid a local heat network. In addition, the reasons need to be communicated as to why a local heat network will lead to the existing gas network in the quarter being dismantled in the long run, due to its dwindling demand. This discussion process is not easy and requires multiple debates not only with the citizens of the quarter but as well with the management of the utility, as they are concerned about the stranded costs of their distribution pipelines for natural gas.

5. Dynamic scenario simulator for interactive participation of residents

An accurate and effective data generation on all buildings and their energy systems is necessary for analysis purposes and for developing strategies as well as quantitative scenarios and a practical pathway to CO_2 neutrality. For these reasons, all district data were collected at a building level based on cadastral extracts and stored in an adapted Energy Application Domain Extension (Energy ADE) of the CityGML standard.[7] Using this database, a dynamic building simulator was developed and used for interactive participation of residents in the quarter.[8]

Based on this detailed inventory, different potentials for emission reduction in the quarter were identified and examined, for example, building renovation, the use of solar power, or heat supply via renewable district heating. To create a detailed forecast for the quarter in 2030 and 2050, the results of the studies were taken together with other possibilities for changing CO_2 emissions in the quarter and were calculated and presented in a *dynamic scenario generator*.[9]

In that respect, two kinds of inputs were distinguished:

- *Fixed inputs:* The quarter and city administration of Trier have no influence on, for example, CO_2 emission factors (German district concept standard BISKO was used; Gugel et al., 2020) or physical limitations (efficiency of the conversion systems, effects of renovation on heat demand, etc.).
- *Variable inputs:* The citizens of the quarter and the city administration of Trier have an active influence and corresponding action is required (Gebauer et al., 2021). These variable inputs can be determined for the period from now until 2030 and between 2030 and 2050. Sometimes yearly rates are used (yearly renovation in % per year—p.a.); sometimes goals are set to achieve the target value.

All this input information was used to calculate the future energy demand for each building as well as the resulting CO_2 output of the whole quarter for 2030 and 2050.

For the purposes of public discussion, any option for energy improvement the citizens of the quarter might wish to undertake was presented in an input sheet for the process of calculation, as shown in Fig. 23.6 (Gebauer et al., 2021).

The possible list of input parameter for the dynamic scenario generator got more extensive over time, as stakeholders wanted "all" options evaluated. So, at the end the study, authors ended up with a complicated and extended list of parameters for the estimation of possible emissions reductions in the quarter as displayed in Fig. 23.6.

In the first section of Fig. 23.6—"Building stock"—the annual renovation rate can be fixed, distinguishing between standard renovation or a "nearly zero energy building" standard, as well as the decision whether landmarked buildings should be considered as well in all actions. In this section, growth in the quarter through more

[7] Energy ADE of the CityGML standard is a data standard for Application Domain Extension (ADE) of the City Geography Markup Language standard used in the city of Trier.

[8] The communal data allowed the function of each building, its 2D geometry as well as landmark status, and number of inhabitants to be addressed. With additional data from the local supplier and existing networks, 458 conversion, 28 distribution, and 1453 demand energy systems could be modeled with real data for the years 2016–18 in 1946 energy flows. More data (floors heated, roof shape and usage, number of post-boxes) were acquired by complete visual inspection of the district combined with evaluation of aerial photographs. Only for 69 buildings the heat demand was estimated (2).

[9] The dynamic scenario generator is a simulation tool developed using Excel to simulate energy demand for each building in the quarter based on different assumptions. The approach and results are shown below in Figs. 23.7–23.9.

year	2030	2050	
Building stock	629 buildings		
Complete renovation	1.0%	1.0%	p.a.
NZEB renovation	1.0%	1.5%	p.a.
also in 137 landmarked buildings	☐ Yes		
Consumption behavior/SmartHome	0.0%	0.0%	target in year
More inhabitants per m²	0.0%	0.0%	target in year
More m² (redensification)	0.0%	0.0%	target in year
Demand for electricty			
More efficiency	0.0%	0.0%	p.a.
More electrification	0.0%	0.0%	p.a.
Heat networks			
Enlargement network "Mutterhaus"	continued exist.	enlargement	implementation
Connection rate	0%	80%	target in year
Enlargement network "Vereinigte Hospitien"	enlargement	enlargement	implementation
Connection rate	80%	80%	target in year
New network "Energiebunker"	without	small	implementation
Connection rate	100%	80%	target in year
Heat losses	12%		
Renewable energies			
Solar fitted roofs	28,531 m²		
PV	30.0%	40.0%	target in year
PVT	0.0%	0.0%	target in year
ST	30.0%	50.0%	target in year
No. of building supplied by networks	147	513	
Heatpumps			
No. of building with demand < 50 kWh/m²	108	145	
Realisation rate	0.0%	0.0%	target in year
No. of buildings with heatpumps	0	0	
Biogas-boiler	0.0%	0.0%	target in year
Solid biomass	0.0%	0.0%	target in year
Remaining energy supply			
Natural gas	24,674,287	19,460,303	
Fuel oil	3,695,625	2,815,440	

Figure 23.6 Variable inputs for the scenario generator.
Source: Gebauer, M., Hill, A., Menke, C., et al., 2021. Integriertes Energetisches Quartierskonzept Trier-Innenstadt Südwest EQTI; Final Report. https://www.trier.de/umwelt-verkehr/klimaschutz/energiequartier/.

inhabitants or redensification can also be inputted. For the "Demand for electricity" section for normal appliances in households, a balanced rate of more efficient devices versus more devices can be assumed. In the third section, "Heat networks," the realization and connection rate of each of the local heat networks described can be assumed. The last input section determines future use of renewables in the quarter, based on the calculated potential for solar, heat pumps, and bioenergy. In the last two lines, the resulting remaining consumption of fossil fuel oil and natural gas is shown, based on the above assumptions.

To display the consequences of assumptions and estimations within the quarter, the scenario generator graphics show the total future energy demand and supply, both combined as well as separately for electricity and heat, as shown in Fig. 23.7 (combined heat and power), Fig. 23.8 (only power), and Fig. 23.9 (only heat) (Gebauer et al., 2021).

Fig. 23.7 shows the results for the *combined energy demand of electricity and heat* for 2020, 2030, and 2050. In addition, results for the set of input parameters currently chosen show the trend of CO_2 emissions until 2050, e.g., 60% reduction.

For each year—2020, 2030, and 2050—the 2 bars display the demand for heat (for space heating and hot water) and electricity and the way these are supplied (by fuel oil, natural gas, etc.) (Gebauer et al., 2021). One can see that the combined demand and the supply is reduced by only 10% over this timespan in the quarter. On the demand side, the distribution between the three energy consuming sectors does not change, while on the production side, the CHP heat and electricity proportion increases from 40% to more than 80% in 2050. However, the reduction of CO_2 emissions was mainly achieved through increasing the proportion of combined heat and power using biomethane-driven CHP units within the quarter and so replacing the use of natural fossil gas for heating and use of electricity from the national grid with higher specific CO_2 emissions.

Fig. 23.8 shows the demand and supply of the electricity sector. It shows how much electricity is produced within the quarter through CHP and PV and how much still comes from the German grid. It displays how much electricity is consumed in the different consuming sectors, like 30% for private households and 35% each for public

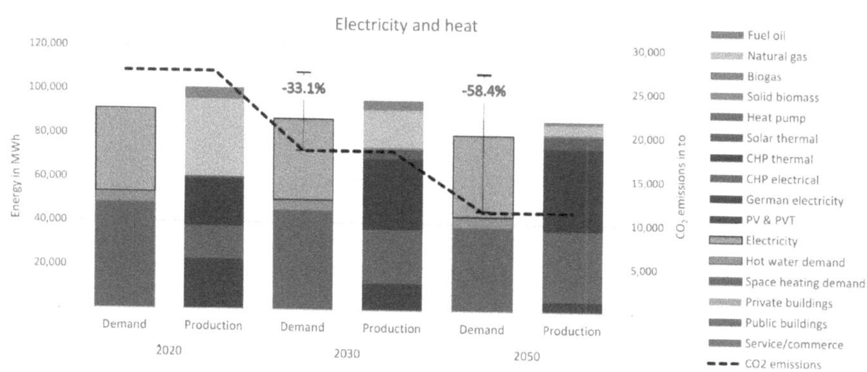

Figure 23.7 Result for energy demand and supply based on inputs in the scenario generator for electricity and heat in the quarter for 2020, 2030, and 2050.
Source: Gebauer, M., Hill, A., Menke, C., et al., 2021. Integriertes Energetisches Quartierskonzept Trier-Innenstadt Südwest EQTI; Final Report. https://www.trier.de/umwelt-verkehr/klimaschutz/energiequartier/.

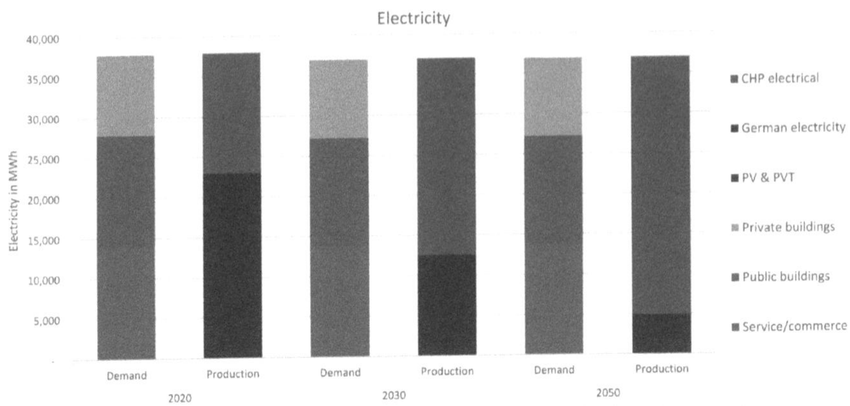

Figure 23.8 Example for electricity demand and supply based on inputs in the scenario generator in the quarter for 2020, 2030, and 2050.
Source: Gebauer, M., Hill, A., Menke, C., et al., 2021. Integriertes Energetisches Quartierskonzept Trier-Innenstadt Südwest EQTI; Final Report. https://www.trier.de/umwelt-verkehr/klimaschutz/energiequartier/.

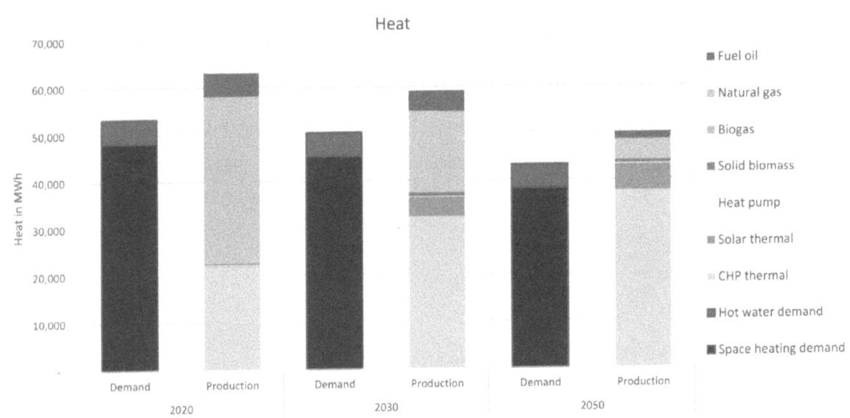

Figure 23.9 Example for heat demand and supply based on inputs in the scenario generator in the quarter for 2020, 2030, and 2050.
Source: Gebauer, M., Hill, A., Menke, C., et al., 2021. Integriertes Energetisches Quartierskonzept Trier-Innenstadt Südwest EQTI; Final Report. https://www.trier.de/umwelt-verkehr/klimaschutz/energiequartier/.

buildings and the service sector/commerce. It shows that in 2050, only 5% of electricity demand in the quarter still comes from outside the quarter, while 95% comes from CHP units and a small proportion from PV systems within the quarter. However, the total electricity demand in the quarter will be more or less the same over time, due to the different assumptions, efficiency, and electrification balancing each other out.

Fig. 23.9 shows the results for the heat sector. One can see that 90% of heat demand is for heating, with the rest for hot water. In addition, one can see that by 2050 the heat demand is reduced only by around 15% due to energy-saving measures. On the supply side, it shows that CHP heat will supply around 80% in 2050, while solar heat can contribute 10% and the remaining 10% will still be fossil gas and heating oil for some remaining buildings not connected to the network. The fact that the supply is higher than the demand results from the heat losses of the local heat network and burners as well.

This dynamic scenario generator was used to discuss intensively with participants (in online workshops due to COVID restrictions) the options for reducing CO_2 emissions in the quarter.

This kind of tool provides instant answers to questions as well as visual results for different assumptions and scenarios. For instance, what happens when the rate of building restoration rises, or more solar thermal systems, heat pumps, etc., are installed. It helps bring much more rationalization into discussions with residents and to focus on realistic and possible solutions and what the consequences are of individual measures.

The scenario generator shows what effects can be achieved with individual measures by house owners in comparison to joint action. The results show as well that although renovating buildings helps, as the building stock is old and is partly landmark restricted, even with PV and solar thermal systems on the roofs, there is a need for a low-carbon heat network to reduce GHG emissions in the quarter. Extending the local heat network in this way can save 60% of GHG compared to the current status and up to 90% compared to base year 1990 (Gebauer et al., 2021).

This kind of online tool proves to be very useful in discussions with citizens and with technical staff of the city administration, as any kind of technical options to reduce emissions in the quarter could be quantified immediately and so its individual contribution could be shown. It is for sure one of the lessons that such kind of online tool helps a lot in public discussions with stakeholders to narrow down the options.

6. Participation process for residents and stakeholders for building up an energy community

The overall participation process for developing a concept for the district was started with an official conference in a historical location within the quarter. The well-attended launch event—in which the process, objectives, and partner consortium were introduced and explained—gained positive feedback from participants, especially inhabitants from the quarter under consideration. This was followed up by capacity-building workshops[10], showing and discussing different technical, social, and economic aspects of the future development. The selection of possible inputs in the specially developed scenario generator was a key result of these discussions. Finally,

[10] Due to COVID pandemic, all workshops were held online.

the generator was presented and explained in two online working sessions—one open discussion with inhabitants and one political discussion in the KUE committee, including all political parties, universities, environmental organizations, and other local stakeholders. Both workshops offered possibilities to interactively "play" with the various options in the scenario generator tool and view their resulting emissions—a real-time trial and error process to achieve climate goals.

These discussions contributed to a mutually agreed and almost identical "quarter scenario" in both sessions by the participants of both workshops. The reason why this process went so smoothly may be attributed to the fact that the KuE committee of the city of Trier—where all relevant political parties and stakeholders are represented—was constantly kept informed of the progress with the results communicated and discussed with all parties. This helped to keep all parties on board during the entire process. Clearly, community engagement and consent building is an important challenge as described in the chapter by Li.

7. Conclusions

The energy transition and the adaptation of cities and communities to climate change require persistence from all parties and stakeholders involved. The implementation of measures calls for immense changes to cultivated habits and hitherto valid concepts on the part of those affected. It is therefore important to offer something to make it easier to take the necessary steps. The transformation of cities is more like a marathon than a sprint, and so the implementation and continuation of measures requires resources—especially personnel—to foster, monitor, control, advance, and document the process.

As illustrated in this chapter, even when economically viable solutions exist for an up to 90% reduction in GHG for local heat and power demand in complex-built environments—such as in the downtown area of an old city—it is a long-term process to bring these solutions to fruition. It requires extensive and lengthy communication efforts on all levels and insistence by climate managers of the city to get local utility management and the city administration moving, and to finally get it started. It is expected that the new heat network will be installed as many public buildings need a new heating system within the next 3 years offering an opportunity to jumpstart the implementation of the heat network.

The current best option for this quarter in Trier is the biomethane used in CHP units, as biogas is available nearby and collected and upgraded. However, other quarters in Trier may be heated by central heat pumps with the river Mosel as heat source. But the key factor for quarters with old building stock is a local or central heat network as it offers the only solution to provide sufficient renewable heat to existing building stock with high heat demands in an efficient way.

The project offers a number of interesting lessons including:

- The usefulness of an online visualization tool—the scenario generator—to instantly demonstrate to residents in workshops the possibilities for focusing on real options in the energy transition of a local energy community;

- The need to communicate from the beginning of the process extensively with the city administration and the utility management to keep all parties engaged and involved;
- The active engagement of the staff within the city administration such as those responsible for climate protection, environment, and energy to remove stumbling blocks and ensure access to data from city and utility;
- The access to technical know-how such as the engagement of the university to develop local knowledge that remains in the city during the next steps of implementation; and
- Finally, it is essential to have a local committee established, like the KuE committee in Trier, where all political parties, relevant environmental groups, and local institutions like gas and power utilities are engaged and discuss the approach and intermediate results at each meeting—in this case every 3 months. That helps to keep all stakeholders on board during the course of the project.

References

Binczyk, J., 2020. Extension and Optimization of the Block Heat and Power Plant of the Vereinigten Hospitien Trier as Part of the Energetic Quarter Concept for the City Center of Trier; Master's Thesis. Trier University of Applied Sciences.

Dau, S., Jesinghaus, M., 2020. Technical and Economic Design of a Local Heating Grid to Supply the Hospital Mutterhaus Trier and the Nearby Urban District by Means of Biomethane CHP in the Context of the EQTI Project; Project Study. Trier University of Applied Sciences.

Gebauer, M., Hill, A., Menke, C., et al., March 2021. Integriertes Energetisches Quartierskonzept Trier-Innenstadt Südwest EQTI; Final Report. https://www.trier.de/umwelt-verkehr/klimaschutz/energiequartier/.

Gugel, B., Hertle, H., Dünnebeil, F., Herhoffer, V., 2020. Institut für Energie- und Umweltforschung Heidelberg gGmbH, Heidelberg: Weiterentwicklung des kommunalen Bilanzierungsstandards für THG-Emissionen – Bilanzierungssystematik kommunal – BISKO; final report. https://www.umweltbundesamt.de/.../cc_19-2020_endbericht_sv-gutachten_bisko.pdf.

Lellinger, D., Lölsberg, T., 2020. Technical and Economic Evaluation of District Heating Designs in the Context of the Trier District Energy Project EQTI; Project Study. Trier University of Applied Sciences.

Establishing energy communities in postcommunist states: the case of Bulgaria

24

Toby D. Couture[1] *and Teodora Stoyanova*[2]
[1]E3 Analytics, Berlin, Germany; [2]Independent Consultant, Varna, Bulgaria

1. Introduction

Bulgaria is poised for a significant transformation of its energy system in the coming decades leading up to 2050. Among the major drivers for this are the rapidly decreasing costs of renewable energy sources, the European Union's (EU) new Renewable Energy Directive (RED II), a steady rise in retail electricity prices due in part to the liberalization of the market, and the influence of bottom-up market forces. In addition, the 2018 Directive on the promotion of the use of energy from renewable sources (RED II) combined with the European Green Deal is fueling a fundamental rethink of both energy and climate policy and investment across the EU.

As a Member State of the EU, Bulgaria is required to implement RED II. As part of this process, it will need to adopt a range of new regulations to provide greater legal and policy certainty for the development of energy communities. As articulated in a number of chapters in this volume, RED II enshrines the right of households and businesses to install their own onsite renewable energy supply and establishes a range of clear guidelines regarding the regulation of the sector. RED II also lays out a number of specific requirements to facilitate the establishment of energy communities.

As further described in this chapter, Bulgaria faces unique challenges due to the legacy of 40+ years under Soviet control, a period in which decision-making was highly centralized and the presence of entrepreneurial activity and of civil society was often actively suppressed. A similar legacy persists in other countries that lived under Soviet control, such as Poland and Romania, and continues to shape how citizens in former Soviet countries respond to efforts to establish energy communities.

This chapter is structured as follows:

- Section 2 provides a brief overview of the electricity sector in Bulgaria including its current state and underlying trends;
- Section 3 looks at the history of the cooperative movement in Bulgaria;
- Section 4 examines the opportunities for developing energy communities in Bulgaria followed by the chapter's conclusions.

Energy Communities. https://doi.org/10.1016/B978-0-323-91135-1.00009-2

2. Overview of the electricity sector in Bulgaria

Bulgaria is one of the newest members of the EU, having joined the EU in 2007 together with Romania. While becoming a Member State has had many positive impacts on the country's social and economic development in recent years, Bulgaria remains the country with the lowest GDP per capita in the EU at €8,680, compared to an EU average of €26,340.[1]

Another important factor is that Bulgaria's population has been steadily declining since the mid-1980s when it reached its peak of nearly 9 million people. In the first 15 years after the collapse of the communism, more than 1.2 million Bulgarian citizens left the country, mainly in search of better employment opportunities and in order to escape poverty.[2] The latest data show the total population at a level of 6.9 million.[3]

Like many other EU Member States, such as Poland, Czech Republic, and Romania, Bulgaria was under communist/socialist rule from roughly 1945 until the fall of the Berlin wall in 1989. The 1990s were a period of a dramatic transition for the Bulgarian economy following nearly 50 years of a planned economy. The electricity sector in Bulgaria was no exception. The power sector was state owned for much of the second half of the 20th century, including for several years after 1989 as the country transitioned to a market economy.

While efforts to liberalize the market and introduce competition started in the early 2000s, the state continues to play a major role in the energy sector, and it retains an ownership stake in a significant share (roughly 60%) of the power generation business. In addition, while electricity prices for business and industrial consumers have been liberalized, prices for households continue to be regulated and are kept artificially low.

As part of the electricity market liberalization process, Bulgaria introduced an electricity exchange in 2014, the Independent Bulgarian Energy Exchange (IBEX), making it one of the last countries in the EU to do so.[4] IBEX was initially established as a subsidiary of the Bulgarian Energy Holding (BEH) company, the former state-owned monopoly supplier. However, due to the fact that BEH also owns some of the largest power plants in the country, the European Commission raised the issue of potential conflict of interests and the Bulgarian government transferred the ownership to the Bulgarian Stock Exchange, which is now the sole owner of the power exchange (OECD, 2019). The legal framework governing the IBEX operations has been subject to improvements and revisions throughout the years. Some of the most important regulations implemented in 2018 and 2019 included the elimination of the "single buyer" role of the Bulgarian National Electricity Company and the inclusion of producers of 1 MW and above on the free market. Centralized trading via the IBEX was also introduced.

An overview of the current structure of the electricity market is illustrated in Fig. 24.1. In addition to owning a substantial share of power generation through

[1] See: https://appsso.eurostat.ec.europa.eu/nui/show.do?dataset=nama_10_pc&lang=en and https://ec.europa.eu/eurostat/databrowser/view/sdg_08_10/default/table?lang=en.

[2] See: https://www.dw.com/en/will-eu-entry-shrink-bulgarias-population-even-more/a-2287183.

[3] See: https://bg.wikipedia.org/wiki/%D0%91%D1%8A%D0%BB%D0%B3%D0%B0%D1%80%D0%B8%D1%8F.

[4] See: https://cms.law/en/bgr/publication/bulgaria-the-drive-for-full-liberalization-of-the-energy-market-and-the-upcoming-changes.

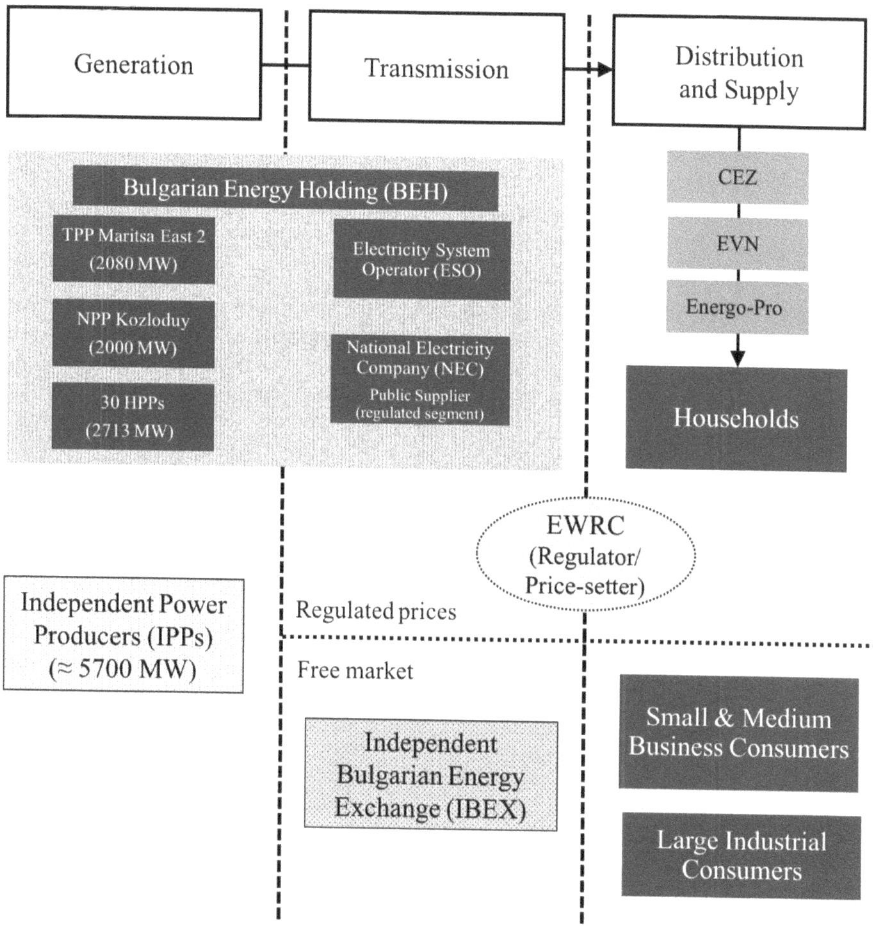

Figure 24.1 High-level structure of the Bulgarian electricity market.
Source: Pavlov, T., 2022. The political economy of coal in Bulgaria: the silent phase-out, (Chapter 3). In: Jakob, M., Steckel, J.C. (Eds.), The Political Economy of Coal: Obstacles to Clean Energy Transitions, forthcoming, Routledge. ISBN 9780367491024.

subsidiaries, the state-owned BEH also owns the high voltage transmission grid. The distribution network and retail supply, by contrast, are privately run.

The country is divided into three main regions, each of which is managed by a different distribution system operator (DSO) with a regional monopoly. As the map shows, there are currently three main DSOs active in the market:

1. CEZ Razpredelenie Bulgaria, part of CEZ Bulgaria (currently in the process of selling its business in Bulgaria)—responsible for the capital Sofia and the western part of the country;
2. Elektrorazpredelenie Yug, part of EVN Bulgaria—responsible for the southeastern part of the country; and
3. Elektrorazpredelenie Sever, part of Energo-Pro Varna holding—responsible for Varna and the northeastern region.

There is also Electricity Distribution Company Zlatni Pyasatsi, which is responsible for the distribution grid in a seaside resort in northeastern Bulgaria, though its share of the market is negligible. CEZ has the biggest market share with 40% (or 2.043.566 clients), followed by EVN with 37% (1.899.531 clients) and ENERGO-PRO with 23% (1.165.052 clients).[5]

The DSOs provide electricity to households who are part of the so-called regulated market, which refers to households whose electricity prices are set by the Energy and Water Regulatory Commission (EWRC). However, starting October 1, 2020, small and medium business consumers are required to purchase their electricity on the open market at market-determined prices. The full liberalization of the electricity market is expected to occur by 2025, at which point electricity prices for households will also be exposed to market prices.

The total number of electricity customers in Bulgaria is approximately 5.1 million. Out of this, 4.5 million (or approximately 90%) are individual household customers, while the remaining 600.000 are nonhousehold customers. However, in terms of demand, commercial and industrial customers make up almost two-thirds of total power demand. As can be seen in Fig. 24.2, the total final electricity demand for households is 11 TWh, while the total final electricity demand for nonhousehold customers (commercial and industrial) is 20 TWh.[6]

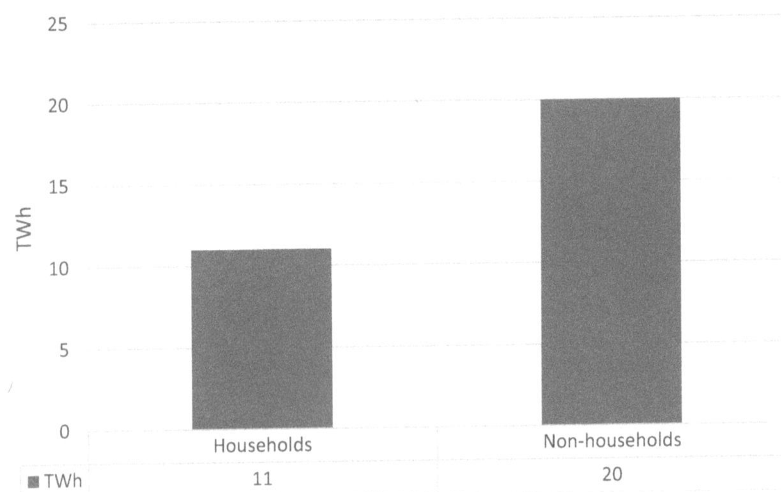

Figure 24.2 Final electricity demand in 2018 (in TWh).
Source: Own illustration, based on Ministry of Energy, 2020. Integrated Energy and Climate Plan of the Republic of Bulgaria, 2021–2030.

[5] See: https://www.dker.bg/uploads/2020/report_EC_2020_EN.pdf.
[6] https://me.government.bg/files/useruploads/files/buletinenergy2018-04.06.2019-finish.pdf.

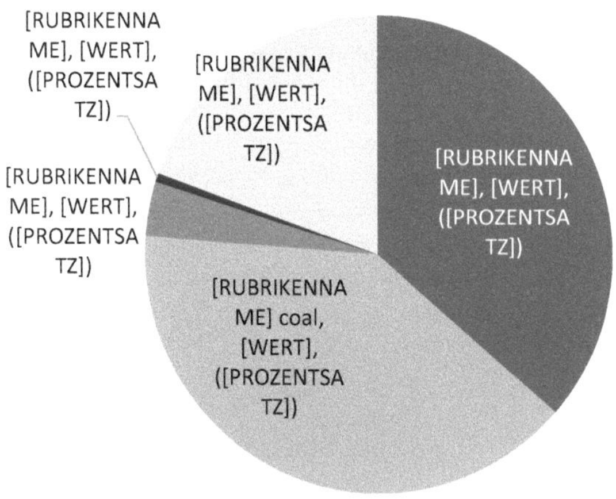

Figure 24.3 Net electricity generation by power plant type in GWh.
Source: Own illustration, based on Energy and Water Regulatory Commission (2019).[8]

Fig. 24.3 shows Bulgaria's current electricity mix.[7]

As can be seen, coal and nuclear power dominate Bulgaria's power mix. Moreover, the latest version of Bulgaria's National Climate and Energy Plan (NCEP) continues to envision a prominent role for both nuclear power and fossil gas in the country, despite the fact that these technologies are costlier, riskier, and slower to build than renewables.[9]

The NCEP sets a modest target for the share of renewable energy in the electricity sector at 30% by 2030, which represents a 7% increase from the current level.[10] The latest plan indicates a target of roughly 2.000 MW of new solar PV by 2030, although recent analyses suggest that Bulgaria has a potential of well over 5.000 MW of cost-effective solar PV, or more than four times current deployment levels.[11] Another key shortcoming is that the current NCEP does not have a separate RES objective or projections for renewable energy produced by cities, energy communities, or by prosumers.

At the same time, some positive regulatory changes have occurred in Bulgaria in recent years as highlighted in Box 24.1.

With regard to electricity rates, Bulgaria currently has the lowest household electricity prices in the EU, with households paying approximately €0.10/kWh. However, prices have increased by 17% between 2009 and 2019, and it is expected that retail

[7] For a more detailed overview of Bulgaria's electricity system and electricity rates, please see Couture et al. (2021).

[8] Note that we have used 2018 data for the time being, because there are some significant discrepancies in the 2019 data available.

[9] See: https://ec.europa.eu/energy/sites/default/files/documents/bg_final_necp_main_en.pdf.

[10] See: https://www.nsi.bg/en/content/5062/electricity-generated-renewable-sources-share-gross-electricity-consumption.

[11] See: https://irena.org/events/2020/Oct/Launch-of-CESEC-Report.

Box 24.1 Positive developments for energy communities in Bulgaria

- From an administrative perspective, amendments to the Spatial Development Act in 2019 removed the requirement for solar PV systems with a total installed capacity of up to 1 MW mounted on rooftops and façades to have an approved investment plan in order to obtain a building permit for the installation.
- Systems up to 1 MW that are not connected to the grid do not require a formal commissioning, which enables them to avoid some administrative hurdles and delays.
- Also, producers with installations that are connected to the grid after 2021 and do not receive feed-in tariffs are no longer be required to pay a 5% levy on their revenues that was introduced on FIT projects in 2015.[12]

[12] See: https://www.capital.bg/biznes/energetika/2020/10/24/4130766_otpada_edna_ot_golemite_prechki_za_novi_vei_moshtnosti.

prices for households will continue to rise in the years ahead.[13] This is likely to make it even more economically attractive to invest in solar power, both for individuals as well as for energy communities.

3. A brief history of the cooperative movement in Bulgaria

Although energy communities have existed in different EU Member States for decades—covered in multiple chapters in this volume—this concept is relatively new to Bulgaria. As such, a host of barriers persists, including a range of obstacles linked to its history.

The cooperative movement in Bulgaria has a long history. The first cooperative in Bulgaria was established in the village of Mirkovo in 1890 following the principles of the Rochdale and Raiffeisen cooperatives. Its roots can be traced back to a group of craftsmen and artisans from the same village organized in a fellowship whose purpose was to mutually support their work and raise funds for the needs of the village. The following decades saw a boom in establishing cooperatives—not only in agriculture but also in manufacturing, banking and credit, construction and housing, as well as fishing. In fact, in certain cities in Bulgaria, cooperatives were the first producers and distributors of electricity.[14]

Historically, the cooperative movement in Bulgaria can be divided into three periods:

[13] See: https://ec.europa.eu/eurostat/cache/metadata/en/nrg_pc_204_esms.htm and https://ec.europa.eu/eurostat/statistics-explained/pdfscache/45239.pdf.

[14] See: https://www.bia-bg.com/magazine/view/28061.

- Presocialism—from the creation of the first cooperative in 1890–1944;
- The socialist period—from 1945 to 1989; and
- Postsocialism—the period of democratic change after 1990 (Sarov, 2021).

Cooperatives in the first period follow the principles and values of Rochdale and Raiffeisen and were focused on providing social and economic benefits to their community; they played a vital role in the life of local communities by establishing school canteens, libraries, bookstores, and theaters.

The word that best describes the second period of the cooperative development is the "collectivization" of private property. At that time, private ownership was not consistent with the collective ownership model preferred by the ruling communist parties in Eastern Europe. While in the Soviet Union the process of collectivization was gradual and took several decades to complete (starting in the early 1920s), that was not the case in countries like Bulgaria where the idea was involuntarily enforced after 1945. Rich landowners in the rural areas were forced to give up their property and enter so called "labor-cooperative agricultural enterprises" (TKZS).[15] Initially, these formations were a variation of the cooperatives following some of the original principles (such as voluntary membership) but over time, they became swept along in a wave of party propaganda.

By the 1960s, the communist party took control of the governance of cooperatives and introduced a range of new principles including the following:

- Compulsory membership;
- State sponsored and state controlled;
- Authoritarian management;
- Party dependence;
- Party control;
- Equity ownership is share of state;
- Net income is distributed to state;
- Benefits are for the state; and
- Nationalized private property (Sarov, 2021).

Historically, this period left deep scars in the image of cooperatives in Bulgaria. During this period, cooperatives lost their independence and autonomous governance structure; landowners were forced to join the cooperative and to give up ownership of their land; income and production were controlled and distributed by the state; the cooperatives were required to supply their goods to the state at administratively set prices.

Over time, the state planning of economic activities including production, distribution, and pricing together with the inefficient use of resources it encouraged led to a host of wide-ranging economic and social issues in Bulgaria. Cooperatives, as a cornerstone of the agricultural sector at the time, were particularly hard hit. As a recent paper on the topic puts it "*After the collapse of communism, the reemergence of capitalism was again associated with the state which proved to be the major factor for private capital accumulation.*

[15] The abbreviation in Bulgarian stands for трудово-кооперативни земеделски стопанства (ТКЗС).

Being an alternative [to] capitalism, the cooperative sector was ignored and was often associated with the communist economic inefficiency." (Marinova and Nenovsky, 2020).

Between 1945 and 1989, the cooperative movement in Bulgaria underwent significant changes—its democratic foundations and governance principles were ignored and largely replaced by party policy. Cooperative members lost their autonomy and their ability to self-govern; income was not proportionately distributed among the members of the cooperative anymore, and instead, people were paid based on the number of days they worked; over time, virtually all economic activities became subject to state planning, including the activities and operations previously conducted by cooperative members. As a result of these and other profound changes, the cooperatives that existed in Bulgaria gradually became dysfunctional and many ceased to operate.

With the collapse of communism in 1989, the cooperative movement faced its third period of development. Slowly but steadily, cooperatives (mainly agricultural and manufacturing) reemerged and regained rebuilt some of the trust they had lost during the Socialist period. However, the harms done during the socialist period still hang like a cloud over the image of cooperatives in Bulgaria: in fact, the word cooperative (кооператив) is still associated with a certain stigma that has proved difficult to shake and this stigma continues to limit the creation and development of new cooperatives in the country (Ignatov, 2021). Although public attitudes and perceptions often take years if not decades to change, it is becoming clear to a growing number of stakeholders in Bulgaria that cooperatives are not solely an artifact of the country's socialist/communist past, but that they can potentially play a more important role in the country's energy transition as well by introducing more bottom-up decision making and enabling greater participation. Some have even argued that energy communities can help play a transformative role in the economic and social life of communities in Bulgaria by introducing greater local control over resources, as well as greater control over local planning and investment.[16]

As a result of this historical development, establishing any form of mutual cooperation in Bulgaria is accompanied with many challenges not only administrative and bureaucratic but also behavioral and philosophical. This is especially difficult for energy communities as by their nature they deal with a very technical and specific product that would require high level of mutual dedication, trust, and expertise.

One easy step would be to promote the development of energy *communities* rather than cooperatives. In the years before the revised RED II in Bulgaria "cooperative" (кооператив) was used as a collective term of energy community projects mainly because many of the examples of successful energy community projects were of energy cooperatives from Germany and Denmark. As a result, "energy cooperatives" became a collective term for all projects involving community cooperation. While there are still cooperatives operating today (mainly in the field of agriculture, tourism, and industry), the word is somewhat antiquated.

By contrast, the word "community" (общност) has more modern connotations: it conveys notions of inclusion, equality, and a shared sense of belonging and could

[16] See: https://csd.bg/fileadmin/user_upload/publications_library/files/2020_12/BRIEF_96_ENG_WEB.pdf.

provide a more appropriate term to guide the further development of local ownership in the country and foster greater participation in energy communities. For the people who have lived during the socialist period, there is an inherited sense of distrust in the co-operatives and any form of cooperative partnerships and projects. Younger generations (Millennials and Generation Z, for example) do not carry many of these negative associations and are showing signs of being more open and willing to engage in activities that benefit and support their own communities.

4. Establishing energy communities in Bulgaria

Energy communities have existed in different forms long before their formal recognition in EU legislation. As a result, energy communities vary widely from country to country. The increase of energy community projects in Europe is largely driven by the introduction of various support mechanisms and incentives such as feed-in tariffs, tax exemptions, and grants. These mechanisms enable energy communities and prosumers to actively participate in the production of renewable energy and benefit from it. Recent studies, however, show that the decline in support schemes negatively impacts the development of energy communities. Cuts in feed-in tariff schemes in Germany and Denmark, for example, have led to a significant slowdown in establishing energy cooperatives in those two countries in recent years. Moreover, the realization is dawning in both Germany and Denmark (see the chapter by Benedettini and Stagnaro) is that in order to maintain the pace of renewable energy development required to achieve energy and climate goals, citizens and community support are critical (Caramizaru and Uihlein, 2020).

Beyond increasing the local acceptance of energy projects, community-owned energy projects have several additional benefits: they enable citizens to be active participants in the energy system, they foster decentralization of the energy system, and reduce the overall reliance on fossil fuels, thus contributing to the reduction of greenhouse gas emissions. In addition, the concept of energy communities can help alleviate energy poverty, a major issue that affects several different regions throughout the EU, not only in eastern and central Europe, but also in countries like Spain, Portugal, Italy, and Ireland. According to EU estimates, approximately 40% of households in Bulgaria struggle to pay their energy (electricity/heating) bills, one of the highest rates in the EU (Center for the Study of Democracy, 2018).

There is currently little information on energy communities in Bulgaria. The idea has been sporadically mentioned by environmental NGOs over the years but has so far failed to receive high-level attention from either policymakers or stakeholders in the country. Discussions with various stakeholders indicate that a combination of factors including the lack of support, of awareness, and of a clear governing framework for energy communities is responsible for the absence of momentum on energy communities in Bulgaria.

However, a few examples exist.

One example that is similar to an energy community is a 117-unit apartment complex in the "Hristo Smirnenski" district of the capital Sofia, the owners of which have pooled their resources to install a 28 kW rooftop PV installation on their building. Organized via the property owners' association (a legal entity requited by law for every condominium building in the country), it took the entity 3 years (from 2010 to 2013) to successfully complete the project.[17] The community-owned PV installation brings a regular monthly income to the entity which is used to pay common bills and building expenses. All of the electricity produced is sold back to the grid under the country's feed-in tariff policy.

In the 7 years since the project was completed, the number of similar projects remains small. And although there have been efforts to reduce some of the administrative obstacles related to renewable energy installations in the country, the near absence of similar projects elsewhere in the country indicates that the measures adopted to date have not been enough to fuel widespread interest. Another example is found in the municipality of Straldzha (see Box 24.2) as well as in Burgas (see Box 24.3).

As part of efforts to encourage more direct citizen involvement and engagement in the energy transition, there are several new business and investment models emerging to make it easier for individuals, businesses, and local municipalities to develop their own energy community projects. Financing renewable energy projects (especially community-based ones) is a challenging task not only in Bulgaria (IRENA, 2020). A range of innovative financing methods has emerged in recent years partly to help overcome these difficulties, such as crowdfunding (Box 24.4).

Box 24.2 How Straldzha municipality is taking the lead

The municipality of Straldzha is located in south-eastern Bulgaria and is part of Yambol Province. It is a small municipality with an established agriculture and food industry, and a population of 12,014. In 2016, the Municipality started a project "Utilization of the local potential of geothermal energy in heating installations for schools and kindergartens in the town of Straldzha."[1] The grant financing is 772,109.23 BGN (€394,773.18). The project is funded by the BG04 Energy Efficiency and Renewable Energy Program as part of the Financial Mechanism of the European Economic Area (EEAFM). The project was completed in 2016.

Additionally, in 2018, the municipality installed a system of street lights that runs entirely on renewable energy. Currently, the municipality continues to work on building on and expanding its portfolio of renewable energy projects. With the right framework conditions, municipalities like Straldzha could position themselves to develop a range of energy community projects, providing a template that other municipalities in the country could follow.

[1] See: http://www.straldzha.net.

[17] See: https://bnt.bg/bg/a/kolko-kilograma-dokumenti-sa-nuzhni-za-rabotata-na-dva-solarni-panela-v-balgariya.

Box 24.3 Municipality-led projects in city of Burgas

The municipality of Burgas is one of the most active regional authorities working on advancing renewables through a range of municipality-led projects. Located in the southern part of Bulgaria along the Black Sea Coast, Burgas is the fourth largest city in the country with a population of 202,694 people. As part of an EU cofunded energy efficiency project in 2019, the renovation of the municipality building also included the installation of a 30 kWp rooftop solar PV system configured for self-consumption. The whole project cost BGN 2.28 million (€1.16 million), with 65% coming from EU's Regional Development Fund (ERDF), and the remaining 35% came from the municipality's own budget. Through other ERDF cofunded projects, five kindergartens across the city were equipped with rooftop solar PV installations entirely for self-consumption. Residential buildings have also benefited from municipality-led projects promoting energy efficiency that includes renewables.

A significant obstacle highlighted by the Burgas Municipality, however, is that the installation of any rooftop system requires the consent of all apartment owners. Interviews suggest that due to the lack of awareness and the inherent skepticism over the benefits of such systems, some apartment owners refuse to give their consent, hindering the realization of more such projects.

In accordance with the Renewable Energy Sources Act, the Burgas Municipality has drafted a plan for the promotion of renewables and biofuels for the 2020–30 period. The plan envisions the share of renewables and biofuels on the territory of the municipality to reach 32% by 2030.[1]

[1] See: https://www.burgas.bg/uploads/posts/2021/burgas_res_longterm1.pdf.

While crowdfunding is popular and accessible method to raise finances, it comes with many administrative challenges and difficulties. It requires significant expertise and know-how to navigate through the network of administrative and regulatory procedures.

Other innovative forms are being created to make it easier for individuals, businesses, and municipalities to form energy communities such as *consumer stock ownership plans* (CSOPs) (see Box 24.5).[18]

Another interesting example comes from Poland, another country that experienced a transition to a planned economy in the decades after the Second World War. Poland's energy system is notorious for its reliance on coal. *Spółdzielnia Nasza Energia,*[19] the first energy cooperative in Poland, was established in 2014. It has 300 members and its main activity is the generation of renewable heat and electricity from local biogas installations (up to 15 small installations between 0.5 and 1 MW each). It was initiated by a private company in a collaboration with another ESCO company and four

[18] See: https://www.score-h2020.eu/about-us.

[19] See: http://biopower.home.pl/consulting.

Box 24.4 How crowdfunding is being used to develop solar projects

The Case of Krizevci, Croatia.

Overview: Križevci is a town located in central Croatia with a total population of 21,122.

In 2018, it started the first pilot project in Croatia for citizens' crowd funding for a renewable energy project.[1] The goal was to install a solar PV power plant on the rooftop of the municipality's Development Center and Technology Park's administrative building. The municipality of Križevci provided administrative and financial support during the preparation phase of the project.

Funding: A fundraising campaign was organized; it resulted in 53 investors with an average investment of €500, raising a total of €31,000 for a 30 kW PV plant. The campaign managed to collect the total amount of money needed in only 10 days. A follow-up campaign was launched for a second PV plan. The €23,000 target budget was raised in only 48 h.

How it works: The PV system, with a capacity of 30 kW, was installed on the rooftop of a municipal business support center. The first goal was to ensure that the electricity produced meets the needs of the users in the building. Any excess energy is sold back to the grid. The municipality covers the cost of the electricity consumption and pays back the citizens that have invested in the project through the monthly savings obtained. The cooperative works with the municipality as the main implementing partner, involving citizens through microloans and paying them back with a fixed interest of 4.5%. In the second campaign, the interest rate was brought to 3%, as the main driver to investments from citizens was contributing to community development rather than profit. The estimated time for the return on investment is 10 years, after which the PV system ownership will be transferred to the municipality.

[1] See: https://energy-cities.eu/best-practice/crowd-investing-for-solar-roofs-in-krizevci.

municipalities in the eastern region of Lubelskie voivodeship.[20] It was established to tackle energy security and high electricity prices by tapping into the potential of a network of agricultural biogas plants. The cooperative is a public–private government initiative created with the aim to supply electricity and heat to meet the needs of public buildings as well as the needs of households in the community.

Another policy that could make it easier to develop energy communities in Bulgaria is Virtual Net Metering a policy that enables customers to develop a renewable energy project (or coinvest in one) and use the generation to offset their own consumption from the grid (Couture et al., 2021). In the case of Virtual Net Metering, customers can generate power elsewhere on the grid and use it to offset their onsite demand in their home or business, a policy option that can be useful in establishing energy communities featuring a large number of different members.

[20] See: https://www.communitypower.eu/en/poland.html.

Box 24.5 Can consumer stock ownership plans (CSOPs) help citizens in Bulgaria set up an energy community?

Consumer stock ownership plans are a consumer-centered investment model that provides both participation in terms of finance and a role in decision-making. CSOPs entail the creation of an intermediary entity that is established to invest into new or existing RES projects and operates on behalf of coowners. They have several attractive characteristics:

1. CSOPs have a low threshold to enter, making it easy for low-income families and vulnerable groups to participate;
2. They are easily scalable, and can grow as the pool of financial resources grows;
3. They offer participants a secure supply of energy and a reliable return on investment;
4. They allow participants to avoid personal liability; and
5. CSOPs are open to coinvestments from municipalities, small-, and medium-sized businesses and other local stakeholders.

Fig. 24.4 provides an overview of how the CSOP model works in practice.

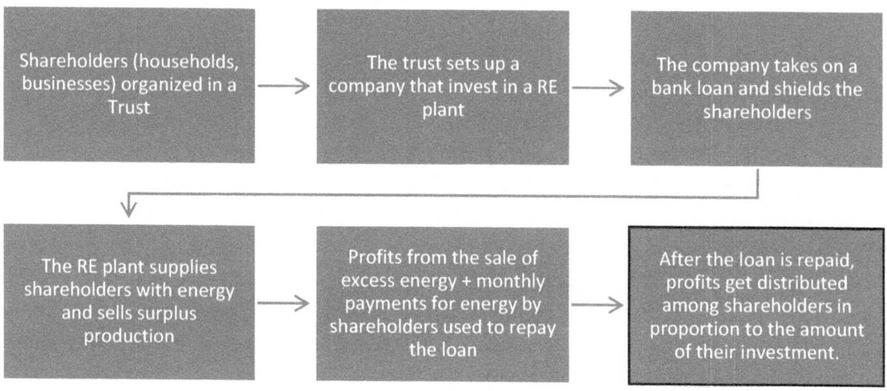

Figure 24.4 Financing structure of a CSOP.
Source: https://www.score-h2020.eu/csop-financing/csop-step-by-step.

5. Conclusion

Bulgaria still has a lot of work to do to create a stable and investable framework for energy communities. While cooperatives and community organizations have a long history in Bulgaria, today they are perceived as outdated structures, artifacts of the communist past. Little has been done over the years to address this misconception, making it difficult for organizations and proponents of energy communities to advocate for the development of community-based renewable energy projects.

Interviews conducted in 2020 and early 2021 with local stakeholders in Bulgaria suggest that energy communities are still seen by many as new and therefore distant from the day-to-day realities of many Bulgarian families. Low levels of disposable income combined with the country's past experience during the socialist period from 1945 to 1989 contribute further to a hesitancy to experiment with new ideas, and new forms of community and citizen engagement such as energy communities. Despite the fact that the socialist period only lasted for just over 40 years in Bulgaria, the hesitancy to engage in citizen-led forms of organization persists.

According to the latest Assessment of the National Energy and Climate Plan (NECP):

> *As regards enabling renewable self-consumers and renewable energy communities, the final plan still provides no details on the form of support it intends to provide. It only refers in general terms to unspecified legislative changes that should be made to optimize the current regulatory framework and to regulate the rights of self-consumer in a more satisfactory way, with a view to encouraging and facilitating the development of self-consumption of renewable energy.*
> *European Commission, 2020. Assessment of the Final National Energy and Climate Plan of Bulgaria.*

Households, small businesses, farmers, and municipalities could all potentially benefit from the development of energy communities in Bulgaria, but the interviews conducted as part of this analysis indicate that it will take time, sustained support, and a clear policy framework to drive change.

As this chapter shows, history and local culture play an important role in shaping the willingness of individuals to initiate or to participate in an energy community. These issues are especially relevant to citizens in postcommunist countries where community and cooperative forms often evoke memories of the forced collectivization that occurred under socialist rule. Countries like Denmark and Germany have long traditions of cooperative ownership and governance, a factor that helps explain the success of community energy projects in these countries.

Thus, beyond the many tax, permitting, grid access, and regulatory barriers to establishing energy communities in Bulgaria, among the most critical barriers to success, remain linked to intangible factors such as a country's culture and history. Seen from this perspective, transposing the renewable energy directive into the national legislation is just one step of a much broader process.

References

Caramizaru, A., Uihlein, A., 2020. Energy Communities: An Overview of Energy and Social Innovation, EUR 30083 EN. Publications Office of the European Union, Luxembourg, ISBN 978-92-76-10713-2. https://doi.org/10.2760/180576. JRC11943.

Center for the Study of Democracy, 2018. Development of Small-Scale Renewable Energy Sources in Bulgaria Legislative and Administrative Challenges.

Couture, T.D., Pavlov, T., Stoyanova, T., 2021. Scaling-up Distributed Solar PV in Bulgaria. E3 Analytics, Berlin. https://www.e3analytics.eu/wp-content/uploads/2021/06/E3A_Bulgaria_Analysis_of_Energy_Communities_EN_FINAL.pdf.
Energy and Water Regulatory Commission, 2019. Annual Report to the European Commission.
European Commission, 2020. Assessment of the Final National Energy and Climate Plan of Bulgaria.
Ignatov, K., 2021. Cooperatives in the System of Public Relations the Bulgarian Experience as a Historical Tradition and Modern Practices.
IRENA, 2020. Renewable Energy Prospects for Central and South-Eastern Europe Energy Connectivity (CESEC). International Renewable Energy Agency, Abu Dhabi.
Marinova, T., Nenovsky, N., 2020. Cooperative Agricultural Farms in Bulgaria (1890–1989).
Ministry of Energy, 2020. Integrated Energy and Climate Plan of the Republic of Bulgaria, 2021–2030.
OECD, 2019. OECD Review of the Corporate Governance of State-Owned Enterprises: Bulgaria.
Pavlov, T., 2022. The political economy of coal in Bulgaria: the silent phase-out (Chapter 3) in. In: Jakob, M., Steckel, J.C. (Eds.), The Political Economy of Coal: Obstacles to Clean Energy Transitions. forthcoming, Routledge. ISBN 9780367491024.
Sarov, A., 2021. Governance transformation of the Bulgarian cooperative movement: values, principles, practices and members' satisfaction. Bulg. J. Agric. Sci. 27 (1), 65–71.

Further reading

Directive (EU) 2018/2001 of the European Parliament and of the Council of 11 December 2018 on the Promotion of the Use of Energy from Renewable Sources.
Interreg Europe, 2018. A Policy Brief from the Policy Learning Platform on Low-Carbon Economy.
IRENA Coalition for Action, 2020. Stimulating Investment in Community Energy: Broadening the Ownership of Renewables. International Renewable Energy Agency, Abu Dhabi.
Ivanov, M., 2019. Governed by tensions: the introduction of renewable energies and their integration in the Bulgarian energy system (2006–2016). Environ. Innov. Societal Transitions 32, 90–106.
PROSEU, 2020. Transposition Guidance for Citizen Energy Policies.
SCORE Project.
Solar United Neighbors, Solar for Municipalities.

Sustainable island energy systems: a case study of Tilos Island, Greece

25

Xin Li
University of East Anglia, Norwich, United Kingdom

1. Introduction

Tilos is a small, remote Greek island in the Aegean Sea with a total area of 61 km^2. It is located between the islands of Kos and Rhodes, both popular tourist destinations. Tilos has around 500 permanent inhabitants who live on the island all year round and 300 temporary workers, tourists, and expatriates who arrive in the summer. The main economic activities are fishing and livestock all year round as well as tourism in summer.

TILOS, or Technology Innovation for the Local Scale, Optimum Integration of Battery Energy Storage, is a European project on Tilos island that demonstrates the potential of local small-scale battery storage systems to integrate with renewable energy technologies, smart metering, and demand-side measures. The project, which started in February 2015 and was completed in January 2019, was funded by the European Union's Horizon 2020 Research and Innovation Program and with industrial cofunding. In 2017, the project won two EU sustainable energy awards for the best energy island project and the citizen's choice award. The system became fully operational by late 2018. The project's success illustrates the potential for a green energy transition on a small and isolated island, which historically had relied on expensive and highly polluting diesel power generation.

This chapter examines the experiences and lessons learned from the implementation of the project. There are two main elements to the success of TILOS project, namely the design and operation of a sophisticated island energy system and the active participation of stakeholders, especially local residents, in a community energy project.[1]

The chapter is organized as follows:

- Section 2 describes the background to the TILOS project;
- Section 3 illustrates the technology solutions adopted in the project;
- Section 4 discusses the community engagement activities; and
- Section 5 examines the possibility of replicating Tilos's experiences in other islands or remote communities, followed by the chapter's conclusions.

[1] The chapter by Rossetto et al. defines community energy projects very broadly. Among the most important issues are: what to share, who shares, and who delivers services to energy communities.

Energy Communities. https://doi.org/10.1016/B978-0-323-91135-1.00023-7

2. The TILOS project—a brief description

With Greece announcing its target to be carbon neutral by 2050, there has been significant progress in the deployment of renewable energy sources and the transition away from fossil fuel–based generation. Between January and May 2020, renewable energy generation from wind and solar reached 4.27 TWh, the second largest generation source in Greece's power mix following natural gas (Argus, 2020). One of the major challenges in Greece's energy transition is its geographically dispersed islands, which rely on diesel oil–fired power plants. These systems are typically expensive, unreliable, and highly polluting. Greece needed to find a new technological solution to support the energy transition of the islands.

Before the TILOS project, the power supply of Tilos island relied primarily on an undersea cable built in 2010. The cable transmits electricity generated from oil-based power plants on the island of Kos via the island of Nisyros. Fig. 25.1 shows the locations of Kos, Nisyros, and Tilos islands. Tilos also had a small diesel genset used to "black start"[2] the Tilos power grid in the event of a power outage and to provide emer-

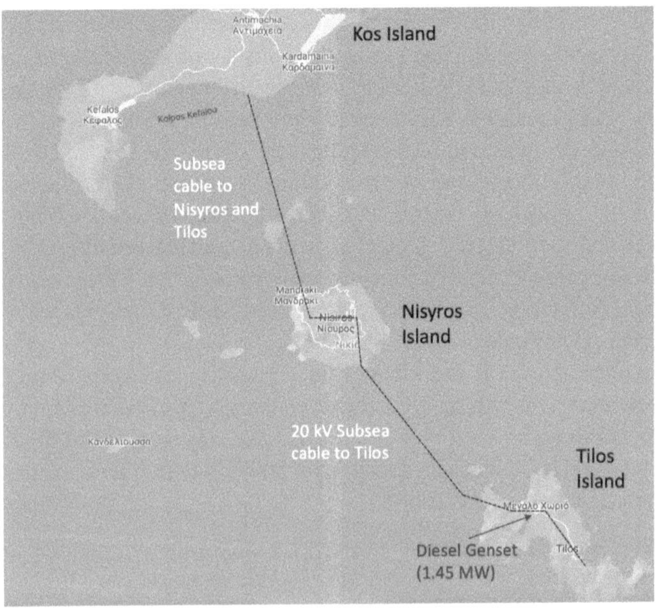

Figure 25.1 Tilos island interconnection with Kos island.
Source: Own illustration, adapted from Kaldellis, J.K., Zafirakis, D., 2020. Prospects and challenges for clean energy in European Islands. The TILOS paradigm. Renew. Energy, 145, 2489–2502.

[2] Black start is the process of restoring a power station to operation in the event of a total or partial shutdown.

gency power. The connection with Kos is unreliable and Tilos had experienced frequent and sometimes long-lasting power supply disruptions. For example, the island relied on its own diesel genset for two continuous weeks in January 2016 due to severe damage on the undersea cable. Between 2015 and early 2019, Tilos experienced over 200 h of black-out. Although the diesel genset provides backup to its power supply, local residents are not licensed to operate the local diesel power unit themselves. Technicians need to be sent from Kos and the delay in arrival depends on the transportation available. Consequently, the power supply on Tilos was neither secure nor sustainable.

Not only was the supply insecure, the cost was also high. Kaldellis and Zafirakis (2020) estimated the cost of power supply in the Aegean islands in 2015. There were significant variations in generation cost among the islands and costs also depend on fluctuating international oil prices. Because of economies of scale, generation costs on bigger islands are lower than on smaller and remote islands. For example, Kos has an average generation cost of 50 euros/MWh, while Agathonisi, a remote island in the north Aegean Sea, has an average generation cost of about 1500 euros/MWh. We have seen no precise estimation for power generation cost on Tilos island, but it is obviously higher than the cost of power on Kos' since one must account for the costs of connection, which are significant. The cost for interconnection between Tilos and Nisyros was 57.9 million euros including undersea cable, substation, and other equipment. A similar amount was spent on the interconnection between Nisyros and Kos (Zafeiratou & Spataru, 2017). The high cost of island electricity generation has been heavily subsidized through taxes paid by all electricity consumers in mainland Greece (Apostolopoulos et al., 2020). For example, for the island of Sifnos, the generation cost was round 300 euros per MWh with oil prices around 27 euros per barrel or 30 US dollars in 2015. Consumers on the island, however, were charged 180 euros/MWh for their electricity consumption (EC, 2019a).

Apart from the high generation cost, reliance on oil-based generation also resulted in emissions of particulate matter, SO_2 and NO_x that led to air pollution on these islands (Kaldellis et al., 2008; Spyropoulos et al., 2005). Furthermore, CO_2 emissions per unit of electricity are high, around 800 g per kWh.

The main source of electricity consumption is air conditioners, which provide heating in winter and cooling in summer for residential and business premises such as hotels and restaurants. On average, each household has 2–3 units of air conditioning. Each hotel room is also equipped with air conditioners. In addition, professional refrigeration and washing machines represent a significant share of electricity consumption at business premises. However, consumption is usually season specific. Most of the tourists arrive in summer when the electrical equipment is used most intensively. For households, besides heating and cooling, cooking and electronics equipment are the main sources of electricity consumption. Prior to the project, local residents already had a good awareness of the energy efficiency classification of their equipment. In some cases, people had installed solar-based water heaters.

At the beginning of the project, one of the project consortium members conducted research on local citizens' views on the project and their expectations from it. The results illustrated that people had been suffering from frequent power outages on the island: over 34% of the local residents considered the electricity supply on the island to be inadequate and 38% believed it was problematic. For example, a hotel owner mentioned that he had to replace the air conditioners and televisions a few times due to damages caused by frequent power outages and service interruptions. There was clear evidence of a wish for more secure power.

On the other hand, there were concerns about the introduction of technologies with which people were not familiar. On many Greek islands, some islanders were not enthusiastic about having wind turbines and solar PVs installed. Residents were concerned that renewable energy systems would obstruct the scenic views of the Aegean Sea, with negative consequences for the tourism industry on which most of them rely. There were no significant concerns with restricted views on Tilos island; only 2% of the population expressed negative views in TILOS (Stephanides et al., 2019b).

2.1 TILOS community energy project

As defined by the EU legislative framework, there are two types of community energy, namely "renewable energy communities" and "citizen energy communities" (Caramizaru & Uihlein, 2020). The chapter by Spasova and Braungardt examines and compares the two definitions in detail. It highlights that the deployment of energy communities in the EU is considerably different among Member States due to social, economic, and legislative factors. Technically, the TILOS project does not fit the description of energy communities defined by the EU legislation. For instance, the energy system is now owned by a private company in Greece, which is specialized in renewable energy development and integrated energy solutions. All the technical aspects of the energy system are centralized and controlled by the company. However, as discussed in the chapter by Kolesar, there is a wide range of different approaches in the building a community energy project. Kolesar highlights that:

> *The principal objectives of an energy community are to encourage engagement in and support for a project or outcome, and the nature and level of community engagement among communities and community members will likely differ .... Too rigid a framework may eliminate consideration of models and approaches to community involvement that have not emerged elsewhere.*

TILOS project was supported by stakeholders and local citizens, which were actively involved in the decision-making process. Benefits are equally shared among citizens, meaning renewable energy generation is mostly used to serve local people. On the Aegean Sea islands, as with most remote islands, local renewable energy—especially in combination with storage and demand flexibility—can improve the security of electricity supply and alleviate air pollution caused by oil-based power generation.

The main objectives of the project included the following:

- Assessing the characteristics and potential of renewable energy on the island and maximizing the use of wind and solar energy.
- Developing and operating a prototype battery storage system that allows optimal integration with the system.

- Designing a smart grid control system that can optimize the operation of the whole system and ensure secure supply.
- Integrating demand-side measures through deployment of smart metering.
- Engaging with local stakeholders to ensure a timely delivery of relevant messages with regard to the project (at different stages) and its impacts on electricity supply and energy bills. The project would also educate citizens about different technologies and help them participate in the operation of the system.
- Creating a platform for other islands in similar circumstances to build similar power systems and enable the transfer of technical solutions to other islands.
- Developing novel community energy-based and self-reliance business models and policy instruments that can stimulate the market uptake of the energy storage system and the integrated energy solution in other places.

Based on these main objectives, the project included, on the one hand, 12 work packages[3] to ensure successful technical installation and operations. The involvement and empowerment of citizens was the other key element of success. The following sections focus on the technical solutions and on public engagement activities.

3. Technology solutions in the TILOS project

The main challenge for the project was the optimal integration of different generation technologies as well as their interaction with consumption, storage, and benefit-sharing among the island community. The technical solution is a hybrid power system (HPS) that integrates local renewable energy sources with a battery storage system and demand-side measures using the existing local grid infrastructure. Detailed load demand information is integrated into the technical solution to formulate a system design layout.

From the generation side, a key challenge was to identify the potential of renewable energy generation on the island. A weather station was installed to measure wind speed, solar irradiance, ambient temperature, and pressure at the beginning of the project. These measurements provided important inputs to the forecasting models that project renewable energy output on the island as well as future local electricity demand. This information is also required to size properly the wind turbine and solar farm, which are the foundation of the HPS.

Another key component of the project is the $NaNiCl_2$ battery storage system. This system included two subsystems each with 1 MWh capacity, 400 kW nominal power, and an inverter. The battery supported operation in both island mode when there was a problem with the interconnection and in microgrid connected mode when there was no problem. Two other important features of the battery system are the minimum maintenance needs and the prolonged battery life; both critical given the remote location of the island and the lack of local technical skills (Li et al., 2018). The battery management

[3] Details of 12 work packages can be found from the project website at: https://www.tiloshorizon.eu/project-overview/work-packages.html.

model considers battery cycling, aging, maintenance needs and it integrates renewable energy generation, local demand, and connections with the island grid.

On the demand side, the main components included in the project are smart metering, demand-side management devices, and energy monitors. Smart metering devices are used to monitor the consumption behavior of residential, commercial, and other public buildings. The energy monitor is then used to better inform consumers on their participation in demand-side management programs. An assessment of deferrable and nondeferrable loads was conducted through data collection from smart meters. The collected data provided useful information for the design of DSM programs on the island. A range of appliances for domestic and business use, as well as public utilities equipment, were involved in providing better management of the loads, including water heaters, air conditioners, refrigerators, and water pumping equipment. The demand-side activities included engagement opportunities for consumers, to introduce each piece of equipment involved as well as the potential benefits of using them. These engagement activities with residential and business users are considered as the key to the eventual success of the TILOS project, further described in Section 4.

An SCADA operations center is installed which integrates communication of various system subcomponents from the supply side, the local grid, and the demand side. Therefore, the entire system is centrally monitored and controlled.

4. Local citizen engagement

The success of the TILOS project does not stop at the building of a hybrid power station in combination with energy storage facilities and demand-side measures. Before the TILOS project, several other islands in the Aegean Sea had developed proposals to explore their own renewable energy sources to reduce emissions and alleviate dependence on energy imports (Oikonomou et al., 2009; Dimitropoulos & Kontoleon, 2009). However, they were not successful due to opposition from the local community. In other words, the technological solutions would not be possible without community involvement and public acceptance, which require that the project address their concerns and reflect their needs and values (Hills et al., 2018).

For the TILOS project to succeed, from the onset of the project, the consortium committed to enhancing energy literacy on the island. This involved dissemination of accessible and tailored information, including the benefits of the new system for local citizens as well as expert-led site visits that helped islanders develop a better understanding of the technologies. These dissemination activities stimulated support for sustainable energy technologies. On the one hand, citizens learned more about the benefits of green energy. On the other hand, project managers could address misconceptions about adverse impacts.

Mikkelson et al. argue that the key to successful consumer engagement activity is the shift from technology push toward a needs-based approach (Mikkelsen et al., 2019). They refer to experience from other EU Horizon projects that demonstrates the importance of considering consumer needs and communicating effectively the benefits of the projects. Needs and values vary significantly among regions and islands (Ioannidis et al., 2019). The evidence from pilot projects suggests that success requires a clear understanding at an early stage of the project of what economic incentives will be effective, the environmental awareness of the local population, as well as their energy context. A number of studies have examined the main challenges posed by community opposition in achieving national goals for a sustainable energy transition (Wüstenhagen et al., 2007; Devine-Wright, 2005).

There are several factors that can hinder the successful deployment of renewable energy technologies, including, but not limited to, restricted views, NIMBYism, and stakeholder influences.[4] Stephanides et al. (2019b) examine the public attitudes toward innovative energy solutions across the islands of the Aegean Sea. Even though most islanders are supportive of a sustainable and clean energy transition, their preferences on specific energy solutions differ considerably. In particular, islanders have questions about the use of technologies that would potentially change their consumption behavior, such as the use of demand-side measures. On the other hand, Braunholtz-Speight et al. consider the inclusion of demand-side flexibility as a key to the future development for community energy projects in Great Britain, since existing efforts mainly focus on renewable energy and energy efficiency (Braunholtz-Speight et al., 2021). Therefore, a more tailored approach is needed to take into account profiles of different energy users.

In terms of how information can be collected from and delivered to consumers, Darby (2020) suggests that a "middle actor" who has contact directly with the program leads (in TILOS, the project teams) and the participants (in TILOS, the local residents) is key to the success of this type of project. Apart from choosing a suitable medium of communication, it is also important to maintain contact with the participants in order to build trust and collect feedback in a timely manner. Other stakeholders may also be important to facilitate and support the development of energy communities, such as the Community Energy Confederation[5] in Great Britain and different levels of governments (Braunholtz-Speight et al., 2021).

The TILOS project has a dedicated work package to emphasize the training, study, and engagement of the local population as noted in the project's objectives.[6]

> *The engagement activities aim at the development of a mindset that will facilitate DSM and smart grid operation and at the same time enable local consumers to gain an understanding of effective energy monitoring and potential energy savings through the proper use of smart meters. In general, the activities concern the long-term observation of the islanders throughout the project duration, in order to both filter their needs in relation to the system operation and also measure their gradual engagement and willingness to actively support the operation of the microgrid.*

[4] These factors are also discussed in the chapter by Kolesar.

[5] The Community Energy Confederation is an organization whose members are local community energy groups. The organization represents its members and enables them to trade renewable energy and demand-side flexibility in regional and national markets.

[6] From TILOS Project Proposal Work Package Eight Public Engagement.

To promote awareness of the project, at the beginning of the project the consortium team prepared the project brochure in Greek and English. The brochure includes a brief introduction to the project and the consortium teams and offers an opportunity to participate in the project by leaving contact details. An info kiosk on the island was hosted and operated by the local municipality. It was a communication hub for the project that was used to convey information related to the project to locals and visitors to the island. The info kiosk includes a touch screen monitor which shows the process of system installation, an architectural model of the local electricity grid, and informative material. The project used a few other approaches to engage with local residents. For example, a series of training seminars was conducted to introduce specific technologies, the energy monitoring system, as well as the operation of the system. The Board of the local municipality, local residents, and island visitors participated in such seminars.

The first seminar included an introduction to technologies (such as smart meters) and programs (such as DSM) and the possible impacts on energy bills, island society, and possibly on the local tourism industry. Local residents had the opportunity to raise questions and discuss their concerns.

The second and other seminars included open discussion between local residents and the project team. The project team addressed concerns of the locals with regard to issues such as the improvement of local electricity supply quality, the potential for energy saving, and the role of demand side measures. Explanations of these topics were usually detailed. For example, the installation of smart meters led to questions and answers concerning how long it would take to install a meter, what the possible interruptions to power supply were, and how people could benefit from using smart meters. These discussions allowed citizens to get familiar with the equipment before installation, ease their concerns on the possible impact of installation, and build trust with the project team members.

The seminars also enabled the team to collect feedback from local residents. For example, in one seminar introducing demand side measures, local residents gave their ideas on critical loads and noncritical loads. They were also encouraged to identify the most energy-consuming activities and the timing of such activities. Together with the data collected from smart meters, these ideas helped formulate the demand-side management program that became operational on the island, without compromising on comfort and convenience. In another seminar, residents reflected on the problem of power outages and the damage to electric appliances; they wanted to know—and the project team explained—how the system could alleviate the problem on the island.

The project team also demonstrated the procedures of EV charging, introduced the different types of EVs available, and discussed the suitability of EVs to a small island (where driving range is not a significant barrier to consider) for local residents. They insisted on the benefits (environmental and system management) of switching from ICE vehicles and addressed the possible G2V and V2G services.

Apart from learning about the technologies and solutions of the project, local residents could also ask questions specific to their potential to benefit from the new system. For example, local hotel and restaurant owners expressed their interest in a net

Table 25.1 A summary of main methods of engagement.

Main methods	Description
Ethnographic and behavioral research	Study day to day practices of the people in the local community. Examine and understand the diverse dynamics in order to adopt the most suitable approach method for each subgroup (e.g., farmers, businessmen, etc.).
Stakeholder mapping	Identify local stakeholders and those who can collaborate and service the scope of the project. Local authorities are the first to be contacted.
Media mapping	Identify key media sources, establish connection, and interaction. When the appropriate liaisons are in place, feed them with news and developments of the project (e.g., articles, press releases, interviews, etc.).
Establishment of a network of influencers	Approach individuals that have increased influence on the community. Form alliances and collaboration strategies. Members of the local authorities and elected representatives are usually individuals with significant influence in local communities.
Social media engagement	Identify popular social media groups and pages that are widely used by members of the community, become a member and interact with them.
Establishment of an affiliated network of people	Identify those individuals who have expressed an interest, want to stay informed, and be involved in order to promote the scope of the project. Delegating affiliated individuals a role in the project can provide multiple short-term and long-term benefits.
Organization of events	Events involving the whole community can be inspiring and motivational to a large group of people.

Source: Adapted from Tilos Project, 2019. TILOS Final Review Meeting: Rhodes—Tilos, p. 29.

metering PV solution, which they believed could help to reduce their electricity bills in a cost-effective manner.

Besides training seminars, there were other engagement activities. Table 25.1 summarizes the main methods and tools used.

5. Summary of results and extended work beyond the TILOS project

Both the wind turbine and solar PV station were commissioned in September 2018, while the battery storage system was commissioned in May 2018. Fig. 25.2 shows the monthly energy generation from renewable sources and Tilos island electricity

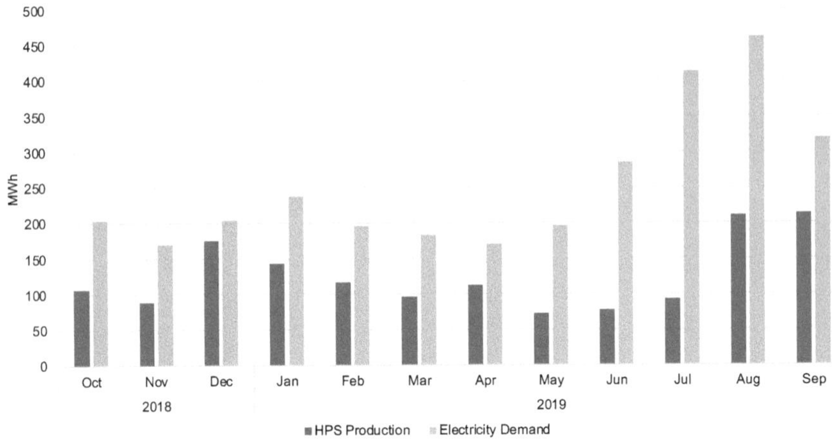

Figure 25.2 Monthly electricity generation of hybrid power system and electricity demand in Tilos between October 2018 and September 2019.
Source: Own illustration, based on data from Hellenic Electricity Distribution Network Operator (HEDNO).[7]

demand during the first year the system HPS was in operation (between October 2018 and September 2019). Renewable energy generation reached its highest level in December 2018, when it accounted for approximately 90% of the local consumption. However, the system saw much lower generation during the summer period due to lower wind energy generation, whereas demand in summer was more than twice the demand in winter (450 MWh vs. 200 MWh). Therefore, renewable energy generation accounted for one quarter of the total generation in summer period. Nevertheless, total renewable energy—based power generation accounted for approximately half of the total electricity consumption on Tilos island (or 1.5 GWh) during the first full year of operation of the HPS, with the rest imported from the main Kos-Kalymnos grid.

In fact, the actual generation from renewable energy is much lower than projected generation. This is mainly due to the curtailment of Tilos renewable energy generation during times that local consumption is low and renewable generation from the Kos-Kalymnos grid is high.[8] However, the level of renewable energy generation in total power generation mix on Tilos is much higher than the other island, which is never higher than 22% (Kaldellis, 2021).

It is estimated that the generation costs of the integrated TILOS energy solution are between 200 and 250 euros per MWh (Kaldellis & Zafirakis, 2020), compared to around 350 euros per MWh before the project. The overall cost will fall further with the advancement in battery technologies and falling battery costs. Savings also

[7] HEDNO publishes monthly statistics in the non-Interconnected Islands in Greece.

[8] The Kos-Kalymnos grid has a total installed capacity of 15.2 MW for wind and 8.8 MW for solar PV, excluding Tilos. Given the proximity of these islands, renewable energy generation is highly homogeneous. The Kos-Kalymnos grid experiences significant renewable energy curtailment. One-third of the total generation is curtailed on average.

derive from the automation of energy system components, for instance, enabling the local diesel genset to be activated—in the event of a black-out—without sending a technician from the Kos island. Thus, the new energy system can offer lower cost energy as well as improved supply security.

The TILOS project has successfully demonstrated the technical potential to increase the share of renewable energy in island energy consumption. The full operation of the existing hybrid power plant could potentially satisfy 70%–75% of electricity consumption in Tilos, despite a lower than expected performance between 2018 and 2019. Different modes of operation were tested toward the end of the project, including load following and black start (Kaldellis & Zafirakis, 2020).

While the main activities were conducted on Tilos island, the project also engaged with the islands of La Graciosa and Corsica with the aim of enabling transfer of technological know-how between islands (Notton et al., 2020). In addition, a geographical study was carried out to compare the views of Tilos and Pellworm populations and also with the views of other island communities around the Aegean Sea (Stephanides et al., 2019a). The study was then extended with a wider geographical coverage. For example, a survey was conducted with a sample size of 1001 households across 15 Aegean Sea islands (Stephanides et al., 2019b). The aim was to examine the possibility for other Aegean islands to replicate the technological solutions adopted in Tilos and to address the concern that local opposition might undermine what appeared to be promising energy transition plans. A range of demographic factors were identified that could increase the likelihood of accepting a TILOS-like system. For instance, support is likely to come from green prosumers, who are usually young, male, well educated, and have a higher level of environmental awareness. The results of the survey only present a prescreening of the islanders, who had no previous experience or adequate knowledge about renewable energy technologies and battery storage systems, among other technology gaps. Nevertheless, the findings may inform future interventions aiming to secure public support for the energy transition in Aegean Sea islands.

Tilos has also encouraged similar energy projects in the Aegean Sea. For instance, the Agios Efstratios Island in the north Aegean Sea has started to construct a similar energy system, the completion of which will include a power system that includes renewables, energy storage, and demand-side measures (Hope, 2021).

In the context of the Clean Energy for EU Island Initiative,[9] several islands in the EU have developed decarbonization pathways tailored to their values and needs (EC, 2019b). Most of these islands focus on the development of energy community projects. For example, Culatra Island in Portugal aims to achieve 100% renewable energy self-generation, which will be owned by the local community. Sifnos Island in Greece also aims to achieve 100% renewable energy self-generation, which will be coowned by the local community and private investors.

[9] Clean Energy for EU Island Initiative provides a long-term framework to help islands to generate their own sustainable and low cost energy.

6. Conclusions

The TILOS project successfully demonstrated the possibility of building an HPS on a remote island. It brings multiple benefits to the island, such as more sustainable energy supply, lower costs, a higher level of energy security, a lower level of pollution, and reduced reliance on expensive and polluting oil-based power generation. The core of the new energy system is the integration of multiple technological solutions (e.g., wind, solar, battery storage, and demand flexibility). The novel integrated energy solutions could also bring new economic opportunities to the island, when more secure and sustainable energy sources are used.

The innovative technological solutions are only part of the successful implementation of the project. The effective communication with local islanders, and their participation, throughout the project has been key. Through training and seminars, local islanders were frequently informed about the progress of the project and involved in the setup of the new energy system. They also actively participated in the provision of demand-side management. The engagement activities ensured the acceptance of the new power system on the island by local islanders, which is now owned and managed by a private company.

Efforts are ongoing to replicate the TILOS experience in other remote and small Aegean islands. Apart from the benefits to the islanders, these community energy projects reduce both the subsidy paid by the rest of the Greek consumers and the carbon footprint of power generation in Greece. However, such efforts on other islands are not always welcome by islanders. The TILOS project demonstrates that it is important to understand local needs, values, and the environmental awareness of citizens before designing a new sustainable energy system.

Bibliography

Apostolopoulos, N., Chalvatzis, K.J., Liargovas, P.G., Newbery, R., Rokou, E., 2020. The role of the expert knowledge broker in rural development: renewable energy funding decisions in Greece. J. Rural Stud. 78, 96–106.

Argus, 2020. Renewables Overtake Lignite in Greek Power Mix. Argus Media. Available: https://www.argusmedia.com/en/news/2115773-renewables-overtake-lignite-in-greek-power-mix. (Accessed 31 May 2021).

Braunholtz-Speight, T., Mclachlan, C., Mander, S., Hannon, M., Hardy, J., Cairns, I., Sharmina, M., Manderson, E., 2021. The long term future for community energy in Great Britain: a co-created vision of a thriving sector and steps towards realising it. Energy Res. Social Sci. 78, 102044.

Caramizaru, A., Uihlein, A., 2020. Energy Communities: An Overview of Energy and Social Innovation. Publications Office of the European Union, Luxembourg, ISBN 978-92-76-10713-2.

Darby, S.J., 2020. Demand response and smart technology in theory and practice: customer experiences and system actors. Energy Pol. 143, 111573.

Devine-Wright, P., 2005. Beyond NIMBYism: towards an integrated framework for understanding public perceptions of wind energy. Wind Energy 8, 125–139.

Dimitropoulos, A., Kontoleon, A., 2009. Assessing the determinants of local acceptability of wind-farm investment: a choice experiment in the Greek Aegean Islands. Energy Pol. 37, 1842–1854.

EC, 2019a. Sifnos (Greece). European Commission: Clean energy for EU islands. Available: https://euislands.eu/island/sifnos. (Accessed 14 July 2021).

EC, 2019b. Six EU Islands Publish Clean Energy Transition Agendas, Putting Citizens First. European Commission. Available: https://www.euislands.eu/index.php/eu-islands-publish-clean-energy-transition-plans. (Accessed 14 July 2021).

Hills, J.M., Michalena, E., Chalvatzis, K.J., 2018. Innovative technology in the Pacific: building resilience for vulnerable communities. Technol. Forecast. Soc. Change 129, 16–26.

Hope, K., 2021. The Greek Island where Renewable Energy and Hybrid Cars Rule, Financial Times. Financial Times. Available: https://www.ft.com/content/2dff3e85-2fb7-4a95-83cc-a38e1558a111. (Accessed 31 May 2021).

Ioannidis, A., Chalvatzis, K.J., Li, X., Notton, G., Stephanides, P., 2019. The case for islands' energy vulnerability: electricity supply diversity in 44 global islands. Renew. Energy 143, 440–452.

Kaldellis, J., Kondili, E., Paliatsos, A., 2008. The contribution of renewable energy sources on reducing the air pollution of Greek electricity generation sector. Fresenius Environ. Bull. 17, 1584–1593.

Kaldellis, J.K., 2021. Supporting the clean electrification for remote islands: the case of the Greek Tilos island. Energies 14, 1336.

Kaldellis, J.K., Zafirakis, D., 2020. Prospects and challenges for clean energy in European Islands. The TILOS paradigm. Renew. Energy 145, 2489–2502.

Li, X., Chalvatzis, K., Stephanides, P., 2018. Innovative energy islands: life-cycle cost-benefit analysis for battery energy storage. Sustainability 10, 3371.

Mikkelsen, T., Chalvatzis, K., Li, X., Tawil-Jamault, D., Paravan, D., Barreto, M., Bijnens, M., Negreira, J., Karg, L., Watson, M., Miralles, A., Darby, S., Fikiin, K., 2019. Cooperation between Horizon 2020 Projects in the Filed of Smart Grids and Energy Storage—Main Findings and Recommendations, Customer Engagement Working Group. Bridge Horizon 2020 European Commission. July 2019.

Notton, G., Nivet, M.L., Voyant, C., Duchaud, J.L., Fouilloy, A., Zafirakis, D., Kaldellis, J., 2020. Tilos, an autonomous Greek island thanks to a PV/Wind/Zebra battery plant and a smart Energy Management System. In: 7th International Conference on Energy Efficiency and Agricultural Engineering (EE&AE), 12–14 November 2020, pp. 1–4.

Oikonomou, E.K., Kilias, V., Goumas, A., Rigopoulos, A., Karakatsani, E., Damasiotis, M., Papastefanakis, D., Marini, N., 2009. Renewable energy sources (RES) projects and their barriers on a regional scale: the case study of wind parks in the Dodecanese islands, Greece. Energy Pol. 37, 4874–4883.

Spyropoulos, G., Chalvatzis, K., Paliatsos, A., Kaldellis, J., 2005. Sulphur dioxide emissions due to electricity generation in the Aegean islands: real threat or overestimated danger? The 9th International Conference on Environmental Science and Technology, Rhodes Island, Greece.

Stephanides, P., Chalvatzis, K., Li, X., Mantzaris, N., Prodromou, M., Papapostolou, C., Zafirakis, D., 2019a. Sustainable energy solutions for the Aegean Archipelago Islands: what is the public attitude? Energy Proc. 159, 243–248.

Stephanides, P., Chalvatzis, K.J., Li, X., Lettice, F., Guan, D., Ioannidis, A., Zafirakis, D., Papapostolou, C., 2019b. The social perspective on island energy transitions: evidence from the Aegean archipelago. Appl. Energy 255, 113725.

Tilos Project, 2019. TILOS Final Review Meeting: Rhodes. Tilos.
Wüstenhagen, R., Wolsink, M., Burer, M.J., 2007. Social acceptance of renewable energy innovation: an introduction to the concept. Energy Pol. 2683–2691.
Zafeiratou, E., Spataru, C., 2017. Potential economic and environmental benefits from the interconnection of the Greek islands. Int. J. Glob. Warming 13, 426.

Epilogue

This book contains a compendium of fascinating responses to the question of whether "energy communities" are citizen-centered, market-driven, and welfare-enhancing.

What is immediately evident from its opening chapters, and which continues to its closing chapters, is the widely differing understanding of "energy communities". Many chapters chop away at taxonomy and narrow the discussion to analyze particular conceptions of what energy community means. Several of the authors advise that we should not be prescriptive or pedantic in our understanding of community energy.

I have found this to be excellent advice, but it nonetheless begs the question of why such a volume on "community energy" has now been compiled. What is it about "community energy" that has stimulated such a rich supply of ideas, case studies, theoretical, and regulatory expositions from so many excellent energy researchers, thinkers, and practitioners scattered around the globe? And just why is it so hard to pin down what "community energy" really is?

Let me advance a thesis in response. Technology change, and specifically solar photovoltaics initially and now also batteries, is making the distributed production and storage of electricity possible and economically attractive. The economies of scale that led to ever greater centralization of electricity supply over the first century of the industry's history are now breaking down and, if anything, going into reverse.

This technology change has opened many possibilities ranging from increasing levels of individual energy self-sufficiency to various forms of communal distributed production and electricity trade. Innovators are responding to technology and institutional changes by developing platforms and other tools to organize such distributed production and trade.

Several authors in this volume have suggested that these technology changes are taking us back to the institutional arrangements from the early years of electricity supply when distributed energy sources—run-of-river hydro, tallow, biogas, wood, coal, and liquid fossil fuels—provided the primary source of energy for conversion into electricity (and sometime also waste heat) used by local communities. Energy communities could not exist without access to locally produced energy.

While in many ways distributed supply might in some ways be taking us back to a place from whence we first came, we do so now in a very different way. As various chapters explain, tools now exist to make all sorts of sharing of distributed production possible. Furthermore, distributed community energy can exist in cooperation with or in competition to centralized supply. The two might be substitutes, but with modern mechanisms for trade and technology control, they can also be complements.

In this volume, some authors rankle at the prospect of inefficiencies that may arise if we duplicate energy infrastructure that already exists, or strand assets in centrally coordinated electricity supply that have been developed at great expense over the preceding decades. Efficiency concerns relate not just to the treatment of sunk costs, but also to the recovery of "policy" costs in variable charges. In the context in which ever more demand may be met through decentralized production, individual prosumers or communities of prosumers may be able to avoid paying their fair share of those policy costs.

Other authors draw attention to the many things that different consumers value: autarky, interdependence, individual control, community, low prices and simple products, sophisticated and bespoke products, low environmental impacts, and so on. This heterogeneity, surely, is the main point. The combination of technology change in electricity production and storage with innovations in trading, information sharing, and control is increasing the different ways that electricity supply might be arranged and customers' heterogenous demands met.

There are now so many possibilities. Those consumers that wish to totally opt-out of grid supply or drastically reduce their dependence on the grid can now, increasingly, do so. Those that wish to find a market for their surplus production and trade those surpluses on their own account or, conceptually, sell them to friends can expect that arrangements will be developed to allow this. Those that wish to club together to form cooperatives to jointly develop their own local production or to share their own resources can increasingly do so. Those that wish to have as little to do with the energy industry as possible and just buy from their existing utility with as little fuss as possible can find suppliers that offer this even if they source much of their supplies from embedded generators.

Faced with such diversity and choice, many of us that are long schooled in the merits of systematic, hierarchical organization find such new possibilities exciting but daunting. The many interesting chapters in this volume should provide confidence that the challenges of these times are being met by fine minds.

Bruce Mountain
Director, Energy Policy Center
University of Victoria
Melbourne, Australia

Index

'*Note*: Page numbers followed by "f" indicate figures, "t" indicate tables and "b" indicate boxes.'

E

H

I

FSC
www.fsc.org
MIX
Papier | Fördert
gute Waldnutzung
FSC® C083411

Kolibri 360 GmbH
Ferdinand-Jühlke-Straße 7
99095 Erfurt, Deutschland
produktsicherheit@kolibri360.de